T0138031

Lecture Notes in Computer Science 12659

More information about this subseries at http://www.springer.com/series/7412

Alessandro Crimi · Spyridon Bakas (Eds.)

Brainlesion: Glioma, Multiple Sclerosis, Stroke and Traumatic Brain Injuries

6th International Workshop, BrainLes 2020
Held in Conjunction with MICCAI 2020
Lima, Peru, October 4, 2020
Revised Selected Papers, Part II

 Springer

Editors
Alessandro Crimi 🆔
University of Zurich
Zurich, Switzerland

Spyridon Bakas 🆔
University of Pennsylvania
Philadelphia, PA, USA

ISSN 0302-9743   ISSN 1611-3349   (electronic)
Lecture Notes in Computer Science
ISBN 978-3-030-72086-5   ISBN 978-3-030-72087-2   (eBook)
https://doi.org/10.1007/978-3-030-72087-2

LNCS Sublibrary: SL6 – Image Processing, Computer Vision, Pattern Recognition, and Graphics

This Springer imprint is published by the registered company Springer Nature Switzerland AG
The registered company address is: Gewerbestrasse 11, 6330 Cham, Switzerland

in loving memory of Prof. Christian Barillot

Preface

This volume contains articles from the Brain Lesion workshop (BrainLes), as well as (a) the International Brain Tumor Segmentation (BraTS) challenge and (b) the Computational Precision Medicine: Radiology-Pathology Challenge on Brain Tumor Classification (CPM-RadPath). All these events were held in conjunction with the Medical Image Computing and Computer Assisted Intervention (MICCAI) conference on October the 4th 2020 in Lima, Peru, though online due to COVID19 restrictions.

The papers presented describe the research of computational scientists and clinical researchers working on glioma, multiple sclerosis, cerebral stroke, trauma brain injuries, and white matter hyper-intensities of presumed vascular origin. This compilation does not claim to provide a comprehensive understanding from all points of view; however the authors present their latest advances in segmentation, disease prognosis, and other applications to the clinical context.

The volume is divided into four parts: The first part comprises invited papers summarizing the presentations of the keynotes during the full-day BrainLes workshop, the second includes the accepted paper submissions to the BrainLes workshop, the third contains a selection of papers regarding methods presented at the BraTS 2020 challenge, and lastly there is a selection of papers on the methods presented at the CPM-RadPath 2020 challenge.

The content of the first chapter with the three invited papers covers the current state-of-the-art literature on radiomics and radiogenomics of brain tumors, gives a review of the work done so far in detecting and segmenting multiple sclerosis lesions, and last but not least illuminates the clinical gold standard in diagnosis and classification of glioma.

The aim of the second chapter, focusing on the accepted BrainLes workshop submissions, is to provide an overview of new advances of medical image analysis in all of the aforementioned brain pathologies. It brings together researchers from the medical image analysis domain, neurologists, and radiologists working on at least one of these diseases. The aim is to consider neuroimaging biomarkers used for one disease applied to the other diseases. This session did not have a specific dataset to be used.

The third chapter focuses on a selection of papers from the BraTS 2020 challenge participants. BraTS 2020 made publicly available a large (n = 660) manually annotated dataset of baseline pre-operative brain glioma scans from 20 international institutions, in order to gauge the current state of the art in automated brain tumor segmentation using multi-parametric MRI sequences and to compare different methods. To pinpoint and evaluate the clinical relevance of tumor segmentation, BraTS 2020 also included the prediction of patient overall survival, via integrative analyses of radiomic features and machine learning algorithms, and evaluated the algorithmic uncertainty in the predicted segmentations, as noted in: www.med.upenn.edu/cbica/brats2020/.

The fourth chapter contains descriptions of a selection of the leading algorithms participating in the CPM-RadPath 2020 challenge. The "Combined MRI and Pathology

Brain Tumor Classification" challenge used corresponding radiographic and pathologic imaging data towards classifying a cohort of diffuse glioma tumors into three categories. This challenge presented a new paradigm in algorithmic challenges, where data and analytical tasks related to the management of brain tumors were combined to arrive at a more accurate tumor classification. Data from both challenges were obtained from The Cancer Genome Atlas/The Cancer Imaging Archive (TCGA/TCIA) repository, and the Hospital of the University of Pennsylvania.

We heartily hope that this volume will promote further exciting research about brain lesions.

March 2021   Alessandro Crimi
                                                                        Spyridon Bakas

Organization

Main BrainLes Organizing Committee

Spyridon Bakas   Center for Biomedical Image Computing and
Analytics, University of Pennsylvania, USA
Alessandro Crimi   African Institute for Mathematical Sciences, Ghana

Challenges Organizing Committee

Brain Tumor Segmentation (BraTS) Challenge

Spyridon Bakas   Center for Biomedical Image Computing and
Analytics, University of Pennsylvania, USA
Christos Davatzikos   Center for Biomedical Image Computing and
Analytics, University of Pennsylvania, USA
Keyvan Farahani   Center for Biomedical Informatics and Information
Technology, National Cancer Institute,
National Institutes of Health, USA
Jayashree Kalpathy-Cramer   Massachusetts General Hospital, Harvard University,
USA
Bjoern Menze   Technical University of Munich, Germany

Computational Precision Medicine: Radiology-Pathology (CPM-RadPath) Challenge on Brain Tumor Classification

Spyridon Bakas   Center for Biomedical Image Computing and
Analytics, University of Pennsylvania, USA
Benjamin Aaron Bearce   Massachusetts General Hospital, USA
Keyvan Farahani   Center for Biomedical Informatics and Information
Technology, National Cancer Institute,
National Institutes of Health, USA
John Freymann   Frederick National Lab for Cancer Research, USA
Jayashree Kalpathy-Cramer   Massachusetts General Hospital, Harvard University,
USA
Tahsin Kurc   Stony Brook Cancer Center, USA
MacLean P. Nasrallah   Neuropathology, University of Pennsylvania, USA
Joel Saltz   Stony Brook Cancer Center, USA
Russell Taki Shinohara   Biostatistics, University of Pennsylvania, USA
Eric Stahlberg   Frederick National Lab for Cancer Research, USA
George Zaki   Frederick National Lab for Cancer Research, USA

Program Committee

Meritxell Bach Cuadra	University of Lausanne, Switzerland
Ujjwal Baid	University of Pennsylvania, USA
Jacopo Cavazza	Italian Institute of Technology, Italy
Keyvan Farahani	Center for Biomedical Informatics and Information Technology, National Cancer Institute, National Institutes of Health, USA
Madhura Ingalhalikar	Symbiosis International University, India
Tahsin Kurc	Stony Brook Cancer Center, USA
Jana Lipkova	Harvard University, USA
Sarthak Pati	University of Pennsylvania, USA
Sanjay Saxena	University of Pennsylvania, USA
Anupa Vijayakumari	University of Pennsylvania, USA
Benedikt Wiestler	Technical University of Munich, Germany
George Zaki	Frederick National Lab for Cancer Research, USA

Sponsoring Institution

Center for Biomedical Image Computing and Analytics, University of Pennsylvania, USA

Contents – Part II

**Computational Precision Medicine: Radiology-Pathology
Challenge on Brain Tumor Classification**

Contents – Part I

Brain Tumor Segmentation

Brain Tumor Segmentation

Lightweight U-Nets for Brain Tumor Segmentation

Tomasz Tarasiewicz, Michal Kawulok, and Jakub Nalepa[(⊠)]

Silesian University of Technology, Gliwice, Poland
{tomasz.tarasiewicz,michal.kawulok,jakub.nalepa}@polsl.pl

Abstract. Automated brain tumor segmentation is a vital topic due to its clinical applications. We propose to exploit a lightweight U-Net-based deep architecture called Skinny for this task—it was originally employed for skin detection from color images, and benefits from a wider spatial context. We train multiple Skinny networks over all image planes (axial, coronal, and sagittal), and form an ensemble containing such models. The experiments showed that our approach allows us to obtain accurate brain tumor delineation from multi-modal magnetic resonance images.

Keywords: Brain tumor · Segmentation · Deep learning · U-Net

1 Introduction

Automated brain tumor segmentation from magnetic resonance images (MRI) is a critical step in precision patient care. Accurate delineation of tumorous tissue is pivotal for further diagnosis, prognosis and treatment, and it can directly influence the treatment. Ensuring the reproducibility, robustness, and quality of segmentations are critical in clinical settings [7,13,19,32].

The state-of-the-art brain tumor delineation techniques are split into the *atlas-based, unsupervised, supervised,* and *hybrid* algorithms. In the *atlas-based* techniques, manually segmented images are used to segment the unseen scans [29] by applying image registration [5]. *Unsupervised* algorithms capture intrinsic characteristics of the *unlabeled* data [8] with the use of clustering [15,30,34], Gaussian modeling [31], and other approaches [27]. Once the labeled data is available, we can utilize the *supervised* techniques. Such classical supervised learners include, among others, decision forests [11,37], conditional random fields [35], support vector machines [18], extremely randomized trees [28], and supervised self-organizing maps [26]. These methods exploit hand-crafted features that are extracted from the raw input data in the process of feature engineering which commonly includes feature extraction and selection [9].

We have been witnessing deep learning-powered breakthroughs in a variety of fields, including the brain tumor detection and segmentation [4,12,17,21,23]. Such systems are built upon various network architectures and techniques, and they encompass holistically nested neural nets [36], ensembles of deep neural

© Springer Nature Switzerland AG 2021
A. Crimi and S. Bakas (Eds.): BrainLes 2020, LNCS 12659, pp. 3–14, 2021.
https://doi.org/10.1007/978-3-030-72087-2_1

nets [16], U-Net-based architectures [14,25], autoencoder networks, convolutional encoder-decoder approaches [6], and more [10], very often coupled with a battery of regularization techniques [24]. Importantly, deep models perform automated representation learning, hence do not require designing manual feature extractors, and can potentially capture features that are unknown for humans. Finally, the *hybrid* algorithms combine and benefit from advantages of various methods, commonly including both "classical" and deep learning.

In this paper, we propose a modified version of our Skinny network [33], being a lightweight U-Net-based architecture (Sect. 3) which was originally designed for human skin segmentation from color images. Such models are trained over 2D images in all image planes—axial, coronal, and sagittal—and form a weighted ensemble. The experiments proved Skinny to be successful in brain tumor detection and segmentation, and allowed us to deliver high-quality deep models that generalize well over the unseen test data (Sect. 4).

2   Data

The BraTS'20 dataset [1–4] includes MRI data of 369 glioma patients—293 high-grade glioblastomas (HGG), and 76 low-grade gliomas (LGG). Each study was annotated by one to four experienced readers. The data comes in co-registered modalities: native pre-contrast (T1), post-contrast T1-weighted (T1Gd), T2-weighted (T2), and T2 Fluid Attenuated Inversion Recovery (T2-FLAIR). All pixels are labeled with one of the following classes: healthy tissue, Gd-enhancing tumor (ET), peritumoral edema (ED), the necrotic and non-enhancing tumor core (NCR/NET) [20]—see an example training patient in Fig. 1.

The data was acquired with different scanners at 19 institutions. The studies were interpolated to the same shape ($240 \times 240 \times 155$ there are 155 images of 240×240 size, with $1\,\mathrm{mm}^3$ voxel size), and they were skull-stripped. There are 125 and 166 patients in the unlabeled validation (V) and test (Ψ) sets, respectively.

3   Methods

In this section, we discuss our algorithm for the brain tumor detection and segmentation from MRI. First, we present our pre-processing (Sect. 3.1), which is followed by a detailed discussion of the Skinny-based model (Sect. 3.2).

3.1   Data Pre-processing

We apply the Z-score normalization to the input images by setting their mean values to zero, and variance to one, in order to ensure that the examples share the same scale. We also performed cropping by 24 pixels from both sides for the second and the third dimensions, and padding by 5 pixels for the first one. That led us to obtain $160 \times 192 \times 192$ images that were compatible with our

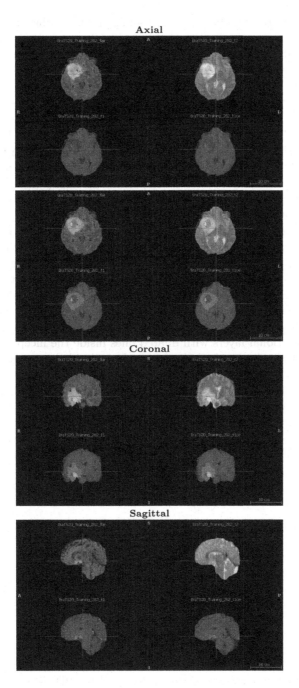

Fig. 1. Example BraTS training scan (T2-FLAIR, T2, T1, and T1Gd), alongside the overlaid ground-truth segmentation (green—ED, yellow—ET, red—NCR/NET). (Color figure online)

architectures (see the number of max-pooling layers in the models). We also made sure that the cropped parts of the images do not contain any significant information that would affect the training—it was not the case for the first dimension, hence we performed its padding.

3.2 Our Deep Network Architecture

We utilize our U-Net-based Skinny architecture with a wide spatial context resulting from the inception modules and dense blocks placed in each level of the network (Fig. 2), where each level consists of the operations performed over features of the same spatial resolution separated by max-pooling and upscaling layers. The number of kernels in subsequent convolutional layers is doubled after each pooling layer, and halved after the upscaling ones.

In this work, we exploit (i) the original Skinny architecture with all modalities concatenated in the very first layer, alongside (ii) Skinny$_{\text{Split}}$, in which we process each modality independently and concatenate them at the beginning of each level in the contracting path. Apart from that, we created a third architecture, (iii) Skinny$_{\text{Dilated}}$, resembling Skinny$_{\text{Split}}$ but taking an additional input, being a difference between T1Gd and T1 modalities. To reduce the number of trainable parameters of this network and enhance its spatial context, we decided to replace standard convolutional layers with dilated ones inside the inception modules.

Even though the data is three-dimensional, we decided not to use 3D convolutions as they would drastically increase the memory requirements, and splitting the images to smaller patches would neglect the most important feature of this architecture which is its wide spatial context. To avoid losing the benefits of exploiting the multi-dimensional information of relations between voxels due to these memory constraints, we train three models of the same architecture, each trained on 2D slices taken from images along different axes (axial, coronal, and sagittal planes). We ensemble the models in the final segmentation pipeline—here, we average the predictions of the base models, and softmax the result.

4 Experimental Study

4.1 Setup

Since the validation data V does not contain labels, we divided the training set T into two subsets from which one was used for training the models (T_1), and the second one (V_1) was employed for local calculation of evaluation metrics. The examples of T_1 were split into 2D slices along each dimension (axial, coronal, and sagittal), corresponding to a model that was being trained, and then shuffled to ensure that there are no relations between images in a single batch.

We fed our models with batches consisting of ten 2D slices, and we trained them for approximately 20 epochs on average. As an optimizer, we used Adam with $\beta_1 = 0.9$ and $\beta_2 = 0.999$ with the learning rate initially set to 10^{-4} for the initial seven epochs with an exponential decay obtained by multiplying the

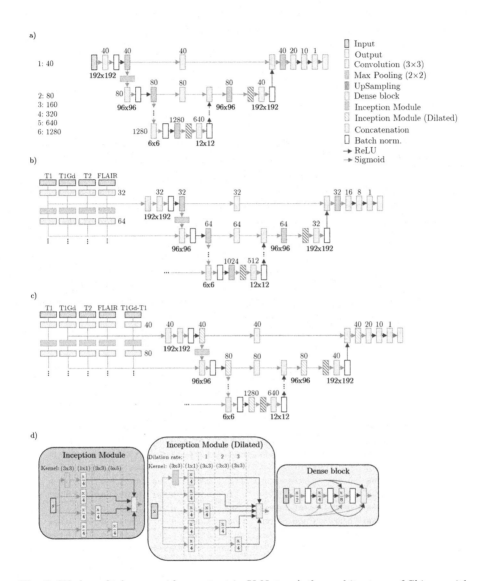

Fig. 2. We benefit from a wider context in U-Nets: a) the architecture of Skinny with all modalities concatenated in the first layer, b) Skinny with modalities processed in parallel before concatenation, c) Skinny with parallel processing of modalities with dilated convolutions in inception modules and additional input feature provided as a difference between T1Gd and T1 modalities, d) the Inception modules and dense blocks used in our models. Here, x denotes the number of kernels in each convolutional layer. For each layer, we present the number of filters used in convolution layers. Note that the impact of the inception modules and dense blocks on the overall performance of our architecture was investigated in detail in [33].

value of the learning rate by a factor of 0.85 at the beginning of each consecutive epoch. The loss function was defined as a sum of the DICE coefficient and categorical cross-entropy. The deep models were implemented using `Python3` with `TensorFlow` 2.2.0 over CUDA 10.1 and CuDNN 7.6. The experiments were run on a machine equipped with NVIDIA RTX 2080Ti GPU (11 GB VRAM).

4.2   The Results

As we trained nine models (three different architectures trained using 2D slices along three different anatomical planes), apart from submitting results for each individual one, we also combined them by adding their outputs with the weights corresponding to the mean DICE score obtained over the validation set V. The ensemble model allowed us to render better results than those obtained by each single model. Table 1 gathers the mean and median of metrics calculated for each

Table 1. The results (mean and median) obtained using our techniques: **Ens.**—an ensemble classifier, (a) $\text{Skinny}^{\text{Axial}}$, (b) $\text{Skinny}^{\text{Sagittal}}$, (c) $\text{Skinny}^{\text{Coronal}}$, (d) $\text{Skinny}^{\text{Axial}}_{\text{Split}}$, (e) $\text{Skinny}^{\text{Sagittal}}_{\text{Split}}$, (f) $\text{Skinny}^{\text{Coronal}}_{\text{Split}}$, (g) $\text{Skinny}^{\text{Axial}}_{\text{Dilated}}$, (h) $\text{Skinny}^{\text{Sagittal}}_{\text{Dilated}}$, (i) $\text{Skinny}^{\text{Coronal}}_{\text{Dilated}}$. Segmentation performance is quantified by DICE, Hausdorff (95%) distance (H95), sensitivity (Sens.), and specificity (Spec.) over the BraTS'20 validation set for the enhancing tumor (ET), tumor core (TC), and whole tumor (WT) classes. The best results are boldfaced.

Mean

Metrics	Ens.	(a)	(b)	(c)	(d)	(e)	(f)	(g)	(h)	(i)
DICE (ET)	**0.703**	0.663	0.641	0.679	0.663	0.662	0.598	0.619	0.616	0.614
DICE (TC)	**0.749**	0.728	0.689	0.722	0.736	0.725	0.712	0.715	0.708	0.667
DICE (WT)	**0.888**	0.868	0.856	0.863	0.862	0.856	0.836	0.836	0.836	0.808
H95 (ET)	**40.132**	48.212	48.514	40.360	44.819	43.017	50.750	53.961	49.897	53.247
H95 (TC)	**10.678**	14.158	15.840	14.114	12.860	15.774	13.976	20.376	17.884	26.843
H95 (WT)	**4.552**	20.397	14.410	19.846	19.064	19.310	19.145	24.071	10.082	29.572
Sens. (ET)	0.748	0.725	0.652	0.695	0.721	0.707	**0.786**	0.722	0.720	0.689
Sens. (TC)	0.740	0.732	0.705	0.729	0.742	0.735	**0.791**	0.747	0.778	0.737
Sens. (WT)	0.886	0.875	0.846	0.898	0.893	0.894	**0.927**	0.868	0.815	0.819
Spec. (ET)	**1.000**	0.999	1.000	1.000	1.000	1.000	0.999	0.999	0.999	0.999
Spec. (TC)	**1.000**	1.000	0.999	0.999	0.999	0.999	0.999	0.999	0.999	0.999
Spec. (WT)	0.999	0.999	0.999	0.998	**0.999**	0.998	0.997	0.998	**0.999**	**0.999**

Median

Metrics	Ens.	(a)	(b)	(c)	(d)	(e)	(f)	(g)	(h)	(i)
DICE (ET)	**0.827**	0.800	0.794	0.809	0.788	0.791	0.718	0.773	0.748	0.758
DICE (TC)	**0.873**	0.857	0.791	0.825	0.835	0.815	0.776	0.819	0.784	0.751
DICE (WT)	**0.916**	0.906	0.899	0.894	0.892	0.881	0.866	0.885	0.888	0.859
H95 (ET)	**2.449**	4.000	3.742	3.000	3.162	3.317	8.062	5.099	5.000	7.874
H95 (TC)	**4.359**	7.681	10.344	8.602	7.874	8.246	9.434	9.695	9.000	16.245
H95 (WT)	**3.000**	5.385	5.196	6.928	6.708	7.348	8.775	11.956	4.330	16.149
Sens. (ET)	0.913	0.895	0.799	0.868	0.887	0.860	**0.957**	0.915	0.910	0.875
Sens. (TC)	0.899	0.883	0.810	0.870	0.877	0.856	**0.934**	0.893	0.902	0.852
Sens. (WT)	0.942	0.928	0.897	0.939	0.939	0.939	**0.964**	0.933	0.897	0.876
Spec. (ET)	1.000	1.000	1.000	1.000	1.000	1.000	0.999	0.999	0.999	1.000
Spec. (TC)	1.000	1.000	1.000	1.000	1.000	1.000	0.999	1.000	0.999	0.999
Spec. (WT)	0.999	0.999	0.999	0.999	0.999	0.999	0.998	0.999	0.999	0.999

variant of Skinny[1]. Here, we compare the results obtained over different image orientations and architectures—we can appreciate that ensembling such models together leads to the best segmentation.

In Figs. 3, 4 and 5, we present example segmentations obtained using our ensemble approach over the validation patients. Here, we selected scans for which we elaborated the highest DICE scores over ET (Fig. 3), WT (Fig. 4), and TC (Fig. 5). This quantitative analysis reveals that accurate segmentation of one

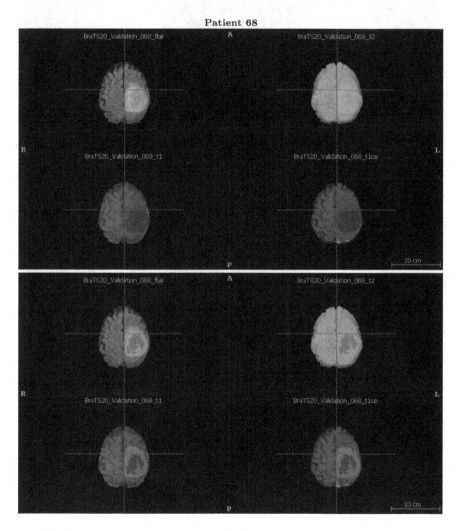

Fig. 3. Example segmentation for which we obtained the best ET DICE within the validation set (green—ED, yellow—ET, red—NCR/NET), and the DICE values amounted to 1.00000 (ET), 0.94528 (WT), and 0.42972 (TC). (Color figure online)

---

[1] Our team is named `ttarasiewicz`.

class, e.g., ET, does not necessarily allow us to obtaining high-quality delineation of other classes (e.g., TC), hence they could be tackled independently, perhaps by applying additional post-processing that would target specific tumor parts.

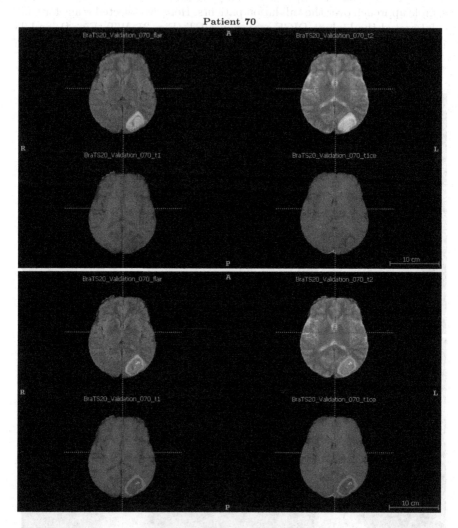

Fig. 4. Example segmentation for which we obtained the best WT DICE within the validation set (green—ED, yellow—ET, red—NCR/NET), and the DICE values amounted to 0.30969 (ET), 0.97788 (WT), and 0.7673 (TC). (Color figure online)

The results obtained over the BraTS'20 test data are gathered in Table 2—the metrics are averaged across 166 test patients, and we report the average (μ), standard deviation (s), and the median values (m). We can observe that our final model (an ensemble of Skinny architectures) can generalize well over the unseen

Patient 40

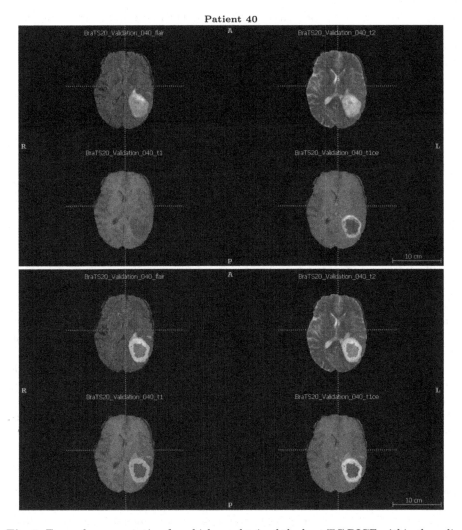

Fig. 5. Example segmentation for which we obtained the best TC DICE within the validation set (green—ED, yellow—ET, red—NCR/NET), and the DICE values amounted to 0.92826 (ET), 0.96111 (WT), and 0.96963 (TC). (Color figure online)

test data, and consistently delivers accurate segmentation. By confronting the average and median metrics, it is evident that there were outlying patients in the dataset, where Skinny failed to obtained high-quality delineations. Investigating such cases constitutes our current research effort (note that in this work we did not exploit any additional "data-level" regularization, e.g., data augmentation, that could help us better capture "extreme" cases with e.g., low-quality scans and different image acquisition settings).

Table 2. Segmentation performance quantified by DICE, sensitivity, specificity, and Hausdorff (95%) distance over the BraTS'20 test set obtained using our algorithm. The scores (average μ, standard deviation s, median m, and the 25 and 75 quantile) are presented for the whole tumor (WT), tumor core (TC), and enhancing tumor (ET).

	DICE ET	DICE WT	DICE TC	Sens. ET	Sens. WT	Sens. TC	Spec. ET	Spec. WT	Spec. TC	Haus. ET	Haus. WT	Haus. TC
μ	0.748	0.873	0.804	0.857	0.913	0.851	0.999	0.999	0.999	21.458	6.196	19.595
s	0.229	0.111	0.244	0.251	0.124	0.247	0.001	0.001	0.001	79.537	9.559	69.063
m	0.822	0.905	0.893	0.953	0.957	0.954	1.000	0.999	1.000	2.236	3.317	3.468
25q	0.719	0.852	0.796	0.869	0.905	0.861	0.999	0.998	0.999	1.414	2.236	1.732
75q	0.886	0.937	0.946	0.989	0.981	0.984	1.000	1.000	1.000	3.464	5.454	9.242

5   Conclusions

In this paper, we exploited lightweight U-Nets, referred to as the Skinny models, for multi-class brain tumor segmentation. Although Skinny was originally designed to perform segmentation of human skin regions from color images, its results are very satisfying especially when taking into account that it operates solely on 2D slices of 3D data in different planes (axial, coronal, and sagittal). The experimental validation showed that our approach generalizes well over the unseen validation and test data, and delivers high-quality delineation.

Currently, we focus on converting this architecture into a model suitable for processing 3D data to ensure relational information is not only preserved but also effectively utilized in a process of segmentation. Also, we aim at exploiting additional regularization techniques that may help improve the abilities of the deep models for heterogeneous and under-represented image data [22,24].

Acknowledgments. This research was supported by the National Science Centre, Poland, under Research Grant DEC-2017/25/B/ST6/00474. MK was supported by the Silesian University of Technology funds through the Rector's Research and Development Grant 02/080/RGJ20/0004. JN was supported by the Silesian University of Technology grant for maintaining and developing research potential, and by the Rector's Research and Development Grant 02/080/RGJ20/0003.

This paper is in memory of Dr. Grzegorz Nalepa, an extraordinary scientist and pediatric hematologist/oncologist at Riley Hospital for Children, Indianapolis, USA, who helped countless patients and their families through some of the most challenging moments of their lives.

References

1. Bakas, S., et al.: Advancing the cancer genome atlas glioma MRI collections with expert segmentation labels and radiomic features. Nat. Sci. Data **4**, 1–13 (2017). https://doi.org/10.1038/sdata.2017.117
2. Bakas, S., et al.: Segmentation labels and radiomic features for the pre-operative scans of the TCGA-GBM collection. Cancer Imaging Arch. (2017). https://doi.org/10.7937/K9/TCIA.2017.KLXWJJ1Q

3. Bakas, S., et al.: Segmentation labels and radiomic features for the pre-operative scans of the TCGA-LGG collection. Cancer Imaging Arch. (2017). https://doi.org/10.7937/K9/TCIA.2017.GJQ7R0EF

4. Bakas, S., et al.: Identifying the best machine learning algorithms for brain tumor segmentation, progression assessment, and overall survival prediction in the BRATS challenge. CoRR abs/1811.02629 (2018). http://arxiv.org/abs/1811.02629

5. Bauer, S., Seiler, C., Bardyn, T., Buechler, P., Reyes, M.: Atlas-based segmentation of brain tumor images using a Markov random field-based tumor growth model and non-rigid registration. In: Proceedings of the IEEE EMBC, pp. 4080–4083 (2010)

6. Bontempi, D., Benini, S., Signoroni, A., Svanera, M., Muckli, L.: Cerebrum: a fast and fully-volumetric convolutional encoder-decoder for weakly-supervised segmentation of brain structures from out-of-the-scanner MRI. Med. Image Anal. **62**, 101688 (2020)

7. Bousselham, A., Bouattane, O., Youss, M., Raihani, A.: Towards reinforced brain tumor segmentation on MRI images based on temperature changes on pathologic area. Int. J. Biomed. Imaging **2019**, 1758948 (2019)

8. Chander, A., Chatterjee, A., Siarry, P.: A new social and momentum component adaptive PSO algorithm for image segmentation. Expert Syst. Appl. **38**(5), 4998–5004 (2011)

9. Chowdhary, C.L., Acharjya, D.: Segmentation and feature extraction in medical imaging: a systematic review. Procedia Comput. Sci. **167**, 26–36 (2020)

10. Estienne, T., et al.: Deep learning-based concurrent brain registration and tumor segmentation. Front. Comput. Neurosci. **14**, 17 (2020)

11. Geremia, E., Clatz, O., Menze, B.H., Konukoglu, E., Criminisi, A., Ayache, N.: Spatial decision forests for MS lesion segmentation in multi-channel magnetic resonance images. Neuroimage **57**(2), 378–390 (2011)

12. Ghafoorian, M., et al.: Transfer learning for domain adaptation in MRI: application in brain lesion segmentation. In: Descoteaux, M., Maier-Hein, L., Franz, A., Jannin, P., Collins, D.L., Duchesne, S. (eds.) MICCAI 2017. LNCS, vol. 10435, pp. 516–524. Springer, Cham (2017). https://doi.org/10.1007/978-3-319-66179-7_59

13. Hasan, S.M.K., Ahmad, M.: Two-step verification of brain tumor segmentation using watershed-matching algorithm. Brain Inform. **5**(2), 8 (2018)

14. Isensee, F., Kickingereder, P., Wick, W., Bendszus, M., Maier-Hein, K.H.: No new-net. In: Crimi, A., Bakas, S., Kuijf, H., Keyvan, F., Reyes, M., van Walsum, T. (eds.) BrainLes 2018. LNCS, vol. 11384, pp. 234–244. Springer, Cham (2019). https://doi.org/10.1007/978-3-030-11726-9_21

15. Ji, S., Wei, B., Yu, Z., Yang, G., Yin, Y.: A new multistage medical segmentation method based on superpixel and fuzzy clustering. Comput. Math. Methods Med. **2014**, 747549:1–747549:13 (2014)

16. Kamnitsas, K., et al.: Ensembles of multiple models and architectures for robust brain tumour segmentation. In: Crimi, A., Bakas, S., Kuijf, H., Menze, B., Reyes, M. (eds.) BrainLes 2017. LNCS, vol. 10670, pp. 450–462. Springer, Cham (2018). https://doi.org/10.1007/978-3-319-75238-9_38

17. Korfiatis, P., Kline, T.L., Erickson, B.J.: Automated segmentation of hyperintense regions in FLAIR MRI using deep learning. Tomogr. J. Imaging Res. **2**(4), 334–340 (2016)

18. Ladgham, A., Torkhani, G., Sakly, A., Mtibaa, A.: Modified support vector machines for MR brain images recognition. In: Proceedings of the CoDIT, pp. 032–035 (2013)

19. Meier, R., et al.: Clinical evaluation of a fully-automatic segmentation method for longitudinal brain tumor volumetry. Sci. Rep. **6**(1), 23376 (2016)

20. Menze, B.H., et al.: The multimodal brain tumor image segmentation benchmark (BRATS). IEEE Trans. Med. Imaging **34**(10), 1993–2024 (2015)
21. Moeskops, P., Viergever, M.A., Mendrik, A.M., de Vries, L.S., Benders, M.J.N.L., Isgum, I.: Automatic segmentation of MR brain images with a convolutional neural network. IEEE Trans. Med. Imaging **35**(5), 1252–1261 (2016)
22. Moradi, R., Berangi, R., Minaei, B.: A survey of regularization strategies for deep models. Artif. Intell. Rev. **53**(6), 3947–3986 (2020). https://doi.org/10.1007/s10462-019-09784-7
23. Myronenko, A.: 3D MRI brain tumor segmentation using autoencoder regularization. In: Crimi, A., Bakas, S., Kuijf, H., Keyvan, F., Reyes, M., van Walsum, T. (eds.) BrainLes 2018. LNCS, vol. 11384, pp. 311–320. Springer, Cham (2019). https://doi.org/10.1007/978-3-030-11726-9_28
24. Nalepa, J., Marcinkiewicz, M., Kawulok, M.: Data augmentation for brain-tumor segmentation: a review. Front. Comput. Neurosci. **13**, 83 (2019)
25. Nalepa, J., et al.: Fully-automated deep learning-powered system for DCE-MRI analysis of brain tumors. Artif. Intell. Med. **102**, 101769 (2020)
26. Ortiz, A., Górriz, J.M., Ramírez, J., Salas-Gonzalez, D.: MRI brain image segmentation with supervised SOM and probability-based clustering method. In: Proceedings of the IWINAC, pp. 49–58 (2011)
27. Ouchicha, C., Ammor, O., Meknassi, M.: Unsupervised brain tumor segmentation from magnetic resonance images. In: Proceedings of the IEEE WINCOM, pp. 1–5 (2019)
28. Pinto, A., Pereira, S., Correia, H., Oliveira, J., Rasteiro, D.M.L.D., Silva, C.A.: Brain tumour segmentation based on extremely randomized forest with high-level features. In: Proceedings of the IEEE EMBC, pp. 3037–3040 (2015)
29. Pipitone, J., et al.: Multi-atlas segmentation of the whole hippocampus and sub-fields using multiple automatically generated templates. Neuroimage **101**, 494–512 (2014)
30. Saha, S., Bandyopadhyay, S.: MRI brain image segmentation by fuzzy symmetry based genetic clustering technique. In: Proceedings of the IEEE CEC, pp. 4417–4424 (2007)
31. Simi, V., Joseph, J.: Segmentation of glioblastoma multiforme from MR images - a comprehensive review. Egypt. J. Radiol. Nuclear Med. **46**(4), 1105–1110 (2015)
32. Sun, L., Zhang, S., Chen, H., Luo, L.: Brain tumor segmentation and survival prediction using multimodal MRI scans with deep learning. Front. Neurosci. **13**, 810 (2019)
33. Tarasiewicz, T., Nalepa, J., Kawulok, M.: Skinny: a lightweight U-Net for skin detection and segmentation. In: Proceedings of the IEEE ICIP, pp. 2386–2390 (2020)
34. Verma, N., Cowperthwaite, M.C., Markey, M.K.: Superpixels in brain MR image analysis. In: Proceedings of the IEEE EMBC, pp. 1077–1080 (2013)
35. Wu, W., Chen, A.Y.C., Zhao, L., Corso, J.J.: Brain tumor detection and segmentation in a CRF (conditional random fields) framework with pixel-pairwise affinity and superpixel-level features. Int. J. Comput. Assist. Radiol. Surg. **9**(2), 241–253 (2014)
36. Zhuge, Y., et al.: Brain tumor segmentation using holistically nested neural networks in MRI images. Med. Phys. **44**, 5234–5243 (2017)
37. Zikic, D., et al.: Decision forests for tissue-specific segmentation of high-grade gliomas in multi-channel MR. In: Ayache, N., Delingette, H., Golland, P., Mori, K. (eds.) MICCAI 2012. LNCS, vol. 7512, pp. 369–376. Springer, Heidelberg (2012). https://doi.org/10.1007/978-3-642-33454-2_46

Efficient Brain Tumour Segmentation Using Co-registered Data and Ensembles of Specialised Learners

Beenitaben Shah$^{(\boxtimes)}$ and Harish Tayyar Madabushi

University of Birmingham, Birmingham, West Midlands, UK
bxs633@alumni.bham.ac.uk, h.tayyarmadabushi.1@bham.ac.uk

Abstract. Gliomas are the most common and aggressive form of all brain tumours, leading to a very short survival time at their highest grade. Hence, swift and accurate treatment planning is key. Magnetic resonance imaging (MRI) is a widely used imaging technique for the assessment of these tumours but the large amount of data generated by them prevents rapid manual segmentation, the task of dividing visual input into tumorous and non-tumorous regions. Hence, reliable automatic segmentation methods are required. This paper proposes, tests and validates two different approaches to achieving this. Firstly, it is hypothesised that co-registering multiple MRI modalities into a single volume will result in a more time and memory efficient approach which captures the same, if not more, information resulting in accurate segmentation. Secondly, it is hypothesised that training models independently on different MRI modalities allow models to specialise on certain labels or regions, which can then be ensembled to achieve improved predictions. These hypotheses were tested by training and evaluating 3D U-Net models on the BraTS 2020 data set. The experiments show that these hypotheses are indeed valid.

Keywords: Segmentation · U-Net · Brain tumour · Ensembles · Co-registration · MRI

1 Introduction

A brain tumour is an abnormal mass of tissue in which cells grow and multiply uncontrollably. Categorised into primary and secondary tumours, the most common type of primary tumours are gliomas. With the highest frequency and mortality rate, they can be categorised into Low Grade Gliomas (LGG) and High Grade Gliomas (HGG) with the latter being more aggressive and infiltrative (Pereira et al. 2016) with treatment highly dependent on early detection.

Accurate segmentation of gliomas is important not only for diagnosis and treatment but also for post-treatment analysis. Manual segmentation techniques used by experts to complete this task are time consuming, difficult (Tjahyaningtijas 2018) and subject to inter- and intra-rater errors (Pereira et al. 2016).

© Springer Nature Switzerland AG 2021
A. Crimi and S. Bakas (Eds.): BrainLes 2020, LNCS 12659, pp. 15–29, 2021.
https://doi.org/10.1007/978-3-030-72087-2_2

Automated methods could save physicians time and provide an accurate and reproducible solution for further tumour analysis and monitoring. However, this is a challenging task as shape, structure and location of these tumours are highly irregular, made harder by intensity inhomogeneity amongst the same MRI sequences and acquisition scanners (Pereira et al. 2016) as well as the imbalance in healthy and tumorous tissues, evident in the resulting data set.

As noted by Menze et al., the number of publications devoted to automated brain tumour segmentation has grown exponentially in the last several decades, emphasising not only the need for a solution but also that this is still an area under development. Deep Learning methods, learning a complex hierarchy of features directly from the data, have proven to be the most effective for this task (Pereira et al. 2016). It is unchallenged now that Deep Learning, specifically Convolutions Neural Networks (CNNs) and its variants, dictate the state-of-the-art in biomedical image segmentation (Pereira et al. 2016).

Co-registration is a method of aligning two volumes such that information is pulled into a single volume. Usually, its application is in the form of aligning anatomical and functional scans so that corresponding image features are spatially aligned (Despotović et al. 2015).

This paper proposes the use of co-registration across multiple MRI modalities to produce a single volume on a per-patient basis, hypothesising that this will produce an informative volume on which a deep learning model can be trained. Modality refers to a particular type or sequence of MRI. This work builds on the work by Uhlich et al., who produce robust segmentation on the basis of a co-registered T1c and Flair atlas (Uhlich 2018), which are different MRI modalities. This paper thus examines the effect of such a data processing method so as to prove that training using such data can result in competitive performance in less time with a lower memory requirement. This is especially crucial given the carbon footprint of deep learning (Anthony et al. 2020).

This paper additionally hypothesises that ensemble models consisting of models trained on different MRI modalities, each specialising on specific labels will lead to a well performing, unbiased, and generic system while simultaneously reducing training time when trained in parallel.

Networks vary in performance, with architectural choices and training settings largely impacting behaviour (Kamnitsas et al. 2018). Following this, Isensee et al. focused on training rather than architectural changes concluding that one may be able to achieve a competitive model from existing architectures purely by focusing on model training. Based on this conclusion, this paper tests the proposed hypotheses over a single, well-tuned model as opposed to multiple architectures. By focusing on tuning a single network, the aim was to thoroughly examine the proposed methodologies under identical conditions; conditions optimised for competitive performance.

This work is evaluated on data from the Multimodal Brain Tumour Segmentation Challenge 2020. The task in relation to this challenge is to segment MRI volumes into three hierarchical regions: whole tumour (whole), tumour core

(core) and enhancing tumour (enh) (Menze et al. 2015; Bakas et al. 2017; Bakas and Reyes et al. 2018).

In summary, this paper aims to test the following hypothesis:

1. Can the use of co-registration of different MRI modalities result in competitive performance while using less computational resources?
2. Can performance be improved by the use of ensemble models consisting of individual models trained on different MRI modalities, each specialising on specific labels?

The remainder of this paper is organised as follows. Section 2 discusses the methodology employed. Evaluation, experimentation and results are detailed in Sects. 3 and 4. Finally, future work and conclusions are presented in Sects. 5 and 6.

2   Related Work

2.1   BraTS: Challenge and Data Set

BraTS 2020 constitutes of multi-institutional pre-operative MRI scans and focuses on a range of tasks. For the purpose of this paper only the task of segmentation of brain tumours, namely gliomas, is considered. The BraTS 2020 training data set is composed of 369 patients' MRIs. Each patient has multi-modal MRI data available – T1, T1c, T2 and Flair. Obtained from various clinical protocols and several institutions, all MRIs were segmented manually following standardised annotation protocols (*MICCAI BRATS - The Multimodal Brain Tumor Segmentation Challenge* 2019). Annotations constitute of four distinct labels:

- Label 0: Background
- Label 1: Necrotic and non-enhancing tumour core (NCR/NET)
- Label 2: Peritumoral edema (ED)
- Label 4: GD-enhancing tumour (ET)

It should be noted that label distribution in this data set is highly unbalanced.

2.2   Related Literature

The original U-Net implementation works with 2D data (Ronneberger et al. 2015); biomedical data is primarily volumetric. Following this, Çiçek et al. 2016 propose an alternate network which takes as input 3D volumes and processes them with equivalent 3D operations, mirroring those found in a 2D U-Net. Like the standard U-Net, the contraction and expansion path each consisted of four resolution steps. Each convolution block was followed by ReLU and max-pooling functions. A weighted cross-entropy loss was used to suppress frequently seen background voxels and increase weights for those labels which appear less frequently (Çiçek et al. 2016). Working with Xenopus kidney data, an average Intersection over Union (IoU) value of 0.863 was achieved. Trained from scratch and not optimised in any way for this particular application, this result indicates great performance of a 3D architecture when working with volumetric data.

Milletari et al. 2016 also propose a 3D segmentation approach, the V-Net architecture, for prostate segmentation. Additionally, they propose that novel objective functions, based on the Dice coefficient, can optimise training. Their results show such an architecture and training loss result in fast and accurate segmentation.

Ranking second place in BraTS 2018, Isensee et al. 2019 demonstrate the power of a well trained model, proposing that a well-trained U-Net, without any significant architectural alterations, is capable of generating state-of-the-art segmentations. A 3D U-Net architecture, very close to the original publication, was implemented and the training procedure was optimised to maximise the performance on the BraTS 2018 data set. All MRI modalities were normalised independently. The training procedure consisted of randomly sampled patches of size $128 \times 128 \times 128$ and a batch size of 2 (Isensee et al. 2019). Training was done for a maximum of 500 epochs using the Adam optimiser with an initial learning rate of 1e−04. Trained with a multi-class Dice loss implementation the model achieved Dice scores of 87.81/80.62/77.88 (whole, core, enh) on the test set.

Competitive performers of the BraTS 2017 challenge included Kamnitsas et al. 2018, who proposed to Ensemble Several Models for Robust Segmentation (EMMA). This method considered the effect of aggregating a range of models in order to lessen the effect of meta-parameters of individual models as well as over-fitting. EMMA took advantage of several independently trained models to produce a final ensembled segmentation result. The DeepMedic architecture (Kamnitsas et al. 2017) and variants of a Fully Connected Network and U-Nets were combined to achieve good generalisation over the 2017 data set. During training of U-Nets specifically, crops of size $64 \times 64 \times 64$ were used with a batch size of 8; ReLUs were used to add non-linearity and cross-entropy was used with AdaDelta and Adam. The ensembled system achieved dice scores of 90.1/79.7/73.8 (whole, core, enh).

Zhou et al. 2019 also proposed the use of an ensemble of different networks. The brain tumour segmentation task was broken down into three different but related sub-tasks to deal with the class imbalance problem. 1: Coarse segmentation to detect the whole tumour. 2: Refined segmentation for whole tumour and its intra-tumoural classes. 3: Precise segmentation for the enhancing tumour. This resulted in a multi-task learning framework that incorporated the three sub-tasks into an end-to-end holistic network consisting of a variation of deep learning networks. Pre-processing consisted only of individual MRI normalisation. Input patches of size $32 \times 32 \times 16 \times 4$ were used. (Zhou et al. 2019). Evaluated on the BraTS 2018 data set, the average Dice Scores were 0.0905/0.8651/0.8136 (whole, core, enh).

3   Methodology

In this section, the methodology employed is discussed. The data set used, data pre-processing, model architecture and training approaches are detailed.

3.1   Data Pre-processing

Bias field signals affect the intensity gradient of MRI scans, making some parts of the images brighter than others, despite them being structurally the same. This can introduce errors in segmentation as incorrect exposure makes it harder to accurately identify, group and classify voxels (Larsen et al. 2014). Owing to this, the N4 bias field correction algorithm (Song et al. 2017) was applied so as to correct this signal.

Following Bias Field Correction, normalisation, the process of changing the range of pixel intensity values, was applied as a means of bringing the numeric values to a common scale without distorting the differences in the ranges of the values.

Following these two pre-processing steps, which were applied to all scans in the data set, co-registration was applied in an iterative manner to multiple MRI modalities allowing for spatial alignment, producing a single volume in return. Typically used for intra-subject scan alignment, applied to two different scan types e.g., MRI and CT scans or to a maximum of two MRI modalities, this method effectively results in multi-scan information being drawn into a single scan volume. It can be speculated that the repeated application of co-registration across all four MRI modalities, on a per subject basis, could result in a single, concise and informative scan where significant information from each modality is maintained and combined to form a 'super MRI' of sorts. As mentioned previously, this paper theorises that co-registering multiple modalities may result in a clear probability atlas with which to do segmentation. The hypothesis looks to prove that bringing together the modalities in such a way will enhance the distinct information captured in each modality whilst simultaneously balancing the uncertainty, achieving a clear probability atlas. Not only could this provide the best possible prior probabilities but also reduce training time and memory requirements as all information can now be presented in a single volume. This method of co-registering multiple modalities is used to test this hypothesis. Figure 1 shows an example of that co-registering multiple modalities results in.

3.2   Model Architecture

As biomedical data is volumetric, a 2D model would process the data on a per slice basis. This is both tedious and inefficient as neighbouring slices almost always show the same information (Çiçek et al. 2016) and handling volumetric data in such a manner results in the loss of spatial information. To deal with such data effectively, a 3D U-Net was implemented based on the 2D U-Net outlined in the original paper (Ronneberger et al. 2015).

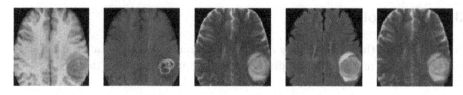

Fig. 1. Starting from the left most image of the figure, T1, T1c, T2, Flair and the result of co-registering T1c, T2 and Flair modalities

Figure 2 shows the final U-Net architecture. This U-Net architecture was subsequently used to form two different models. Firstly, it was trained using co-registered data to explore such data's effectiveness and enable the testing of the first hypothesis. Such models are referred to as co-reg models for the remainder of this paper.

Secondly, it was trained on single MRI modalities following which the individual models were combined to produce an ensemble. Ensembles were used in this manner to examine if training models on different MRI modalities allow them to specialise on different tumour labels or sub-regions so that when combined, each predicts its specialization accurately, producing an overall well performing model, thus allowing for the testing of the second hypothesis. *Specialisation* refers to a model's ability to outperform other models on certain labels. The effect of such a model on the data bias and variance trade-off is also examined. If across models each label is predicted confidently, then the class imbalance problem may be overcome. Ensembles are a combination of a set of weak base learners (Lutins 2017). Usually, these learners are trained on data of the same format. Instead here, each learner was trained on a different modality so as to test the stated hypotheses. Hence, instead of a combination of weak learners, ensembles of a set of learners specialised on certain labels are considered, creating a multi-expert system.

Input to this network were patches of size $128 \times 128 \times 128$ with a batch size of 1. The ground truth was one- hot encoded, producing a $5 \times 128 \times 128 \times 128$ map, mirrored in the output shape. The model was trained for a maximum of 250 epochs, where training was terminated early, using the early stopping mechanism, if validation loss did not improve over the last 100 epochs. The training was carried out using the Adam optimiser with an adaptive learning rate, starting at 1e−04 and shrinking by a factor of 0.1 if validation loss did not improve in the last 10 epochs. Appropriate losses were used to train the models. To avoid over-fitting the aforementioned strategies of training for an appropriate number of epochs, randomisation of data points during training and the splitting of the data into training, validation and test sets were employed.

The BraTS 2020 data set (Menze et al. 2015; Bakas et al. 2017; Bakas and Reyes et al. 2018; Bakas et al. 2017a, 2017b) was split into a training (75%) and held-out (25%) set during training. The held-out set provided an unbiased evaluation of a model whilst training, allowing a model's generalisation ability to increase and its use aided hyperparameter training. The BraTS Validation set

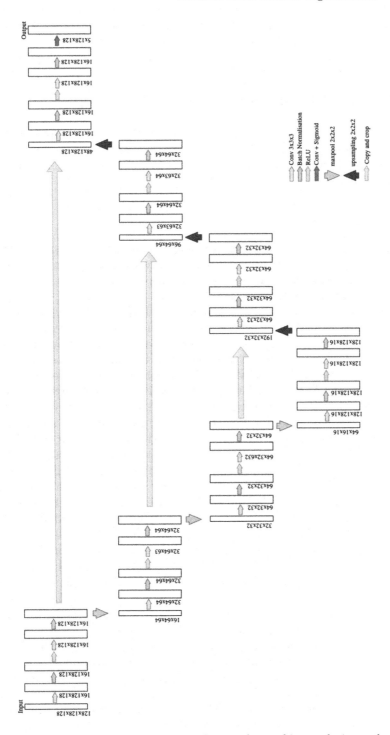

Fig. 2. Implemented 3D U-Net architecture. It is made up of 3 convolution and upsampling blocks connected by max pooling, upsampling and skip connections (copy and concatenate).

was then used to gauge the performance of the fully trained models. It should be noted that a held-out set was kept aside for rapid testing and development and the inclusion of that data could lead to an improvement in performance.

4   Empirical Evaluation

Results are presented and examined in this section, both in relation to key variables of the experiments carried out and the original hypotheses stated earlier in the paper. Performance of models is also compared to prior participants of the BraTS challenge.

4.1   Results

The baseline model was a U-Net with randomly chosen hyperparameters trained on a single modality. Comparing its performance with the co-reg and ensemble methods, it is clear to see the sizable gains in performance and accuracy with the use of these methods.

The success of co-reg models, especially with the whole tumour region, as visible in Tables 1 and 2, proves that the hypothesis stated earlier in this paper does in fact hold. Applying co-registration to volumes does result in a single, concise and informative volume. Dice scores of 0.66 on well represented labels proves that this method can prove effective. Furthermore, Dice scores of 0.6 were achieved over the Validation data set, as shown in Table 2, further evidencing the fact that the use of co-registered data is a viable and effect method. This is a key result as handling data in such a manner could result in an exponential speed up in training time as the input data is reduced by a factor of 4 whilst not increasing the data pre-processing step massively either. The limiting factor for these models, much like for all the models tested, was the class imbalance issue. The severity of this issue can be seen in the Dice scores of the enhancing tumour label (Label 4). Although class imbalance is a limiting factor, the impact of it is similar to that on the ensemble models, proving that the reduced accuracy in this model's prediction capability isn't consequent or increased due to the employment of co-registration. Furthermore, the Dice scores of the core and enhancing labels are many-fold higher than the base-line model confirming, to an extent, that important information is maintained through the use of co-registration, leading to the conclusion that the use of this method does result in a 'super MRI' which can be used to effectively train a model in less time and with a lowered memory requirement. Overall, the Dice scored achieved by the co-reg models for the held-out set and validation set prove that the use of co-registered data can result in an effective and well-performing model which has the ability to consistently perform on data outside the training data set.

This paper also hypothesised that training models on separate modalities would result in robust performance. This is confirmed by the results presented in Tables 1 and 2. High accuracy values of 77% on labels which are well presented

Table 1. Whole, core and enhancing dice scores for the ablation study of different models used in experiments. Results from BraTS 2019 are not directly comparable.

Model	Dice_WT	Dice_TC	Dice_ET
Brats 2019 - 1st Place	0.8879	0.8369	0.8326
Baseline	0.3146	1.63E−05	1.11E−05
Co-reg (4 modalities, soft Dice)	0.5993	**0.4513**	0.086
Co-reg (3 modalities, soft Dice)	0.6116	0.2936	0.095
Co-reg (4 modalities, Balanced avg loss)	0.6206	0.2051	0.00937
Co-reg (3 modalities, Multi-label loss)	**0.6609**	**0.3290**	**0.1733**
Ensemble (4 modalities, soft Dice, Aggregation)	0.7248	0.6092	0.5625
Ensemble (4 modalities, soft Dice, Voting)	**0.6858**	**0.6279**	**0.5835**
Ensemble (3 modalities, soft Dice, Aggregation)	0.7515	0.5058	0.4100
Ensemble (3 modalities, soft Dice, Voting)	**0.7713**	**0.5353**	**0.4363**

Table 2. Whole, core and enhancing dice scores for the ablation study of different models over BraTS 2020 Validation data set.

Model	Dice_WT	Dice_TC	Dice_ET
Co-reg (4 modalities, Balanced avg loss)	0.5933	0.1865	0.05314
Co-reg (3 modalities, Multi-label loss)	0.6080	0.2954	0.1432
Ensemble (3 modalities, soft Dice, Aggregation)	0.7290	0.4061	0.3391
Ensemble (3 modalities, soft Dice, Voting)	**0.7592**	**0.4686**	**0.3237**
Ensemble (4 modalities, soft Dice, Aggregation)	0.6835	0.3075	0.2667
Ensemble (4 modalities, soft Dice, Voting)	0.6760	0.3117	0.2776

in the data set show that individual models were able to specialise depending on modalities and an ensemble of these individuals with an appropriate method resulted in an accurate segmentation map. Enhancing tumours gained Dice scores of 43% and 58%; not as high as the whole tumour label however high enough to prove the effectiveness of specialisation. This leads to the conclusion that ensembles of such models are effective, with specialisation occurring within individual models, where a more precise ensemble mechanism and better management of data bias can lead to more robust and accurate segmentation. Again, such performance can also be seen over the validation data set proving that the models are indeed robust and generalised so that performance is constant even for unseen data.

Finally, ensembles did consistently outperform the co-reg models indicating that the training of models on individual modalities did provide a performance gain. There was a significant difference in the Dice scores of these models, indicating that this mechanism may have the ability to reduce the effect of data bias in a more successful manner than the strategies applied to the co-reg models.

Furthermore, it indicates that this may be a more effective method for leveraging the different sorts of information captured in the various MRI modalities so that it can be used in the most effective way.

4.2   Performance Comparison

The performance of this paper's best performing model, at the task of brain tumour segmentation, in comparison to other participants to the BraTS 2020 challenge can be seen in Table 3 (below). Other models of this paper's ablation study, whilst having a similar Dice score for the whole tumour, do differ more greatly in performance for the core and enhancing tumours in comparison to the paper's best model.

The results in Table 4 show this paper's best performing model's performance over the BraTS 2020 Test set. The mean Dice values across the three labels over the test set are not only in line with the performance seen during validation, but the values for both the tumour core and enhancing tumour are actually higher. Furthermore, where the values in the 1st quartile lie close the averages seen during validation and testing, the values in the third quartile are significantly higher, getting close to top performers presented in Table 3. These values indicate that this approach of pre-processing and ensembling of models can achieve competitive results across all labels and holds the potential to perform competitively across the data set given further experimentation.

Table 3. Results over the BraTS 2020 Validation data set from other participants (taken from the leader board).

Team name	Dice_WT	Dice_TC	Dice_ET
MIC-DKFZ	0.9124	0.85044	0.7918
lescientifik	0.8986	0.84383	0.7968
Anonymity	0.9088	0.8543	0.7927
...	...	...	...
BeenitaShah	**0.7592**	**0.4686**	**0.3237**

Table 4. Averages for whole, core and enhancing dice scores of best performing model (Ensemble - 3 modalities, soft Dice, Voting) over BraTS 2020 Test set.

Model	Dice_WT	Dice_TC	Dice_ET
Mean	0.7555	0.5319	0.4331
25 quantile	0.7021	0.3139	0.2649
75 quantile	0.8919	0.7590	0.6310

4.3   Analysis

Impact of Number and Types of MRI Modalities. Both co-reg and ensemble models were trained and evaluated over combinations of MRI modalities. Considering the co-registered data, results from models trained on co-registered data of 3 vs 4 modalities did have differences however there was no significant difference in the Dice scores of these two models; the differences seen here were not as a direct result of the modalities used.

For the ensemble models, there was a statistically significant difference between the Dice scores of the different tumours proving that the 3 modality ensembles were in fact the better performing models overall. Therefore, the number of modalities that are most effective for ensemble models are 3 using the following modality types: T1c, T2 and Flair.

Impact of Ensemble Techniques. Two simple but effective ensemble techniques were tested as a means to maximise performance. The first was max voting. The second technique involved the use of an aggregation algorithm which calculates class probabilities by summing model predictions and calculating an average probability for each voxel. Both algorithms perform similarly across label groups. Although consistent performance improvement is seen with the use of max voting, the differences in Dice scores is statistically insignificant. Despite the improvements being insignificant, they were consistent and in light of this max voting is accepted as a better ensemble technique.

4.4   Impact and Implications of Loss Functions

Initially, the standard Dice coefficient metric and soft Dice loss were used whilst training. As a means of dealing with the class imbalance problem, a number of different multi-class and weighted loss implementations were tested.

For co-reg models experimentation led to the conclusion that the choice of loss functions was a trade-off between better predicting the whole and enhancing tumour vs the identification of the core tumour. Given that the non-standard implementations better predict both label groups compared to soft Dice, they are considered the preferable choice for co-reg models.

With regards to the models trained using individual modalities, under all conditions, soft Dice consistently outperformed the others. This can be taken as an indication that ensembles of such models may be handling the class imbalance problem in some intrinsic way, as the improvements found when using adapted losses with co-reg models was not seen here. Hence, these models were trained using the standard soft Dice implementation.

4.5   Performance Evaluation

In comparison to top performers of the BraTS 2020 challenge, as presented in Table 3 in Sect. 3.2, the models and methods tested in this paper could be improved so that competitive results can be achieved across all labels.

With regards to the co-reg models a more tailored pre-processing regime could be been applied so that specific information from the different modalities could be combined in a more effective manner such that performance on par with, and possibly greater than, the ensemble models could be achieved.

For the ensemble models more complex and informative ensemble techniques could have been applied so that the results from individual models could have been combined in a more successful manner so as to improve the accuracy of the core and enhancing tumours further.

Finally, a range of architectures could have been tested as they may have proven better at maximising results with the use of the methods outlined in this paper.

5   Discussion

Following the results seen the Dice scores achieved the following statements can be made:

1. The proposed method of training models on co-registered data does achieve competitive results on well represented labels, proving that the use of co-registration does produce a 'super MRI' and training models on such data does produce competitive models, achieving Dice scores of 0.62 and greater on certain labels. Hindered by data bias, an issue for deep learning models in general, these results indicate that if this data imbalance issue is dealt with effectively prior to or whilst training, competitive performance across all labels can be achieved in significantly less time.
2. Performance of ensemble models proves that models trained on individual modalities can become specialised and combined to produce a well-performing, general model which is more capable of dealing with data bias than the co-reg models. With a Dice score of 0.77 on well represented labels, competitive performance can be expected across all labels under a balanced data set.

Other successful optimisation techniques included the use of a weighted loss as a means of handling the class imbalance issue. Despite its persistence, the weighted loss functions used did manage to counteract the imbalance issue, albeit not completely. Although the loss functions were not sufficiently effective in solving the class imbalance problem, their use did result in gains in prediction accuracy, especially for enhancing tumours, indicating to possible performance gains for the co-reg models if class imbalance is dealt with fully. Finally, performance of co-reg models indicates such models' ability to meet the state-of-the-art, proving that co-registration of modalities to reduce data dimensionality is a viable method, one which can result in sizable training time and memory requirement reductions whilst maintaining prediction capability.

6   Conclusions and Future Work

Two main hypotheses were stated and explored in this study.

1. Combining multiple MRI modalities through the use of co-registration can produce a competitive model whilst simultaneously reducing time and memory requirements.
2. Training models on a single MRI modality can result in their specialisation over certain labels, producing a more informed ensemble model.

Findings supporting both these statements are presented. Ensembles of models trained independently on T1c, T2 and Flair achieved Dice scores of 0.7713/0.5353/0.4363 (whole, core, enh). Not as high as state-of-the-art models in all labels, these results still show that such a training process does produce specialised individuals whose combination results in a strong model.

The proposed method of training a model on co-registered data constructed using T1, T1c, T2 and Flair achieved Dice scores of 0.6206/0.3290/0.00937 (whole, core, enh); the one trained on T1c, T2 and Flair co-registered data achieving 0.6609/0.3290/0.1733 (whole, core, enh), proving the success of this method. Despite ensembles performing better, the training of 3 or 4 individual models and following ensembling computations could prove more time consuming than training a single model on co-registered data. This, in line with the possible performance gains of a balanced data set and the co-reg model's potentially unexplored benefits, as indicated by the highest core tumour Dice score of 0.45, indicates such method's potential to match the performance of highly complex and specialised networks. The use of co-registration could lead to exponential speed up in training time, making it feasible to use larger data sets whilst maximally leveraging the information contained within and more importantly, quickly train and use models in the medical field where time is of the essence.

Due to its impact on model performance, more effective measures which balance data and increase data quality should be considered as a means to address the class imbalance problem. Moreover, the performance of other network architectures under co-registered data can be tested in the future. Furthermore, methods of pre-processing modalities, so as to enhance the retainment of the important information they contain prior to co-registration, may be explored. Data augmentation can be employed as a means of diversifying the data set. Finally, weighted ensemble methods could be tested so as to optimally leverage individual models' strengths.

References

Anthony, L.F.W., Kanding, B., Selvan, R.: Carbontracker: tracking and predicting the carbon footprint of training deep learning models. arXiv:2007.03051 [cs.CY] (2020)

Bakas, S., Akbari, H., Sotiras, A., Bilello, M., Rozycki, M., Kirby, J., et al.: Segmentation labels for the pre-operative scans of the TCGA-GBM collection (2017a). https://wiki.cancerimagingarchive.net/x/KoZyAQ

Bakas, S., Akbari, H., Sotiras, A., Bilello, M., Rozycki, M., Kirby, J., et al.: Segmentation Labels for the Pre-operative Scans of the TCGA-LGG collection (2017b). https://doi.org/10.7937/K9/TCIA.2017.GJQ7R0EF. https://wiki.cancerimagingarchive.net/x/LIZyAQ

Bakas, S., Akbari, H., Sotiras, A., Bilello, M., Rozycki, M., Kirby, J.S., et al.: Advancing the cancer genome atlas glioma MRI collections with expert segmentation labels and radiomic features. Sci. Data **4**(1), 170117 (2017). https://doi.org/10.1038/sdata.2017.117. ISSN 2052-4463

Bakas, S., Reyes, M., et al.: Identifying the best machine learning algorithms for brain tumor segmentation, progression assessment, and overall survival prediction in the brats challenge. arXiv:1811.02629 [cs, stat], November 2018

Çiçek, Ö., Abdulkadir, A., Lienkamp, S.S., Brox, T., Ronneberger, O.: 3D U-Net: learning dense volumetric segmentation from sparse annotation. In: Ourselin, S., Joskowicz, L., Sabuncu, M.R., Unal, G., Wells, W. (eds.) MICCAI 2016. LNCS, vol. 9901, pp. 424–432. Springer, Cham (2016). https://doi.org/10.1007/978-3-319-46723-8_49

Despotović, I., Goossens, B., Philips, W.: MRI segmentation of the human brain: challenges, methods, and applications. Comput. Math. Methods Med. 1–23 (2015). https://doi.org/10.1155/2015/450341. ISSN 1748-670X, 1748-6718

Isensee, F., Kickingereder, P., Wick, W., Bendszus, M., Maier-Hein, K.H.: No new-net. In: Crimi, A., Bakas, S., Kuijf, H., Keyvan, F., Reyes, M., van Walsum, T. (eds.) BrainLes 2018. LNCS, vol. 11384, pp. 234–244. Springer, Cham (2019). https://doi.org/10.1007/978-3-030-11726-9_21

Kamnitsas, K., et al.: Ensembles of multiple models and architectures for robust brain tumour segmentation. In: Crimi, A., Bakas, S., Kuijf, H., Menze, B., Reyes, M. (eds.) BrainLes 2017. LNCS, vol. 10670, pp. 450–462. Springer, Cham (2018). https://doi.org/10.1007/978-3-319-75238-9_38

Kamnitsas, K., et al.: Efficient multi-scale 3D CNN with fully connected CRF for accurate brain lesion segmentation. Med. Image Anal. **36**, 61–78 (2017). https://doi.org/10.1016/j.media.2016.10.004. ISSN 13618415

Larsen, C.T., Iglesias, J.E., Van Leemput, K.: N3 bias field correction explained as a Bayesian modeling method. In: Cardoso, M.J., Simpson, I., Arbel, T., Precup, D., Ribbens, A. (eds.) BAMBI 2014. LNCS, vol. 8677, pp. 1–12. Springer, Cham (2014). https://doi.org/10.1007/978-3-319-12289-2_1

Lutins, E.: Ensemble methods in machine learning: What are they and why use them? August 2017. https://towardsdatascience.com/ensemble-methods-in-machine-learning-what-are-they-and-why-use-them-68ec3f9fef5f

Menze, B.H., et al.: The multimodal brain tumor image segmentation benchmark (BRATS). IEEE Trans. Med. Imaging **34**(10), 1993–2024 (2015). https://doi.org/10.1109/TMI.2014.2377694. ISSN 0278-0062, 1558-254X

MICCAI BRATS - The Multimodal Brain Tumor Segmentation Challenge (2019). http://braintumorsegmentation.org/

Milletari, F., Navab, N., Ahmadi, S.-A.: V-net: fully convolutional neural networks for volumetric medical image segmentation. In: 2016 Fourth International Conference on 3D Vision (3DV), pp. 565–571. IEEE, October 2016. https://doi.org/10.1109/3DV.2016.79. http://ieeexplore.ieee.org/document/7785132/. ISBN 978-1-5090-5407-7

Tjahyaningtijas, H.P.A.: Brain tumor image segmentation in MRI image. IOP Conf. Ser. Mater. Sci. Eng. **336**, 012012 (2018). https://doi.org/10.1088/1757-899X/336/1/012012. ISSN 1757-8981, 1757-899X

Pereira, S., et al.: Brain tumor segmentation using convolutional neural networks in MRI images. IEEE Trans. Med. Imaging **35**(5), 1240–1251 (2016). https://doi.org/10.1109/TMI.2016.2538465. ISSN 0278-0062, 1558-254X

Ronneberger, O., Fischer, P., Brox, T.: U-Net: convolutional networks for biomedical image segmentation. arXiv:1505.04597 [cs], May 2015

Song, S., Zheng, Y., He, Y.: A review of methods for bias correction in medical images. Biomed. Eng. Rev. **3**(1) (2017). https://doi.org/10.18103/bme.v3i1.1550. http://journals.ke-i.org/index.php/bme/article/view/1550. ISSN 23759143, 23759151

Uhlich, M., et al.: Improved brain tumor segmentation via registration-based brain extraction. Forecasting **1**(1), 59–69 (2018). https://doi.org/10.3390/forecast1010005. ISSN 2571-9394

Zhou, C., Chen, S., Ding, C., Tao, D.: Learning contextual and attentive information for brain tumor segmentation. In: Crimi, A., Bakas, S., Kuijf, H., Keyvan, F., Reyes, M., van Walsum, T. (eds.) BrainLes 2018. LNCS, vol. 11384, pp. 497–507. Springer, Cham (2019). https://doi.org/10.1007/978-3-030-11726-9_44

Efficient MRI Brain Tumor Segmentation Using Multi-resolution Encoder-Decoder Networks

Mohammadreza Soltaninejad, Tony Pridmore, and Michael Pound[✉]

Computer Vision Lab, School of Computer Science, University of Nottingham, Nottingham, UK
{m.soltaninejad,tony.pridmore,michael.pound}@nottingham.ac.uk

Abstract. In this paper, we propose an automated three dimensional (3D) deep learning approach for the segmentation of gliomas in pre-operative brain MRI scans. We introduce a state-of-the-art multi-resolution architecture based on encoder-decoder which comprise of separate branches to incorporate local high-resolution image features and wider low-resolution contextual information. We also used a unified multi-task loss function to provide end-to-end segmentation training. For the task of survival prediction, we propose a regression algorithm based on random forests to predict the survival days for the patients. Our proposed network is fully automated and designed to take input as patches that can work on input images of any arbitrary size. We trained our proposed network on the BraTS 2020 challenge dataset that consists of 369 training cases, and then validated on 125 unseen validation datasets, and tested on 166 unseen cases from the testing dataset using a blind testing approach. The quantitative and qualitative results demonstrate that our proposed network provides efficient segmentation of brain tumors. The mean Dice overlap measures for automatic brain tumor segmentation of the validation dataset against ground truth are 0.87, 0.80, and 0.66 for the whole tumor, core, and enhancing tumor, respectively. The corresponding results for the testing dataset are 0.78, 0.70, and 0.66, respectively. The accuracy measures of the proposed model for the survival prediction tasks are 0.45 and 0.505 for the validation and testing datasets, respectively.

Keywords: Convolutional neural network · Encoder-decoder · Deep learning · MRI · Brain tumor segmentation

1 Introduction

Magnetic resonance imaging (MRI) is the most common modality which is used for diagnosis, treatment planning, and survival prediction [1]. Accurate segmentation of brain tumors plays an important role in measuring the tumor features to aid the clinical tasks. MR images are acquired using different acquisition protocols such as fluid-attenuated inversion recovery (FLAIR), T1-weighted (with and without contrast agent), and T2-weighted to distinguish between different tumor tissue types.

Segmentation of human brain tumors, traditionally performed manually by physicians in medical images, is a vital and crucial task. The clinical manual delineation procedures are subjective and inherently prone to misinterpretation and human error.

© Springer Nature Switzerland AG 2021
A. Crimi and S. Bakas (Eds.): BrainLes 2020, LNCS 12659, pp. 30–39, 2021.
https://doi.org/10.1007/978-3-030-72087-2_3

Therefore, developing fast, reliable and automated algorithms may aid a more accurate diagnosis. This also may provide remarkable advances in the long term in treatment planning for the patients. Accurate brain tumor segmentation algorithms become even more highlighted when having three-dimensional observation of the volumetric MRI images by the machine instead of a natural two-dimensional view of a human interpreter.

Many efforts addressed inventing such an automatic segmentation system. Powerful computing processors as well as the availability of big datasets have emerged recently. The recent advances in deep learning have revolutionized many aspects of the technology related to various data types, i.e. text, speech, and image. Deep learning has been widely used in the medical imaging domain for various clinical tasks such as diagnosis, detecting, and segmenting lesions and tumors. Segmentation of multimodal MR images for tumor segmentation is inherently a challenging task due to the high dimensionality of the input data and variations in size, shape, and texture of the tumors.

Most of the recent tumor segmentations have focused on deep learning methods due to the recent developments in deep neural networks for image analysis [2], such as fully convolutional networks [3] and U-Net [4–6]. In this paper, we propose a volumetric deep network based on encoder-decoder architecture [7]. We applied significant architectural changes to provide semantic segmentation with an end-to-end training strategy [8]. Our contribution is designing a downsampling path to take a wider field of view and help the network to learn more contextual information.

The segmentation results from our proposed method are then used for the survival prediction task. The statistical features from the segmented tumor tissue types are extracted and then applied to a random forest (RF) [9] for survival prediction. RF has been widely used as an ensemble learning algorithm for both classification and regression purposes. We use the regression RF to predict the number of days for the survival task.

2 Materials and Methods

This section describes our segmentation and survival prediction approaches as well as the dataset. We first describe the dataset that is used for training and validation of our proposed network. We then outline the multi-resolution encoder-decoder network followed by the random forest based method for survival prediction.

2.1 Dataset

The Multimodal Brain Tumor Segmentation Challenge (BraTS) 2020 [10–14] training dataset, including 369 patient cases, are used to train the proposed network. The ground truth is manually annotated by experts and provided on the Center for Biomedical Image Computing and Analytics (CBICA) portal. The BraTS 2020 dataset also includes 125 validation and 166 testing datasets, which will be later used to evaluate the network with a blind testing approach. For the task of survival prediction, CBICA provides 236 of the training patient cases as well as their relevant survival information. There are also 29 validation cases and 107 testing cases for validation of the survival task.

We perform pre-processing on the MRI dataset, i.e. intensity normalization and bias filed correction. The intensities of each protocol are first normalized by subtracting each pixel from the average intensity and then dividing by their standard deviation. Then, the toolbox provided in [15] is used for bias field correction.

2.2 Segmentation

Our proposed segmentation approach comprises two main branches with different resolution levels. The native resolution branch is an encoder-decoder with residual blocks based on the method proposed in [7]. We modified the encoder-decoder for semantic segmentation of volumetric images by using the appropriate 3D convolution and pooling layers. We also replaced the activation function in the final layer and utilized a sigmoid function. The native branch deals with higher resolution but small patches of the image. The residual feature size is 128 throughout the network. The convolution filters are $3 \times 3 \times 3$, with stride and paddings of $1 \times 1 \times 1$. The max-pooling kernels are $2 \times 2 \times 2$ in the encoder and bilinear upsampling is used for the decoder.

To consider the contextual information around the small patch for segmentation, we introduce another branch with a wider field of view but lower resolution. Features from the wider receptive field are then cropped so that they can be aligned with the native branch. The features from both branches are then concatenated and applied to the final layer for segmentation. For each path, two convolution layers of size $1 \times 1 \times 1$ are used to provide the segmentation heatmaps from the features.

Figure 1 shows the architecture of the proposed network. Multiple loss functions are used to train the network so that each branch learns useful features as well as the final layer. L_1 and L_2 are the loss functions of each pipeline and L_{end} is the loss function corresponding to the final segmentation. The network uses auxiliary loss function on each branch and is trained end-to-end using a multi-task criterion, i.e. $L = L_1 + L_2 + L_{end}$. The network is trained using stochastic gradient descent and the output is passed through a Sigmoid function. Binary classification error (BCE) is used as loss function for each class. A separate network is trained for each class against other labels and the background.

2.3 Survival Prediction

We utilize the statistical features of the segmented tumor regions from the previous segmentation stage for the task of survival prediction. The volume sizes for the whole tumor, tumor core, and enhanced tumor that is moralized compared to the whole brain volume size are considered as a set of features. Another set of features is the mean intensity value for each tumor tissue type, i.e. the whole tumor, tumor core, and enhanced tumor. Therefore, 8 features were used for the task of survival prediction. The feature vectors for each volume of interest (VOI) are then applied to an RF. Random forests parameters, i.e. tree depth and the number of trees, are tuned using 5-fold cross-validation on the training dataset. The corresponding parameters, 50 trees with a depth of 10 provides optimum generalization and accuracy. Utilizing the random forest in regression mode generates the predictions as the number of survival days.

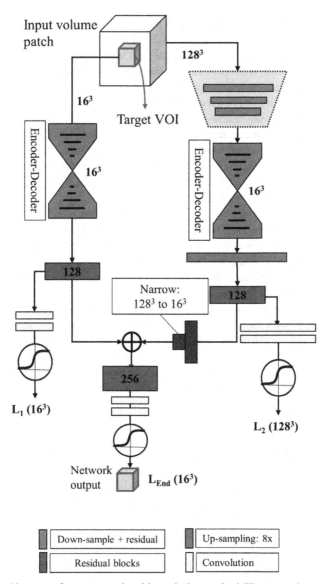

Fig. 1. The architecture of our proposed multi-resolution method. The network uses two branches, a native resolution branch with smaller volumetric input, and a low-resolution branch with a larger field of view.

3   Results

3.1   Segmentation Task

The network for segmentation is implemented using Pytorch [16] on an Nvidia Titan X with 12 GB RAM with the operating system Linux. During training, patch sizes of

128^3 were cropped form the volumes and 5 mini-batches were used. The network was trained using rmsprop with a learning rate of 2.5×10^{-4} for a maximum of 100 epochs. The network is trained using the ground truth that is provided for the BraTS training. However, the ground truth is not provided for the evaluation set and a blind testing system is applied. The evaluation measures (provided by the CBICA's Image Processing Portal) are Dice score, sensitivity, specificity, Hausdorff distance.

Table 1 provides the segmentation results of the proposed segmentation method on the BraTS 2020 validation and testing datasets according to the CBICA blind testing evaluation system. The mean Dice overlap measures for automatic brain tumor segmentation of the validation dataset against ground truth are 0.87, 0.80, and 0.66 for the whole tumor, core, and enhancing tumor, respectively. The corresponding results for the testing dataset are 0.78, 0.70, and 0.66, respectively.

Figure 2 shows segmentation regions obtained from our proposed multi-resolution method for two sample cases of the BraTS 2020 validation dataset. The tumor tissue types are overlaid on T2 modality and depicted in axial, sagittal, and coronal views. The original FLAIR and T1-CE for each corresponding case is also presented in Fig. 2 for visual comparison. Figure 3 shows the corresponding segmentation results for two sample cases from the testing dataset.

The limitation of our proposed method underlies in the margins of the images or volumes. A margin with the size of the difference between the high and low receptive fields of the network, i.e. 56 pixels around the image, will not be used during the training. This is not an issue in the specific case of the BraTS dataset since they have enough black margin around the brain. However, to use our proposed method for other datasets, zero padding might be necessary to include this margin if it contains important content.

3.2 Survival Prediction

For the survival prediction task, 29 patient cases are specified by the CBICA portal for the validation phase. The number of patient cases for the testing phase is 107 and the predictions are provided as the number of survival days. The random forest for survival prediction task is implemented using MATLAB 2019a. The evaluation metrics for this survival prediction are as follows: accuracy for classification mode and mean square error (MSE), median and standard deviation of SE, and Spearman R for the regression model.

The survival predictions are also categorized into three classes: short (less than 10 months), medium (between 10 to 15 months), and long (more than 15 months). Table 2 shows the survival prediction results that are evaluated by the CBICA system. Our proposed method obtained the classification accuracy measures of 0.45 and 0.505 for the validation and testing datasets, respectively.

Table 1. Segmentation accuracy metrics for our proposed network on the validation and testing dataset. The results are evaluated and provided by CBICA portal blind testing system. ET: enhancing tumor, WT: whole tumor, TC: tumor core.

Dataset		Dice			Sensitivity			Specificity			Hausdorff (95%)		
		ET	WT	TC	ET	WT	TC	ET	WT	TC	ET	WT	TC
Validation	Mean	0.66	0.87	0.80	0.68	0.86	0.78	1.00	1.00	1.00	47.33	6.91	7.80
	STD	0.31	0.10	0.18	0.32	0.12	0.21	0.00	0.00	0.00	116.5	9.08	11.97
	Median	0.82	0.91	0.86	0.81	0.89	0.87	1.00	1.00	1.00	2.83	3.74	4.12
	25-quantile	0.57	0.86	0.72	0.59	0.82	0.72	1.00	1.00	1.00	1.62	2.45	2.24
	75-quantile	0.87	0.93	0.92	0.89	0.94	0.93	1.00	1.00	1.00	9.49	5.92	7.68
Testing	Mean	0.66	0.78	0.70	0.70	0.78	0.74	1.00	1.00	1.00	35.49	11.95	29.44
	STD	0.32	0.26	0.34	0.34	0.27	0.32	0.00	0.00	0.00	96.31	19.58	80.05
	Median	0.80	0.89	0.88	0.85	0.88	0.88	1.00	1.00	1.00	2.34	4.42	4.16
	25-quantile	0.63	0.81	0.63	0.66	0.80	0.72	1.00	1.00	1.00	1.41	3.00	2.24
	75-quantile	0.87	0.92	0.93	0.93	0.93	0.94	1.00	1.00	1.00	8.78	7.81	11.04

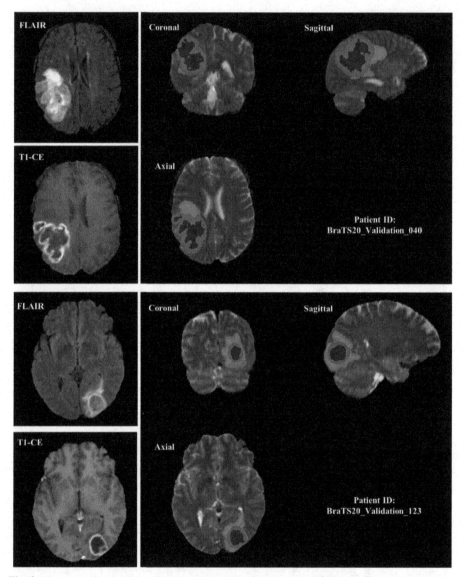

Fig. 2. Segmentation results using our proposed method for two patient cases from the validation dataset (patient IDs: Brats20_Validation_040 and Brats20_Validation _123). The tumor tissue types are overlaid on the corresponding T2 image; blue: necrosis and on-enhancing, green: edema, red: enhancing. (Color figure online)

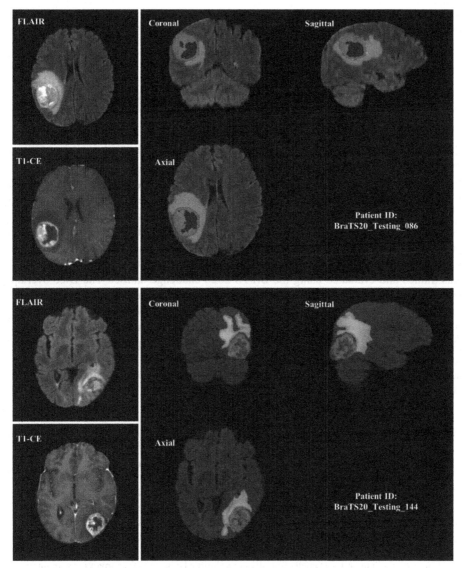

Fig. 3. Segmentation results using our proposed method for two patient cases from the testing dataset (patient IDs: Brats20_Testing_086 and Brats20_Testing_144). The tumor tissue types are overlaid on the corresponding T2 image; blue: necrosis and on-enhancing, green: edema, red: enhancing. (Color figure online)

Table 2. The evaluation results for the survival prediction task on the validation dataset. The measures are provided by the CBICA evaluation system.

Dataset	Accuracy	MSE	Median SE	STD SE	Spearman R
Validation	0.45	109564	39601	192755	0.28
Test	0.505	400914	58564	1154203	0.306

4 Conclusion

In this paper, a novel multi-resolution method based on encoder-decoder was proposed for the volumetric segmentation of brain tumors in MRI images. The proposed network comprises two encoder-decoder paths with different resolutions and receptive fields. The native resolution path has a smaller field of view and handles the higher resolution and provides fine segmentation boundaries. The downsampling path considers a coarse but wider field of view and takes more contextual information as input. The proposed network was trained and evaluated using the BraTS 2020 challenge dataset. The Dice overlap measures for brain tumor segmentation using our proposed method are 0.87, 0.80, and 0.66 for the whole tumor, tumor core, and enhancing tumor on the validation dataset, and 0.78, 0.70, and 0.66 on the testing dataset, respectively. The statistical intensity-based features extracted from the segmentation masks are then applied to an RF classifier to predict the survival days. Our proposed method acquired MSE and classification accuracy of 109564 and 0.45 for the validation datasets, and corresponding measures of 400914 and 0.505 for the testing dataset.

References

1. Gordillo, N., Montseny, E., Sobrevilla, P.: State of the art survey on MRI brain tumor segmentation. Magn. Reson. Imaging **31**, 1426–1438 (2013)
2. Soltaninejad, M., Zhang, L., Lambrou, T., Yang, G., Allinson, N., Ye, X.: MRI brain tumor segmentation and patient survival prediction using random forests and fully convolutional networks. In: Crimi, A., Bakas, S., Kuijf, H., Menze, B., Reyes, M. (eds.) BrainLes 2017. LNCS, vol. 10670, pp. 204–215. Springer, Cham (2018). https://doi.org/10.1007/978-3-319-75238-9_18
3. Long, J., Shelhamer, E., Darrell, T.: Fully convolutional networks for semantic segmentation. In: Presented at the Proceedings of the IEEE Conference on Computer Vision and Pattern Recognition (2015)
4. Ronneberger, O., Fischer, P., Brox, T.: U-net: convolutional networks for biomedical image segmentation. In: Navab, N., Hornegger, J., Wells, W.M., Frangi, A.F. (eds.) MICCAI 2015. LNCS, vol. 9351, pp. 234–241. Springer, Cham (2015). https://doi.org/10.1007/978-3-319-24574-4_28
5. Feng, X., Tustison, N., Meyer, C.: Brain tumor segmentation using an ensemble of 3D U-nets and overall survival prediction using radiomic features. In: Crimi, A., Bakas, S., Kuijf, H., Keyvan, F., Reyes, M., van Walsum, T. (eds.) BrainLes 2018. LNCS, vol. 11384, pp. 279–288. Springer, Cham (2019). https://doi.org/10.1007/978-3-030-11726-9_25

6. Chen, W., Liu, B., Peng, S., Sun, J., Qiao, X.: S3D-UNet: separable 3D U-net for brain tumor segmentation. In: Crimi, A., Bakas, S., Kuijf, H., Keyvan, F., Reyes, M., van Walsum, T. (eds.) BrainLes 2018. LNCS, vol. 11384, pp. 358–368. Springer, Cham (2019). https://doi.org/10.1007/978-3-030-11726-9_32

7. Newell, A., Yang, K., Deng, J.: Stacked hourglass networks for human pose estimation. In: Leibe, B., Matas, J., Sebe, N., Welling, M. (eds.) ECCV 2016. LNCS, vol. 9912, pp. 483–499. Springer, Cham (2016). https://doi.org/10.1007/978-3-319-46484-8_29

8. Soltaninejad, M., Sturrock, C.J., Griffiths, M., Pridmore, T.P., Pound, M.P.: Three dimensional root CT segmentation using multi-resolution encoder-decoder networks. IEEE Trans. Image Process. **29**, 6667–6679 (2020)

9. Breiman, L.: Random forests. Mach. Learn. **45**, 5–32 (2001)

10. Menze, B.H., et al.: The multimodal brain tumor image segmentation benchmark (BRATS). IEEE Trans. Med. Imaging **34**(10), 1993–2024 (2015). https://doi.org/10.1109/TMI.2014.2377694

11. Bakas, S., et al.: Advancing The cancer genome atlas glioma MRI collections with expert segmentation labels and radiomic features. Nat. Sci. Data **4**, 170117 (2017). https://doi.org/10.1038/sdata.2017.117

12. Bakas, S., et al.: Identifying the Best Machine Learning Algorithms for Brain Tumor Segmentation, Progression Assessment, and Overall Survival Prediction in the BRATS Challenge (2018). arXiv preprint arXiv:1811.02629

13. Bakas, S., et al.: Segmentation labels and radiomic features for the pre-operative scans of the TCGA-GBM collection. Cancer Imag. Arch. (2017). https://doi.org/10.7937/K9/TCIA.2017.KLXWJJ1Q

14. Bakas, S., et al.: Segmentation labels and radiomic features for the pre-operative scans of the TCGA-LGG collection. Cancer Imag. Arch. (2017). https://doi.org/10.7937/K9/TCIA.2017.GJQ7R0EF

15. Tustison, N.J., et al.: N4ITK: improved N3 bias correction. IEEE Trans. Med. Imaging **29**, 1310–1320 (2010)

16. Paszke, A., et al.: Automatic differentiation in pytorch (2017)

Trialing U-Net Training Modifications for Segmenting Gliomas Using Open Source Deep Learning Framework

David G. Ellis[(⊠)] [ID] and Michele R. Aizenberg [ID]

Department of Neurosurgery, University of Nebraska Medical Center, Omaha, NE, USA
david.ellis@unmc.edu

Abstract. Automatic brain segmentation has the potential to save time and resources for researchers and clinicians. We aimed to improve upon previously proposed methods by implementing the U-Net model and trialing various modifications to the training and inference strategies. The trials were performed and tested on the Multimodal Brain Tumor Segmentation dataset that provides MR images of brain tumors along with manual segmentations for hundreds of subjects. The U-Net models were trained on a training set of MR images from 369 subjects and then tested against a validation set of images from 125 subjects. The proposed modifications included predicting the labeled region contours, permutations of the input data via rotation and reflection, grouping labels together, as well as creating an ensemble of models. The ensemble of models provided the best results compared to any of the other methods, but the other modifications did not demonstrate improvement. Future work will look at reducing the level of the training augmentation so that the models are better able to generalize to the validation set. Overall, our open source deep learning framework allowed us to quickly implement and test multiple U-Net training modifications. The code for this project is available at https://github.com/ellisdg/3DUnetCNN.

Keywords: Deep learning · Brain tumor segmentation · U-Net

1 Introduction

The automatic segmentation of brain tumors from MR imaging has the potential to save clinicians and researchers time by providing accurate labeling of tumor regions without manual editing. The Multimodal Brain Tumor Segmentation (BraTS) challenge evaluates automatic methods for segmenting glioma type brain tumors by hosting an annual event that invites participants to train and test their segmentation methods on the BraTS dataset [1]. This dataset consists of MR images from hundreds of subjects along with images containing manually labeled tumor regions for those same subjects.

Previous BraTS challenges have shown deep learning to be the most accurate method to segment tumor regions [2–5]. The U-Net convolutional neural network model is a common approach chosen by many of the top-performing teams [3–6]. This model architecture employs convolution layers that encode the information stored in the images

© Springer Nature Switzerland AG 2021
A. Crimi and S. Bakas (Eds.): BrainLes 2020, LNCS 12659, pp. 40–49, 2021.
https://doi.org/10.1007/978-3-030-72087-2_4

at progressively smaller resolutions and then decodes the output of those layers to create a segmentation map at the original resolution. Inspired by previous challenges, our project implemented the U-Net model and trialed various modifications to the training and inference strategy.

2 Methods

2.1 Data

As a part of the challenge, BraTS provided sets of training images from 369 subjects with T1w, contrast-enhanced T1w, T2w, and T2w-FLAIR images along with manually labeled tumor segmentation maps (Fig. 1) [1, 2, 7–9]. The segmentation maps were labeled 1 for the necrotic center and non-enhancing tumor, 2 for edema, and 4 for enhancing tumor. BraTS also provided 125 subjects with sets of images without the segmentation maps as a validation group. We evaluated the performance on the validation set through submissions of segmentation maps to the BraTS challenge online portal. We cropped and resampled all input images to a size of $160 \times 160 \times 160$ voxels.

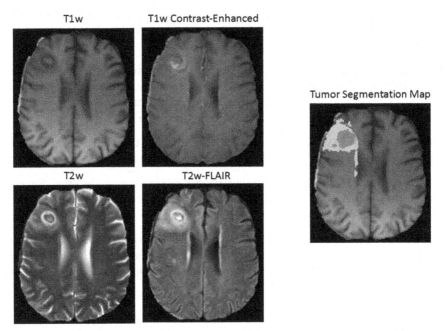

Fig. 1. Example subject part of the BraTS training dataset. Shown are the T1w, T1w with gadolinium contrast-enhancing agent, T2w, and T2w-FLAIR images used as the inputs for training and the manually segmented label map of the tumor used as the ground truth.

2.2 Model Architecture

We trained a U-Net style convolutional neural network with five layers, two ResNet style blocks per encoding and decoding layer, a base width of 32 channels, 20% channel dropout between the two encoding residual blocks of the first layer, and a receptive field of $160 \times 160 \times 160$ voxels [4, 10–12]. Each residual block consisted of two convolutional blocks performing group normalization, followed by rectified linear unit activation, and a $3 \times 3 \times 3$ convolution. The outputs of each residual block were summed with the outputs of a $1 \times 1 \times 1$ convolution of the inputs to that block. Between each encoding layer the images were downsampled by a factor of two using a strided convolution. The number of channels was doubled at each consecutive level of the model. The decoding layers concatenated the outputs of the encoding layer of the same depth to the upsampled output of the previous decoding layer. A final $1 \times 1 \times 1$ convolution linearly resampled the outputs from the 32 channels of the last decoding layer, and a sigmoid activation function was applied to predict the target segmentation maps.

2.3 Augmentation

We employed six different types of augmentation at training time, as detailed below.

Noise. Random Gaussian noise with a mean of zero and a standard deviation of 0.1 multiplied by the standard deviation of the input images was added to the input images randomly with a 50% probability per training iteration.

Blurring. The input images were blurred using a Gaussian kernel randomly with a 50% probability per training iteration. The full-width half-maximum (FWHM) of the kernel was randomly generated independently for each direction according to a normal distribution with a mean of 1.5 mm and a standard deviation of 0.5 mm.

Left–right Mirroring. The input images were mirrored so that the left and right sides of the images were randomly flipped on 50% of the training iterations.

Scale Distortion. The scale of the input images was randomly distorted for each axis independently, with a standard deviation of 0.1. The scale distortion was applied randomly on 50% of the training iterations.

Translation. The input images were randomly translated with a standard deviation of 0.05 times the extent of the cropped images. This translation was performed independently for each direction. This augmentation was applied randomly to 50% of the training iterations.

2.4 Training

Two Nvidia V100 GPUs with 32 gigabytes of memory each were used for training. One minus the average of the dice score per channel was used as the loss function during training. The initial learning rate was 10^{-4} and was decreased by a factor of 0.5 every time the validation loss plateaued for 20 epochs. Each model was trained for seven days (due to the allocated time limit on computational resources) or until the validation loss plateaued for 50 epochs.

3   Experiments

3.1   Effects of Thresholding

To perform an initial test on thresholding methods, we trained a single model on half of the training data. We then examined the effect of summing the label activations before thresholding and experimented with various threshold values.

3.2   Contours, Permutations, and Grouped Labels

In order to test different modifications to the network training, we trained a set of 4 models on the full training set with various modifications as detailed below.

Contours. The per-channel dice loss weighs all tumor voxels of the same label equally, but tumor segmentation can intuitively be considered an estimation of the boundaries of the labeled tumor regions. A high-quality segmentation will accurately estimate the boundaries between the brain and the labeled tumor, the tumor core and the edema, and the enhancing core and the necrotic center voxels. Segmentation methods often accurately segment the center of the labeled tumor regions accurately but fail to segment the boundary between two labels correctly. We tested the effect of teaching the model to focus on the region boundaries by adding the estimated contour of each label as a separate channel during training. Contours of the segmentation maps were generated by performing a binary erosion on each labeled region and then subtracting the eroded labeled region from the original labeled region.

Permutations. Combinations of rotations and reflections allow for 48 unique lossless transformations of the input feature input volumes. These permutations were performed randomly with a 50% probability for each training input. After training a model to predict the tumor labels for all permutations, we also tested the effect of permuting the validation images and then averaging the predictions over all 48 permutations.

Grouped Labels. The BraTS competition scores segmentation maps not based on the accuracy of the individual labels but rather the accuracy of the following label groups: whole tumor (WT), tumor core (TC), and enhancing tumor (ET). The WT group includes labels 1, 2, and 4, the TC group includes labels 1 and 4, and the TC group includes only label 4. We trained a model to predict the grouped labels rather than the individual labels themselves to test if this would result in higher scoring predictions [3].

3.3   Ensembles

Ten models were trained via 10-fold cross-validation on the training set. The models were trained with weighted contours and randomly permutated input volumes on every iteration. The cross-validation dice loss was measured for each fold, and the models performing the best were saved along with the models that completed the full training. The fully trained models were then used to predict the validation set, and the predictions were averaged. This ensemble of predictions was compared to those from the ensemble of the best performing cross-validation models. The fully trained ensemble predictions were also compared to the averaged predictions of both the best cross-validation models and the fully trained models.

3.4 Test Set

The ensemble of the ten fully trained models was used to predict the BraTS 2020 test set consisting of 166 sets of images provided without segmentation maps. The predicted segmentation maps were submitted through BraTS challenge online portal and the results were reported back to the authors after the completion of the challenge.

4 Results

4.1 Effects of Thresholding

The summation of the label activations prior to thresholding with a 0.5 threshold resulted in improved dice scores for all three regions as well as shorter Hausdorff95 distances for the WT and TC regions as compared to no summation before thresholding (Table 1). Interestingly, summing the activations and then thresholding at 0.9 resulted in the highest ET region scores.

Table 1. Mean Dice and Hausdorff95 measurements of the predictions resulting from various thresholding of the sigmoid label activations on the validation set. T refers to the threshold that differentiates between unlabeled and labeled voxels. Sum (Y/N) refers to whether or not the responses for each label were summed before thresholding.

T	Sum	Dice			Hausdorff95		
		ET	WT	TC	ET	WT	TC
0.3	N	0.7149	0.8938	0.8089	36.5840	4.9753	**7.9247**
0.5	N	0.7238	0.8943	0.8071	33.5743	5.3918	10.0016
0.7	N	0.7294	0.8581	0.7726	30.3896	8.7654	12.5598
0.9	N	0.6535	0.7536	0.6768	36.5133	9.6495	13.2661
0.5	Y	0.7231	**0.9002**	**0.8090**	33.6038	**4.8244**	9.8996
0.7	Y	0.7307	0.8949	0.8080	30.6517	5.2314	10.0038
0.9	Y	**0.7322**	0.8659	0.8015	**27.6895**	6.1277	10.2072

4.2 Contours, Permutations, and Grouped Labels

Weighting the contours of the labeled regions slightly improved the whole tumor dice score, but resulted in lower dice scores for the other regions compared to the baseline model trained without the weighted contours as shown in Table 2. Adding random permutations during training did not increase the dice scores but did result in lower Hausdorff95 distances for the whole tumor and tumor core regions. Permuting the data and averaging the predictions for all possible permutations did not improve results. Predicting the grouped labels rather than the individual labels resulted in better ET and TC dice scores but worse TC Hausdorff95 distance. Overall, none of the proposed alterations demonstrated enhanced validation scores over any of the other methods for every evaluation metric.

Table 2. Mean Dice and Hausdorff95 measurements of the predictions from models with the proposed modifications on the BraTS validation dataset. The proposed modifications were models that predicted the contours and as well as the labels, augmentation via permutations, averaging the predictions from each permutation, and predicting the WT, TC, and ET label groups.

	Dice			Hausdorff95		
Trial	ET	WT	TC	ET	WT	TC
Baseline	**0.7412**	0.8988	0.8086	**28.2891**	5.0059	13.9990
w/contours	0.7296	**0.9032**	0.8046	32.3023	5.8643	13.8081
w/contours & permutations	0.7263	0.8995	0.8083	36.3551	**4.7713**	**7.2312**
w/contours, permutations, and permuted predictions	0.7278	0.9014	0.8057	36.2750	4.8782	8.7185
w/contours, permutations, & grouped labels	0.7392	0.9003	**0.8136**	35.9019	4.9854	10.3612

4.3 Ensembles

Ground Truth	Prediction	Ground Truth	Prediction

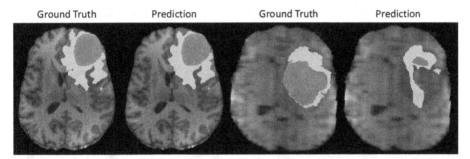

Fig. 2. Example predictions from a single of the cross-validated model on the training set compared to the ground truth segmentation maps overlayed onto the T1w image from two separate subjects. The segmentation maps are colored yellow for edema, blue for enhancing tumor, and green for the necrotic center. The predicted map on the left demonstrates a highly accurate predicted segmentation. In contrast, the predicted map on the right shows a very poor predicted segmentation, likely due to the poor-quality imaging. (Color figure online)

Overall, the baseline models that trained for the full seven days performed the best on the validation set as shown in Table 3. Figure 2 shows example predictions from a single model alongside the ground truth segmentations on the cross-validated training data. Surprisingly, the models that performed the best on the held-out cross-validation data during training had worse scores on the validation set. Also, the combined predictions from all the models scored worse than the baseline models on the validation set.

4.4 Test Set Results

The baseline ensemble consisting of the ten fully trained models was used to predict the test set cases. The results are listed in Table 4 and example predictions are shown in Fig. 3.

Table 3. Mean Dice and Hausdorff95 measurements of the predictions for the validation set from ensembles of cross-validated models. The baseline models were trained for seven days. During training, the models that performed the best on the cross-validation hold-out data were saved. The predictions from both the baseline and best models were also averaged and evaluated.

Trial	Dice			Hausdorff95		
	ET	WT	TC	ET	WT	TC
Baseline (10 models)	**0.7530**	**0.9071**	0.8292	**32.6782**	**4.5609**	9.2570
Best models (10 models)	0.7451	0.9067	**0.8318**	35.6361	4.5914	**9.1908**
Combined (20 models)	0.7444	0.9070	0.8303	35.6515	4.5695	9.2499

Table 4. Results of the ten model U-Net ensemble on the BraTS 2020 test set.

	Dice			Hausdorff95		
	ET	WT	TC	ET	WT	TC
Mean	0.8162	0.8779	0.8475	11.2353	11.0939	19.2312
Std Dev	0.1863	0.1512	0.2372	57.1383	49.5001	74.6450
Median	0.8528	0.9217	0.9248	1.4142	3.0000	2.2361
25th quantile	0.8003	0.8839	0.8727	1.0000	1.7321	1.4142
75th quantile	0.9170	0.9493	0.9595	2.4495	5.3852	3.9354

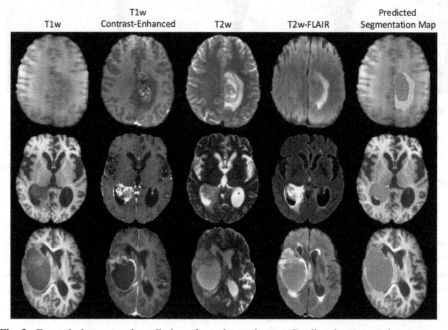

Fig. 3. Example images and predictions from the testing set. Predicted segmentation maps are shown on the far right overlayed on top of the T1 image. Predicted segmentation of the edema is shown in yellow, enhancing tumor in blue, and necrotic center in green. (Color figure online)

5 Discussion

Using an ensemble of models trained via cross-validation offered the best predictions out of all the modifications proposed. This result aligns with results from previous challenges showing that ensembles of models often out-perform the individual models [2, 3]. However, combining more models saved during training did not further increase the scores on the validation set, indicating that bigger ensembles do not always yield better results. It is likely that the models that scored the best on the cross-validation were slightly overfitted to the training set and thus decreased the scores of the fully trained models when ensembled together. Furthermore, creating ensembles with different style of models and approaches is likely to yield more diverse predictions across models in the ensemble. This diversity of predictions may produce better results than the ensembles used in this project that consisted only of identically trained U-Net models.

The median scores of the model were much higher than the average scores, especially for the Hausdorff95 distances. This indicates that the average scores are being heavily skewed by poor predictions of outlying cases. It is possible that post-processing the predictions to account for these outlier cases as performed in [3] may provide enhanced results.

Though much of the segmentation errors occur at the boundaries between labeled tumor regions, weighting the contours of the regions did not improve the results on the validation set. Training with the weighted contours produced predictions with worse dice scores for the ET and TC regions than a model trained without the weighted contours.

Applying random permutations to the input data produced mixed results. The scores on the model trained with the permuted data were not sizably better in any of the scores except for the TC Hausdorff 95 distance, which was much lower than the same model trained without permutations. Surprisingly, averaging the predictions from all possible permutations did not improve the accuracy of the combined predictions on the validation set. It is possible that the rotation of the volumes hurts the model's ability to learn information that is specific to a given axis. For example, slices in the z-axis are typically thicker than the slices in the x and y axes, prior to resampling. Furthermore, the BraTS protocol describes annotations being performed on the axial slices [2]. By rotating the volumes during training, the model is forced to treat all axes equally and cannot learn to mimic the patterns that may exist due to the manual annotation methodology. This can be avoided by only doing mirroring rotation that flip axes without rotating the volumes.

Another advantage of not performing rotation permutations is that receptive field is not required to be the shape of a cube. We used a receptive field that was $160 \times 160 \times 160$ voxels. This allowed any one dimension to be rotated and switched with another dimension. If no rotations are used, then the receptive field can be changed to better match the data. Brain images tend to be longer in the x and y dimensions than in the z dimension. Therefore, a smaller number of voxels along the z axis and a larger number of voxels along the y axis may produce better results than our cube shaped receptive field.

Overall, the augmentation strategy appears to have been too aggressive. A training set size of 369 subjects is likely big enough to train generalizable models without such heavy augmentation. When using datasets this big, permutating the input data via rotations should likely be avoided as the model may be missing valuable information encoded only in a given direction.

6 Conclusion

We trialed several different modifications to U-Net model training, including weighting the region contours, grouping the labels, permuting the input data, and combining models into ensembles. Overall, creating the ensembles of models was the only modification that demonstrated a clear improvement over other methods. Our open source code for this project is available at https://github.com/ellisdg/3DUnetCNN.

Acknowledgements. This work was completed utilizing the Holland Computing Center of the University of Nebraska, which receives support from the Nebraska Research Initiative.

References

1. Menze, B.H., et al.: The multimodal brain tumor image segmentation benchmark (BRATS). IEEE Trans. Med. Imaging **34**(10), 1993–2024 (2014)
2. Bakas, S., et al.: Identifying the best machine learning algorithms for brain tumor segmentation, progression assessment, and overall survival prediction in the BRATS challenge. arXiv preprint arXiv:1811.02629 (2018)
3. Jiang, Z., Ding, C., Liu, M., Tao, D.: Two-stage cascaded U-Net: 1st place solution to BraTS challenge 2019 segmentation Task. In: Crimi, A., Bakas, S. (eds.) Brainlesion: Glioma, Multiple Sclerosis, Stroke and Traumatic Brain Injuries: 5th International Workshop, BrainLes 2019, Held in Conjunction with MICCAI 2019, Shenzhen, China, October 17, 2019, Revised Selected Papers, Part I, pp. 231–241. Springer International Publishing, Cham (2020). https://doi.org/10.1007/978-3-030-46640-4_22
4. Myronenko, A.: 3D MRI brain tumor segmentation using autoencoder regularization. In: Crimi, A., Bakas, S., Kuijf, H., Keyvan, F., Reyes, M., van Walsum, T. (eds.) Brainlesion: Glioma, Multiple Sclerosis, Stroke and Traumatic Brain Injuries: 4th International Workshop, BrainLes 2018, Held in Conjunction with MICCAI 2018, Granada, Spain, September 16, 2018, Revised Selected Papers, Part II, pp. 311–320. Springer International Publishing, Cham (2019). https://doi.org/10.1007/978-3-030-11726-9_28
5. Isensee, F., Kickingereder, P., Wick, W., Bendszus, M., Maier-Hein, K.H.: No new-net. In: Crimi, A., Bakas, S., Kuijf, H., Keyvan, F., Reyes, M., van Walsum, T. (eds.) Brainlesion: Glioma, Multiple Sclerosis, Stroke and Traumatic Brain Injuries: 4th International Workshop, BrainLes 2018, Held in Conjunction with MICCAI 2018, Granada, Spain, September 16, 2018, Revised Selected Papers, Part II, pp. 234–244. Springer International Publishing, Cham (2019). https://doi.org/10.1007/978-3-030-11726-9_21
6. Isensee, F., Kickingereder, P., Wick, W., Bendszus, M., Maier-Hein, K.H.: Brain tumor segmentation and radiomics survival prediction: contribution to the brats 2017 challenge. In: Crimi, A., Bakas, S., Kuijf, H., Menze, B., Reyes, M. (eds.) Brainlesion: Glioma, Multiple Sclerosis, Stroke and Traumatic Brain Injuries, pp. 287–297. Springer International Publishing, Cham (2018). https://doi.org/10.1007/978-3-319-75238-9_25
7. Bakas, S., et al.: Segmentation labels and radiomic features for the pre-operative scans of the TCGA-LGG collection. Cancer Imaging Arch. **286** (2017)

8. Bakas, S., et al.: Segmentation labels and radiomic features for the pre-operative scans of the TCGA-GBM collection. The cancer imaging archive. Nat. Sci. Data. **4**, 170117 (2017)

9. Bakas, S., et al.: Advancing the cancer genome atlas glioma MRI collections with expert segmentation labels and radiomic features. Sci. Data **4**, 170117 (2017)

10. Ronneberger, O., Fischer, P., Brox, T.: U-Net: convolutional networks for biomedical image segmentation. In: Navab, N., Hornegger, J., Wells, W.M., Frangi, A.F. (eds.) Medical Image Computing and Computer-Assisted Intervention—MICCAI 2015. Lecture Notes in Computer Science, vol. 9351, pp. 234–241. Springer, Cham (2015). https://doi.org/10.1007/978-3-319-24574-4_28

11. He, K., et al.: Deep residual learning for image recognition. CoRR abs/1512.03385 (2015)

12. Ellis, D.G., Aizenberg, M.R.: Structural brain imaging predicts individual-level task activation maps using deep learning. bioRxiv, p. 2020.10.05.306951 (2020)

HI-Net: Hyperdense Inception $3D$ UNet for Brain Tumor Segmentation

Saqib Qamar[1,3], Parvez Ahmad[2(✉)], and Linlin Shen[1,3]

[1] Computer Vision Institute, School of Computer Science and Software Engineering,
Shenzhen University, Shenzhen, China
sqbqamar@hust.edu.cn, llshen@szu.edu.cn
[2] National Engineering Research Center for Big Data Technology and System,
Services Computing Technology and System Lab, Cluster and Grid Computing Lab,
School of Computer Science and Technology, Huazhong University of Science
and Technology, 430074 Wuhan, China
parvezamu@hust.edu.cn
[3] AI Research Center on Medical Image Analysis and Diagnosis,
Shenzhen University, Shenzhen, China

Abstract. The brain tumor segmentation task aims to classify tissue into the whole tumor (WT), tumor core (TC) and enhancing tumor (ET) classes using multimodel MRI images. Quantitative analysis of brain tumors is critical for clinical decision making. While manual segmentation is tedious, time-consuming, and subjective, this task is at the same time very challenging to automatic segmentation methods. Thanks to the powerful learning ability, convolutional neural networks (CNNs), mainly fully convolutional networks, have shown promising brain tumor segmentation. This paper further boosts the performance of brain tumor segmentation by proposing hyperdense inception $3D$ UNet (HI-Net), which captures multi-scale information by stacking factorization of 3D weighted convolutional layers in the residual inception block. We use hyper dense connections among factorized convolutional layers to extract more contexual information, with the help of features reusability. We use a dice loss function to cope with class imbalances. We validate the proposed architecture on the multi-modal brain tumor segmentation challenges (BRATS) 2020 testing dataset. Preliminary results on the BRATS 2020 testing set show that achieved by our proposed approach, the dice (DSC) scores of ET, WT, and TC are 0.79457, 0.87494, and 0.83712, respectively.

Keywords: Brain tumor · 3D UNet · Dense connections · Factorized convolutional · Deep learning

1 Introduction

Primary and secondary are two types of brain tumors. Primary brain tumors originate from brain cells, whereas secondary tumors metastasize into the brain

---

S. Qamar and P. Ahmad—Equal contribution.

A. Crimi and S. Bakas (Eds.): BrainLes 2020, LNCS 12659, pp. 50–57, 2021.
https://doi.org/10.1007/978-3-030-72087-2_5

from other organs. Gliomas are primary brain tumors. Gliomas can be further sub-divided into two parts: low-grade (LGG) and high-grade (HGG). High-grade gliomas are an aggressive type of malignant brain tumor that proliferates, usually requires surgery and radiotherapy, and has a poor survival prognosis. Magnetic resonance imaging (MRI) is a critical diagnostic tool for brain tumor analysis, monitoring, and surgery planning. Usually, several complimentary $3D$ MRI modalities - such as $T1$, $T1$ with contrast agent ($T1c$), $T2$, and fluid attenuation inversion recover ($FLAIR$) are required to emphasize different tissue properties and areas of tumor spread. For example, the contrast agent, usually gadolinium, emphasizes hyperactive tumor subregions in $T1c$ MRI modality.

Deep learning techniques, especially CNNs, are prevalent for the automatic segmentation of brain tumors. CNN can learn from examples and demonstrate state-of-the-art segmentation accuracy both in 2D natural images and in 3D medical image modalities. The information of segmentation provides an accurate, reproducible solution for further tumor analysis and monitoring. Multimodal brain tumor segmentation challenge (BRATS) aims to evaluate state-of-the-art methods for the segmentation of brain tumors by providing a 3D MRI dataset with ground truth labels annotated by physicians [1–4,14]. A $3D$ UNet is a popular CNN architecture for automatic brain tumor segmentation [8]. The multi-scale contextual information of the encoder-decoder sub-networks is effective for the accurate brain tumor segmentation task. Several variations of the encoder-decoder architectures were proposed for MICCAI BraTS 2018 and 2019 competitions. The potential of several deep architectures [12,13,17] and their ensembling procedures for brain tumor segmentation was discussed by a top-performing method [11] for MICCAI BRATS 2017 competition. Wang et al. [18] proposed architectures with factorized weighted layers to save the GPU memory and the computational time. At the same time, the majority of these architectures used either the bigger input sizes [16] or cascaded training [10] or novel pre-processing [7] and post-processing strategies [9] to improve the segmentation accuracy. In contrast, few architectures demonstrate the important memory consumption of 3D convolutional layers. Chen et al. [5] used an important concept in which each weighted layer was split into three branches in a parallel fashion, each with a different orthogonal view, namely axial, sagittal, and coronal. However, more complex combinations exist between features within and in-between different orthogonal views, which can significantly increase the learning representation [6]. Inspired by the S3D UNet architecture [5,19], we propose a variant encoder-decoder based architecture for the brain tumor segmentation. The key contributions of our study are as follows:

- A novel hyperdense inception 3D UNet (HI-Net) architecture is proposed by stacking factorization of 3D weighted convolutional layers in the residual inception block.
- In each residual inception block, hyper-dense connections are used in-between different orthogonal views to learn more complex feature representation.
- Our network achieves state-of-the-art performance as compared to other recent methods.

2   Proposed Method

Figure 1 shows the proposed HI-Net architecture for brain tumor segmentation. The network's left side works as an encoder to extract the features of different levels, and the right component of the network acts as a decoder to aggregate the features and the segmentation mask. The modified residual inception blocks of the encoder-decoder sub-networks have two $3D$ convolutional layers, and each layer has followed the structure of Fig. 2(b). In contrast, traditional residual inception blocks are shown in Fig. 2(a). This study employed inter-connections of dense connections within and in-between different orthogonal views to learn more complex feature representation. In the stage of encoding, the encoder extracts feature at multiple scales and create fine-to-coarse feature maps. Fine feature maps contain low-level features but more spatial information, while coarse feature maps provide the opposite. Skip connection is used to combine coarse and fine feature maps for accurate segmentation. Unlike standard residual UNet, the encoder sub-network uses a self-repetition procedure on multiple levels to generate semantic maps for fine feature maps and thus select relevant regions in the fine feature maps to concatenate with the coarse feature maps.

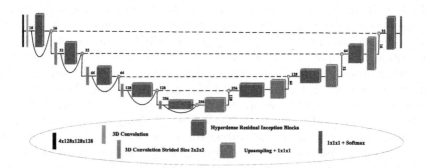

Fig. 1. Proposed HI-Net architecture. The element-wise addition operations (+ symbol with the oval shape) are employed to design the proposed architecture. The modified residual inception blocks (violet), known as hyperdense residual inception blocks, are used in the encoder-decoder paths. The length of the encoder path is longer than the decoder part by performing repetitions on several levels. The maximum repetition is 4 on the last level of the encoder part to draw the semantic information from the lowest input resolution. Finally, the softmax activation is performed for the outcomes. (Color figure online)

3   Implementation Details

3.1   Dataset

The BRATS 2020 [1–4,14] training dataset included 369 cases (293 HGG and 76 LGG), each with four rigidly aligned 3D MRI modalities ($T1$, $T1c$, $T2$, and

$FLAIR$), resampled to $1 \times 1 \times 1$ mm isotropic resolution and skull-stripped. The input image size is $240 \times 240 \times 155$. The data were collected from 19 institutions, using various MRI scanners. Annotations include 3 tumor subregions: WT, TC, and ET. Two additional datasets without the ground truth labels are provided for validation and testing. These datasets required participants to upload the segmentation masks to the organizers' server for evaluations. In validation (125 cases) and testing (166) datasets, each subject includes the same four modalities of brain MRI scans but no ground truth. In our experiment, the training set is applied to optimize the trainable parameters in the network. The validation and testing sets are utilized to evaluate the performance of the trained network.

3.2 Experiments

The network is implemented by Keras and trained on Tesla V100–SXM2 32 GB GPU card with a batch size of 1. Adam optimizer with an initial learning rate 3×10^{-5} is employed to optimize the parameters. The learning rate is reduced by 0.5 per 30 epochs. The network is trained for 350 epochs. During network training, augmentation techniques such as random rotations and mirroring are employed. The size of the input during the training of the network is $128 \times 128 \times 128$. The multi-label dice loss function [15] addressed the class imbalance problem. Equation 1 shows the mathematical representation of loss function.

$$Loss = -\frac{2}{D} \sum_{d \in D} \frac{\sum_j P_{(j,d)} T_{(j,d)} + r}{\sum_j P_{(j,d)} + \sum_j T_{(j,d)} + r} \tag{1}$$

where $P_{(j,d)}$ and $T_{(j,d)}$ are the prediction obtained by softmax activation and ground truth at voxel j for class d, respectively. D is the total number of classes.

Table 1. BRATS 2020 training, validation and testing results. Mean average scores on different metrics.

Dataset	Metrics	WT	TC	ET
BRATS 2020 training	DSC	92.967	90.963	80.009
	Sensitivity	93.004	91.282	80.751
	Specificity	99.932	99.960	99.977
BRATS 2020 validation	DSC	90.673	84.293	74.191
	Sensitivity	90.485	80.572	73.516
	Specificity	99.929	99.974	99.977
BRATS 2020 testing	DSC	87.494	83.712	79.457
	Sensitivity	91.628	85.257	82.409
	Specificity	99.883	99.962	99.965

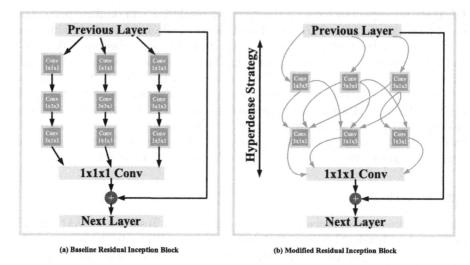

(a) Baseline Residual Inception Block   (b) Modified Residual Inception Block

Fig. 2. Difference between baseline and modified residual inception blocks. (a) represent a baseline residual inception block with a separable 3D convolutional layer, while the proposed block with inter-connected dense connections is shown in (b).

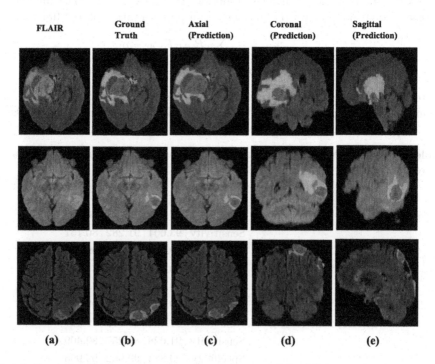

(a)   (b)   (c)   (d)   (e)

Fig. 3. Segmentation results on the training dataset of the BRATS 2020. From left to right: Ground-truth and predicted results on FLAIR modality; WT (brown), TC (red) and ET (blue). (Color figure online)

Table 2. Performance evaluation of different methods on the BRATS 2020 validation dataset. For comparison, only DSC scores are shown. All scores are evaluated online.

Methods	ET	WT	TC
Baseline work	70.616	90.670	82.136
Proposed work	74.191	90.673	84.293

3.3 Evaluation Metrics

Multiple criteria are computed as performance metrics to quantify the segmentation result. Dice coefficient (DSC) is the most frequently used metric for evaluating medical image segmentation. It measures the overlap between the segmentation and ground truth with a value between 0 and 1. The higher the Dice score, the better the segmentation performance. *Sensitivity* and *specificity* are also commonly used statistical measures. The sensitivity called true positive rate is defined as the proportion of positives that are correctly predicted. It measures the portion of tumor regions in the ground truth that is also predicted as tumor regions by the segmentation method. The specificity, called true negative rate, is defined as the proportion of correctly predicted negatives. It measures the portion of normal tissue regions in the ground truth that is also predicted as normal tissue regions by the segmentation method.

3.4 Results

The performance of our proposed architecture is evaluated on training, validation, and the testing sets provided by BRATS 2020. Table 1 presents the quantitative analysis of the proposed work. We have secured mean DSC scores of ET, WT, and TC as 0.74191, 0.90673, and 0.84293, respectively, on the validation dataset, while 0.80009, 0.92967, and 0.90963 on the training dataset. At the same time, our proposed approach obtained mean DSC scores of ET, WT, and TC as 0.79457, 0.87494, and 0.83712, respectively, on the testing dataset. In Table 1, sensitivity and specificity are also presented on training, validation, and the testing datasets. Table 2 shows the comparable study of proposed work with the baseline work [5]. Our proposed HI-Net achieves higher scores for each tumor than the baseline work. Furthermore, ablation studies are conducted to assess the modified residual inception blocks' influence with and without the inter-connected dense connections. The influence of these connections on DSCs of ET, WT, and TC is shown in Table 2. To provide qualitative results of our method, three-segmented images from training data are shown in Fig 3. In summary, modified inception blocks significantly improve the DSCs of ET, WT, and TC against the baseline inception blocks.

4 Conclusion

We proposed a HI-Net architecture for brain tumor segmentation. Each 3D convolution is splitted into three parallel branches in the residual inception block, each with different orthogonal views, namely axial, sagittal and coronal. We also proposed hyperdense connections among factorized convolutional layers to extract more contextual information. The HI-Net architecture secures high DSC scores for all types of tumors. This network has been evaluated on the BRATS 2020 Challenge testing dataset and achieved average DSC scores of 0.79457, 0.87494, and 0.83712 for the segmentation of ET, WT, and TC, respectively. Compared with the performance of the validation dataset, the scores on the testing set are higher. In the future, we will work to enhance the robustness of the network to improve the segmentation performance by using some post-processing methods such as a fully connected conditional random field (CRF).

Acknowledgment. This work is supported by the National Natural Science Foundation of China under Grant No. 91959108.

References

1. Bakas, S., et al.: Segmentation labels and radiomic features for the pre-operative scans of the TCGA-GBM collection. The Cancer Imaging Archive (2017) (2017)
2. Bakas, S., et al.: Segmentation labels and radiomic features for the pre-operative scans of the TCGA-LGG collection. Cancer Imaging Archive **286** (2017)
3. Bakas, S., et al.: Advancing The Cancer Genome Atlas glioma MRI collections with expert segmentation labels and radiomic features. Scientific Data **4**, 170117 (2017). https://doi.org/10.1038/sdata.2017.11710.0.4.14/sdata.2017.117
4. Bakas, S., et al.: Identifying the Best Machine Learning Algorithms for Brain Tumor Segmentation, Progression Assessment, and Overall Survival Prediction in the BRATS Challenge. CoRR abs/1811.0 (2018), http://arxiv.org/abs/1811.02629
5. Chen, W., Liu, B., Peng, S., Sun, J., Qiao, X.: S3D-UNet: separable 3D U-Net for brain tumor segmentation. In: Crimi, A., Bakas, S., Kuijf, H., Keyvan, F., Reyes, M., van Walsum, T. (eds.) BrainLes 2018. LNCS, vol. 11384, pp. 358–368. Springer, Cham (2019). https://doi.org/10.1007/978-3-030-11726-9_32
6. Dolz, J., Gopinath, K., Yuan, J., Lombaert, H., Desrosiers, C., Ayed, I.B.: HyperDense-Net: a hyper-densely connected CNN for multi-modal image segmentation. CoRR abs/1804.0 (2018), http://arxiv.org/abs/1804.02967
7. Feng, X., Tustison, N., Meyer, C.: Brain tumor segmentation using an ensemble of 3D U-Nets and overall survival prediction using radiomic features. In: Crimi, A., Bakas, S., Kuijf, H., Keyvan, F., Reyes, M., van Walsum, T. (eds.) BrainLes 2018. LNCS, vol. 11384, pp. 279–288. Springer, Cham (2019). https://doi.org/10.1007/978-3-030-11726-9_25
8. Isensee, F., Kickingereder, P., Wick, W., Bendszus, M., Maier-Hein, K.H.: Brain Tumor Segmentation and Radiomics Survival Prediction: Contribution to the BRATS 2017 Challenge. CoRR abs/1802.1 (2018), http://arxiv.org/abs/1802.10508
9. Isensee, F., Kickingereder, P., Wick, W., Bendszus, M., Maier-Hein, K.H.: No New-Net. In: Crimi, A., Bakas, S., Kuijf, H., Keyvan, F., Reyes, M., van Walsum, T. (eds.) BrainLes 2018. LNCS, vol. 11384, pp. 234–244. Springer, Cham (2019). https://doi.org/10.1007/978-3-030-11726-9_21

10. Jiang, Z., Ding, C., Liu, M., Tao, D.: Two-stage cascaded U-Net: 1st place solution to BraTS challenge 2019 segmentation task. In: Crimi, A., Bakas, S. (eds.) BrainLes 2019. LNCS, vol. 11992, pp. 231–241. Springer, Cham (2020). https://doi.org/10. 1007/978-3-030-46640-4_22

11. Kamnitsas, K., et al.: Ensembles of Multiple Models and Architectures for Robust Brain Tumour Segmentation. CoRR abs/1711.0 (2017), http://arxiv.org/abs/1711. 01468

12. Kamnitsas, K., Ledig, C., Newcombe, V.F.J., Simpson, J.P., Kane, A.D., Menon, D.K., Rueckert, D., Glocker, B.: Efficient multi-scale 3D CNN with fully connected CRF for accurate brain lesion segmentation. Med. Image Anal. **36**, 61–78 (2017)

13. Long, J., Shelhamer, E., Darrell, T.: Fully Convolutional Networks for Semantic Segmentation. CoRR abs/1411.4 (2014), http://arxiv.org/abs/1411.4038

14. Menze, B.H., et al.: The multimodal brain tumor image segmentation benchmark (BRATS). IEEE Trans. Med. Imaging 34(10), 1993–2024 (2015). https://doi.org/ 10.1109/TMI.2014.2377694

15. Milletari, F., Navab, N., Ahmadi, S.A.: V-Net: fully convolutional neural networks for volumetric medical image segmentation. CoRR abs/1606.0 (2016), http://arxiv. org/abs/1606.04797

16. Myronenko, A.: 3D MRI brain tumor segmentation using autoencoder regularization. CoRR abs/1810.1 (2018), http://arxiv.org/abs/1810.11654

17. Ronneberger, O., Fischer, P., Brox, T.: U-Net: convolutional networks for biomedical image segmentation. CoRR abs/1505.0 (2015), http://arxiv.org/abs/1505. 04597

18. Wang, G., Li, W., Ourselin, S., Vercauteren, T.: Automatic Brain Tumor Segmentation using Cascaded Anisotropic Convolutional Neural Networks. CoRR abs/1709.0 (2017), http://arxiv.org/abs/1709.00382

19. Xie, S., Sun, C., Huang, J., Tu, Z., Murphy, K.: Rethinking Spatiotemporal Feature Learning For Video Understanding. CoRR abs/1712.0 (2017), http://arxiv.org/ abs/1712.04851

H²NF-Net for Brain Tumor Segmentation Using Multimodal MR Imaging: 2nd Place Solution to BraTS Challenge 2020 Segmentation Task

Haozhe Jia[1,2,4], Weidong Cai[3], Heng Huang[4,5], and Yong Xia[1,2(✉)]

[1] Research and Development Institute of Northwestern Polytechnical University Shenzhen, Shenzhen 518057, China
yxia@nwpu.edu.cn

[2] National Engineering Laboratory for Integrated Aero-Space-Ground-Ocean Big Data Application Technology, School of Computer Science and Engineering, Northwestern Polytechnical University, Xi'an 710072, China

[3] School of Computer Science, University of Sydney, Sydney, NSW 2006, Australia

[4] Department of Electrical and Computer Engineering, University of Pittsburgh, Pittsburgh, PA 15261, USA

[5] JD Finance America Corporation, Mountain View, CA 94043, USA

Abstract. In this paper, we propose a Hybrid High-resolution and Non-local Feature Network (H²NF-Net) to segment brain tumor in multimodal MR images. Our H²NF-Net uses the single and cascaded HNF-Nets to segment different brain tumor sub-regions and combines the predictions together as the final segmentation. We trained and evaluated our model on the Multimodal Brain Tumor Segmentation Challenge (BraTS) 2020 dataset. The results on the test set show that the combination of the single and cascaded models achieved average Dice scores of 0.78751, 0.91290, and 0.85461, as well as Hausdorff distances (95%) of 26.57525, 4.18426, and 4.97162 for the enhancing tumor, whole tumor, and tumor core, respectively. Our method won the second place in the BraTS 2020 challenge segmentation task out of nearly 80 participants.

Keywords: Brain tumor · Segmentation · Single and cascaded HNF-Nets

1 Introduction

Brain gliomas are the most common primary brain malignancies, which generally contain heterogeneous histological sub-regions, i.e. edema/invasion, active tumor structures, cystic/necrotic components, and non-enhancing gross abnormality. Accurate and automated segmentation of these intrinsic sub-regions using multimodal magnetic resonance (MR) imaging is critical for the potential diagnosis and treatment of this disease. To this end, the multimodal brain tumor

© Springer Nature Switzerland AG 2021
A. Crimi and S. Bakas (Eds.): BrainLes 2020, LNCS 12659, pp. 58–68, 2021.
https://doi.org/10.1007/978-3-030-72087-2_6

segmentation challenge (BraTS) has been held for many years, which provides a platform to evaluate the state-of-the-art methods for the segmentation of brain tumor sub-regions [1–4,11].

With deep learning being widely applied to medical image analysis, fully convolutional network (FCN) based methods have been designed for this segmentation task and have shown convincing performance in previous challenges. Kamnitsas *et al.* [9] constructed a 3D dual pathway CNN, namely DeepMedic, which simultaneously processes the input image at multiple scales with a dual pathway architecture so as to exploit both local and global contextual information. DeepMedic also uses a 3D fully connected conditional random field to remove false positives. In [6], Isensee *et al.* achieved outstanding segmentation performance using a 3D U-Net with instance normalization and leaky ReLU activation, in conjunction with a combination loss function and a region-based training strategy. In [12], Myronenko *et al.* incorporated a variational auto-encoder (VAE) based reconstruction decoder into a 3D U-Net to regularize the shared encoder, and achieved the 1st place segmentation performance in BraTS 2018. In BraTS 2019, Jiang *et al.* [8] proposed a two-stage cascaded U-Net to segment the brain tumor sub-regions from coarse to fine, where the second-stage model has more channel numbers and uses two decoders so as to boost performance. This method achieved the best performance in the BraTS 2019 segmentation task. In our previous work [7], we proposed a High-resolution and Non-local Feature Network (HNF-Net) to segment brain tumor in multimodal MR images. The HNF-Net is constructed based mainly on the parallel multi-scale fusion (PMF) module, which can maintain strong high-resolution feature representation and aggregate multi-scale contextual information. The expectation-maximization attention (EMA) module is also introduced to the model to enhance the long-range dependent spatial contextual information at the cost of acceptable computational complexity.

In this paper, we further propose a Hybrid High-resolution and Non-local Feature Network (H^{2}NF-Net) for this challenging task. Compared to original HNF-Net, the proposed H^{2}NF-Net adds a two-stage cascaded HNF-Net and uses the single and cascaded models to segment different brain tumor sub-regions. We evaluated the proposed method on the BraTS 2020 challenge dataset. In addition, we also introduced the detailed implementation information of our second place solution to BraTS 2020 challenge segmentation task.

2   Dataset

The BraTS 2020 challenge dataset [1–4,11] contains 369 training, 125 validation and 166 test multimodal brain MR studies. Each study has four MR images, including T1-weighted (T1), post-contrast T1-weighted (T1ce), T2-weighted (T2), and fluid attenuated inversion recovery (Flair) sequences, as shown in Fig. 1. All MR images have the same size of $240 \times 240 \times 155$ and the same voxel spacing of $1 \times 1 \times 1\,\mathrm{mm}^3$. For each study, the enhancing tumor (ET), peritumoral edema (ED), and necrotic and non-enhancing tumor core (NCR/NET) were annotated on a voxel-by-voxel basis by experts. The annotations for training studies are publicly available, and the annotations for validation and test studies are withheld for online evaluation and final segmentation competition, respectively.

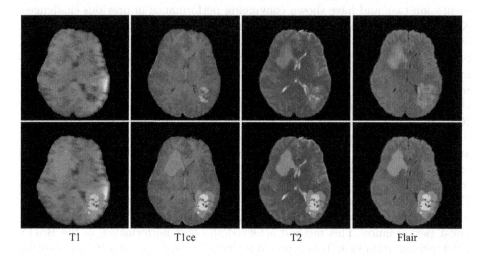

| T1 | T1ce | T2 | Flair |

Fig. 1. Example scans and corresponding annotations of different modalities. The NCR/NET, ED, and ET regions are highlighted in red, green, and yellow, respectively. (Color figure online)

3   Method

In this section, we introduce the structure of our H^2NF-Net. Despite the details have been introduced in our previous work [7], to make the proposed H^2NF-Net self-consistent, here we still give a brief introduction of HNF-Net and its two key modules. Also we provide the details of the two-stage cascaded HNF-Net. We now delve into the details of each part.

3.1   Single HNF-Net

The single HNF-Net has an encoder-decoder structure. For each study, four multimodal brain MR sequences are first concatenated to form a four-channel input and then processed at five scales i.e., r, $1/2r$, ... $1/16r$, highlighted in green, yellow, blue, pink, and orange in the upper figure of Fig. 2, respectively. At the original scale r, there are four convolutional blocks, two for encoding and the other two for decoding. The connection from the encoder to the decoder skips the processing at other scales so as to maintain the high resolution and spatial information in a long-range residual fashion. At other four scales, four

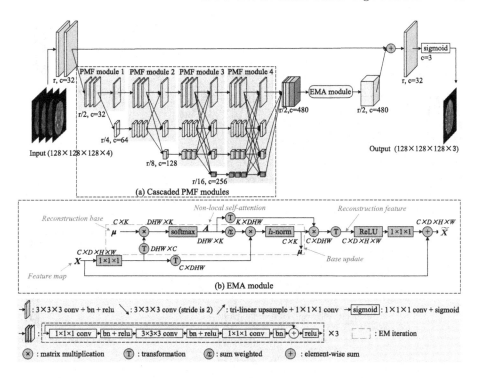

Fig. 2. Architecture of the single HNF-Net, PMF module and EMA module. r denotes the original resolution and c denotes the channel number of feature maps. All down-sample operations are achieved with 2-stride convolutions, and all upsample operations are achieved with joint $1 \times 1 \times 1$ convolutions and tri-linear interpolation. It is noted that, since it is inconvenient to show 4D feature maps $(C \times D \times H \times W)$ in the figure, we show all feature maps without depth information, and the thickness of each feature map reveals its channel number.

PMF modules are jointly used as a high-resolution and multi-scale aggregated feature extractor. At the end of the last PMF module, the output feature maps at four scales are first recovered to the $1/2r$ scale and then concatenated as mixed features. Next, the EMA module is used to efficiently capture long-range dependent contextual information and reduce the redundancy of the obtained mixed features. Finally, the output of the EMA module is recovered to original scale r and 32 channels via $1 \times 1 \times 1$ convolutions and upsampling and then added to the full-resolution feature map produced by the encoder for the dense prediction of voxel labels. All downsample operations are achieved with 2-stride convolutions, and all upsample operations are achieved with joint $1 \times 1 \times 1$ convolutions and tri-linear interpolation.

PMF Module. Given that learning strong high-resolution representation is essential for small object segmentation, high-resolution features are maintained

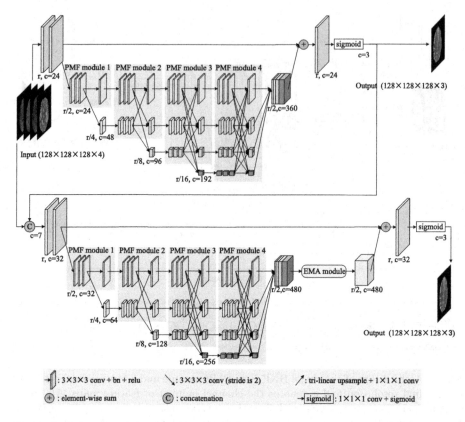

Fig. 3. Architecture of the cascaded HNF-Net. r denotes the original resolution and c denotes the channel number of feature maps.

throughout the segmentation process in recent solutions [13,14,16], and it has been shown that this strategy contributes to convincing performance. Based on this, the PMF module is designed to have two parts, the parallel multi-scale convolutional block and fully connected fusion block. The former has a set of parallel branches similar to the group convolution, and each branch is built with repeated residual convolutional blocks at a specific scale. The latter fuses all output features of the parallel multi-scale convolutional block in a parallel but fully connected fashion, where each branch is the summation of the output features of all resolution branches. Thus, in each PMF module, the parallel multi-scale convolutional block can fully exploit multi-resolution features but maintain high-resolution feature representation, and the fully connected fusion block can aggregate rich multi-scale contextual information. Moreover, we cascade multiple PMF modules, in which the number of branches increases progressively with depth, as shown in Fig. 2(a). As a result, from the perspective of the highest resolution stage, its high-resolution feature representation is boosted with repeated fusion of multi-scale low-resolution representations. Meanwhile, the cascade PMF modules can be regarded as an ensemble of several U-shape sub-networks with different depths and widths, which tends to further reduce the semantic gap of the features at different depths.

EMA Module. It has been proved that Non-local self-attention mechanism can help to aggregate contextual information from all spatial positions and capture long-range dependencies [5,17,19]. However, the potential high computational complexity makes it hard to be applied to 3D medical image segmentation tasks. Hence, we introduce the EMA module [10] to our tumor segmentation model, aiming to incorporate a lightweight Non-local attention mechanism into our model. The main concept of the EMA module is operating the Non-local attention on a set of feature reconstruction bases rather than directly achieving this on the high-resolution feature maps. Since the reconstruction bases have much less elements than the original feature maps, the computation cost of the Non-local attention can be significantly reduced. As shown in the lower figure of Fig. 2(b), the shape of the input feature maps X are $C \times D \times H \times W$, where C is the channel number. The dotted box denotes the EM algorithm operation, where the Non-local spatial attention maps A and the bases μ are generated alternately as the E step and M step, respectively. After convergence, we can use the obtained A and μ to generate reconstructed feature maps $\widetilde{X}$. At both the beginning and ending of the EM algorithm operation, the $1 \times 1 \times 1$ convolutions are adopted to change the channel number. In addition, to avoid overfitting, we further sum $\widetilde{X}$ with X in a residual fashion.

3.2 Cascaded HNF-Net

Inspired by the previous work [8,18], we further construct a two-stage cascaded version of our HNF-Net, with the structure shown in Fig. 3. Compared to the single HNF-Net, the first stage network of the cascaded HNF-Net is narrower and has no EMA module. The second network has the same structure with the single HNF-Net, but receives the concatenation of the original image block and the prediction of the first network as the input. The two-stage cascaded HNF-Net is trained in an end-to-end fashion, and the deep supervision is also added for the output of the first stage network for a stable training. Given that (1) the segmentation performance of three sub-regions, namely ET, tumor core (TC, the union of ET and NCR/NET), and whole tumor (WT, the union of all sub-regions) is evaluated separately in BraTS 2020 and (2) the single HNF-Net shows better segmentation performance on ET and WT, while the cascaded HNF-Net has better results on TC on the validation set, we use the single HNF-Net to segment ED and ET and use the cascaded HNF-Net to segment NCR/NET in the practical use of our H²NF-Net.

Table 1. Segmentation performances of our method on the BraTS 2020 validation set. DSC: dice similarity coefficient, HD95: Hausdorff distance (95%), WT: whole tumor, TC: tumor core, ET: enhancing tumor core. *: ensemble of models trained with 5-fold cross-validation, $^+$: ensemble of models trained with entire training set.

Method	Dice(%)				95%HD(mm)			
	ET	WT	TC	Mean	ET	WT	TC	Mean
Single*	0.78492	0.91261	0.83532	0.84428	26.60476	4.17888	5.41503	12.06622
Single$^+$	**0.78908**	0.91218	0.84887	0.85004	**26.50355**	**4.10500**	5.14468	11.91774
Cascaded*	0.77647	0.91084	0.85631	0.84787	26.68954	4.38397	**4.93158**	12.00169
Cascaded$^+$	0.77338	0.91022	**0.85701**	0.84687	29.71248	4.30247	4.93369	12.98288
H^{2}NF-Net	0.78751	**0.91290**	0.85461	**0.85167**	26.57525	4.18426	4.97162	**11.91038**

Table 2. The parameter numbers and FLOPs of both single and cascaded HNF-Net.

Method	Params (M)	FLOPs(G)
Single HNF-Net	16.85	436.59
Cascaded HNF-Net	26.07	621.09

4 Experiments and Results

4.1 Implementation Details

Pre-processing. Following our previous work [7], we performed a set of pre-processing on each brain MR sequence independently, including brain stripping, clipping all brain voxel intensity with a window of [0.5%–99.5%], and normalizing them into zero mean and unit variance.

Training. In the training phase, we randomly cropped the input image into a fixed size of $128 \times 128 \times 128$ and concatenated four MR sequences along the channel dimension as the input of the model. The training iterations were set to 450 epochs with a linear warmup of the first 5 epochs. We trained the model using the Adam optimizer with a batch size of 4 and betas of (0.9, 0.999). The initial learning rate was set to 0.0085 and decayed by multiplied with $(1 - \frac{current\_epoch}{max\_epoch})^{0.9}$. We also regularized the training with an l_2 weight decay of $1e - 5$. To reduce the potential overfitting, we further employed several online data augmentations, including random flipping (on all three planes independently), random rotation ($\pm10°$ on all three planes independently), random per-channel intensity shift of [±0.1] and intensity scaling of [0.9–1.1]. We empirically set the base number $K = 256$ in all experiments. We adopted a combination of generalized Dice loss [15] and binary cross-entropy loss as the loss function. All experiments were performed based on PyTorch 1.2.0. Since the training of the cascaded HNF-Net requires more than 11 Gb GPU memory, we used 4 NVIDIA Tesla P40 GPUs and 4 NVIDIA Geforce GTX 2080Ti GPUs to train the cascaded HNF-Net and single HNF-Net, respectively.

Inference. In the inference phase, we first cropped the original image with a size of $224 \times 160 \times 155$, which was determined based on the statistical analysis across the whole dataset to cover the whole brain area but with minimal redundant background voxels. Then, we segmented the cropped image with sliding patches instead of predicting the whole image at once, where the input patch size and sliding stride were set to $128 \times 128 \times 128$ and $32 \times 32 \times 27$, respectively. For each inference patch, we adopted test time augmentation (TTA) to further improve the segmentation performance, including 7 different flipping $((x), (y), (z), (x, y), (x, z), (y, z), (x, y, z)$, where x, y, z denotes three axes, respectively). Then, we averaged the predictions of the augmented and partly overlapped patches to generate the whole image segmentation result. At last, suggested by the previous work [6,7], we performed a post-processing by replacing enhancing tumor with NCR/NET when the volume of predicted enhancing tumor is less than the threshold. Based on the results on the validation set, the threshold value was set to 300 and 500 for the single and cascaded models, respectively.

Table 3. Segmentation performances of our method on the BraTS 2020 test set. DSC: dice similarity coefficient, HD95: Hausdorff distance (95%), WT: whole tumor, TC: tumor core, ET: enhancing tumor core.

Method	Dice(%)			95%HD(mm)		
	ET	WT	TC	ET	WT	TC
H^{2}NF-Net	0.82775	0.88790	0.85375	13.04490	4.53440	16.92065

4.2 Results on the BraTS 2020 Challenge Dataset

We first evaluated the performance of our method on the validation set, with the results shown in Table 1. All segmentation results were evaluated by the Dice score and 95% Hausdorff distance (%95HD) and directly obtained from the BraTS 2020 challenge leaderboard. Here Single* and Cascaded* denote the ensemble of 5 single models and 5 cascaded models trained with 5-fold cross-validation, respectively. Single$^+$ and Cascaded$^+$ denote the ensemble of 7 single models and 10 cascaded models trained with entire training set, respectively. We can find that the single models show better segmentation performance on ET and WT, while the cascaded models have better results on TC. Besides, our H^{2}NF-Net which has an ensemble of all above 27 models, achieved the best performance on validation set, with a mean Dice score of 0.85167 and a mean %95HD of 11.91038. We also provide the parameter numbers and FLOPs of both Single and Cascaded HNF-Net, shown in Table 2.

At last, we further used the ensemble H^{2}NF-Net to segment the test set and the results are shown in Table 3. We can observe that our approach achieved average Dice scores of 0.78751, 0.91290, and 0.85461, as well as Hausdorff distances (95%) of 26.57525, 4.18426, and 4.97162 for ET, WT, and TC, respectively,

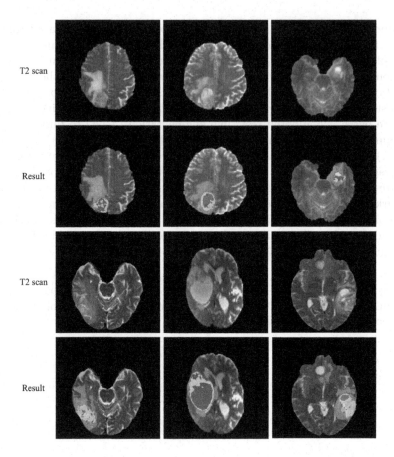

Fig. 4. Some visual results of our method on the test set. The NCR/NET, ED, and ET regions are highlighted in red, green, and yellow, respectively. (Color figure online)

which won the second place out of nearly 80 participants in the BraTS 2020 challenge segmentation task.

Figure 4 visualizes the segmentation results on some scans of the test set. It can be observed that our H^2NF-Net can generate convincing results, even on some small regions.

5 Conclusion

In this paper, we propose a H^2NF-Net for brain tumor segmentation using multimodal MR imaging, which simultaneously utilizes the single and cascaded HNF-Nets to achieve accurate and robust segmentation. We evaluated our method on the BraTS 2020 validation sets and obtained competitive results. In addition, our H^2NF-Net won the second place in the BraTS 2020 challenge segmentation task among near 80 participants, which further demonstrates its superiority and effectiveness.

Acknowledgement. Haozhe Jia and Yong Xia were partially supported by the Science and Technology Innovation Committee of Shenzhen Municipality, China under Grant JCYJ20180306171334997, the National Natural Science Foundation of China under Grant 61771397, and the Innovation Foundation for Doctor Dissertation of Northwestern Polytechnical University under Grant CX202042. The authors would like to appreciate the efforts devoted to collect and share the BraTS 2020 dataset.

References

1. Bakas, S., Akbari, H., Sotiras, A., et al.: Segmentation labels for the pre-operative scans of the TCGA-GBM collection (2017)
2. Bakas, S., et al.: Segmentation labels and radiomic features for the pre-operative scans of the TCGA-LGG collection. Cancer Imaging Arch. **286** (2017)
3. Bakas, S., et al.: Advancing the cancer genome atlas glioma MRI collections with expert segmentation labels and radiomic features. Sci. Data **4**, 170117 (2017)
4. Bakas, S., et al.: Identifying the best machine learning algorithms for brain tumor segmentation, progression assessment, and overall survival prediction in the BRATS challenge. arXiv preprint arXiv:1811.02629 (2018)
5. Fu, J., et al.: Dual attention network for scene segmentation. In: Proceedings of the IEEE Conference on Computer Vision and Pattern Recognition, pp. 3146–3154 (2019)
6. Isensee, F., Kickingereder, P., Wick, W., Bendszus, M., Maier-Hein, K.H.: No new-net. In: Crimi, A., Bakas, S., Kuijf, H., Keyvan, F., Reyes, M., van Walsum, T. (eds.) BrainLes 2018. LNCS, vol. 11384, pp. 234–244. Springer, Cham (2019). https://doi.org/10.1007/978-3-030-11726-9_21
7. Jia, H., Xia, Y., Cai, W., Huang, H.: Learning high-resolution and efficient non-local features for brain glioma segmentation in MR images. In: Martel, A.L., Abolmaesumi, P., Stoyanov, D., Mateus, D., Zuluaga, M.A., Zhou, S.K., Racoceanu, D., Joskowicz, L. (eds.) MICCAI 2020. LNCS, vol. 12264, pp. 480–490. Springer, Cham (2020). https://doi.org/10.1007/978-3-030-59719-1_47
8. Jiang, Z., Ding, C., Liu, M., Tao, D.: Two-stage cascaded U-Net: 1st place solution to BraTS challenge 2019 segmentation task. In: Crimi, A., Bakas, S. (eds.) BrainLes 2019. LNCS, vol. 11992, pp. 231–241. Springer, Cham (2020). https://doi.org/10.1007/978-3-030-46640-4_22
9. Kamnitsas, K., et al.: Efficient multi-scale 3D CNN with fully connected CRF for accurate brain lesion segmentation. Med. Image Anal. **36**, 61–78 (2017)
10. Li, X., Zhong, Z., Wu, J., Yang, Y., Lin, Z., Liu, H.: Expectation-maximization attention networks for semantic segmentation. In: Proceedings of the IEEE International Conference on Computer Vision, pp. 9167–9176 (2019)
11. Menze, B.H., et al.: The multimodal brain tumor image segmentation benchmark (BRATS). IEEE Trans. Med. Imaging **34**(10), 1993–2024 (2014)
12. Myronenko, A.: 3D MRI brain tumor segmentation using autoencoder regularization. In: Crimi, A., Bakas, S., Kuijf, H., Keyvan, F., Reyes, M., van Walsum, T. (eds.) BrainLes 2018. LNCS, vol. 11384, pp. 311–320. Springer, Cham (2019). https://doi.org/10.1007/978-3-030-11726-9_28

13. Pohlen, T., Hermans, A., Mathias, M., Leibe, B.: Full-resolution residual networks for semantic segmentation in street scenes. In: Proceedings of the IEEE Conference on Computer Vision and Pattern Recognition, pp. 4151–4160 (2017)
14. Saxena, S., Verbeek, J.: Convolutional neural fabrics. In: Advances in Neural Information Processing Systems, pp. 4053–4061 (2016)
15. Sudre, C.H., Li, W., Vercauteren, T., Ourselin, S., Jorge Cardoso, M.: Generalised dice overlap as a deep learning loss function for highly unbalanced segmentations. In: Cardoso, M.J., et al. (eds.) DLMIA/ML-CDS-2017. LNCS, vol. 10553, pp. 240–248. Springer, Cham (2017). https://doi.org/10.1007/978-3-319-67558-9_28
16. Sun, K., et al.: High-resolution representations for labeling pixels and regions. arXiv preprint arXiv:1904.04514 (2019)
17. Wang, X., Girshick, R., Gupta, A., He, K.: Non-local neural networks. In: Proceedings of the IEEE Conference on Computer Vision and Pattern Recognition, pp. 7794–7803 (2018)
18. Wu, Y., Xia, Y., Song, Y., Zhang, Y., Cai, W.: Multiscale network followed network model for retinal vessel segmentation. In: Frangi, A.F., Schnabel, J.A., Davatzikos, C., Alberola-López, C., Fichtinger, G. (eds.) MICCAI 2018. LNCS, vol. 11071, pp. 119–126. Springer, Cham (2018). https://doi.org/10.1007/978-3-030-00934-2_14
19. Zhao, H., et al.: Psanet: point-wise spatial attention network for scene parsing. In: Proceedings of the European Conference on Computer Vision (ECCV), pp. 267–283 (2018)

2D Dense-UNet: A Clinically Valid Approach to Automated Glioma Segmentation

Hugh McHugh[1,2]($\boxtimes$), Gonzalo Maso Talou[3], and Alan Wang[1,3]

[1] Faculty of Medical and Health Sciences, University of Auckland, Auckland, New Zealand
[2] Radiology Department, Auckland City Hospital, Auckland, New Zealand
[3] Auckland Bioengineering Institute, University of Auckland, Auckland, New Zealand

Abstract. Brain tumour segmentation is a requirement of many quantitative MRI analyses involving glioma. This paper argues that 2D slice-wise approaches to brain tumour segmentation may be more compatible with current MRI acquisition protocols than 3D methods because clinical MRI is most commonly a slice-based modality. A 2D Dense-UNet segmentation model was trained on the BraTS 2020 dataset. Mean Dice values achieved on the test dataset were: 0.859 (WT), 0.788 (TC) and 0.766 (ET). Median test data Dice values were: 0.902 (WT), 0.887 (TC) and 0.823 (ET). Results were comparable to previous high performing BraTS entries. 2D segmentation may have advantages over 3D methods in clinical MRI datasets where volumetric sequences are not universally available.

Keywords: 2D vs. 3D · BraTS 2020 · Glioma · GBM · MRI · Segmentation · UNet · Convolutional neural network · Artificial intelligence · Radiology

1 Introduction

Gliomas are the most common primary central nervous system malignancy [1]. Gliomas arise from glial cells that surround and support neurons. Glioblastoma multiforme (GBM) is an aggressive glioma subtype which is a devastating diagnosis with a median survival of 15 months despite best treatment [2]. Treatment of GBM typically consists of surgical resection, chemotherapy and radiotherapy.

The Brain Tumour Segmentation Challenge (BraTS) runs in conjunction with the MICCAI Conference and provides MRI data with ground truth segmentation labels of patients with glioma [3–7]. Data is leveraged from the Cancer Imaging Archive [8]. The availability of segmentation labels is a prerequisite for training supervised automated segmentation models. BraTS has arguably resulted in the advancement of this technology, to the extent that performance is approaching that of expert manual segmentation [9].

Manual segmentation is time-consuming and expensive as expert radiologist knowledge is required. Potential clinical applications of automated segmentation include radiotherapy planning, evaluation of treatment response and disease progression. Segmentation forms a key part of the radiomic workflow and allows quantitative assessment of brain tumours. Research applications of segmentation include genomic classification

© Springer Nature Switzerland AG 2021
A. Crimi and S. Bakas (Eds.): BrainLes 2020, LNCS 12659, pp. 69–80, 2021.
https://doi.org/10.1007/978-3-030-72087-2_7

[10], quantifying tumour size and relationships to structures known to affect survival such as the ventricles [11].

The segmentations provided in BraTS delineate enhancing tumour (ET), non-enhancing tumour and necrotic core (NC), and infiltrative/vasogenic oedema (OE). ET demonstrates increased T1 signal following contrast administration. NC is low signal on T1 and intermediate-to-high signal on T2 with no enhancement. Enhancement and necrosis are high-grade features which reflect a poor prognosis [12–14]. OE is the region of high T2 signal that surrounds the tumour. OE contains viable tumour cells and the extent of oedema is a poor prognostic indicator [15, 16].

MRI is an inherently heterogeneous modality making generalisation across institutions and datasets difficult. Furthermore, the use of MRI sequences such as perfusion-weighted imaging is variable and is usually a clinical decision made by the radiologist. The imaging sequences used in BraTS are standard in glioma imaging and are compatible with consensus recommendations [16, 17]. A clinically useful segmentation model ought to account for imaging factors and use routinely available sequences only.

Most of the high performing segmentation models in recent BraTS competitions use 3D approaches. An overlooked aspect is that MRI is not inherently a 3D modality and there are factors that make 2D methods attractive. Consensus statements in brain tumour imaging recommend 2D slice-wise acquisition of T2 and T2 fluid-attenuated inversion recovery (FLAIR) sequences (T1 images are acquired volumetrically) [16, 17]. Although most sequences can be acquired volumetrically, volumetric FLAIR acquisition may take up to three times as long [18], and there are imaging factors such as increased signal-to-noise ratio that make thicker 2D slices preferable in some clinical situations [19]. In the authors' institution, there are typically 25 axial slices per volume for T2 and FLAIR imaging. This results in anisotropy of resolution, and 3D reconstruction of non-volumetric imaging involves interpolation between slices. For non-volumetric imaging, resolution in the coronal and sagittal planes is much less than the axial plane. This is visually apparent in figure one below (Fig. 1).

The major difference between 2D and 3D convolutional neural networks is the additional convolution in the 3rd dimension, increasing the dimensionality of all latent spaces. This results in significantly increased model complexity and memory demands. Due to the reduced sagittal and coronal resolution in clinical MRI, we argue that a 2D approach may be better suited to clinical segmentation than 3D models.

Here we present a fully automated segmentation model utilising a 2D UNet architecture with dense-blocks. This model is trained on multi institutional data with manual segmentations provided by the BraTS organisers and evaluated on independent validation and test datasets. We demonstrate comparable performance to state-of-the-art 3D methods and argue that a 2D approach should be considered due to the slice-based nature of MRI in current clinical practice.

2 Methods

The BraTS 2020 data consists of brain MRIs of patients with high- and low-grade glioma. The provided sequences are T2, FLAIR, T1 pre- and post-contrast. These studies were acquired across 19 separate institutions using a variety of MRI scanners and scanning protocols.

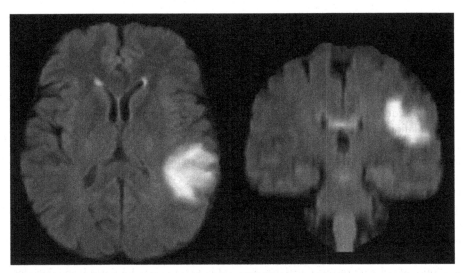

Fig. 1. BraTS axial (left) and coronal (right) FLAIR images demonstrating anisotropy of resolution with reduced resolution in the coronal plane (validation case 58).

Manual segmentation of "ground truth" labels was performed according to a set protocol by experienced neuroradiologists and provided by the BraTS organisers for the training dataset. The segmented labels delineate enhancing tumour (ET), necrotic core (NC) and vasogenic/infiltrative oedema (OE). Segmentation precision scores are calculated for the validation and test datasets. Segmentation scores are provided for: whole tumour (WT), tumour core (TC) and enhancing tumour (ET). Whole tumour consists of oedema, necrotic and non-enhancing core and enhancing tumour (WT = OE + NC + ET). Tumour core consists of necrotic and non-enhancing core and enhancing tumour (TC = NC + ET).

The training, validation and test datasets were released sequentially by the BraTS organisers through May to September 2020. Model development and training occurred over this period. The training dataset consists of 369 cases with images and labels in the training dataset. The validation dataset consists of 125 cases without segmentation labels. Final model performance is evaluated on a test dataset of 166 cases. External MRI data was not utilised.

A multisequence 2D Dense-UNet [20] was trained to generate multilabel tumour segmentations. Inputs are 240 × 240 × 4-channel slices consisting of the four MRI sequences provided. The output is a 240 × 240 × 4-channel soft-max segmentation map corresponding to the labels no tumour, ET, NC and OE. Volumetric segmentations are reconstructed from individual slices by iterating over the provided MRI volumes.

Pre-processing: The BraTS MRI data is provided in Nifti file format and is pre-processed. Preprocessing by the BraTS organisers included rigid co-registration of sequences to an anatomical format, resampling to isotropic 1 mm^3 voxels with interpolation, and skull stripping. Image normalisation was the only additional pre-processing step. Each sequence was normalised by dividing the intensity of all voxels by the 95th percentile of non-zero voxels. This method was chosen to keep the value of non-brain regions set at zero and to prevent the normalisation factor from being affected by non-brain or outlier intensity values. Bias correction was not utilised.

Model: The model consists of an encoder, a bottleneck and a decoder. The encoder contains four sequential levels that iteratively downscale the resolution of the intermediate representations. Each of these levels consists of two dense-blocks [21] and a dense-block transition layer (composed of batch normalization, convolution and max-pooling layers) in series. The last transition layer in the encoder is connected with the so-called bottleneck composed of two dense-blocks and a transpose convolution layer to up-scale the representation resolution. Lastly, the decoder contains four sequential levels that upscale the bottleneck resolution yielding the same resolution as the network input. Each of these levels is composed of two dense-blocks and a transpose convolution layer in series. As proposed in U-Net architecture, skip connections between the encoder and decoder are added, concatenating outputs of the encoder into the decoder for levels with the same resolution. Each unit of two dense-blocks used in the encoder, bottleneck and decoder, has skip connections from the input of each dense-block to the output of the second dense block increasing the number of features in its input by a factor of 3. The first level computes 32 features for each convolution and the number of convolution features in each successive level is increased by a factor of two. The dropout value is fixed at 0.2.

Training: The loss function was weighted cross-entropy (WCE). WCE was chosen over Dice loss as WCE informs gradient descent across a range of soft-max values [22]. Initial weights for the cross-entropy (inter-class weights) were derived from the relative distribution of labels within the training dataset. Following a period of training, the inter-class weights were updated based on visual inspection of the predicted segmentations to increase the specificity of the model. The model was trained using the ReLu activation function and Adam optimiser with a learning rate of 10^{-4}. A batch size of 8 was used. Online data augmentation included random shifting of intensity values, random translation and random rotation. The model was trained for 140 epochs using the initial inter-class weights and then for a further 80 epochs using the updated weights.

Post-processing: The predicted segmentation labels were post-processed by computing the largest connected component (LCC) to remove unconnected labels. The second LCC was also extracted if it was above a volume threshold to account for multifocal glioblastoma [23]. A further step involved changing ET within volumes to NC if the ET volume was below a threshold. This method has been used previously to increase Dice coefficient [22] (Fig. 2).

Uncertainty Task: Uncertainty maps were obtained for WT, TC and ET using an entropy function applied to the soft-max output of the segmentation model. The area under the curve (AUC) of the Dice coefficient was calculated from confidence scores on the segmentations using a function determined by the BraTS organisers.

Implementation: The model was implemented in Python 3.7 using the TensorFlow 2.0 library [24]. Training was performed using a 32GB NVIDIA Tesla V100 GPU. Each

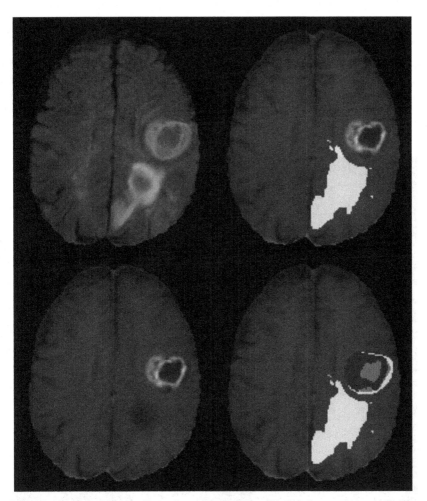

Fig. 2. Example of multifocal glioma from the BraTS validation dataset (case 116). Top and bottom left are the raw FLAIR and T1 post-contrast images respectively. Top right is the segmentation following extraction of the first LCC. The second LCC is also extracted in the bottom right example demonstrating the benefits of this method in multifocal glioma. Segmentation labels are ET (red), NC (blue) and OE (green) – segmentation key continues. (Color figure online)

training epoch required 50 min of processing time, equating to just under 8 days of training total. Prediction of segmentation and certainty maps took less than 3 s per case. Post-processing required negligible computational resources.

3 Results

The primary precision measures for the segmentation task were the Dice coefficient and the 95th percentile Hausdorff distance. Performance in the certainty task was assessed based on the Dice AUC. Mean and median precision measures are provided for the training, validation and test datasets.

Segmentation: Mean Dice coefficient values for the training dataset were: 0.929 (WT), 0.912 (TC) and 0.887 (ET). Median values were: 0.937 (WT), 0.927 (TC) and 0.908 (ET). Standard deviations were: 0.035 (WT), 0.057 (TC) and 0.118 (ET). Mean Housdorff distances were: 2.63 (WT), 2.61 (TC) and 5.58 (ET), median values were: 2.24 (WT), 2.00 (TC) and 1.00 (ET).

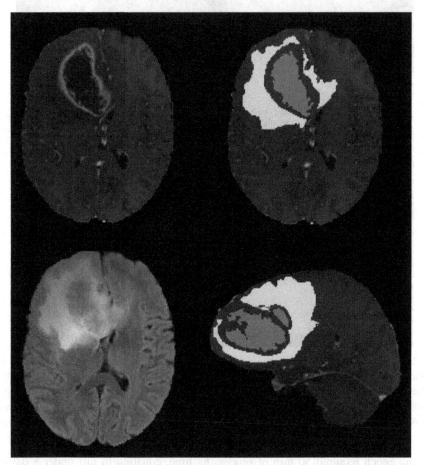

Fig. 3. Example segmentation from the BraTS validation set (case 15). Images are T1 post-contrast (top-left), FLAIR (bottom-left), axial segmentations (top-right) and sagittal segmentation (bottom-right). Dice scores on this case were 0.885 (WT), 0.958 (TC) and 0.901 (ET).

Validation mean Dice coefficient scores were: 0.881 (WT), 0.789 (TC) and 0.712 (ET). Median Dice values were: 0.909 (WT), 0.861 (TC) and 0.825 (ET). Standard deviations were: 0.089 (WT), 0.196 (TC) and 0.285 (ET). Mean Housdorff distances were: 6.72 (WT), 10.2 (ET) and 40.6 (ET). Median Hausdorff distances were: 3.61 (WT), 5.20 (TC) and 3.00 (ET).

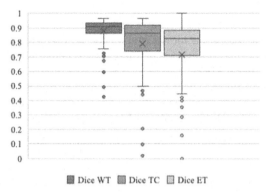

Fig. 4. Box plot demonstrating the distribution of Dice scores for individual cases on the validation dataset for each label.

Test mean Dice coefficient values were: 0.859 (WT), 0.788 (TC) and 0.766 (ET). Median Dice scores were: 0.902 (WT), 0.887 (TC) and 0.823 (ET). Standard deviations were: 0.129 (WT), 0.258 (TC) and 0.212 (ET). Mean Hausdorff distances were: 7.65 (WT), 24.6 (TC) and 19.4 (ET). Median Hausdorff distances were: 3.74 (WT), 4.00 (TC) and 2.24 (ET) (Table 1).

Table 1. Segmentation results on the training, validation and test data.

		Dice			Hausdorff distance (95th Percentile)		
		WT	TC	ET	WT	TC	ET
Training	Mean	0.929	0.912	0.887	2.64	2.61	5.58
	Median	0.937	0.927	0.908	2.24	2.00	1.00
	SD	0.035	0.057	0.118	3.31	3.92	38.7
Validation	Mean	0.881	0.789	0.712	6.72	10.2	40.6
	Median	0.909	0.861	0.825	3.61	5.20	3.00
	SD	0.089	0.196	0.285	10.9	14.6	109
	Mean	0.859	0.788	0.766	7.65	24.6	19.4
Test	Median	0.902	0.887	0.823	3.74	4.00	2.24
	SD	0.129	0.258	0.212	13.1	79.2	74.7

Uncertainty: Mean Dice AUC values for the training dataset were: 0.944 (WT), 0.926 (TC) and 0.823 (ET). Mean Dice AUC values on the validation dataset were: 0.924 (WT), 0.885 (TC) and 0.744 (ET). Test dataset mean Dice AUC values were: 0.898 (WT), 0.837 (TC) and 0.782 (ET).

4 Discussion

We demonstrate accurate segmentations can be generated with a 2D segmentation method with comparable segmentation scores to high-performing 3D methods. The best performing segmentation model in BraTS 2018 achieved test Dice scores of 0.884

(WT), 0.815 (TC) and 0.766 (ET) using a 3D ensemble method [25]. Although the method described here has not achieved state-of-the-art performance, we argue that a 2D segmentation model may be favourable where there is non-volumetric imaging which is often the case in clinical MRI.

There was a large difference between mean and median precision scores for the validation dataset. Performance was generally accurate, however a small number of very poor segmentations reduced the mean by a large amount. The skewed distribution of Dice scores is visually apparent in Fig. 4. The effect of poor performance on mean values was most pronounced for ET. Inspection of individual cases in the validation dataset revealed 12/125 cases where the ET Dice was zero with correspondingly high Hausdorff distances. In low-grade glioma, where there is often no enhancement, a single incorrectly labelled enhancing voxel may result in a Dice of zero for ET. It is recognised that Dice tends to extreme values for small volumes [22]. The model did not perform as well in smaller tumours, which may consist of a small region of oedema only. Poor performance was also observed in cases of cystic GBM. Approximately 20% of GBMs form cysts, which are associated with favourable mutation profiles and better survival [26, 27]. Radiologically cysts are T2 hyperintense and low signal on FLAIR - isointense to cerebrospinal fluid (CSF). It can be inferred the segmentation model tends to misclassify cysts as normal CSF due to the similar signal characteristics. Interpretive errors such as this reinforce that automated segmentation is likely to fit into the radiological workflow mainly as an adjunctive tool, and radiologist oversight may be required to avoid errors requiring higher level image interpretation. From a radiological standpoint, it is fast and easy to tacitly assess the quality of a segmentation. The observation that median Dice and Hausdorff distances remained stable across the validation and test datasets supports the utility of automated segmentation with radiologist oversight in poorly segmented cases.

During model design, the authors experimented with a 3D UNet [28] to compare the effectiveness of different approaches. Despite the use of a high-end 32Gb GPU, the 3D model required a patch-based approach to fit into memory, with patch sizes smaller than $64 \times 64 \times 64 \times 4$ channels to allow for a reasonable batch size and feature count. Using a 2D model it was possible to train using whole slices ($240 \times 240 \times 4$ channels). An advantage of a large field of view is that this contextualises tumour location relative to other structures in the brain, and long-range information may aid segmentation [29]. A recognised problem is the misclassification of age/hypertension related white matter hyperintensities as oedema [9]. These hyperintensities are recognised by their characteristic periventricular location and a large field of view may reduce this error. Interestingly the segmentation model tended to misclassify periventricular white matter hyperintensities as oedema early in training but was better able to discriminate these entities as training progressed. It is possible that the interpretive errors described above may be overcome with sufficient data and training (Fig. 5).

Given the computational restraints, another possible advantage of a 2D model is that 3D convolutions split the limited number of features across an additional degree of freedom, potentially limiting overall feature complexity given the computational constraints. This problem is compounded if image resolution in the 3rd dimension is reduced. For example, if the filter kernel is $3 \times 3 \times 3$ and the slice above and below are interpolated,

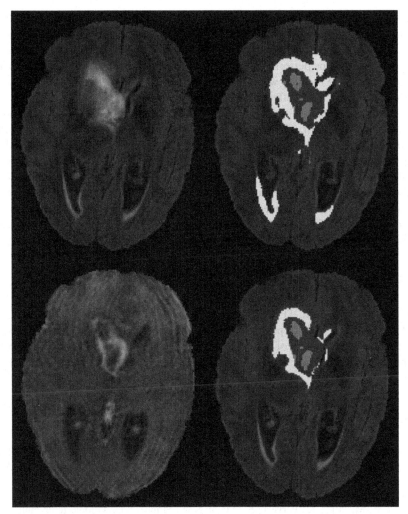

Fig. 5. Images from the validation dataset (case 8) demonstrating erroneous segmentation of periventricular white matter hyperintensity as oedema and choroid plexus as enhancement which improves during training. Images are FLAIR (top-left), T1 post-contrast (bottom-left), segmentation at 50 epochs (top-right) and final segmentation at 140 epochs (bottom-right). LCC extraction has not been performed in this example.

the filter is effectively convolving on itself in the 3rd dimension. We argue this makes the trade-off of a reduced field of view for increased dimensional information less attractive. This trade-off is likely to become less important as computing technology improves and GPU memory increases.

A possible disadvantage of the 2D approach is discontinuity of predictions between adjacent slices. We found the 2D segmentations were generally congruous in the coronal and sagittal planes when 3D reconstructed (see Fig. 3). The presented arguments in favour of a 2D approach related to anisotropy of resolution rely on the segmentation model being "in-plane" with image acquisition. There was at least one case in the validation dataset where the T2 sequence was acquired in the coronal plane. Our 2D segmentation model based in the axial plane performed poorly on this case, although it is not certain whether this was because the imaging was "out of plane". It is likely that a 3D model would be more robust to this kind of effect.

It must be recognised that despite reasonably high performance, there were better performing models in BraTS 2020. 3D segmentation models have been demonstrated to outperform their 2D counterparts for isotropic CT imaging [30]. It was observed that a proportion of the T2 and FLAIR sequences in the BraTS dataset were volumetrically acquired and it is possible that this has contributed to the high performance of 3D models in previous competitions. It remains unclear whether this performance will be upheld in clinical settings where volumetric T2 and FLAIR imaging is not routinely performed for glioma imaging. There would be value in comparing the performance of equivalent 2D and 3D segmentation models to further tease out the arguments presented above.

5 Conclusion

In this paper we present a 2D Dense-UNet brain tumour segmentation model with comparable performance to high precision 3D models in previous BraTS competitions. We argue the potential benefits of 2D segmentation given current MRI acquisition protocols and advocate the consideration of 2D neural network architectures in MRI based research.

References

1. Ostrom, Q.T., Gittleman, H., Liao, P., Rouse, C., Chen, Y., Dowling, J., et al.: CBTRUS statistical report: primary brain and central nervous system tumors diagnosed in the United States in 2007–2011. Neuro. Oncol. **16**(Suppl 4), iv1–iv63 (2014). https://pubmed.ncbi.nlm.nih.gov/25304271
2. Hanif, F., Muzaffar, K., Perveen, K., Malhi, S.M., Simjee, S.U.: Glioblastoma multiforme: a review of its epidemiology and pathogenesis through clinical presentation and treatment. Asian Pac. J. Cancer Prev. **18**(1), 3–9 (2017). https://pubmed.ncbi.nlm.nih.gov/28239999
3. Menze, B.H., et al.: the multimodal brain tumor image segmentation benchmark (BRATS). IEEE Trans. Med. Imaging **34**(10), 1993–2024 (2015)
4. Bakas, S., Akbari, H., Sotiras, A., Bilello, M., Rozycki, M., Kirby, J.S., et al.: Advancing the cancer genome atlas glioma MRI collections with expert segmentation labels and radiomic features. Sci. Data **4**, 170117 (2017)
5. Bakas, S., Reyes, M., Jakab, A., Bauer, S., Rempfler, M., Crimi, A., et al.: Identifying the best machine learning algorithms for brain tumor segmentation, progression assessment, and overall survival prediction in the BRATS challenge (2018)
6. Bakas, S., et al.: Segmentation labels for the pre-operative scans of the TCGA-GBM collection. Cancer Imaging Arch. (2017). https://doi.org/10.7937/K9/TCIA.2017.KLXWJJ1Q
7. Bakas, S., et al.: Segmentation labels and radiomic features for the pre-operative scans of the TCGA-LGG collection. Cancer Imaging Arch. (2017). https://doi.org/10.7937/K9/TCIA.2017.GJQ7R0EF

8. Clark, K., et al.: The cancer imaging archive (TCIA): maintaining and operating a public information repository. J. Digit. Imaging **26**(6), 1045–1057 (2013). https://doi.org/10.1007/s10278-013-9622-7
9. Rudie, J.D., Weiss, D.A., Saluja, R., Rauschecker, A.M., Wang, J., Sugrue, L., et al.: Multi-disease segmentation of gliomas and white matter hyperintensities in the BraTS data using a 3D convolutional neural network. Front. Comput. Neurosci **13**, 84 (2019)
10. Yogananda, C.G.B., et al.: A novel fully automated MRI-based deep learning method for classification of IDH mutation status in brain gliomas. Neuro. Oncol. **22**, 402–411 (2019)
11. van Dijken, B.R.J., van Laar, P.J., Li, C., Yan, J.-L., Boonzaier, N.R., Price, S.J., et al.: Ventricle contact is associated with lower survival and increased peritumoral perfusion in glioblastoma. J. Neurosurg. JNS **131**(3), 717–723 (2018). https://thejns.org/view/journals/j-neurosurg/131/3/article-p717.xml
12. Louis, D.N., et al.: The 2016 world health organization classification of tumors of the central nervous system: A summary. Acta Neuropathol. **131**(6), 803–820 (2016). https://doi.org/10.1007/s00401-016-1545-1
13. Leon, S.P., Folkerth, R.D., Black, P.M.: Microvessel density is a prognostic indicator for patients with astroglial brain tumors. Cancer **77**(2), 362–372 (1996)
14. Wesseling, P., van der Laak, J.A., Link, M., Teepen, H.L., Ruiter, D.J.: Quantitative analysis of microvascular changes in diffuse astrocytic neoplasms with increasing grade of malignancy. Hum. Pathol. **29**(4), 352–358 (1998)
15. Lin, Z.-X.: Glioma-related edema: new insight into molecular mechanisms and their clinical implications. Chin. J. Cancer **32**(1), 49–52 (2013). https://pubmed.ncbi.nlm.nih.gov/23237218
16. Wu, C.-X., Lin, G.-S., Lin, Z.-X., Zhang, J.-D., Liu, S.-Y., Zhou, C.-F.: Peritumoral edema shown by MRI predicts poor clinical outcome in glioblastoma. World J. Surg. Oncol. **11**(13), 97 (2015). https://pubmed.ncbi.nlm.nih.gov/25886608
17. Ellingson, B.M., Bendszus, M., Boxerman, J., Barboriak, D., Erickson, B.J., Smits, M., et al.: Consensus recommendations for a standardized Brain Tumor Imaging Protocol in clinical trials. Neuro. Oncol. **17**(9), 1188–1198 (2015). https://pubmed.ncbi.nlm.nih.gov/26250565
18. Chagla, G.H., Busse, R.F., Sydnor, R., Rowley, H.A., Turski, P.A.: Three-dimensional fluid attenuated inversion recovery imaging with isotropic resolution and nonselective adiabatic inversion provides improved three-dimensional visualization and cerebrospinal fluid suppression compared to two-dimensional flair at 3 tesla. Invest. Radiol. **43**(8), 547–551 (2008). https://pubmed.ncbi.nlm.nih.gov/18648253
19. Hausmann, D., Liu, J., Budjan, J., Reichert, M., Ong, M., Meyer, M., et al.: Image quality assessment of 2D versus 3D T2WI and evaluation of ultra-high b-Value (b=2,000 mm/s(2)) DWI for response assessment in rectal cancer. Anticancer Res. **38**(2), 969–978 (2018)
20. Ronneberger, O., Fischer, P., Brox, T.: U-Net: convolutional networks for biomedical image segmentation. In: Navab, N., Hornegger, J., Wells, W.M., Frangi, A.F. (eds.) MICCAI 2015. LNCS, vol. 9351, pp. 234–241. Springer, Cham (2015). https://doi.org/10.1007/978-3-319-24574-4_28
21. Huang, G., Liu, Z., van der Maaten, L., Weinberger, K.Q.: Densely Connected Convolutional Networks (2016)
22. Isensee, F., Kickingereder, P., Wick, W., Bendszus, M., Maier-Hein, K.H.: No new-net. In: Crimi, A., Bakas, S., Kuijf, H., Keyvan, F., Reyes, M., van Walsum, T. (eds.) BrainLes 2018. LNCS, vol. 11384, pp. 234–244. Springer, Cham (2019). https://doi.org/10.1007/978-3-030-11726-9_21
23. Pérez-Beteta, J., Molina-García, D., Villena, M., Rodríguez, M.J., Velásquez, C., Martino, J., et al.: Morphologic features on MR imaging classify multifocal glioblastomas in different prognostic groups. Am. J. Neuroradiol. **40**, 634–640 (2019). https://www.ajnr.org/content/early/2019/03/28/ajnr.A6019.abstract

24. Abadi, M., Agarwal, A., Barham, P., Brevdo, E., Chen, Z., Citro, C., et al.: TensorFlow: Large-Scale Machine Learning on Heterogeneous Systems (2015). https://tensorflow.org/
25. Myronenko, A.: 3D MRI brain tumor segmentation using autoencoder regularization. In: Crimi, A., Bakas, S., Kuijf, H., Keyvan, F., Reyes, M., van Walsum, T. (eds.) BrainLes 2018. LNCS, vol. 11384, pp. 311–320. Springer, Cham (2019). https://doi.org/10.1007/978-3-030-11726-9_28
26. Curtin, L., Whitmire, P., Rickertsen, C.R., Mazza, G.L., Canoll, P., Johnston, S.K., et al.: Assessment of prognostic value of cystic features in glioblastoma relative to sex and treatment with standard-of-care. medRxiv 19013813 (2020). https://medrxiv.org/content/early/2020/07/07/19013813.abstract
27. Zhou, J., Reddy, M.V., Wilson, B.K.J., Blair, D.A., Taha, A., Frampton, C.M., et al.: MR imaging characteristics associate with tumor-associated macrophages in glioblastoma and provide an improved signature for survival prognostication. Am. J. Neuroradiol. 39(2), 252–259 (2018). https://www.ajnr.org/content/39/2/252.abstract
28. Çiçek, Ö., Abdulkadir, A., Lienkamp, S.S., Brox, T., Ronneberger, O.: 3D U-Net: learning dense volumetric segmentation from sparse annotation. In: Ourselin, S., Joskowicz, L., Sabuncu, M.R., Unal, G., Wells, W. (eds.) MICCAI 2016. LNCS, vol. 9901, pp. 424–432. Springer, Cham (2016). https://doi.org/10.1007/978-3-319-46723-8_49
29. Mlynarski, P., Delingette, H., Criminisi, A., Ayache, N.: 3D convolutional neural networks for tumor segmentation using long-range 2D context. Comput. Med. Imaging Graph. 73, 60–72 (2019). https://www.sciencedirect.com/science/article/pii/S0895611118304221
30. Zhou, X.: Automatic segmentation of multiple organs on 3D CT images by using deep learning approaches. Adv. Exp. Med. Biol. 1213, 135–147 (2020)

Attention U-Net with Dimension-Hybridized Fast Data Density Functional Theory for Automatic Brain Tumor Image Segmentation

Zi-Jun Su[1], Tang-Chen Chang[1,2], Yen-Ling Tai[1], Shu-Jung Chang[1],
and Chien-Chang Chen[1(✉)] (iD)

[1] Bio-Microsystems Integration Laboratory, Department of Biomedical
Sciences and Engineering, National Central University, Taoyuan, Taiwan
`gettgod@ncu.edu.tw`
[2] Multimedia Information System Laboratory, Department of Computer Science,
National Tsing Hua University, Hsinchu, Taiwan

Abstract. In the article, we proposed a hybridized method for brain tumor image segmentation by fusing topological heterogeneities of images and the attention mechanism in the neural networks. The three-dimensional image datasets were first pre-processed using the histogram normalization for the standardization of pixel intensities. Then the normalized images were parallel fed into the procedures of affine transformations and feature pre-extractions. The technique of fast data density functional theory (fDDFT) was adopted for the topological feature extractions. Under the framework of fDDFT, 3-dimensional topological features were extracted and then used for the 2-dimensional tumor image segmentation, then those 2-dimensional significant images are reconstructed back to the 3-dimensional intensity feature maps by utilizing physical perceptrons. The undesired image components would be filtered out in this procedure. Thus, at the pre-processing stage, the proposed framework provided dimension-hybridized intensity feature maps and image sets after the affine transformations simultaneously. Then the feature maps and the transformed images were concatenated and then became the inputs of the attention U-Net. By employing the concept of gate controlling of the data flow, the encoder can perform as a masked feature tracker to concatenate the features produced from the decoder. Under the proposed algorithmic scheme, we constructed a fast method of dimension-hybridized feature pre-extraction for the training procedure in the neural network. Thus, the model size as well as the computational complexity might be reduced safely by applying the proposed algorithm.

Keywords: fDDFT · Attention U-Net · Dimensional hybridization

1 Introduction

Neural network (NN) based models exhibit a superior framework for automatic brain tumor image segmentation through learning from data [1–6]. Among these techniques,

© Springer Nature Switzerland AG 2021
A. Crimi and S. Bakas (Eds.): BrainLes 2020, LNCS 12659, pp. 81–92, 2021.
https://doi.org/10.1007/978-3-030-72087-2_8

the convolutional neural network (CNN) based methods successfully acquire more attention than others. The emergence of fully convolutional neural networks (FCN) further elevates the performance of CNN in modern medical image segmentation [6, 7]. However, as compared to the availability of general image datasets, the medical image datasets usually have sparse data so that the technical direction of development is often guided by models that handle small size datasets. Thus, the evolution of neural networks, starting from FCN over V-net [6] and U-net [8] then to the approach of ensemble several models for robust segmentation (EMMA) [9], gradually become the developing core in brain tumor image segmentation. Yet NN based models rely on high-cost hardware and high computational complexity during the training processes [2, 5]. For instance, the occupation of GPU memory and the computational time is proportional to the size of convolutional kernels [1]. Large numbers of hyper-parameters in the procedure of model training also challenge the memory size and requirement of hardware [2]. Also, the labor and time spent on data labeling might become heavy burdens to the technical progression.

Therefore, to inherit the properties of compact modeling form and affordable computational complexity from interactive approaches [10, 11], also to employ the characteristics of learning procedures from NN based models, we proposed a hybridized method by introducing the fast data density functional theory (fDDFT) [12, 13]. The fDDFT is established by fusing the mathematical characteristics from sophisticated machine learning methods and a technique from quantum mechanics known as density functional theory [13]. In this article, we introduced dimension-hybridized structures into the theoretical framework of fDDFT to simultaneously preserve local and global features of images then the topological features extracted from fDDFT were fed into the attention U-Net model. The tumor images in the BraTS 2020 MRI dataset [14–18] were subsequently segmented by considering their topological heterogeneities in specific energy space.

2   Methods

In the first phase of the proposed algorithmic framework, the MRI datasets [14–18] were first processed with a histogram transformation for the intensity normalization. Then the normalized image datasets were individually and parallel fed into the blocks of fDDFT [12, 13] and affine transformations for the feature pre-extraction and image augmentation. It should be emphasized that in the block of fDDFT, the 3-dimensional image set of each sample was treated as a frozen physical system and the voxels within the system were treated as physical particles [19]. According to the theoretical frame of the fDDFT, the Lagrangian density functional was estimated to define the cluster boundaries by means of finding the topological heterogeneities between the tumor components and the normal tissue components. To reduce the computational complexity, the technique of 2-dimensional fast Fourier transforms (2D-FFTs) was then employed in the fDDFT scheme. The structure of the employed kernel is illustrated in Fig. 1. Then then relevant framework is illustrated in Fig. 2.

In general, the voxel components consisted of tumors reveal obvious topological heterogeneity compared to that of normal tissues and skull. Thus, these voxel components would be extracted to be as global intensity features and then used to inspect the

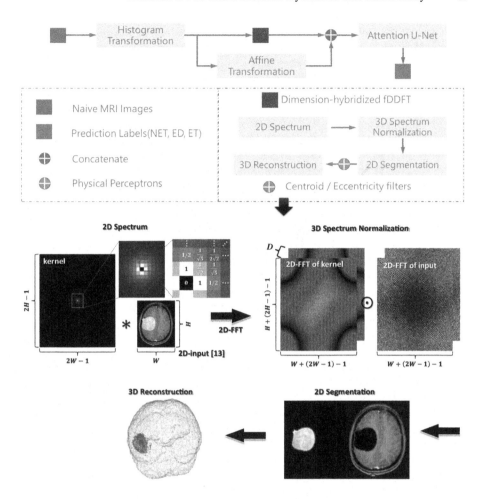

Fig. 1. The framework of the first phase. A scheme of feature pre-extraction constructed by the affine transformation and fDDFT is illustrated. The well-normalized image sets are individually fed into these two paths, wherein the fDDFT provides a different image configuration by employing the dimension-hybridized estimation. The global intensity features will be utilized for the 2-dimensional brain tumor image segmentation. Then the 2-dimensional images are reconstructed bask the 3-dimensional image sets. During the reconstructions, physical perceptrons are applied for avoiding the undesired image components. The parameters H, W, and D are respectively the height, width, and depth of adopted image sets. The mathematical symbols $*$ and $\odot$ are convolution and Hadamard product, respectively. It also noted that the input illustrated in the block of fDDFT was adopted from Ref. [13].

regions of tumor images in 2-dimensional space for local image recognition and segmentation. Then the 2-dimensional segmented images were reconstructed back to the 3-dimensional intensity feature maps. Most boundary estimations of the tumor images were calculated in 2-dimensional, thus the fDDFT can offer very low computational complexity. Meanwhile, the undesired image components would also be filtered out in

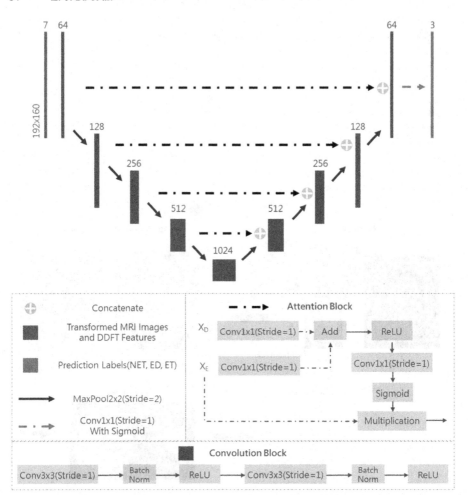

Fig. 2. Visualization of the attention U-Net. The input feature maps contain seven channels: four of transformed images and the rest are fDDFT features. Each attention block consists of two inputs X_E and X_D respectively extracted from encoder and decoder and used sigmoid on the X_E. Convolution block is a common CNN-based architecture and contains the functions of convolution, BatchNorm, and ReLU, repeatedly. The outputs of segmentation correspond to the three types of tumor labels: Non-Enhancing Tumor (NET), Peritumoral Edema (ED), and Enhancing Tumor (ET).

this procedure. Furthermore, in the dimension-hybridized structure of fDDFT, the global and local features extracted by the topological heterogeneity of images could be simultaneously preserved. The framework of the first phase is illustrated in Fig. 1. The feature maps extracted from the fDDFT and the augmented images produced from the affine transformations were concatenated and became the inputs of the attention U-Net.

In the second phase as illustrated in Fig. 2, the attention U-Net, a general encoder-decoder architecture based CNN with attention mechanism [20, 21], was utilized for the brain tumor image segmentation. In our proposed framework, we replaced the original branches of multi-convolution layers in U-Net to be as attention blocks for creating masks which can filter out the information unrelated to the tumor images. The motivation for applying the attention mechanism, which is commonly used in natural language preprocessing and computer vision, is stemmed from its good capability of capturing the relevant information from image datasets.

2.1 Normalization and Augmentation

Since the BraTS MRI intensities between samples might have significantly various due to being collected from different imaging equipment, the normalization process is inevitable for the tumor image recognition. In the first step, we utilized the method of histogram normalization [22] implemented by TorchIO [23] for the image intensity normalization in each either training or validation sample. Also in order to exclude the suspected artifacts or outliers, we only considered the image intensities enclosed between the 1st and the 99th percentiles in each sample. The comparison between before and after executing the histogram normalization is illustrated in Fig. 3. It is obvious that the employed normalization method successfully corrected whole the image intensity distributions to relatively stable states.

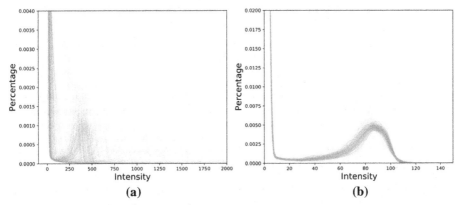

Fig. 3. (a) illustrates the intensity distributions of whole image datasets and (b) shows the results of intensity distributions after processing the histogram transformation. The intensity of each sample was mapped to a similar interval, thus the results of intensity normalizations would benefit the procedure in the first phase.

Furthermore, we applied the image augmentation for each sample in each training step. We processed a random mirror flip for all dimensions of image sets with a probability of 0.5. We also processed a random shift $(-0.1, 0.1)$, a random zoom $(0.8, 1.1)$, and a random shear $(-0.1, 0.1)$ for the sake of emphasizing on significant features of images.

2.2 Feature Pre-extraction Using Fast Data Density Functional Theory

Theoretically, the Lagrangian density functional of a pixel space is defined by the combination of kinetic and potential energy density functionals. By assigning sufficient boundary information from the energy property of a frozen physical system [24, 25], the relationship between image intensity distribution ρ and the expectation value of kinetic energy density functional (KEDF) can then be derived as:

$$t[\rho] = \frac{2\pi^2 D}{D+2} \cdot (D\rho)^{2/D}. \tag{1}$$

In a 2-dimensional pixel space ($D = 2$), $t[\rho] = 2\pi^2 \rho(\mathbf{r})$, where $\mathbf{r}$ represents the position vector of an observed point. Then, in the article we adopted a Coulombic potential form to imitate long-distance interaction and thus the data potential energy density functional (PEDF) has a simple Coulombic form:

$$u[\rho] = \frac{1}{2} \int \frac{\rho(\mathbf{r}')}{|\mathbf{r} - \mathbf{r}'|} d^D \mathbf{r}', \tag{2}$$

where $\mathbf{r}'$ represents the position vector of a source point. Thus, the Lagrangian density functional with fixing data size can be well defined in a scale-free energy system:

$$\mathcal{L}[\rho] = \gamma^2 t[\rho] - \gamma u[\rho]. \tag{3}$$

The adaptive scaling factor γ in Eq. (3) is the core of fDDFT and is used to prevent an imbalance between KEDF and PEDF when scaling or normalizing them:

$$\gamma = \frac{1}{2} \frac{\langle u[\rho] \rangle}{\langle t[\rho] \rangle}, \tag{4}$$

where $\langle u[\rho] \rangle$ and $\langle t[\rho] \rangle$ are the global means of PEDF and KEDF, respectively. it should be emphasized that the 2D-FFT was adopted to reduce the computational complexity caused by the integration of PEDF and the mapping structure has been shown in the 3-dimensional spectrum normalization of fDDFT in Fig. 1.

2.3   Encoder, Decoder, and Attention Block

As shown in Fig. 2, attention blocks were employed to replace the parts of the copy and crop of the original U-net structure. Additionally, an encoder and a decoder consist of the main core of each attention block. In the encoder part, each convolution block consists of two convolution layers and each of which was followed by a normalization function and an activation function ReLU. In the normalization step, we used batch normalization to reduce internal covariance shifts and the dependency of the scales of parameters and gradients. At the end of each convolution block, we added MaxPool, a down-sampling layer that halved the spatial sizes. In the decoder part, we added an up-sampling layer with bilinear interpolation to increase spatial contents before the convolution blocks. Convolution blocks reduce channels by a factor of 2 in comparison to the last layer. All sizes of convolution kernels are 3×3, except for the final convolution layer which has a size of 1×1 in the decoder, and are followed by the activation function, ReLU.

We then computed the input gates from the attention block by utilizing convolution blocks in the encoder and decoder with the same spatial sizes and number of chan-nels. Each attention block began with two convolution layers to obtain two features

X_E (encoder feature) and X_D (decoder feature) respectively extracted from encoder and decoder features, which were then added together to produce a new feature. The production of the new feature was then followed by a convolution layer and a sigmoid function which was like a gate controlling the data flow from the encoder. The gate finally multiplied with the feature of the encoder as a masked feature which then concatenates the feature from the decoder.

2.4 Optimization

An Adam optimizer was employed in the training process for extracting features via a specified loss function and a justified learning rate. As listed in Eq. (5), the adopted loss function contents two terms:

$$L = L_{dice} + \alpha \cdot L_{L2} \tag{5}$$

As listed in Eq. (6), the L_{dice} is defined as the Jaccard dice loss used for evaluating the interaction between the ground truth A_{true} parts and the decoder prediction outputs A_{pred}, wherein the parameter ε is a small constant for avoiding the zero division:

$$L_{dice} = \frac{2 \cdot \left(\sum A_{true} * A_{pred}\right) + \varepsilon}{\sum A_{true}^2 + \sum A_{pred}^2 + \varepsilon} \tag{6}$$

The second term L_{L2} is the two-norm sum of overall weighting factors $\{W\}$ for reducing dependencies between tumor features and particular weighting factors:

$$L_{L2} = \sum_{w \in \{W\}} ||w||^2 \tag{7}$$

The parameter α used in Eq. (5) is a small floating weighting factor to maintain the balance between L_{dice} and L_{L2}. To maintain the robustness of model segmentation, we also added the L2 regularization and spatial dropouts [26] with a probability of 0.5 at the end of the encoder.

3 Results

Figure 4 and 5 respectively illustrate the visualization outcomes of training and validation sets of the BraTS 2020 dataset. In each figure, (a) and (b) are respectively the selected 3- and 2-dimensional segmented brain tumor configurations. The left column in Fig. 4(b) and 5(b) are the pre-extracted features provide by the fDDFT. The ground truth as shown in Fig. 4(c) verifies the feasibility of the proposed algorithm. The dice score of this case is 0.9142. Furthermore, as expected, the undesired image components were filtered out so that the 3-dimensional tumor topology can be well defined as shown in Fig. 4(a) and 5(a).

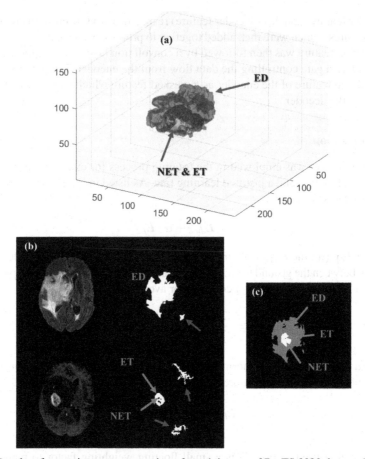

Fig. 4. Results of tumor image segmentation of a training set of BraTS 2020 dataset. Figure (a) shows the 3-dimensional segmented component of ED, NET, and ET. Figure (b) depicts the 2-dimensional segmented results and (c) is the ground truths. As expected, the undesired components as shown in (b) were filtered out in the procedure of fDDFT. The Dice of this sample is 0.9142.

Following the instructions of the BraTS challenge, we uploaded the prediction labels, which are inclusive of ED, ET, and NET, for retrieving the results (Dice similarity coefficient, sensitivity, and Hausdorff distance) provided by CBICA Image Processing Portal. For the convenience of comparison, we built the naive U-Net as the baseline benchmark for adjusting the proposed model. Table 1 and 2 respectively show the scores obtained by utilizing the proposed method and naive U-Net on the BraTS 2020 validation dataset, which consists of 125 cases. From the evaluation results of naive U-net, it is

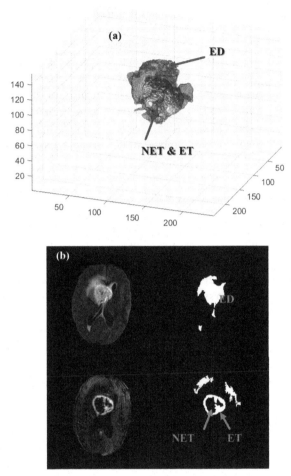

Fig. 5. Results of tumor image segmentation of a validation set of the BraTS 2020 dataset. Figure (a) shows the 3-dimensional segmented component of ED, NET, and ET. Figure (b) depicts the 2-dimensional segmented results. As expected, the undesired components also were filtered out in the procedure of fDDFT.

obvious that the proposed architecture provided a more robust brain segmentation method especially on the DSC score of ET and TC as well as Hausdorff 95 distance of WT, TC, and ET. Table 3 shows the score of the proposed method on the BraTS 2020 testing data, which includes 166 cases.

Table 1. Evaluation results of proposed method on 2020 validation data

	DSC			Sensitivity			Hausdorff 95		
	ET	WT	TC	ET	WT	TC	ET	WT	TC
Mean	0.754	0.896	0.821	0.749	0.886	0.788	32.659	5.358	6.576
StdDev	0.270	0.069	0.154	0.286	0.099	0.201	100.883	8.802	11.036
Median	0.848	0.916	0.883	0.851	0.918	0.876	2.000	3.000	3.606
25quantile	0.757	0.879	0.784	0.742	0.875	0.713	1.414	2.236	2.000
75quantile	0.893	0.939	0.929	0.918	0.949	0.936	4.000	4.899	7.071

Table 2. Evaluation results of naive U-Net on 2020 validation data

	DSC			Sensitivity			Hausdorff 95		
	ET	WT	TC	ET	WT	TC	ET	WT	TC
Mean	0.690	0.886	0.773	0.700	0.896	0.754	36.614	16.366	12.251
StdDev	0.311	0.073	0.219	0.325	0.087	0.246	100.935	24.803	18.158
Median	0.827	0.906	0.855	0.847	0.915	0.851	2.449	4.359	6.000
25quantile	0.685	0.867	0.697	0.637	0.866	0.647	1.414	2.828	2.449
75quantile	0.885	0.936	0.924	0.924	0.959	0.932	6.403	17.912	11.874

Table 3. Evaluation results of proposed method on 2020 testing data

	DSC			Sensitivity			Hausdorff 95		
	ET	WT	TC	ET	WT	TC	ET	WT	TC
Mean	0.775	0.873	0.808	0.798	0.891	0.825	22.690	6.399	20.778
StdDev	0.216	0.129	0.257	0.234	0.113	0.230	84.279	11.044	74.668
Median	0.837	0.909	0.906	0.878	0.922	0.908	2.000	3.162	2.828
25quantile	0.753	0.870	0.827	0.750	0.868	0.781	1.414	2.236	1.732
75quantile	0.890	0.939	0.948	0.933	0.955	0.960	2.828	5.385	6.041

4   Conclusion

In this article, we provide a feature pre-extraction scheme by combining the affine transformation and the fast data density functional theory. Additionally, the dimension-hybridized procedure of the fDDFT integrates the global and local intensity features with a significant low computational complexity. Then the intensity features extracted by the fDDFT and the topological features collected in the affine transformation are combined and to be as the input of the attention U-Net. The segmented results verified the feasibility of the proposed algorithm.

Acknowledgments. This work was supported by the Ministry of Science and Technology, Taiwan, with a grant number of MOST 108-2221-E-008-081-MY3.

References

1. Myronenko, A.: 3D MRI brain tumor segmentation using autoencoder regularization. In: Crimi, A., Bakas, S., Kuijf, H., Keyvan, F., Reyes, M., van Walsum, T. (eds.) BrainLes 2018. LNCS, vol. 11384, pp. 311–320. Springer, Cham (2019). https://doi.org/10.1007/978-3-030-11726-9_28

2. Ma, C., Luo, G., Wang, K.: Concatenated and connected random forests with multiscale patch driven active contour model for automated brain tumor segmentation of MR images. IEEE Trans. Med. Imaging **37**, 1943–1954 (2018)

3. Zhang, W., et al.: Deep convolutional neural networks for multi-modality isointense infant brain image segmentation. Neuroimage **108**, 214–224 (2015)

4. Xu, Y., Wang, Y., Yuan, J., Cheng, Q., Wang, X., Carson, P.L.: Medical breast ultrasound image segmentation by machine learning. Ultrasonics **91**, 1–9 (2019)

5. Smistad, E., Falch, T.L., Bozorgi, M., Elster, A.C., Lindseth, F.: Medical image segmentation on GPUs–a comprehensive review. Med. Image Anal. **20**, 1–8 (2015)

6. Milletari, F., Navab, N., Ahmadi, S.-A.: V-net: fully convolutional neural networks for volumetric medical image segmentation. In: 2016 fourth international conference on 3D vision (3DV), pp. 565–571. IEEE (2016)

7. Dasgupta, A., Singh, S.: A fully convolutional neural network based structured prediction approach towards the retinal vessel segmentation. In: 2017 IEEE 14th International Symposium on Biomedical Imaging (ISBI 2017), pp. 248–251. IEEE (2017)

8. Ronneberger, O., Fischer, P., Brox, T.: U-net: convolutional networks for biomedical image segmentation. In: Navab, N., Hornegger, J., Wells, W.M., Frangi, A.F. (eds.) MICCAI 2015. LNCS, vol. 9351, pp. 234–241. Springer, Cham (2015). https://doi.org/10.1007/978-3-319-24574-4_28

9. Kamnitsas, K., et al.: Ensembles of multiple models and architectures for robust brain tumour segmentation. In: Crimi, A., Bakas, S., Kuijf, H., Menze, B., Reyes, M. (eds.) BrainLes 2017. LNCS, vol. 10670, pp. 450–462. Springer, Cham (2018). https://doi.org/10.1007/978-3-319-75238-9_38

10. Pratondo, A., Chui, C.-K., Ong, S.-H.: Integrating machine learning with region-based active contour models in medical image segmentation. J. Vis. Commun. Image Represent. **43**, 1–9 (2017)

11. Li, C., Huang, R., Ding, Z., Gatenby, J.C., Metaxas, D.N., Gore, J.C.: A level set method for image segmentation in the presence of intensity inhomogeneities with application to MRI. IEEE Trans. Image Process. **20**, 2007–2016 (2011)

12. Chen, C.-C., Tsai, M.-Y., Kao, M.-Z., Lu, H.H.-S.: Medical image segmentation with adjustable computational complexity using data density functionals. Appl. Sci. **9**, 1718 (2019)

13. Chen, C.-C., Juan, H.-H., Tsai, M.-Y., Lu, H.H.-S.: Unsupervised learning and pattern recognition of biological data structures with density functional theory and machine learning. Sci. Rep. **8**, 1–11 (2018)

14. Bakas, S., et al.: Identifying the best machine learning algorithms for brain tumor segmentation, progression assessment, and overall survival prediction in the BRATS challenge. arXiv preprint arXiv:.02629 (2018)

15. Bakas, S., et al.: Advancing the cancer genome atlas glioma MRI collections with expert segmentation labels and radiomic features. Sci. Data **4**, 170117 (2017)

16. Menze, B.H., et al.: The multimodal brain tumor image segmentation benchmark (BRATS). IEEE Trans. Med. Imaging **34**, 1993–2024 (2014)
17. Bakas, S., et al.: Segmentation labels and radiomic features for the pre-operative scans of the TCGA-GBM collection. The cancer imaging archive. Nat. Sci. Data **4**, 170117 (2017)
18. Bakas, S., et al.: Segmentation labels and radiomic features for the pre-operative scans of the TCGA-LGG collection. Cancer Imaging Arch. **286** (2017)
19. Chen, C.-C., Juan, H.-H., Tsai, M.-Y., Lu, H.-S.: Bridging density functional theory and big data analytics with applications. In: Härdle, W.K., Lu, H.-S., Shen, X. (eds.) Handbook of Big Data Analytics. SHCS, pp. 351–374. Springer, Cham (2018). https://doi.org/10.1007/978-3-319-18284-1_15
20. Vinyals, O., Toshev, A., Bengio, S., Erhan, D.: Show and tell: a neural image caption generator. In: Proceedings of the IEEE Conference on Computer Vision and Pattern Recognition, pp. 3156–3164 (2015)
21. Oktay, O., et al.: Attention u-net: learning where to look for the pancreas. arXiv preprint arXiv:.03999 (2018)
22. Nyúl, L.G., Udupa, J.K., Zhang, X.: New variants of a method of MRI scale standardization. IEEE Trans. Med. Imaging **19**, 143–150 (2000)
23. Pérez-García, F., Sparks, R., Ourselin, S.: TorchIO: a Python library for efficient loading, preprocessing, augmentation and patch-based sampling of medical images in deep learning. arXiv preprint arXiv:.04696 (2020)
24. Langreth, D.C., Mehl, M.: Beyond the local-density approximation in calculations of ground-state electronic properties. Phys. Rev. B **28**, 1809 (1983)
25. Zaiser, M.: Local density approximation for the energy functional of three-dimensional dislocation systems. Phys. Rev. B **92**, 174120 (2015)
26. Tompson, J., Goroshin, R., Jain, A., LeCun, Y., Bregler, C.: Efficient object localization using convolutional networks. In: Proceedings of the IEEE Conference on Computer Vision and Pattern Recognition, pp. 648–656 (2015)

MVP U-Net: Multi-View Pointwise U-Net for Brain Tumor Segmentation

Changchen Zhao, Zhiming Zhao, Qingrun Zeng, and Yuanjing Feng[✉]

Zhengjiang University of Technology, Hangzhou 310023, China
fyjing@zjut.edu.cn

Abstract. It is a challenging task to segment brain tumors from multi-modality MRI scans. How to segment and reconstruct brain tumors more accurately and faster remains an open question. The key is to effectively model spatial-temporal information that resides in the input volumetric data. In this paper, we propose Multi-View Pointwise U-Net (MVP U-Net) for brain tumor segmentation. Our segmentation approach follows encoder-decoder based 3D U-Net architecture, among which, the 3D convolution is replaced by three 2D multi-view convolutions in three orthogonal views (axial, sagittal, coronal) of the input data to learn spatial features and one pointwise convolution to learn channel features. Further, we modify the Squeeze-and-Excitation (SE) block properly and introduce it into our original MVP U-Net after the concatenation section. In this way, the generalization ability of the model can be improved while the number of parameters can be reduced. In BraTS 2020 testing dataset, the mean Dice scores of the proposed method were 0.715, 0.839, and 0.768 for enhanced tumor, whole tumor, and tumor core, respectively. The results show the effectiveness of the proposed MVP U-Net with the SE block for multi-modal brain tumor segmentation.

Keywords: Multi-View Pointwise U-Net · Brain tumor segmentation · BraTS 2020 · Spatial-temporal network · SE block

1 Introduction

Qualitative and quantitative assessment of brain tumors is the key to determine whether medical images can be used in clinical diagnosis and treatment. Researchers began to explore faster and more accurate methods for brain tumor segmentation. However, due to the fuzziness of the boundaries of each tumor subregion, the complete automatic segmentation of brain tumors remains challenging.

Brain Tumor Segmentation (BraTS) Challenge [1–4,17] has always been focusing on the evaluation of state-of-the-art methods for the segmentation of brain tumors in multimodal magnetic resonance imaging (MRI) scans. BraTS 2020 utilizes multi-institutional pre-operative MRI scans and primarily focuses on the segmentation of intrinsically heterogeneous brain tumors, namely gliomas.

A. Crimi and S. Bakas (Eds.): BrainLes 2020, LNCS 12659, pp. 93–103, 2021.
https://doi.org/10.1007/978-3-030-72087-2_9

The BraTS 2020 dataset is annotated manually by one to four raters, following the same annotation protocol, and their annotations are approved by experienced neuro-radiologists. Annotations comprise the background (label 0), the enhancing tumor (ET, label 4), the peritumoral edema (ED, label 2), and the necrotic and non-enhancing tumor (NCR/NET, label 1). Each patient's MRI scan consists of four modalities, i.e., native T1 weighted, post-contrast T1-weighted (T1ce), T2-weighted (T2), and T2 Fluid Attenuated Inversion Recovery (T2-FLAIR).

Since the U-Net network was first proposed by Ronneberger et al. in 2015 [18], the neural network represented by U-Net and its variants has been shining brightly in the field of medical image segmentation. It is a specialized convolutional neural network (CNN) with a down-sampling encoding path and an up-sampling decoding path similar to an auto-encoder architecture. However, because the U-Net network takes as input two-dimensional data while medical images are usually three-dimensional, using the U-Net network will lose the spatial details of the original data. As a result, the image segmentation accuracy is not satisfying. Alternatively, 3D U-Net [8] was proposed and has been widely used for segmentation in medical image segmentation due to its outstanding performance. However, 3D U-Net network is prone to overfitting and difficult to train because of its huge number of model parameters, which greatly limits its application. Both 2D and 3D U-Net models have their own advantages and disadvantages. One question naturally arises, is it possible to build a neural network that computes as low cost as a 2D network, but performs as well as a 3D network?

Researchers have been investigating this question for a long time and numerous approaches have been proposed. Haquer et al. [9] proposed 2.5D U-Net which consists of three 2D U-Net trained with axial, coronal, and sagittal slices, respectively. Although it achieves the goal of lower computation cost of 2D U-Net and the effectiveness of 3D U-Net, it does not make full use of the spatial information of volumetric medical image data. Chen et al. [6] proposed S3D U-Net which uses separable 3D convolutions instead of 3D convolutions. Although its segmentation of medical images is efficient, the number of model parameters is still large, which greatly limits its application in practical scenarios. It is a challenging task to achieve both low computational cost and high performance. The key is how to explore the spatial-temporal modeling for the input volumetric data. Recently, spatial-temporal 3D networks have received more and more attention [15]. It performs 2D convolution along three orthogonal views of volumetric video data to learn the spatial appearance and temporal motion cues, respectively, and fuses together to obtain the final output. Inspired by this, we propose Multi-View Pointwise U-Net (MVP U-Net) for brain tumor segmentation. The proposed MVP U-Net follows the encoder-decoder-based 3D U-Net architecture in which we use three multi-view convolutions and one pointwise convolution to reconstruct the 3D convolution.

Meanwhile, the Squeeze-and-Excitation (SE) block, proposed by Hu et al. [11] in 2017, can be incorporated into existing state-of-the-art deep CNN architecture such as ResNet and DenseNet as a subunit structure, and it can further improve the generalization ability of the original network by explicitly modeling the interdependencies between channels and adaptively calibrating the characteristic responses of channel correlation. In view of this, we incorporate this block into our MVP U-Net after appropriate modification.

2   Methods

2.1   Preprocessing

The images are preprocessed according to the following three steps before fed into the proposed MVP U-Net. First, each image is cropped to the region of nonzero values, and, at the same time, the image is normalized to the [2.0, 98.0] percentiles of the intensity values of the entire image. Second, the brain regions of images for each modality are normalized by Z-score normalization. The region outsides the brain is set to 0. Third, batch generators (a python package maintained by the Division of Medical Image Computing at the German Cancer Research Center) are applied to do data augmentation, including random elastic deformation, rotation, scaling, and mirroring [12].

2.2   Network Architecture

MVP Convolution Block. The architecture of the proposed MVP convolution is shown in Fig. 1(a), where we divide a 3D convolution into three orthogonal views (axial, sagittal, coronal) in a parallel fashion, followed by a pointwise convolution. The pointwise convolution is part of the Depthwise Separable Convolution network first proposed by Google [7], which consists of a depthwise convolution and a pointwise convolution. Figure 1(b) shows the traditional 3D convolution.

Figure 2 shows our MVP convolution block, which includes an activation function and an instance normalization, where our activation function is LeakyReLU (leakiness = 0.01). At the same time, in order to solve the problem of gradient disappearance caused by the increase of depth, we add the residual structure on the basis of the original structure. MVP convolution block is the main contribution of our proposed method. Each level of the network comprises two MVP convolution blocks of different resolutions.

MVP U-Net Architecture. The proposed MVP U-Net follows the encoder-decoder based 3D U-Net architecture. Instead of traditional 3D convolution [19], we employ the multi-view convolution to learn spatial-temporal features and the pointwise convolution to learn channel features. The multi-view convolution performs 2D convolutions in three orthogonal views of the input data, i.e., axial, sagittal, coronal. The pointwise convolution [10] is used to merge the antecedent outputs. In this way, the generalization ability of the model can be improved while the number of parameters can be reduced.

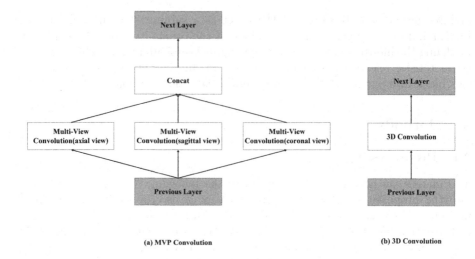

Fig. 1. Comparison of MVP convolution and 3D convolution.

The sketch map of the proposed network is shown in Fig. 3. Just like the original 3D U-Net [8], our network consists of three parts: 1) the left part corresponds to the contracting path that encodes the increasingly abstract representation of the input. Different layers are connected through an encoder module which consists of a $3 \times 3 \times 3$ convolution with stride 2, padding 1 instead of max pooling; 2) the right part corresponds to the expanding path that restores the original resolution, and 3) the jump connection which corresponds to connecting the encoder results to the output of submodules with the same resolution in the encoder as input to the next submodule in the decoder.

MVP U-Net with the SE Block Architecture. SE block consists of three operation modules: Squeeze, Exception, and Reweight. It is a new subunit structure which focuses on the characteristic channel. Among them, the Squeeze operation is to compress each feature channel in the spatial dimension and transform each two-dimensional feature channel into a real number. The real number has the global receptive field to some extent, and the output dimension matches the number of input feature channels. The Exception operation is a mechanism similar to the gate in recurrent neural networks. It can generate weights for each feature channel, which is learned to explicitly model the correlation between feature channels. The Reweight operation regards the output weight of the exception operation as the importance of each feature channel and then weights it to the previous feature channel by channel through multiplication. After the above steps, the recalibration of the original features in the channel dimension is completed.

However, it was originally proposed to improve the classification performance of two-dimensional images. We modify it properly so that it can be used in the

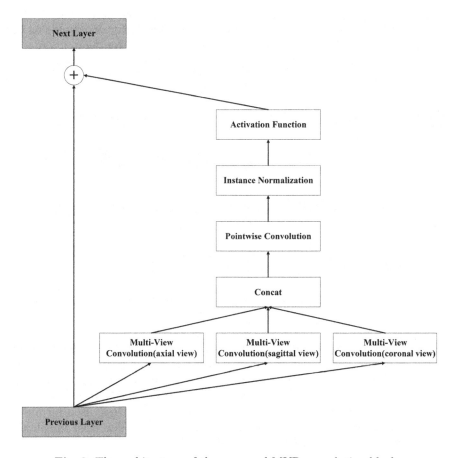

Fig. 2. The architecture of the proposed MVP convolution block.

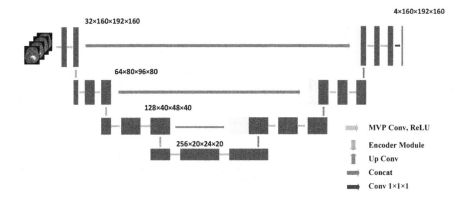

Fig. 3. The architecture of the proposed MVP U-Net.

classification of 3D feature map, and introduce it into our MVP U-Net after the concatenation section, as is shown in Fig. 4. It gives different weights to the features of different channels in the feature map after concatenation, in order to enhance those related features and suppress those less related features.

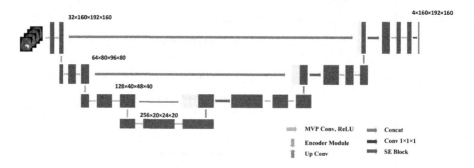

Fig. 4. The architecture of the proposed MVP U-Net with SE block.

2.3 Loss

The performance of a neural network depends not only on the choice of the network structure but also on the choice of the loss function, especially in the case of class imbalance. It holds for the task of brain tumor segmentation, in which the dataset varies in the size of classes [5,14]. In this paper, a hybrid loss function is employed that combines a multiclass Dice loss, used for multi-classification segmentation, and a focal loss aimed to alleviate class imbalance. Our loss function can be expressed as follows,

$$L = L_{Dice} + L_{focal} \tag{1}$$

The Dice loss is defined as,

$$L_{Dice} = \left(1 - \frac{2}{K} \sum_{k \in K} \frac{\sum_i u_i^k v_i^k}{\sum_i u_i^k + \sum_i v_i^k}\right) \tag{2}$$

where u is the softmax of the output map, v is the one-hot encoding of the corresponding ground truth label, i is the number of voxels of the output map and the corresponding ground truth label, k represents the current class, and K is the total number of classes.

The focal loss [16] is defined as,

$$L_{focal} = \begin{cases} -\alpha(1 - y')^\gamma \log y' & , \quad y = 1 \\ -(1 - \alpha)y'^\gamma \log(1 - y') , & y = 0 \end{cases} \tag{3}$$

where α and γ are constants. In our experiments, they are 0.25 and 2, respectively. y is the voxel value of the output map, and correspondingly, y' is the voxel value of the ground truth label.

2.4 Optimization

We use Adam optimizer to train our model [13]. The learning rate decreases as the epoch increases, which can be expressed as

$$lr = lr_0 * (1 - \frac{i}{N_i})^{0.9} \tag{4}$$

where i represents the current number of epochs, N_i is total number of epochs. The initial learning rate lr_0 is set as 10^{-4}.

3 Experiments and Results

We use the data provided by Brain Tumor Segmentation (BraTS) Challenge 2020 to evaluate the proposed network. The training dataset consists of 369 cases with accompanying ground truth labels by expert board-certified neuroradiologists. Our model is trained on one GeForce GTX 1080Ti GPU in a Pytorch environment. The batch size is 1 and the patch size is set to $160 \times 192 \times 160$. We concatenate four modalities into a four-channel feature map as input where each channel represents one modality. The results of our MVP U-Net on the BraTS 2020 training dataset are shown in Table 1, and the BraTS 2020 Training 013 case in training dataset with groundtruth and predicted labels are shown in Fig. 5.

Table 1. Mean Dice, Hausdorff95, Sensitivity and Specificity on BraTS 2020 training dataset of the proposed method: original MVP U-Net. ET: enhancing tumor, WT: whole tumor, TC: tumor core.

	Dice			Sensitivity			Specificity			Hausdorff95		
	ET	WT	TC	ET	WT	TC	ET	WT	TC	ET	WT	TC
Original MVP U-Net	0.600	0.799	0.635	0.676	0.909	0.716	0.999	0.997	0.999	56.655	29.831	26.878

The validation dataset and testing dataset contain 125 and 166 cases with unknown glioma grade and unknown segmentation, respectively. Ground truth segmentations for them are unknown and the evaluation is carried out via an online CBICA portal for the BraTS 2020 challenge. The models we have trained on the training dataset, including the original 3D U-Net, the original MVP U-Net, and MVP U-Net with SE block, are respectively used to predict the validation dataset of BraTS 2020, and the quantitative evaluation is obtained as shown in Table 2. As can be seen from the results, compared with the original 3D U-Net, the original MVP U-Net and the MVP U-Net with SE block has improved performance in most metrics. Meanwhile, the segmentation effect of the MVP U-Net with SE block is better than the original MVP U-Net.

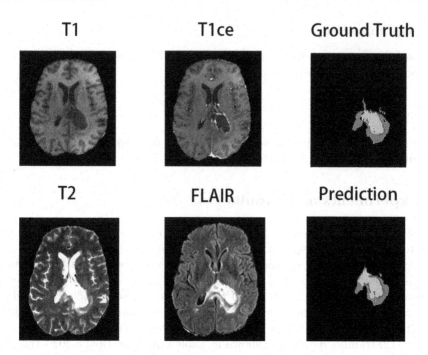

Fig. 5. The BraTS 2020 Training 013 case of training dataset with groundtruth and predicted labels (yelow:NCR/NET, green:ED, red:ET). (Color figure online)

Table 2. Mean Dice, Hausdorff95, Sensitivity and Specificity on BraTS 2020 validation dataset of the proposed methods: original MVP U-Net and MVP U-Net with SE block. ET: enhancing tumor, WT: whole tumor, TC: tumor core.

	Dice			Sensitivity			Specificity			Hausdorff95		
	ET	WT	TC	ET	WT	TC	ET	WT	TC	ET	WT	TC
Original 3D U-Net	0.585	0.762	0.604	0.686	0.868	0.742	0.999	0.997	0.998	78.429	43.598	44.543
Original MVP U-Net	0.601	0.785	0.639	0.659	0.901	0.715	0.993	0.970	0.997	56.653	29.837	26.870
MVP U-Net with the SE block	0.671	0.862	0.623	0.675	0.885	0.634	1.000	0.998	0.999	47.333	12.581	50.149

Finally, we used the MVP U-Net with the SE block to predict the testing dataset, and the results are shown in Table 3. Our method achieves average Dice scores of 0.715, 0.839, and 0.768 for enhancing tumor, whole tumor, and tumor core, respectively. The results are similar to those in the validation dataset, indicating that the model we designed has achieved desirable results in the automatic segmentation of multimodal brain tumors and the generalization ability of this model is also relatively powerful.

Table 3. Dice, Hausdorff95, Sensitivity and Specificity on BraTS 2020 testing dataset of the proposed method: MVP U-Net with SE block. ET: enhancing tumor, WT: whole tumor, TC: tumor core.

	Dice			Sensitivity			Specificity			Hausdorff95		
	ET	WT	TC	ET	WT	TC	ET	WT	TC	ET	WT	TC
Mean	0.715	0.839	0.768	0.756	0.912	0.800	0.999	0.998	0.999	33.147	10.362	33.577
StdDev	0.272	0.160	0.292	0.301	0.141	0.318	0.002	0.003	0.003	96.501	16.592	92.627
Median	0.818	0.893	0.892	0.903	0.961	0.962	1.000	0.998	1.000	2.000	4.243	3.317
25quantile	0.654	0.835	0.742	0.665	0.900	0.816	0.999	0.997	0.999	1.414	2.871	2.000
75quantile	0.895	0.927	0.940	0.957	0.985	0.988	1.000	0.999	1.000	4.472	7.729	9.312

4   Conclusion

In this paper, we propose a novel CNN-based neural network called Multi-View Pointwise (MVP) U-Net for brain tumor segmentation from multi-model 3D MRI. We use three multi-view convolutions and one pointwise convolution to reconstruct the 3D convolution in conventional 3D U-Net, in which the purpose of multi-view convolution is to learn spatial-temporal features while pointwise convolution to learn channel features. In this way, the proposed architecture can not only improve the generalization ability of the network but also reduce the number of parameters. Further, we modify the SE block properly and introduce it into our original MVP U-Net after the concatenation section. Experiments showed that the performance of this method was improved compared with the original MVP U-Net.

During the experiment, we tried a variety of approaches. We found that the model performance could be improved by changing the encoders of the U-shaped network from max pooling to 3D convolution, and the results could also be improved by increasing the number of channels. Finally, the trained MVP U-Net with SE block was used to predict the testing dataset, and achieved mean Dice scores of 0.715, 0.839, and 0.768 for enhancing tumor, whole tumor, and tumor core, respectively. The results showed the effectiveness of the proposed MVP U-Net with the SE block for multi-modal brain tumor segmentation.

In the future, we will make further efforts in data preprocess and network architecture design to alleviate the imbalance of tumor categories and improve the accuracy of tumor segmentation.

Acknowledgement. This work was supported by the National Natural Science Foundation of China under Grant No. 61903336, 61703369, 61976190, Natural Science Foundation of Zhejiang Province under Grant No. LY21F030015, Key Research and Development Program of Zhejiang Province under Grant No. 2020C03070, Major Science & Technology Projects of Wenzhou under Grant No ZS2017007.

References

1. Bakas, S., et al.: Segmentation labels and radiomic features for the pre-operative scans of the TCGA-GBM collection. The cancer imaging archive. Nat. Sci. Data **4**, 170117 (2017)
2. Bakas, S., et al.: Segmentation labels and radiomic features for the pre-operative scans of the TCGA-LGG collection. Cancer Imaging Arch. **286** (2017)
3. Bakas, S., et al.: Advancing the cancer genome atlas glioma MRI collections with expert segmentation labels and radiomic features. Sci. Data **4**, 170117 (2017)
4. Bakas, S., et al.: Identifying the best machine learning algorithms for brain tumor segmentation, progression assessment, and overall survival prediction in the brats challenge. arXiv preprint arXiv:1811.02629 (2018)
5. Berman, M., Triki, A.R., Blaschko. , M.BT:he lovász-softmax loss: a tractable surrogate for the optimization of the intersection-over-union measure in neural networks. In: Proceedings of the IEEE Conference on Computer Vision and Pattern Recognition, pp. 4413–4421 (2018)
6. Chen, W., Liu, B., Peng, S., Sun, J., Qiao, X.: S3D-UNet: separable 3D U-Net for brain tumor segmentation. In: Crimi, A., Bakas, S., Kuijf, H., Keyvan, F., Reyes, M., van Walsum, T. (eds.) BrainLes 2018. LNCS, vol. 11384, pp. 358–368. Springer, Cham (2019). https://doi.org/10.1007/978-3-030-11726-9_32
7. Chollet, F.: Xception: deep learning with depthwise separable convolutions. In: Proceedings of the IEEE Conference on Computer Vision and Pattern Recognition, pp. 1251–1258 (2017)
8. Çiçek, Ö., Abdulkadir, A., Lienkamp, S.S., Brox, T., Ronneberger, O.: 3D U-Net: learning dense volumetric segmentation from sparse annotation. In: Ourselin, S., Joskowicz, L., Sabuncu, M.R., Unal, G., Wells, W. (eds.) MICCAI 2016. LNCS, vol. 9901, pp. 424–432. Springer, Cham (2016). https://doi.org/10.1007/978-3-319-46723-8_49
9. Haque, H., Hashimoto, M., Uetake, N., Jinzaki, M.: Semantic segmentation of thigh muscle using 2.5 d deep learning network trained with limited datasets. arXiv preprint arXiv:1911.09249 (2019)
10. Howard, A.G., et al.: Mobilenets: efficient convolutional neural networks for mobile vision applications. arXiv preprint arXiv:1704.04861 (2017)
11. Hu, J., Shen, L., Sun, G.: Squeeze-and-excitation networks. In: Proceedings of the IEEE Conference on Computer Vision and Pattern Recognition, pp. 7132–7141 (2018)d
12. Huang, C., Han, H., Yao, Q., Zhu, S., Zhou, S.K.: 3D U^2-Net: a 3D universal u-net for multi-domain medical image segmentation. In: Shen, D., et al. (eds.) MICCAI 2019. LNCS, vol. 11765, pp. 291–299. Springer, Cham (2019). https://doi.org/10.1007/978-3-030-32245-8_33
13. Isensee, F., Kickingereder, P., Wick, W., Bendszus, M., Maier-Hein, K.H.: No new-net. In: Crimi, A., Bakas, S., Kuijf, H., Keyvan, F., Reyes, M., van Walsum, T. (eds.) BrainLes 2018. LNCS, vol. 11384, pp. 234–244. Springer, Cham (2019). https://doi.org/10.1007/978-3-030-11726-9_21
14. Kamnitsas, K., et al.: Ensembles of multiple models and architectures for robust brain tumour segmentation. In: Crimi, A., Bakas, S., Kuijf, H., Menze, B., Reyes, M. (eds.) BrainLes 2017. LNCS, vol. 10670, pp. 450–462. Springer, Cham (2018). https://doi.org/10.1007/978-3-319-75238-9_38
15. Li, C., Zhong, Q., Xie, D., Pu, S.: Collaborative spatiotemporal feature learning for video action recognition. In: Proceedings of the IEEE Conference on Computer Vision and Pattern Recognition, pp. 7872–7881 (2019)

16. Lin, T.-Y., Goyal, P., Girshick, R., He, K., Dollár, P.: Focal loss for dense object detection. In: Proceedings of the IEEE International Conference on Computer Vision, pp. 2980–2988 (2017)
17. Menze, B.H., et al.: The multimodal brain tumor image segmentation benchmark (brats). IEEE Trans. Med. Imaging **34**(10), 1993–2024 (2014)
18. Ronneberger, O., Fischer, P., Brox, T.: U-Net: convolutional networks for biomedical image segmentation. In: Navab, N., Hornegger, J., Wells, W.M., Frangi, A.F. (eds.) MICCAI 2015. LNCS, vol. 9351, pp. 234–241. Springer, Cham (2015). https://doi.org/10.1007/978-3-319-24574-4_28
19. Xie, S., Sun, C., Huang, J., Tu, Z., Murphy, K.: Rethinking spatiotemporal feature learning: speed-accuracy trade-offs in video classification. In: Proceedings of the European Conference on Computer Vision (ECCV), pp. 305–321 (2018)

Glioma Segmentation with 3D U-Net Backed with Energy-Based Post-Processing

Richard Zsamboki[✉], Petra Takacs[✉], and Borbala Deak-Karancsi[✉]

GE Healthcare, Budapest, Hungary
{Richard.Zsamboki,Petra.Takacs,Borbala.Deak-Karancsi}@ge.com

Abstract. This paper proposes a glioma segmentation method based on neural networks. The base of the network is a UNet, expanded by residual blocks. Several preprocessing steps were applied before training, such as intensity normalization, high intensity cutting, cropping, and random flips. 2D and 3D solutions are implemented and tested, and results show that the 3D network outperforms 2D directions, therefore we stayed with 3D directions.

The novelty of the method is the energy-based post-processing. Snakes [10], and conditional random fields (CRF) [11] were applied to the neural network's predictions. Snake or active contour needs an initial outline around the object – e.g. the network's prediction outline - and it can correct the contours of the tumor based on calculating the energy minimum, based on the intensity values at a given area. CRF is a specific type of graphical model, it uses the network's prediction and the raw image features to estimate the posterior distribution (the tumor contour) using energy function minimization.

The proposed methods are evaluated within the framework of the BRATS 2020 challenge. Measured on the test dataset the mean dice scores of the whole tumor (WT), tumor core (TC) and enhancing tumor (ET) are 86.9%, 83.2% and 81.8% respectively. The results show high performance and promising future work in tumor segmentation, even outside of the brain.

Keywords: Tumor segmentation · Medical imaging · Neural networks · Supervised segmentation · UNet · ResNet · Radiation therapy

1 Introduction

Cancer is one of the leading causes of deaths nowadays, malignant tumors take more than 30 000 people's life in a year just in a small country like Hungary. Gliomas are the most common type of brain tumors for adults and the treatment process is far from optimal.

Glioma is a type of tumor which originates from a type of cells – glial cells – that make up approximately half of the brain. About 33% of all brain and CNS (central nervous system) tumors, and 80% of all malignant brain tumors are gliomas. Gliomas contain three cell types: astrocyta, epedyma and oligodendroglia cells.

There are no proven risk factors connected to the origin of these types of tumors, but gliomas are slightly more common in men than women, and Caucasian people than

© Springer Nature Switzerland AG 2021
A. Crimi and S. Bakas (Eds.): BrainLes 2020, LNCS 12659, pp. 104–117, 2021.
https://doi.org/10.1007/978-3-030-72087-2_10

African-American people. Although brain tumors are not the most common tumors, the patients sadly suffer from severe symptoms and often even in the case of recovering they remain with permanent health damage. Since the brain is maybe the most important organ in the human body, a tumor that disrupts its integrity affects the whole body. Symptoms such as nausea, vomiting, blindness or visual loss, cachexy, loss or impairment in motoric and sensory functions render the patient unable to work and live a normal life [12].

Since the recovery is a possibility it is essential to discover the tumor as soon as possible to start the treatment. Contouring tumors and organs to protect for radiation therapy is a slow, mechanical process for the doctors. Automating these processes is an active research area, however tumor and anomaly segmentation is not solved because of their wide range of intensity, size, shape and position inside the body. Properly contoured data for the whole body is not publicly available and hard to obtain. BRATS datasets are proper for implementing and testing algorithms with four modalities of MR images, T1, T1c, T2 and FLAIR, all of which are co-registered. However, BRATS datasets focus on gliomas only.

The most popular state-of-the-art solutions for organ and anomaly – like glioma – segmentation are neural networks, especially 2D or 3D U-Nets, with residual connections supplemented by basic preprocessing steps. Our model fits into the state-of-the-art methods, using intensity normalization and several augmentations as preprocessing and building a 3D (and for comparison also a 2D) network with residual blocks to the segmentation task. Using features from more than 300 training data ensures that most of the intensity and shape variations of the gliomas are covered and the network can adapt to the contouring protocol of the dataset.

BRATS is well unified, but real-world problems include inconsistent image or contouring protocols for example. This research includes two, energy-based post-processing steps in order to fix errors introduced by inconsistent contouring. Their working mechanism is based on modifying the neural network's prediction in an unsupervised way (without the target contours) so that the edges of the predicted contours will match the pixels corresponding to the tumors. Since the correction is expected be to be a tighter fit on the contours than the average clinical contour, they should be dilated for radiation therapy if needed.

1.1 Related Work

Organ and tumor segmentation are active branches in medical imaging research. Convolutional neural networks are the most important elements of the deep learning algorithms designed for applications like glioma segmentation in brain MRI scans. Recent state-of-the-art works have significantly increased segmentation accuracy through various developments in neural network architectures. Most of the implementations are based on the U-Net structure, often boosted with residual blocks, as it is simple and effective, and the design can be easily adapted to numerous tasks.

An outstanding example is the work of Andriy Myronenko, where a 3D U-Net with residual connections was built, and VAE regularization was applied to the encoder to increase the network's generalization ability. The obtained DICE score for the Whole

Tumor (WT) was 0.91 which is the highest among BRATS challenge works [1]. Furthermore, Fabian Isensee et al. built a 3D U-Net with instance normalization and negative log-likelihood loss and reached the accuracy of the previous algorithm [2].

Not far behind the work of Shaocheng Wu et al. with 3D U-Net also with residual connections containing cascaded networks for each subtask [3]. Indrajit Mazumdar hit the top-level score with a 2D U-Net based idea with weighted DICE loss and using three orthogonal views for all patches [4].

Raghav Mehta et al. used more preprocessing steps, such as rescaling, cropping, random translation, scaling and shearing besides the standard intensity normalization and random axis flips and built a 3D U-Net with instance normalization and cross-entropy loss [5]. Li Sun et al. applied a 2D hierarchical, cascaded U-Net with tumor localization at first, and evaluated the algorithm successfully on a different brain tumor dataset [6].

Xiaoyue Zhang et al. applied elastic deformation before training the 2D U-Net with dense blocks and reached 0.88 DICE [7]. Another interesting approach belongs to Malathi et al. training and testing on a less preprocessed brain tumor dataset, applying n4ITK bias correction, noise reduction with median filtering and reached 0.77 DICE score with a 2D U-Net [8].

2 Data

The BRATS 2020 train image data consists of 369 MR scans of glioma patients with ground truth masks made by expert clinicians. The dataset contains four contrast types of MRI: T1-weighted, native image, sagittal or axial 2D acquisitions, with 1–6 mm slice thickness; T1-weighted, contrast-enhanced image, with 3D acquisition and 1 mm isotropic voxel size for most patients; T2-weighted image, axial 2D acquisition, with 2–6 mm slice thickness; T2-weighted FLAIR image, axial, coronal, or sagittal 2D acquisitions, 2–6 mm slice thickness. The validation data consists of 125 MR scans containing the same contrasts as the training set.

All image data were co-registered rigidly to the T1-weighted, contrast-enhanced images, which has the highest resolution spatially for most of the cases, then resampling was made to 1 mm resolution in axial orientation with linear interpolation. The dataset was skull stripped, so anonymity of the patients is ensured. Image volumes were synthesized and degraded with Gaussian noise and polynomial bias fields. The size of the final version of the training data considering one contrast of a single patient is $240 \times 240 \times 155$ pixels.

The different visual structures available in the ground truth are the following: the edema (label 2), the enhancing tumor core (label 4), the necrotic tumor core (label 1), and the non-enhancing tumor core (label 3) [9, 13–17]. The following abbreviations will be used later in the article regarding the different classes in the data: Whole Tumor – WT, Tumor Core – TC, edema – ED, Necrotic Tumor Core – NCR, Enhancing tumor core – ET.

3   Methods

3.1   Preprocessing

In order to improve the performance and robustness of the deep learning model and to prevent overfitting, several normalization and augmentation preprocessing steps were applied to the BRATS images before sending them to the training process.

In case of the 3D model training the first step is an intensity normalization using mean and standard deviation values. It is important to point out that these mean and standard deviation values are calculated only considering the voxels corresponding to the brain itself while ignoring the outlying intensity values for a more stable result. After normalization the extremely high or low values are clipped, followed by two stochastic intensity augmentations, specifically random scaling and shifting of the intensity values. Finally, the image is randomly flipped along all three axes with a 50% probability for each axis.

The 2D training process incorporates even more steps - besides the 3D process described above - without increasing the algorithm's total processing time too much, including affine transformations (zooming, shifting, rotating), and adding Gaussian noise to the image. Example original and preprocessed images can be seen on Fig. 1.

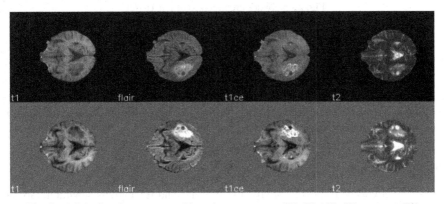

Fig. 1. Original and preprocessed imaging sequences (T1, FLAIR, T1 contrast, T2)

3.2   Model

A 2D and a 3D network architecture was implemented with similar structures. The base of the network is a U-Net with residual connections like in [1].

Besides convolutions with kernel size of 3, ReLu activation, group normalization with groups of 8 and dropout with 25% probability were applied simultaneously. Upsamplings and simple additions led through the encoder section of the U-Net as Fig. 2 shows. The difference between the 2D and the 3D version of our model is the kernel shape of the convolution and using the specific 2D or 3D versions of different layers.

The input of the network is a 4-channel 3D image created by concatenating the 4 input modalities, each with size of $160 \times 192 \times 128$. The reason for size decrease from the original is that we cropped the brain area from the background in order to keep high quality through reducing the ratio of low information value voxels. The first block on Fig. 2. Gets the concatenated and convolved input images. At the encoder part of the U-Net, convolutions with stride 2 are applied at the place of the black arrows as down sampling operation, and at the decoder side, there are bilinear upsizing operations at the corresponding levels. The smallest resolution at the middle is $20 \times 24 \times 16$ with 256 filters. The implementation was done using Tensorflow 2.1.

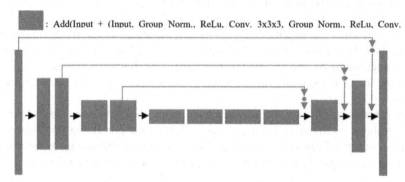

Fig. 2. Network architecture of our U-Net

We tested two different approaches regarding the output of the network. In one case the output consists of 3 channels with sigmoid activation applied to each of them, the other approach uses 4 output channels with softmax activation function. The former predicts the WT, TC and ET regions independently, the latter is a multiclass probability distribution output for background, NCR, ED and ET classes. Figure 3 shows an example result of the training process.

3.3 Loss Function

When the network is trained to predict the WT, TC and ET regions separately, loss is directly calculated from the DICE score for each region. In case of multiclass probability output over the background, NCR, ED and ET regions, categorical cross-entropy loss is applied with label smoothing.

In addition to the former mention loss calculated for the network's prediction we also applied deep supervision (DS) loss. Before each block in the decoder part of the network a separate branch is created with independent convolutional layers and the output of these branches are treated as downscaled versions of the expected output, and the same loss function is applied to them. The experiments showed that this method improved the smallest target class (ET) accuracy significantly without increasing the model size (the branches are only created during training).

Although the Dice scores measured on the training set were lower than without deep supervision, based on validation dice scores we chose to include deep supervision to our baseline model (see Table 1 for comparison).

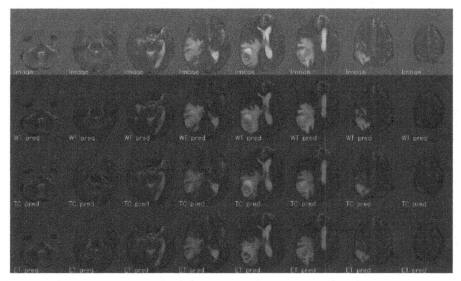

Fig. 3. Training results: the first row contains the T2 contrast, the second, third and fourth rows show the prediction results of the whole tumor area (WT), tumor core (TC), and enhancing tumor (ET)

3.4 Training

The training was done through 300 epochs using Adam optimizer with 0.0001 starting learning rate and polynomial decreasing through the whole training to a small learning rate value, and batch size of 20 in 2D case and 1 in 3D case.

3.5 Post Processing

Basic post processing contains a few steps in order to get binary tumor masks. In case of separate prediction of WT, TC and ET, each output channel undergoes a binary thresholding at 0.5 value. When the network predicts multiclass probability output, argmax operation is executed to get the most probable class for each pixel.

Over Contouring: According to oncologists, it is common for the manual annotation to be slightly over contoured for some images (as an example shows in Fig. 4.) because of the lack of time or because of the common practice to over contour, since the tumor cells do not stop at the visible border. The result is the same, the tumor masses are usually slightly over contoured clinically. As it can be seen in Fig. 5 or Fig. 6 the visible tumor border fits more with the algorithm than with the manual contour, and the neural net using said manual contour as the learning base.

It can be useful to use an algorithm that demarcates the tumor mass as accurately as possible, because with that, more healthy tissue can be spared from the radiative therapy. Of course, it does not substitute the CTV (Clinical Tumor Volume) or PTV (Planning Tumor Volume) in calculating the correct radiation doses, it simply gives the doctors a more accurate GTV (Gross Tumor Volume).

Since the manual contours obviously differ as the doctors and the contouring guidelines they use also differ, it could be a great help, if the GTV would be a standard volume, defined by a standard algorithm. It would make the contours less subjective, and the therapies more accurate.

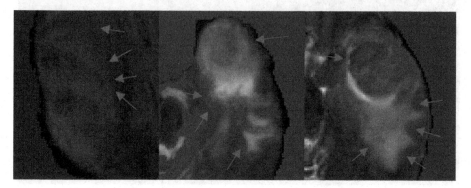

Fig. 4. Over contouring of tumors in BRATS dataset

Energy-Based Post Processing: As brain tumor outlines can be separated from the white and grey matter of the brain in intensity values, using snake - an energy minimizing spline guided by external constraint forces and influenced by image forces pulling it towards lines and edges - can help the prediction map fit better on the tumor outlines [10].

Contour models describe the boundaries of shapes on images. Snakes are designed to solve problems where the approximate shape of the boundary is known, which is fulfilled in our case with the prediction outline of the network. Compared to common feature extraction algorithms, snakes have multiple advantages. For example, they autonomously and adaptively search for the minimum state, and external image forces act upon the contour line in an intuitive manner. This can help categorize the questionable parts during tumor segmentation. Furthermore, incorporating Gaussian smoothing in the image energy function introduces scale sensitivity. As glioma's sizes are quite varied, it also adds value to the segmentation.

For the implementation python's "skimage" active contour method was used with the following parameters: alpha is 0.1, beta is 5, gamma is 0.001, line is 0, edge is 1, max move is 1, and max iteration is 10. The drawbacks of the traditional snakes are their sensitivity to local minima states, and their accuracy, which strongly depends on the convergence policy. Additionally, the method can find illusory contours in the image, ignoring missing boundary information - that can help, but also can lead to false directions, so the modification the active contour method is allowed to make on the network's prediction shape area is 15% [10].

The other energy-based approach in our model's post-processing is conditional random fields, or CRF. The CRF is a specific type of graphical model. CRFs are a class of statistical modeling methods often applied in machine learning which takes context into

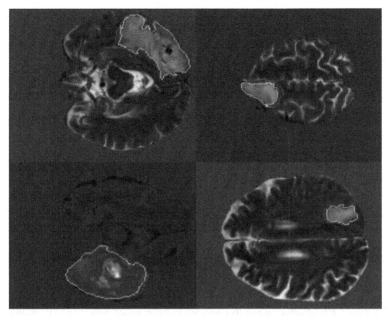

Fig. 5. Blue: Target, Red: Neural network prediction, Green: Snake post processed contour (fitting the GTV better, without over contouring) (Color figure online)

account when making predictions. The exact graph used depends on the application. For details and the theoretical and mathematical background please refer to [11]. In our case we can use CRF to estimate the posterior distribution (the tumor contour) given prediction from our network and the raw image features in our input. It achieves this goal by minimizing a user defined energy function. In our case the effect is similar to bilateral filter which takes into account the spatial closeness of pixels and their similarity in intensity space.

CRF-s are useful when the target object border is featured by a big intensity change compared to the background. It also penalizes small segmentation regions, because the target objects are usually defined by big spatially adjacent regions.

Since pixel similarity is expected over a class region, using the complete network probability prediction and all the input modalities is not advised, since not all class regions appear different on every image modality. Based on consultation with experts, we identified pairs of tumor region class and modality to be used in CRF postprocess: flair + ED, t2 + TC, t1ce + NCR, ET. Example results can be seen on Fig. 6.

Energy based methods - as they are working based on visual features - can help reduce the over contoured borders and concentrate on the tumor mass (which can be dilated later according to the specific planning protocol). However, at BRATS challenge it is important to adjust to their contouring guidelines, so for that, the post-processing steps are expected to perform poorly compared to skipping them.

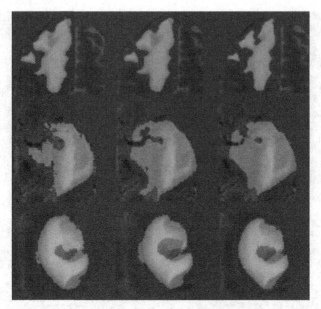

Fig. 6. Example CRF postprocess results (from left to right: image with ground truth, image with network prediction, image with CRF optimized prediction)

4 Performance Evaluation

For each tumor region a binary map has been generated with the network's post processed predictions (P) consisting 0 (non-tumor) and 1 (tumor) values, and the ground truth images (T) also consisting 0 or 1 pixels, and we calculated the Dice score.

$$\text{Dice}(P, T) = \frac{|P_1 \wedge T_1|}{(|P_1| + |T_1|)/2}$$

Table 1. DICE values of different models for WT, TC, ET on training and validation set

Method	Train			Validation		
	WT	TC	ET	WT	TC	ET
2D Baseline	0.936	0.936	**0.862**	0.897	0.796	0.706
3D Baseline - DS	0.949	**0.947**	0.853	0.900	**0.831**	0.710
3D Baseline	0.927	0.908	0.850	**0.902**	0.830	**0.780**
3D Multiclass	**0.950**	0.931	0.810	0.899	0.801	0.707
3D Baseline + Snake	0.932	0.931	0.830	0.886	0.820	0.695
3D Multiclass + CRF	0.948	0.931	0.823	0.879	0.790	**0.714**

The evaluation on BRATS 2020 data was done using the CBICA Image Processing Portal [1–4]. Tables 1 and 2 contains all the training, validation and test results obtained from the portal. Note, that baseline means the network predicted the WT, TC and ET regions separately, the multiclass means the network predicted a probability distribution over background, NCR, ED and ET classes.

Table 2. Results with 3D Baseline model on BRATS 2020 final test set

	DICE			Sensitivity			Specificity		
	WT	TC	ET	WT	TC	ET	WT	TC	ET
Mean	0.869	0.832	0.818	0.910	0.857	0.862	0.999	1.000	1.000
StdDev	0.159	0.253	0.171	0.137	0.234	0.182	0.001	0.001	0.000
Median	0.917	0.921	0.850	0.942	0.943	0.917	0.999	1.000	1.000
25 quantile	0.868	0.861	0.788	0.913	0.869	0.833	0.998	1.000	0.999
75 quantile	0.945	0.955	0.912	0.970	0.970	0.954	1.000	1.000	1.000

$$\text{sensitivity} = \frac{\text{number of true positives}}{\text{number of true positives} + \text{number of false negatives}}$$

$$\text{sensitivity} = \frac{\text{number of true negatives}}{\text{number of true negatives} + \text{number of false positives}}$$

The following pictures show prediction visualization for some examples from the train, validation and test dataset respectively (Figs. 7, 8 and 9).

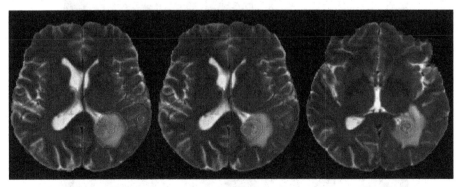

Fig. 7. Example Train results (green – target, red – prediction) for ED, TC and ET (Color figure online)

5   Conclusion

In the previous sections a state-of-the-art U-Net with novel post-processing ideas has been introduced for brain tumor segmentation. The results are promising, fitting into the top results based on earlier BRATS datasets and evaluated with a top DICE score of 0.900 for WT, 0.831 for TC and 0.714 for ET on the BRATS 2020 validation data. The former two belong to the chosen best model, the 3D UNet with separate prediction channels for WT, TC and ET.

As a novelty, two energy-based post-processing steps have been implemented and tested, Conditional Random Fields (CRF) and active contour, alias snake. As they are working without learning, using the image features only, they can mitigate the over contoured area around tumors along the contrasts on the edge of tumors, giving more proper results of the tumor area which can be dilated later according to the current application.

Because this challenge is also a competition, where adjusting to the dataset is needed, the post-processing steps caused inferior results (because the inherent over contouring of the given training data). Further research and tuning of these processes will be a future work on different datasets (Figs. 8 and 9).

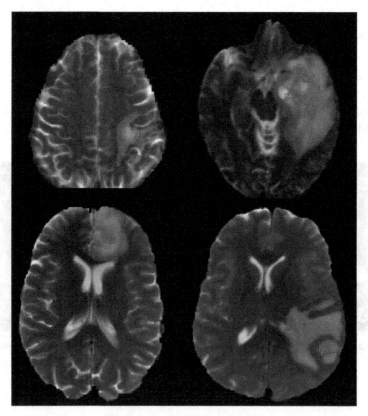

Fig. 8. Example validation results

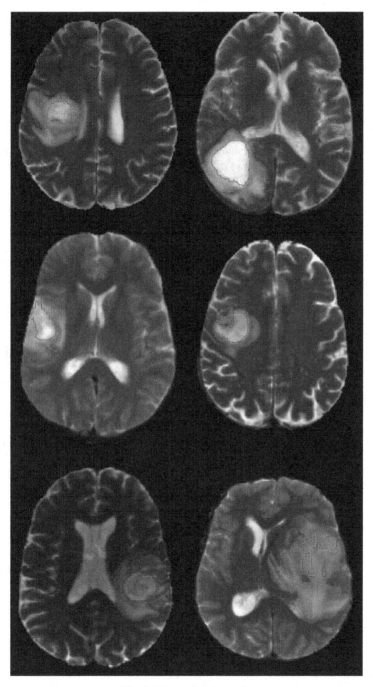

Fig. 9. Example test results

Acknowledgement. This research is part of the Deep MR-only Radiation Therapy activity (project numbers: 19037, 20648) that has received funding from EIT Health. EIT Health is supported by the European Institute of Innovation and Technology (EIT), a body of the European Union receives support from the European Union´s Horizon 2020 Research and innovation programme.

References

1. Myronenko, A.: 3D MRI brain tumor segmentation using autoencoder regularization. In: Crimi, A., Bakas, S., Kuijf, H., Keyvan, F., Reyes, M., van Walsum, T. (eds) BrainLes 2018. LNCS, vol. 11384. Springer, Cham (2019). https://doi.org/10.1007/978-3-030-11726-9_28
2. Isensee, F., Kickingereder, P., Wick, W., Bendszus, M., Maier-Hein, K.H.: No New-Net. In: Crimi, A., Bakas, S., Kuijf, H., Keyvan, F., Reyes, M., van Walsum, T. (eds.) BrainLes 2018. LNCS, vol. 11384, pp. 234–244. Springer, Cham (2019). https://doi.org/10.1007/978-3-030-11726-9_21
3. Wu, S., Li, H., Guan, Y.: Multimodal brain tumor segmentation using u-net. In: MICCAI BraTS, pp. 508–515 (2018)
4. Mazumdar, I.: Automated Brain tumour segmentation using deep fully residual convolutional neural networks. arXiv (2019): arXiv-1908
5. Mehta, R., Arbel, T.: 3D U-Net for brain tumour segmentation. In: Crimi, A., Bakas, S., Kuijf, H., Keyvan, F., Reyes, M., van Walsum, T. (eds) BrainLes 2018. LNCS, vol. 11384. Springer, Cham (2019). https://doi.org/10.1007/978-3-030-11726-9_23
6. Sun, L., et al.: Brain tumor segmentation and survival prediction using multimodal MRI scans with deep learning. Front. Neurosci. **13**, 810 (2019)
7. Zhang, X., Jian, W., Cheng, K.: 3D dense U-nets for brain tumor segmentation. Springer (2018)
8. Malathi, M., Sinthia, P.: Brain tumour segmentation using convolutional neural network with tensor flow. Asian Pacific J. Cancer Prevention: APJCP **20**(7), 2095 (2019)
9. Menze, B.H., et al.: The multimodal brain tumor image segmentation benchmark (BRATS). IEEE Trans. Med. Imaging **34**(10), 1993–2024 (2014)
10. Kass, M., Witkin, A., Terzopoulos, D.: Snakes: active contour models. Int. J. Comput. Vision **1**(4), 321–331 (1988)
11. Sutton, C., McCallum, A.: An introduction to conditional random fields. Found. Trends Mach. Learn. 4(4), 267–373 (2011)
12. Gladson, C.L., Prayson, R.A., Liu, W.M.: The pathobiology of glioma tumors. Ann. Rev. Pathol. Mech. Disease **5**, 33–50 (2010)
13. Menze, B.H., Jakab, A., Bauer, S., Kalpathy-Cramer, J., Farahani, K., Kirby, J., et al.: The multimodal brain tumor image segmentation benchmark (BRATS). IEEE Trans. Med. Imaging **34**(10), 1993–2024 (2015). https://doi.org/10.1109/TMI.2014.2377694
14. Bakas, S., Akbari, H., Sotiras, A., Bilello, M., Rozycki, M., Kirby, J.S., et al.: Advancing The Cancer Genome Atlas glioma MRI collections with expert segmentation labels and radiomic features. Nature Sci. Data **4**, 170117 (2017). https://doi.org/10.1038/sdata.2017.117
15. Bakas, S., Reyes, M., Jakab, A., Bauer, S., Rempfler, M., Crimi, A., et al.: Identifying the Best Machine Learning Algorithms for Brain Tumor Segmentation, Progression Assessment, and Overall Survival Prediction in the BRATS Challenge, arXiv preprint arXiv:1811.02629 (2018)

16. Bakas, S., Akbari, H., Sotiras, A., Bilello, M., Rozycki, M., Kirby, J., et al.: segmentation labels and radiomic features for the pre-operative scans of the TCGA-GBM collection. Cancer Imaging Archive (2017). https://doi.org/10.7937/K9/TCIA.2017.KLXWJJ1Q
17. Bakas, S., Akbari, H., Sotiras, A., Bilello, M., Rozycki, M., Kirby, J., et al.: Segmentation labels and radiomic features for the pre-operative scans of the TCGA-LGG collection. Cancer Imaging Archive (2017). https://doi.org/10.7937/K9/TCIA.2017.GJQ7R0EF

nnU-Net for Brain Tumor Segmentation

Fabian Isensee[1,3]([✉]), Paul F. Jäger[1], Peter M. Full[1], Philipp Vollmuth[2],
and Klaus H. Maier-Hein[1]

[1] Division of Medical Image Computing, German Cancer Research Center (DKFZ),
Heidelberg, Germany
[2] Section for Computational Neuroimaging, Department of Neuroradiology,
Heidelberg University Hospital, Heidelberg, Germany
[3] HIP Applied Computer Vision Lab, DKFZ, Heidelberg, Germany
`f.isensee@dkfz-heidelberg.de`

Abstract. We apply nnU-Net to the segmentation task of the BraTS
2020 challenge. The unmodified nnU-Net baseline configuration already
achieves a respectable result. By incorporating BraTS-specific modifications regarding postprocessing, region-based training, a more aggressive data augmentation as well as several minor modifications to the
nnU-Net pipeline we are able to improve its segmentation performance
substantially. We furthermore re-implement the BraTS ranking scheme
to determine which of our nnU-Net variants best fits the requirements
imposed by it. Our method took the first place in the BraTS 2020 competition with Dice scores of 88.95, 85.06 and 82.03 and HD95 values
of 8.498,17.337 and 17.805 for whole tumor, tumor core and enhancing
tumor, respectively.

Keywords: nnU-Net · Brain tumor segmentation · U-Net · Medical
image segmentation

1 Introduction

Brain tumor segmentation is considered one of the most difficult segmentation
problems in the medical domain. At the same time, widespread availability of
accurate tumor delineations could significantly improve the quality of care by
supporting diagnosis, therapy planning and therapy response monitoring [1].
Furthermore, segmentation of tumors and associated subregions allows for identification of novel imaging biomarkers, which in turn enable more precise and
reliable disease stratification [2] and treatment response prediction [3].

The Brain Tumor Segmentation Challenge (BraTS) [4,5] provides the largest
fully annotated and publicly available database for model development and is the
go-to competition for objective comparison of segmentation methods. The BraTS
2020 dataset [5–8] comprises 369 training and 125 validation cases. Reference
annotations for the validation set are not made available to the participants.

P. Vollmuth—né Kickingereder.

© Springer Nature Switzerland AG 2021
A. Crimi and S. Bakas (Eds.): BrainLes 2020, LNCS 12659, pp. 118–132, 2021.
https://doi.org/10.1007/978-3-030-72087-2_11

Instead, participants can use the online evaluation platform[1] to evaluate their models and compare their results with other teams on the online leaderboard[2]. Besides the segmentation task, the BraTS 2020 competition features a survival prediction task as well as an uncertainty modelling task. In this work, we only participate in the segmentation task.

Recent successful entries in the BraTS challenge are exclusively based on deep neural networks, more specifically on encoder-decoder architectures with skip connections, a pattern which was first introduced by the U-Net [9]. Numerous architectural improvements upon the U-Net have been introduced, many of which are also used in the context of brain tumor segmentation, for example the addition of residual connections [10–14], densely connected layers [15–17,17,18] and attention mechanisms [18,19]. In the context of network architectures, the winning contributions of 2018 [11] and 2019 [12] should be highlighted as they extend the encoder-decoder pattern with a second decoder trained on an auxiliary task for regularization purposes. Training schemes are usually adapted to cope with the particular challenges imposed by the task of brain tumor segmentation. The stark class imbalance, for instance, necessitates appropriate loss functions for optimization: Dice loss [20,21] and focal loss [22] are popular choices [11–13,16,18,23,24]. Since BraTS evaluates segmentations using the partially overlapping whole tumor, tumor core and enhancing tumor regions [4], optimizing these regions instead of the three provided class labels (edema, necrosis and enhancing tumor) can be beneficial for performance [12,14,16,23]. Quantifying the uncertainty in the data has also been shown to improve results [18].

The methods presented above are highly specialized for brain tumor segmentation and their development required expertise as well as extensive experimentation. We recently proposed nnU-Net [25], a general purpose segmentation method that automatically configures segmentation pipelines for arbitrary biomedical datasets. nnU-Net set new state of the art results on the majority of the 23 datasets it was tested on, underlining the effectiveness of this approach. In the following, we will investigate nnU-Net's suitability for brain tumor segmentation. We use nnU-Net both as a baseline algorithm and as a framework for model development.

2 Method

2.1 Rankings Should Be Used for Model Selection

Optimizing a model for a competition is often mistakenly treated equivalently to optimizing the model towards the segmentation metrics used in that competition. Segmentation metrics, however, only tell half the story: they describe the model on a per-image level whereas the actual ranking is based on a consolidation of these the metrics across all test cases. Ranking schemes can be differentiated in 'aggregate then rank' and 'rank then average' approaches. In the former, some

---

[1] https://ipp.cbica.upenn.edu/.
[2] https://www.cbica.upenn.edu/BraTS20/lboardValidation.html.

aggregated metric (for example the average) is computed and then used to rank the participants. In the latter, the participants are ranked on each individual training case and then their ranks are accumulated across all cases. Different algorithm characteristics may be desired depending on the ranking scheme that is used to evaluate them. For example, in an 'aggregate then rank' scheme, median aggregation (as opposed to the mean) would be more forgiving to algorithms that produce badly predicted outliers. We refer to [26] for a comprehensive analysis on the impact of ranking on challenge results.

BraTS uses a 'rank then aggregate' approach, most likely because it is well suited to combine different types of segmentation metrics (such as HD95 and Dice). Effectively, each submission obtains six ranks per test case (one for each of the 3 evaluated regions times the 2 segmentation metrics) and the ranks are then averaged across all cases and metrics (see[3]). The final rank is normalized by the number of participating algorithms to form the ranking score, which ranges from 0 (best) to 1 (worst).

The BraTS evaluation of cases with empty reference segmentations for enhancing tumor is tailored for the ranking scheme used by the competition: if a participant predicts false positive voxels in these cases, BraTS assigns the worst possible value for both metrics (Dice 0, HD95 373.13) whereas a correctly returned empty segmentation yields perfect values (Dice 1, HD95 0). Thus, the enhancing tumor label essentially becomes binary for all participants: they either achieve the (shared) first or (shared) last rank.

As a consequence, optimizing models to maximize the mean-aggregated Dice scores and HD95 values returned by the BraTS evaluation platform may not be an ideal surrogate for optimal performance on the BraTS competition. We therefore reimplemented the ranking used by the BraTS competition and used it to rank our models against each other in order to select the best performing variant(s).

2.2   nnU-Net Baseline

We base our method on nnU-Net, our recently proposed fully automatic framework for the configuration of segmentation methods. We refer to [25] for a detailed description of nnU-Net.

First, we apply nnU-Net without any modifications to provide a baseline for later modifications. The design choices made by nnU-Net are described in the following.

nnU-Net normalizes the brain voxels of each image by subtracting their mean and dividing by their standard deviation. The non-brain voxels remain at 0. The network architecture generated by nnU-Net is displayed in Fig. 1. It follows a 3D U-Net [9,27] like pattern and consists of an encoder and a decoder which are interconnected by skip connections. nnU-net does not use any of the recently proposed architectural variations and only relies on plain convolutions for feature

---

[3] https://zenodo.org/record/3718904.

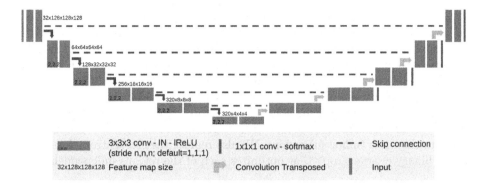

Fig. 1. Network Architecture as generated by nnU-Net. nnU-Net uses only plain U-Net-like architectures. For BraTS 2020, an input patch size of $128 \times 128 \times 128$ was selected. Downsampling is done with strided convolutions, upsampling is implemented as convolution transposed. Feature map sizes are displayed in the encoder part of the architecture. The feature maps in the decoder mirror the encoder. Auxiliary segmentation outputs used for deep supervision branch off at all but the two lowest resolutions in the decoder.

extraction. Downsampling is performed with strided convolutions and upsampling is performed with convolution transposed. Auxiliary segmentation outputs, which are used for deep supervision, branch off at all but the two lowest resolutions in the decoder. The input patch size is selected to be $128 \times 128 \times 128$ with a batch size of 2. A total of five downsampling operations are performed, resulting in a feature map size of $4 \times 4 \times 4$ in the bottleneck. Initial number of convolutional kernels is set to 32, which is doubled with each downsampling up to a maximum of 320. The number of kernels in the decoder mirrors the encoder. Leaky ReLUs [28] are used as nonlinearities. Instance normalization [29] is used for feature map normalization.

Training objective is the sum of Dice [20, 21] and cross-entropy loss. The loss operates on the three class labels edema, necrosis and enhancing tumor. nnU-Net uses stochastic gradient descent with an initial learning rate of 0.01 and a Nesterov momentum of 0.99. Training runs for a total of 1000 epochs, where one epoch is defined as 250 iterations. The learning rate is decayed with a polynomial schedule as described in [30]. Training patches are cropped from randomly selected training cases. Data augmentation is applied on the fly during training (see also Supplementary Information D in [25]).

2.3 BraTS-Specific Optimizations

Besides its role as a high quality standardized baseline and out-of-the-box segmentation tool, we also advertise nnU-Net as a framework for method development. To underline this aspect of nnU-Net's broad functionality, we select promising BraTS-specific modifications and integrate them into nnU-Net's configuration.

Region-Based Training. The provided labels for training are 'edema', 'non-enhancing tumor and necrosis' and 'enhancing tumor'. The evaluation of the segmentations is, however, performed on three partially overlapping *regions*: whole tumor (consisting of all 3 classes), tumor core (non-enh. & necrosis + enh. tumor) and enhancing tumor. It has been shown previously [11,12,14,16,18,23] that directly optimizing these regions instead of the individual classes can improve performance on the BraTS challenge. To this end, we replace the softmax non-linearity in our network architecture with a sigmoid and change the optimization target to the three tumor subregions. We also replace the crossentropy loss term with a binary cross-entropy that optimizes each of the regions independently.

Postprocessing. When the reference segmentation for enhancing tumor is empty, BraTS evaluation awards zero false positive predictions with a Dice score of 1 (the Dice would otherwise be undefined due to division by 0), thus placing the corresponding algorithm in the (shared) first rank for this test case, region and metric. This can be exploited with the goal of improving the average rank of the submitted method and thus its overall ranking in the challenge. By removing enhancing tumor entirely if the predicted volume is less than some threshold, one can accumulate more perfect rankings at the expense of some additional cases with a Dice score of 0 (and corresponding worst rank). Even though this strategy has the side effect of removing some true positive predictions, the net gain can out-weigh the losses. Removed enhancing tumor is replaced with necrosis to ensure that these voxels are still considered part of the tumor core. We optimize the threshold for postprocessing on our training set cross-validation twice, once via maximizing the mean Dice score and once via minimizing the ranking score in our internal BraTS-like ranking. Whenever we present postprocessed results, we select the best value even if that value was achieved by the opposing selection strategy.

2.4   Further nnU-Net Modifications

Increased Batch Size. Over the past years, the BraTS dataset has continued to grow in size. The low batch size used by nnU-Net results in noisier gradients, which potentially reduces overfitting but also constrains how accurately the model can fit the training data. With a larger dataset, it may be beneficial to increase the batch size (bias variance trade-off). To this end, we modify nnU-Net's configuration to increase the batch size from 2 to 5 in an attempt to improve the model accuracy.

More Data Augmentation. Data augmentation can effectively be used to artificially enlarge the training set. While nnU-Net already uses a broad range of aggressive data augmentation techniques, we make use of even more aggressive augmentations in an attempt to increase the robustness of our models. All augmentations are applied on the fly during training using the *batchgenerators* framework[4]. Relative to the nnU-Net baseline, we make the following changes:

- increase the probability of applying rotation and scaling from 0.2 to 0.3.
- increase the scale range from (0.85, 1.25) to (0.65, 1.6)
- select a scaling factor for each axis individually
- use elastic deformation with a probability of 0.3
- use additive brightness augmentation with a probability of 0.3
- increase the aggressiveness of the Gamma augmentation (gamma parameter sampled uniformly from (0.5, 1.6) instead of (0.7, 1.5))

Batch Normalization. In our participation in the M&Ms challenge[5] we noticed that more aggressive data augmentation could be used to effectively close the domain gap to other scanners, but only when used in conjunction with batch normalization (instead of instance normalization) (results available here[6], paper is now available [31]). In BraTS, Dice scores for the test cases are often lower than the reported values on the training and validation dataset, which makes us believe that there may be a domain gap between the test set and the training and validation sets. This suggests that pursuing this strategy for BraTS as well may be beneficial.

Batch Dice. The standard implementation of the Dice loss computes the loss for every sample in the minibatch independently and then averages the loss over the batch (we call this the *sample Dice*). Small errors in samples with few annotated voxels can cause large gradients and dominate the parameter updates during training. If these errors are caused by imperfect predictions of the model, these large gradients are desired to push the model towards better predictions. However, if the models predictions are accurate and the reference segmentation is imperfect, these large gradients will be counterproductive during training. We therefore implement a different Dice loss computation: instead of treating the samples in a minibatch independently, we compute the dice loss over all samples in the batch (pretending they are just a single large sample, we call this the *batch Dice*). This effectively regularizes the Dice loss because samples with previously only few annotated voxels are now overshadowed by other samples in the same batch.

Abbreviations. We will represent our models using the following abbreviations:

- **BL/BL*:** baseline nnU-Net without modifications. * indicates a batch size of 5 instead of 2.
- **R:** Region-based training (see Sect. 2.3)

---

[4] https://github.com/MIC-DKFZ/batchgenerators.

[5] https://www.ub.edu/mnms/.

[6] https://www.ub.edu/mnms/results.html.

- **DA/DA*:** more aggressive data augmentation as described in Sect. 2.4. * indicates that the brightness augmentation is only applied with a probability of 0.5 for each input modality.
- **BD:** model trained with batch Dice (as opposed to sample Dice, see Sect. 2.4)

We denote our models by the modifications that were applied to it, for example BL*+R+BD. Note that all models shown will have postprocessing applied to them.

3   Results

3.1   Aggregated Scores

Table 1. Training set results (n = 369). All experiments were run as 5-fold cross-validation. See Sect. 2.4 for decoding the abbreviations

Model	Training set (Dice, 5-fold CV)			
	Whole	Core	Enh.	Mean
BL	91.6	87.23	80.83	86.55
BL*	91.85	86.24	80.18	86.09
BL*+R	91.75	87.24	82.21	86.73
BL*+R+DA	91.87	87.97	81.37	87.07
BL*+R+DA+BN	91.57	87.59	81.29	86.82
BL*+R+DA+BD	91.76	87.67	80.94	86.79
BL*+R+DA+BN+BD	91.7	87.21	81.7	86.87
BL*+R+DA*+BN	91.6	87.51	80.94	86.68
BL*+R+DA*+BN+BD	91.47	87.13	81.33	86.64

We train each configuration as a five-fold cross-validation on the training cases (no external data is used). This not only provides us with a performance estimate on the 369 training cases but also enables us to select the threshold for postprocessing on the training set rather than the validation set. All trainings were done on single GPUs (either RTX 2080 ti, Titan RTX or V100). Training a single model took between 2 and 3 d for batch size 2 and 4–5 days for batch size 5. Training the batch size 2 models required 8 GB of GPU memory whereas training with batch size 5 required 17 GB.

The results for the different configurations are presented in Table 1.

We use the five models obtained from the cross-validation on the training cases as an ensemble to predict the validation set. The aggregated Dice scores and HD95 values as computed by the online evaluation platform are reported in

Table 2. Validation set results. Predictions were made using the 5 models from the training cross-validation as an ensemble. Metrics computed by the validation platform. See Sect. 2.4 for decoding the abbreviations

Model	Dice				HD95			
	Whole	Core	Enh.	Mean	Whole	Core	Enh.	Mean
BL	90.6	84.26	77.67	84.18	4.89	5.91	35.10	15.30
BL*	90.93	83.7	76.64	83.76	4.23	6.01	41.06	17.10
BL*+R	90.96	83.76	77.65	84.13	4.41	8.80	29.82	14.34
BL*+R+DA	90.9	84.61	78.67	84.73	4.70	5.62	29.50	13.28
BL*+R+DA+BN	91.24	85.04	79.32	85.2	3.97	5.17	29.25	12.80
BL*+R+DA+BD	90.97	83.91	77.48	84.12	4.11	8.60	38.06	16.93
BL*+R+DA+BN+BD	91.15	84.19	79.99	85.11	3.72	7.97	26.28	12.66
BL*+R+DA*+BN	91.18	85.71	79.85	85.58	3.73	5.64	26.41	11.93
BL*+R+DA*+BN+BD	91.19	85.24	79.45	85.29	3.79	7.77	29.23	13.60

Table 2. Again we provide averages of the HD95 values and Dice scores across the three evaluated regions for clarity.

As discussed in Sect. 2.1, these aggregated metrics should not be used to run model selection for the BraTS challenge because they may not directly optimize the ranking that will be used to evaluate the submissions. This becomes evident when looking at the HD95 values of the enhancing tumor HD95. The HD95 averages are artificially inflated because false positive predictions in cases where no reference segmentation for enhancing tumor is present are hard coded to receive a HD95 value of 373.13. Given that the vast majority of HD95 for this class is much smaller (even the 90th percentile is just 11.32 for our final validation set submission), these outlier values dominate the mean aggregation and result in mostly uninformative mean HD95: Switching the metric a single case from 373.13 to 0 (due to the removal of false positive predictions) can have a huge effect on the mean. The same is true for the Dice score, but to a lesser extend: a HD95 of 373.13 is basically impossible to occur even if the predicted segmentation is of poor quality whereas a low Dice score is likely to happen.

3.2 Internal BraTS-Like Ranking

To overcome the shortcomings of metric aggregation-based model selection we create an internal leaderboard where we rank our nnU-Net variants presented above (as well as others that we omitted for brevity) against each other using a reimplementation of the BraTS ranking. Results of both ranking schemes are summarized in Table 3. While the ranks between the mean Dice and the BraTS ranking scheme are different, both still seem to follow similar trends. On the training set, for example, the BL and BL* models constitute the worst contenders while BL*+R+DA+BN+BD performs quite well. Within the validation

Table 3. Rankings of our proposed nnU-Net variants on the training and validation set. For each set, we rank twice: once based on the mean Dice and once using our reimplementation of the BraTS ranking.

Model	Training set				Validation set			
	Rank based on mean Dice		BraTS ranking		Rank based on mean Dice		BraTS ranking	
	Value	Rank	Value	Rank	Value	Rank	Value	Rank
BL	86.55	8	0.3763	8	84.18	6	0.4079	8
BL*	86.09	9	0.3767	9	83.76	9	0.4236	9
BL*+R	86.73	5	0.3393	5	84.13	7	0.4005	7
BL*+R+DA	87.07	1	0.3243	3	84.73	5	0.3647	5
BL*+R+DA+BN	86.82	3	0.3377	4	85.20	3	0.3577	4
BL*+R+DA+BD	86.79	4	0.3231	2	84.12	8	0.3726	6
BL*+R+DA+BN+BD	86.87	2	0.3226	1	85.11	4	0.3487	3*
BL*+R+DA*+BN	86.68	6	0.3521	6	85.58	1	0.3125	1*
BL*+R+DA*+BN+BD	86.64	7	0.3595	7	85.29	2	0.3437	2*

set the correlation between the ranking schemes is even more pronounced. When comparing the model performance between the training and validation sets (any of the ranking schemes) the rankings differ drastically: the DA* models perform exceptionally well on the validation set (rank 1 and 2) whereas they underperform on the training set (rank 6 and 7). This creates a difficult situation. How do we choose the models used for the test set submission? Do we trust more in the 125 validation cases because we do not have access to the reference labels and thus can not overfit as easily or do we favor the training set because is has almost 3 times as many cases and should therefore provide more reliable performance estimates?

We opted for trusting the validation set over the training set and thus selected the three top performing models to build our final ensemble: BL* + R + DA* + BN + BD, BL*+R+DA*+BN and BL*+R+DA+BN+BD. Note that additionally to the 5 models from the cross-validation we had 10 more models of the BL*+R+DA*+BN configuration (each trained on a random 80:20 split of the training cases). Thus, our final ensemble consisted of 5 + 5 + 15 = 25 models. Note that ensembling was implemented by first predicting the test cases individually with each configuration, followed by averaging the sigmoid outputs to obtain the final prediction. Therefore, the 15 models of the BL*+R+DA*+BN configuration had the same influence on the final prediction as the other configurations.

Our final ensemble achieved mean Dice scores of 91.24, 85.06 and 79.89 and HD95 of 3.69, 7.82 and 23.50 for whole tumor, tumor core and enhancing tumor, respectively (on the validation set). With a mean Dice score of 85.4 the ensemble appears to perform worse in terms of the 'aggregate then rank' approach. However, since it sits comfortably on the first place of our internal 'rank then aggregate' ranking with a score of 0.2963 we are confident in its segmentation performance.

3.3   Qualitative Results

Figure 2 provides a qualitative overview of the segmentation performance. It shows results generated by our ensemble on the validation set. To rule out cherry picking we standardized the selection of presented validation cases: They were selected as best, worst, median and 75th and 25th percentiles based on their Dice scores (averaged over the three validation regions). As can be seen in the figure, the segmentation quality is high overall. The low tumor core score for the *worst* example hints at one of the potential issues with the definition of this class in the reference segmentations (see Discussion). The enhancing tumor score of 0 in the absence of predicted enhancing tumor voxels indicates either that our model missed a small enhancing tumor lesion or that it was removed as a result of our postprocessing. An inspection of the non-postprocessed segmentation mask reveals that the enhancing tumor lesion was indeed segmented by the model and must have been removed during postprocessing.

3.4   Test Set Results

Table 4. Quantitative test set results. Values were provided by the challenge organizers.

	Dice			HD95		
	Enh	Whole	Core	Enh	Whole	Core
Mean	82.03	88.95	85.06	17.805	8.498	17.337
StdDev	19.71	13.23	24.02	74.886	40.650	69.513
Median	86.27	92.73	92.98	1.414	2.639	2.000
25th percentile	79.30	87.76	88.33	1.000	1.414	1.104
75th percentile	92.25	95.17	96.19	2.236	4.663	3.606

Table 4 provides quantitative test set results. Our submission took the first place in the BraTS 2020 competition (see https://www.med.upenn.edu/cbica/brats2020/rankings.html).

4   Discussion

This manuscript describes our participation in the segmentation task of the BraTS 2020 challenge. We use nnU-Net [25] not only as our baseline algorithm but also as a framework for method development. nnU-Net again proved its generalizability by providing high segmentation accuracy in an out-of-the-box fashion. On the training set cross-validation, the performance of the nnU-Net baseline is very close to the configurations that were specifically modified for the BraTS challenge. On the validation set, however, the proposed BraTS-specific

128   F. Isensee et al.

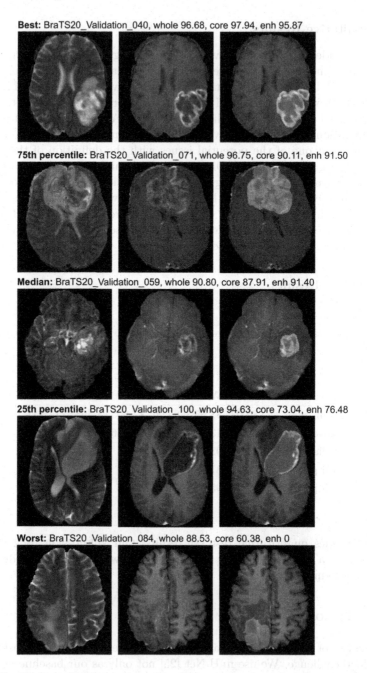

Fig. 2. Qualitative validation set results. Cases were selected as best, worst, median and 75th and 25th percentile. Within each row, the raw T2 image is shown to the left, the T1c image in the middle and on overlay with the generated segmentation on the T1c image is shown on the right. Edema is shown in violet, enhancing tumor in yellow and necrosis/non-enhancing tumor on turquoise. (Color figure online)

modifications provide substantially higher segmentation performance. Our final ensemble was selected based on the three best models on the validation set (as determined with our reimplementation of the BraTS ranking). It obtained validation Dice scores of 91.24, 85.06 and 79.89 as well as HD95 of 3.69, 7.82 and 23.5 for whole tumor, tumor core and enhancing tumor, respectively.

Our approach won the BraTS 2020 competition. On the test set, our model obtained Dice scores of 88.95, 85.06 and 82.03 and HD95 values of 8.498, 17.337 and 17.805 for whole tumor, tumor core and enhancing tumor, respectively.

We should note that this manuscript only spans a small number of modifications and lacks sufficiently extensive experimental validation thereof. While most of our nnU-Net configurations achieve good performance, our results do not allow us to accurately pinpoint which aspects contributed the most. Some design choices may even have reduced our overall performance. Increasing the batch size from 2 to 5 gave worse results according to both evaluated ranking schemes and on both the training and validation set (see Table 3). This is a clear signal that the batch size should have been reverted to 2 for the remaining experiments. We also believe that a more thorough optimization of hyperparameters regarding both the training scheme as well as the data augmentation could result in further performance gains.

Based on our observations in Fig. 2, one major source of error appears to be the tumor core prediction. In particular the *worst* presented example has a rather low Dice score for this region, even though the result seems visually plausible. We believe that this failure mode is not necessarily an issue inherent to our model but potentially originates from an inconsistency in the definition of the *non-enhancing tumor and necrosis* label, particularly in LGG cases. While the necrosis part of this label is easy to recognize, the non-enhancing tumor region often has little evidence in the image and the associated annotations may be subjective.

The enhancing tumor class is arguably the most difficult to segment in this dataset. What makes this class particularly challenging is the way its evaluation is handled when the reference segmentation of an image does not contain this class. The BraTS evaluation scheme favors the removal of small enhancing tumor lesions and thus encourages such postprocessing. In a clinical scenario where the accurate detection of small enhancing tumors could be critical, this property is not necessarily desired and we recommend to omit the postprocessing presented in this manuscript.

Just like many other BraTS participants, we have used the aggregated Dice scores and HD95 values for model selection in previous BraTS challenges [13,23]. As discussed in this manuscript, this strategy may not be ideal because the ranking that is used to determine the winning contributions works on a different principle. Indeed, when comparing the rankings of our models relative to each other on the training and validation set, we observed disparities between 'aggregate then rank' and 'rank then aggregate'. However, the overall trends were similar between the ranking methods indicating that a model selection based on metric aggregation is still going into the right direction.

While experimenting with the 'rank then aggregate' ranking on our model configurations we noted some instability in the ranks obtained by our models: when models are added or removed from the pool of methods, the ranking of other methods relative to each other could change. This seems to be an inherent property of this ranking [26]. Even though this might appear alarming at first glance, we attribute most of the instability to the similarity of the methods within the pool which has certainly overemphasized this effect. We expect the ranking to be more stable the more diverse the evaluated methods are (especially on the test set where all methods are compared against each other).

Given that aggregated metrics are at least in part disconnected from the actual evaluation of the challenge, it is frustrating that the online leaderboards for training and validation sets only display mean values for the competing methods and therefore do not allow for a perfectly accurate comparison with other participants. Especially in cases where the enhancing tumor class is absent from the reference segmentation, the dichotomy of the returned metric values (0 or 1 for the Dice score and 0 or 373.13 for the HD95) can overshadow the more minute (but still very meaningful) differences between the teams on the validation cases that do actually contain this label. We would love to see the 'rank then aggregate' ranking scheme implemented in the online leaderboards as well so that they become more informative about a models performance on the BraTS challenge.

The source code for our submission is publicly available as part of the nnU-Net repository (https://github.com/MIC-DKFZ/nnUNet). A docker image for reproducing our test set predictions is available on Docker Hub[7].

Achnowledgements. This work was funded by the Helmholtz Imaging Platform (HIP), a platform of the Helmholtz Incubator on Information and Data Science.

References

1. Kickingereder, P., Isensee, F., Tursunova, I., Petersen, J., Neuberger, U., Bonekamp, D., Brugnara, G., Schell, M., Kessler, T., Foltyn, M., et al.: Automated quantitative tumour response assessment of MRI in neuro-oncology with artificial neural networks: a multicentre, retrospective study. Lancet Oncol. **20**(5), 728–740 (2019)
2. Kickingereder, P., et al.: Radiomic subtyping improves disease stratification beyond key molecular, clinical, and standard imaging characteristics in patients with glioblastoma. Neuro Oncol. **20**(6), 848–857 (2018)
3. Kickingereder, P., et al.: Large-scale radiomic profiling of recurrent glioblastoma identifies an imaging predictor for stratifying anti-angiogenic treatment response. Clin. Cancer Res. **22**(23), 5765–5771 (2016)
4. Menze, B.H., Jakab, A., Bauer, S., Kalpathy-Cramer, J., Farahani, K., Kirby, J., Burren, Y., Porz, N., Slotboom, J., Wiest, R., et al.: The multimodal brain tumor image segmentation benchmark (brats). IEEE Trans. Med. Imaging **34**(10), 1993–2024 (2014)

[7] https://hub.docker.com/repository/docker/fabianisensee/isen2020.

5. Bakas, S., et al.: Identifying the best machine learning algorithms for brain tumor segmentation, progression assessment, and overall survival prediction in the brats challenge. arXiv preprint arXiv:1811.02629 (2018)

6. Bakas, S., Akbari, H., Sotiras, A., Bilello, M., Rozycki, M., Kirby, J.S., Freymann, J.B., Farahani, K., Davatzikos, C.: Advancing the cancer genome atlas glioma MRI collections with expert segmentation labels and radiomic features. Sci. Data **4**, 170117 (2017)

7. Bakas, S., et al.: Segmentation labels and radiomic features for the pre-operative scans of the TCGA-LGG collection. The Cancer Imaging Archive (2017)

8. Bakas, S., et al.: Segmentation labels and radiomic features for the pre-operative scans of the TCGA-GBM collection. The cancer imaging archive (2017)

9. Ronneberger, O., Fischer, P., Brox, T.: U-Net: convolutional networks for biomedical image segmentation. In: Navab, N., Hornegger, J., Wells, W.M., Frangi, A.F. (eds.) MICCAI 2015. LNCS, vol. 9351, pp. 234–241. Springer, Cham (2015). https://doi.org/10.1007/978-3-319-24574-4_28

10. He, K., Zhang, X., Ren, S., Sun, J.: Deep residual learning for image recognition. In: Proceedings of the IEEE Conference on Computer Vision and Pattern Recognition, pp. 770–778 (2016)

11. Myronenko, A.: 3D MRI brain tumor segmentation using autoencoder regularization. In: Crimi, A., Bakas, S., Kuijf, H., Keyvan, F., Reyes, M., van Walsum, T. (eds.) BrainLes 2018. LNCS, vol. 11384, pp. 311–320. Springer, Cham (2019). https://doi.org/10.1007/978-3-030-11726-9_28

12. Jiang, Z., Ding, C., Liu, M., Tao, D.: Two-stage cascaded U-Net: 1st place solution to BraTS challenge 2019 segmentation task. In: Crimi, A., Bakas, S. (eds.) BrainLes 2019. LNCS, vol. 11992, pp. 231–241. Springer, Cham (2020). https://doi.org/10.1007/978-3-030-46640-4_22

13. Isensee, F., Kickingereder, P., Wick, W., Bendszus, M., Maier-Hein, K.H.: Brain tumor segmentation and radiomics survival prediction: contribution to the BRATS 2017 challenge. In: Crimi, A., Bakas, S., Kuijf, H., Menze, B., Reyes, M. (eds.) BrainLes 2017. LNCS, vol. 10670, pp. 287–297. Springer, Cham (2018). https://doi.org/10.1007/978-3-319-75238-9_25

14. Wang, G., Li, W., Ourselin, S., Vercauteren, T.: Automatic brain tumor segmentation using cascaded anisotropic convolutional neural networks. In: Crimi, A., Bakas, S., Kuijf, H., Menze, B., Reyes, M. (eds.) BrainLes 2017. LNCS, vol. 10670, pp. 178–190. Springer, Cham (2018). https://doi.org/10.1007/978-3-319-75238-9_16

15. Huang, G., Liu, Z., Van Der Maaten, L., Weinberger, K.Q.: Densely connected convolutional networks. In: Proceedings of the IEEE Conference on Computer Vision and Pattern Recognition, pp. 4700–4708 (2017)

16. Zhao, Y.-X., Zhang, Y.-M., Liu, C.-L.: Bag of tricks for 3D MRI brain tumor segmentation. In: Crimi, A., Bakas, S. (eds.) BrainLes 2019. LNCS, vol. 11992, pp. 210–220. Springer, Cham (2020). https://doi.org/10.1007/978-3-030-46640-4_20

17. McKinley, R., Meier, R., Wiest, R.: Ensembles of densely-connected CNNs with label-uncertainty for brain tumor segmentation. In: Crimi, A., Bakas, S., Kuijf, H., Keyvan, F., Reyes, M., van Walsum, T. (eds.) BrainLes 2018. LNCS, vol. 11384, pp. 456–465. Springer, Cham (2019). https://doi.org/10.1007/978-3-030-11726-9_40

18. McKinley, R., Rebsamen, M., Meier, R., Wiest, R.: Triplanar ensemble of 3D-to-2D CNNs with label-uncertainty for brain tumor segmentation. In: Crimi, A., Bakas, S. (eds.) BrainLes 2019. LNCS, vol. 11992, pp. 379–387. Springer, Cham (2020). https://doi.org/10.1007/978-3-030-46640-4_36

19. Vaswani, A., et al.: Attention is all you need. In: Advances in Neural Information Processing Systems, pp. 5998–6008 (2017)

20. Milletari, F., Navab, N., Ahmadi, S.-A.: V-net: Fully convolutional neural networks for volumetric medical image segmentation. In: International Conference on 3D Vision (3DV). IEEE, pp. 565–571 (2016)
21. Drozdzal, M., Vorontsov, E., Chartrand, G., Kadoury, S., Pal, C.: The importance of skip connections in biomedical image segmentation. In: Carneiro, G., Mateus, D., Peter, L., Bradley, A., Tavares, J.M.R.S., Belagiannis, V., Papa, J.P., Nascimento, J.C., Loog, M., Lu, Z., Cardoso, J.S., Cornebise, J. (eds.) LABELS/DLMIA -2016. LNCS, vol. 10008, pp. 179–187. Springer, Cham (2016). https://doi.org/10.1007/978-3-319-46976-8_19
22. Lin, T.-Y., Goyal, P., Girshick, R., He, K., Dollár, P.: Focal loss for dense object detection. In: Proceedings of the IEEE International Conference on Computer Vision, pp. 2980–2988 (2017)
23. Isensee, F., Kickingereder, P., Wick, W., Bendszus, M., Maier-Hein, K.H.: No new-net. In: Crimi, A., Bakas, S., Kuijf, H., Keyvan, F., Reyes, M., van Walsum, T. (eds.) BrainLes 2018. LNCS, vol. 11384, pp. 234–244. Springer, Cham (2019). https://doi.org/10.1007/978-3-030-11726-9_21
24. Kamnitsas, K., Bai, W., Ferrante, E., McDonagh, S., Sinclair, M., Pawlowski, N., Rajchl, M., Lee, M., Kainz, B., Rueckert, D., Glocker, B.: Ensembles of multiple models and architectures for robust brain tumour segmentation. In: Crimi, A., Bakas, S., Kuijf, H., Menze, B., Reyes, M. (eds.) BrainLes 2017. LNCS, vol. 10670, pp. 450–462. Springer, Cham (2018). https://doi.org/10.1007/978-3-319-75238-9_38
25. Isensee, F., Jaeger, P.F., Kohl, S.A., Petersen, J., Maier-Hein, K.H.: nnU-Net: a self-configuring method for deep learning-based biomedical image segmentation. Nat. Methods, **18**(2), 203–211 (2021)
26. Maier-Hein, L., Eisenmann, M., Reinke, A., Onogur, S., Stankovic, M., Scholz, P., Arbel, T., Bogunovic, H., Bradley, A.P., Carass, A., et al.: Why rankings of biomedical image analysis competitions should be interpreted with care. Nat. Commun. **9**(1), 5217 (2018)
27. Çiçek, Ö., Abdulkadir, A., Lienkamp, S.S., Brox, T., Ronneberger, O.: 3D U-Net: learning dense volumetric segmentation from sparse annotation. In: Ourselin, S., Joskowicz, L., Sabuncu, M.R., Unal, G., Wells, W. (eds.) MICCAI 2016. LNCS, vol. 9901, pp. 424–432. Springer, Cham (2016). https://doi.org/10.1007/978-3-319-46723-8_49
28. Maas, A.L., Hannun, A.Y., Ng, A.Y.: Rectifier nonlinearities improve neural network acoustic models. In: Proceedings of the ICML, vol. 30, no. 1, p. 3 (2013)
29. Ulyanov, D., Vedaldi, A., Lempitsky, V.: Instance normalization: The missing ingredient for fast stylization. arXiv preprint arXiv:1607.08022 92016)
30. Chen, L.-C., Papandreou, G., Kokkinos, I., Murphy, K., Yuille, A.L.: Deeplab: semantic image segmentation with deep convolutional nets, atrous convolution, and fully connected CRFs. IEEE Trans. Pattern Anal. Mach. Intell. **40**(4), 834–848 (2017)
31. Full, P.M., Isensee, F., Jäger, P.F., Maier-Hein, K.: Studying robustness of semantic segmentation under domain shift in cardiac MRI. arXiv preprint arXiv:2011.07592 (2020)

A Deep Random Forest Approach for Multimodal Brain Tumor Segmentation

Sameer Shaikh and Ashish Phophalia[✉]

Indian Institute of Information Technology, Vadodara,
Gandhinagar 382028, Gujarat, India
{201861010,ashish_p}@iiitvadodara.ac.in

Abstract. Locating brain tumor and its various sub-regions are crucial for treating tumor in humans. The challenge lies in taking cues for identification of tumors having different size, shape, and location in the brain using multimodal data. Numerous work has been done in the recent past in BRATS challenge [16]. In this work, an ensemble based approach using Deep Random Forest [23] in incremental learning mechanism is deployed. The proposed approach divides data and features into disjoint subsets and learn in chunk as cascading architecture of multi layer RFs. Each layer is also a combination of RFs to use sample of the data to learn diversity present. Given the huge amount of data, the proposed approach is fast and paralleled. In addition, we have proposed new kind of Local Binary Pattern (LBP) features with rotation. Also, few more handcrafted are designed primarily texture based features, appearance based features, statistical based features. The experiments are performed only on MICCAI BRATS 2020 dataset.

Keywords: Brain tumor segmentation · Magnetic Resonance Imaging · Random Forest

1 Introduction

One of the tasks in Biomedical Image Analysis is Tumor detection and Brain segmentation (segmentation of brain tissues). Brain tumors are of two types: primary and secondary. Primary brain tumors originate in brain cells and proliferate whereas Secondary brain tumors arise from other organ and metastasize to the brain. Glioma, a type of primary brain tumor, is classified into Low grade glioma (LGG) and High grade glioma (HGG). LGG is benign in nature and is the slowest growing glioma in adults, while HGG is highly malignant and tends to burgeon rather quickly. According to World Health Organisation (WHO), LGG is of grade 1 and grade 2 while HGG is of Grade 3 and grade 4. The treatment of brain tumors generally includes surgery, radiotherapy, chemotherapy, etc., yet the patient's survival rate is quite low. Being a non-invasive technique, Magnetic Resonance Imaging (MRI) has been proven to quite useful in automated analysis

© Springer Nature Switzerland AG 2021
A. Crimi and S. Bakas (Eds.): BrainLes 2020, LNCS 12659, pp. 133–147, 2021.
https://doi.org/10.1007/978-3-030-72087-2_12

where multi modal (T1, T2, FLAIR etc.) data is also available [16]. Automated brain segmentation can provide accurate solutions and also aids in further analysis and monitoring of tumor. However, tumors in different patients highly vary with respect to its intensity, texture, appearance and location, etc. Therefore, one needs to take multi-parametric approach into account for classification and segmentation task [8]. Various machine learning based segmentation techniques have been developed using Deep Convolutions Neural Networks (DCNN), Support Vector Machines (SVM), Random Forests (RF) and Conditional Random Fields (CRF) etc. [2,9,11,14,16,18,21]. Figure 1 shows different modalities of MRI.

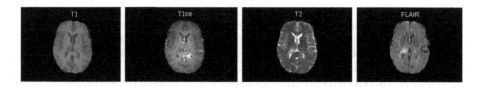

Fig. 1. An example of brain MRI modalities.

Myronenko [17] proposed an encoder-decoder based CNN architecture where a variational auto-encoder (VAE) branch added to reconstruct the input image in order to regularize the shared encoder and introduce additional constraints on its layers. Isensee et al. [10] proposed a well trained U-Net with modifications. Kamnitsas et al. [12] proposed an Ensembles of Multiple Models and Architectures (EMMA), to advantage of being ensemble of several independently trained architectures. Specifically, EMMA is an ensemble of DeepMedic [13], FCN [15] and U-Net [20] models and combines their segmentation predictions. CNN, being a black-box model, has its own demerits whose decision processes are hard to understand, and the learning behaviors are very difficult for theoretical analysis [23].

Decision Tree is a supervised machine learning algorithm that can be used to solve regression as well as classification problems. For classification task, it is used to predict the class label of the test input by applying set of decision rules present at each internal node, learnt during the training time. The Leaf node of decision tree stores the class probability distribution which is used to decide the class of the test sample falling in that leaf node. Splitting criteria is required to find the attribute with maximum impurity drop, at each internal node.

Random Forest (RF) [7] is an ensemble method based on Decision Tree model where output is depending on aggregation of individual tree in the forest. Phophalia and Maji [19] proposed an Ensemble of Forest (EoF) approach which defines relevance between the test input and the training dataset. Zikic et al. [24] proposed a decision forest based method to classify active cells, necrotic core, and edema, by using context-aware spatial features and the probabilities obtained by tissue-specific Gaussian mixture models. In recent years, Zhou and feng have proposed gcForest architecture (also called as Deep Random Forest), which generates a deep forest having the characteristics of a typical neural network [23]. It have 2 parts, multi-grain scanning to detect spatial or sequential relationships among the data, and a cascade structure for layer-wise representation learning, as found in a neural network. It has less hyper-parameters than deep neural networks, and number of layers in cascade structure is automatically determined in a data-dependent way.

Inspired by previous methods, we proposed A Deep Random Forest approach for Brain Tumor Segmentation using data and feature sampling, where each layer (having multiple RFs) learns from disjoint subset of data and features for training. In addition, output of one layer is passed to next layer as class weight feature as input. The present work tries to mimic learning mechanism of neural network using subset of data.

The rest of the paper is organized as follows: Section 2 presents the proposed architecture followed by Experimental results in Sect. 3. The paper is concluded in Sect. 4.

2 Proposed Architecture

In this section, an architecture is proposed to mimic learning capabilities of Neural Network architecture using Random Forest. Being a rule based method, decision trees and hence RF are easy to understand and highly realizable in parallel fashion. The idea of this work is to divide data and features into multiple disjoint subsets and learn them individually. Given the large dataset $D \in \mathbb{R}^{M \times N}$, where M is number of instances and N is number of features, disjoint subsets are created with respect to instances and features in small chunks as $D_i \in \mathbb{R}^{m \times n}, i \in 1, 2, \ldots, k$, where k is an integer whose value is taken experimentally and $m << M$ and $n << N$.

Each layer in proposed architecture consist of F number of Random Forest each with T trees. Before giving the training data D_i to a particular layer $L_j, j \in 1, 2, \ldots$, this D_i is passed to all the previous layers to get their class weights. Let's assume there are C number of classes present in the dataset. In this process each previous layer will generate a class weight (as feature vector) as vector $\mathbb{R}^{C \times F \times (j-1)}$. Each forest will generate class vector of length C and hence one layer will generate vector of length $C \times F$. The new class weight vector is used as augmented features for the training data D_i. Hence, at layer L_j, input size would be now $D_i \in \mathbb{R}^{m \times (n + C \times F \times (j-1))}$. The proposed architecture is shown in Fig. 3.

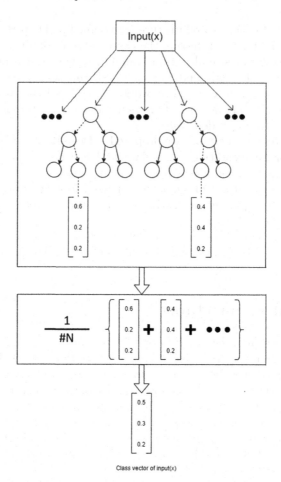

Fig. 2. Illustration of class vector generation. For an input sample, each random forest will produce a class probability distribution, by counting the percentage of different classes of training examples at the leaf node where the concerned sample falls, and then averaging across all trees in the same forest. The class probability distribution generated is treated as features, which is then concatenated with the original features to be input to the next layer. N is number of trees in a forest.

The augmented feature vector encompasses the representation of the past layers in it and used as feature in next layer. In essence, data can be divided in such a way that set of features used to learn particular characteristics of underlying data under rule based regime as done in CNN. Also, for the brain tumor segmentation, training data generated is quite large so learning a single RF is not sufficient. Hence, sampling of data will also provide diversity to the proposed model.

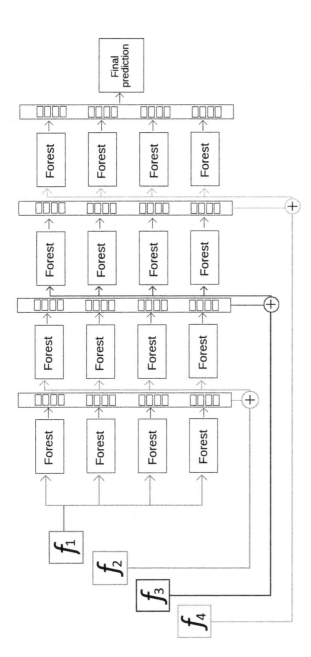

Fig. 3. Overall architecture of proposed model. In training phase, first f_1 features are used to train first layer of multi-layer cascaded RFs. Then, test samples of D_2 with f_1 features are passed to layer 1 and class probability distribution vectors generated which augmented with f_2 features of D_2 for training in layer 2 and so on. As the dataset has 4 classes, each forest will generate 4×1 class vector, and each layer will generate four 4×1 ($4 \times 4 = 16$ features) class vectors. This will be concatenated with test input generating train input for next layer having 86 features. This process will be carried out till last layer. In test phase, last layer's class vectors are aggregated and final prediction is made by max pooling. The subset of features are defined as f_1, f_2, f_3, f_4 where $\bigcup\limits_{i=1}^{4} f_i = F$ and $f_i \cap f_j = \emptyset$, where F is the set of all features.

3   Experimental Results

3.1   Dataset

MICCAI BRATS 2020 dataset [3–6] as provided by the organizers [16] is used to train and test the proposed model. The dataset contains total of 369 patient cases. Each patient has 4 modalities, namely, FLAIR, T1, T1-Contrast enhanced, T2 and ground truth label volume of size $240 \times 240 \times 155$. The sub-regions are: (1) the "Enhancing Tumor" (ET), (2) the "Tumor Core" (TC), and (3) the "Whole Tumor" (WT). The labels in the data are: 1 for Necrotic (NCR) and the Non-Enhancing Tumor (NET), 2 for Edema (ED), 4 for ET, and 0 for everything else. The TC entails the ET, as well as the necrotic (fluid-filled) and the non-enhancing (solid) parts of the tumor. The WT describes the complete extent of the disease, as it entails the TC and the peritumoral edema (ED). All the imaging data-sets have been segmented manually, by one to four raters, following the same annotation protocol. The validation data consists of total 125 patient cases with the 4 modalities, which are, FLAIR, T1, T1-Contrast enhanced, and T2. The dimensions are same as training data ($240 \times 240 \times 155$) for ease of processing. The provided data are distributed after their pre-processing, i.e. co-registered to the same anatomical template, interpolated to the same size ($1 \, \text{mm}^3$) and skull-stripped [16].

3.2   Preprocessing

The Bias field correction for all images in the dataset was carried out using a python script based on N4ITK package [22]. A Region of Interest (ROI) is defined from extracting cube of $51 \times 51 \times 25$ around first voxel found with label 4, as occurrence of label 4 is fewer. For validation data, the voxels of flair modality with intensity value greater than 110 as tumor voxels tends to brighten up in flair modality are considered.

3.3   Feature Generation

Following data pre-processing and extracting initial ROI for each patient, appearance - based features, texture-based features and statistical-based features are extracted, which gives 224 features in total.

Appearance-Based Features: Appearance-based features are calculated in each ROI. Intensity of each modality (T1, T1c, T2 and FLAIR) is treated as features ($1 \times 4 = 4$ features).

Texture-Based Features: Image gradient in three directions in each modality is treated as features ($3 \times 4 = 12$ features) along with the minimum, maximum, mean, median, mode intensities of images after Gaussian smoothing at scale 0.5, 1.0, 1.5 and window size of $3 \times 3 \times 3$ and $5 \times 5 \times 5$ ($5 \times 3 \times 2 \times 4 = 120$ features).

Statistical-Based Features: Maximum, minimum, mean, median, mode, standard deviation, kurtosis, skewness, and three central order moments in each modality is taken as features with window size of $3 \times 3 \times 3$ and $5 \times 5 \times 5$ ($11 \times 4 \times 2 = 88$ features). Refer to [19] for more details. In addition to these set of features, another set of features are extracted called Linear Binary Pattern (LBP). The local binary pattern is a simple, fast, and efficient method for extracting feature from image texture. In this method, a neighborhood of an image is selected. Then, the intensity of the points existing in this neighborhood is compared with the intensity of the pixel located in the center of the patch. If the intensity of the neighbouring pixel is smaller than the intensity of the center pixel, it is marked as 0, otherwise 1. In this way, a binary code is considered for each pixel. In case of three dimensional images, binary codes are extracted from all voxels in XY, YZ, and XZ planes and specified as $XY-LBP$, $YZ-LBP$, and $XZ-LBP$ [1]. The local binary pattern for each voxel is separately extracted based on its location on each plane. The final feature vector is achieved by arranging the values of the local binary pattern. Figure 4 illustrates the LBP feature extraction technique. The local binary pattern algorithm has been successful in recognizing edges, flat regions, and the texture structure. The LBP features are adopted with some modification. Once a local binary pattern of a voxel is generated, the local binary pattern is calculated by considering the right neighbour (clock-wise) as MSB. This will be repeated for each neighbouring voxel in clockwise and then anti-clockwise rotation, producing 16 distinct decimal numbers. This 16 features are generated for all three planes, for each modality, resulting in 192 features ($16 \times 3 \times 4 = 192$).

3.4 Implementation Details

Proposed model is a multi-layered cascaded RFs, having $L = 4$ layers and each layer consists of $F = 4$ RFs. Note that, number of classes present in dataset is $C = 4$. The output class vector of previous layer is augmented as feature in the training data of next layer. Each layer is an ensemble of forests. Being a large amount of data, it is balanced and combined as a set of few patient's data into one file, producing 35 such files. Because the size of the training data is very large, randomly 25,000 samples is taken from each file to trained 100 decision trees with max height capped at 50, providing subset of features. The subset of features are defined as f_1, f_2, f_3, f_4 where $\bigcup_{i=1}^{4} f_i = \mathscr{F}$ and $f_i \cap f_j = \emptyset$, where $\mathscr{F}$ is the set of all features and divided with f_1, f_2, f_3 consisting of 100 features each, whereas f_4 has 116 features at initial. First 100 features are used to train first layer of multi-layer cascaded RFs. Then, test samples of D_2 with f_1 features are passed to layer 1 and class probability distribution vectors generated which augmented with f_2 features of D_2 for training in layer 2 and so on. As the dataset has 4 classes, each forest will generate 4×1 class vector, and each layer will generate four 4×1 ($4 \times 4 = 16$ features) class vectors. This will be concatenated with test input generating train input for next layer having 116 features. Note that, layer 3 will have feature length as 132 and layer 4 will have length of 164 features. In test phase, last layer's class vectors are aggregated and final prediction is made by max pooling (Refer to Fig. 3).

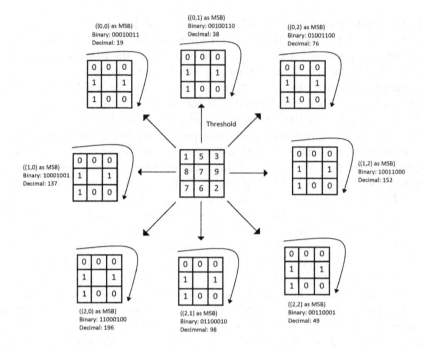

Fig. 4. Binary code of each pixel in local binary patterns.

3.5   Performance

N trees are trained per file and total files are k. so Nk trees are trained in each forest. Each layer has l such forest. If number of classes are m, each layer will produce l $m \times 1$ vectors. These vectors are treated as features and augmented in training for next layer. Last layer will also produce l $m \times 1$ vectors and the vectors are aggregated and decide the class label by max pooling. N is set to 100 experimentally, k is 35 after data balancing and combining.

Table 1 presents the Dice scores produced by proposed model on validation data, while Table 2, Table 3, and Table 4 presents the sensitivity, specificity, and hausdorff scores produced by proposed model, respectively. Figure 6, Fig. 7 and Fig. 8 shows the histograms of validation data for Dice scores for Tumor core, Enhanced Tumor and Whole Tumor respectively. Figures 9, 10, 11, and 12 shows the box plot of Dice Scores, Sensitivity Scores, Specificity Scores, and Hausdorff Scores obtained on validation data. Tables 5, 6, 7, and 8 shows the box plot of Dice Scores, Sensitivity Scores, Specificity Scores, and Hausdorff Scores obtained on test data.

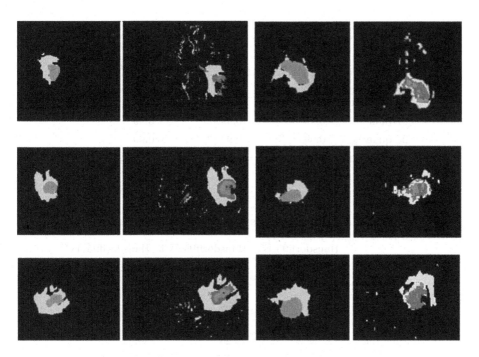

Fig. 5. First and Third columns are ground truth, second and fourth columns are segmented image by model for patient number 4 (Slice number is 79) and 13 respectively (Slice number is 79). Row wise: Axial View, Coronal View and Sagittal view. Edema (Yellow), necrotic core (Green), enhancing core (blue). (Color figure online)

Table 1. Dice Scores obtained on validation data.

	Dice_ET	Dice_WT	Dice_TC
Mean	0.20	0.38	0.27
Std. deviation	0.21	0.21	0.20
Median	0.13	0.36	0.23
25 quantile	0.02	0.22	0.11
75 quantile	0.33	0.54	0.39

Table 2. Sensitivity scores obtained on validation data.

	Sensitivity_ET	Sensitivity_WT	Sensitivity_TC
Mean	0.26	0.46	0.35
Std. deviation	0.26	0.26	0.26
Median	0.17	0.48	0.29
25 quantile	0.03	0.25	0.13
75 quantile	0.42	0.64	0.50

Table 3. Specificity scores obtained on validation data.

	Specificity_ET	Specificity_WT	Specificity_TC
Mean	0.99	0.99	0.99
Std. deviation	0.01	0.01	0.01
Median	0.99	0.99	0.99
25 quantile	0.99	0.98	0.99
75 quantile	0.99	0.99	0.99

Table 4. Hausdorff scores obtained on validation data.

	Hausdorff95_ET	Hausdorff95_WT	Hausdorff95_TC
Mean	99.02	60.06	66.24
Std. deviation	103.10	25.83	29.18
Median	77.16	64.38	68.99
25 quantile	44.08	41.89	48.34
75 quantile	97.88	79.28	89.92

Table 5. Dice Scores obtained on test data.

	Dice_ET	Dice_WT	Dice_TC
Mean	0.16	0.26	0.16
Std. deviation	0.19	0.23	0.20
Median	0.07	0.22	0.07
25 quantile	0.01	0.01	0.01
75 quantile	0.29	0.41	0.24

Table 6. Sensitivity scores obtained on test data.

	Sensitivity_ET	Sensitivity_WT	Sensitivity_TC
Mean	0.12	0.22	0.14
Std. deviation	0.16	0.23	0.20
Median	0.04	0.16	0.04
25 quantile	0.01	0.01	0.01
75 quantile	0.19	0.35	0.21

Table 7. Specificity scores obtained on test data.

	Specificity_ET	Specificity_WT	Specificity_TC
Mean	0.99	0.99	0.99
Std. deviation	0.01	0.01	0.01
Median	0.99	0.99	0.99
25 quantile	0.99	0.99	0.99
75 quantile	0.99	0.99	0.99

Table 8. Hausdorff scores obtained on test data.

	Hausdorff95_ET	Hausdorff95_WT	Hausdorff95_TC
Mean	61.83	43.15	61.76
Std. deviation	91.55	25.41	75.15
Median	32.18	39.13	41.55
25 quantile	14.08	21.16	24.53
75 quantile	67.17	58.77	70.08

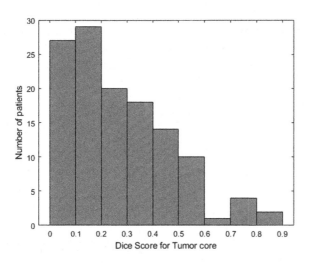

Fig. 6. Histogram for validation data according to dice score for tumor core.

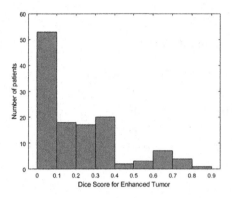

Fig. 7. Histogram for validation data according to dice score for enhanced tumor.

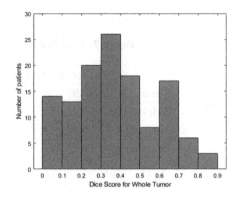

Fig. 8. Histogram for validation data according to dice score for whole tumor.

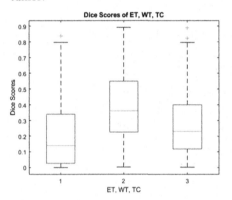

Fig. 9. Box plot of dice Scores

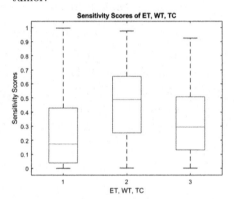

Fig. 10. Box plot of sensitivity scores

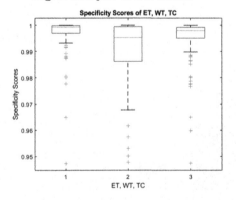

Fig. 11. Box plot of specificity scores

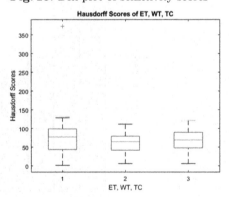

Fig. 12. Box plot of hausdorff scores

4   Conclusion

In this work, a multi-layered cascaded Random Forest based architecture for brain tumor segmentation is proposed. The architecture tries to learn about tumor and non-tumor region in incremental fashion. In each layer, multiple forests are build to classify patterns with subset of features so that every forest can learn particular representation of the data. In addition, the number layer can be made adaptive. Yet the proposed method can easily be extended to any number of layers by allowing non-disjoint set of features. To deal with large amount of data and also imbalance in nature, it is combined as a set of ten patient's data and a small forest with 100 decision trees are trained based on the extracted features. Only subset of extracted feature is provided for training each layer of the model. The proposed model does not perform well in comparison to CNN based methods and lacks with significant margin in terms of accuracy. Hence, a detailed analysis is required to tweak feature generation and construction of such large architecture.

References

1. Abbasi, S., Tajeripour, F.: Detection of brain tumor in 3d MRI images using local binary patterns and histogram orientation gradient. Neurocomputing **219**, 526–535 (2017). https://doi.org/10.1016/j.neucom.2016.09.051, http://www.sciencedirect.com/science/article/pii/S0925231216310864
2. Akkus, Z., Galimzianova, A., Hoogi, A., Rubin, D.L., Erickson, B.J.: Deep learning for brain MRI segmentation: State of the art and future directions. J. Digital Imaging **30**(4), 449–459 (2017). https://doi.org/10.1007/s10278-017-9983-4
3. Bakas, S., et al.: Advancing the cancer genome atlas glioma mri collections with expert segmentation labels and radiomic features. Scientific Data 4 (09 2017). https://doi.org/10.1038/sdata.2017.117
4. Bakas, S., et al.: Segmentation labels and radiomic features for the pre-operative scans of the TCGA-GBM collection, July 2017. https://doi.org/10.7937/K9/TCIA.2017.KLXWJJ1Q
5. Bakas, S., et al.: Segmentation labels and radiomic features for the pre-operative scans of the TCGA-LGG collection, July 2017. https://doi.org/10.7937/K9/TCIA.2017.GJQ7R0EF
6. Bakas, S., et al.: Identifying the best machine learning algorithms for brain tumor segmentation, progression assessment, and overall survival prediction in the BRATS challenge. CoRR abs/1811.02629 (2018). http://arxiv.org/abs/1811.02629
7. Breiman, L.: Random forests. Mach. Learn. **45**(1), 5–32 (2001)
8. Gordillo, N., Montseny, E., Sobrevilla, P.: State of the art survey on mri brain tumor segmentation. Magnetic Resonance Imaging **31**(8), 1426–1438 (2013). https://doi.org/10.1016/j.mri.2013.05.002. http://www.sciencedirect.com/science/article/pii/S0730725X13001872
9. Havaei, M., et al.: Brain tumor segmentation with deep neural networks. Med. Image Anal. **35**, 18–31 (2017). https://doi.org/10.1016/j.media.2016.05.004, http://www.sciencedirect.com/science/article/pii/S1361841516300330

10. Isensee, F., Kickingereder, P., Wick, W., Bendszus, M., Maier-Hein, K.H.: No new-net. In: Crimi, A., Bakas, S., Kuijf, H., Keyvan, F., Reyes, M., van Walsum, T. (eds.) BrainLes 2018. LNCS, vol. 11384, pp. 234–244. Springer, Cham (2019). https://doi.org/10.1007/978-3-030-11726-9_21

11. Işin, A., Direkolu, C., Şah, M.: Review of MRI-based brain tumor image segmentation using deep learning methods. Procedia Comput. Sci. **102**, 317–324 (2016). https://doi.org/10.1016/j.procs.2016.09.407,http://www.sciencedirect.com/scienc e/article/pii/S187705091632587X, 12th International Conference on Application of Fuzzy Systems and Soft Computing, ICAFS: 29–30 August 2016. Austria, Vienna (2016)

12. Kamnitsas, K., et al.: Ensembles of multiple models and architectures for robust brain tumour segmentation. In: Crimi, A., Bakas, S., Kuijf, H., Menze, B., Reyes, M. (eds.) BrainLes 2017. LNCS, vol. 10670, pp. 450–462. Springer, Cham (2018). https://doi.org/10.1007/978-3-319-75238-9_38

13. Kamnitsas, K., et al.: Efficient multi-scale 3D CNN with fully connected CRF for accurate brain lesion segmentation. Med. Image Anal. **36**, 61–78 (2017). https://doi.org/10.1016/j.media.2016.10.004, http://www.sciencedirect. com/science/article/pii/S1361841516301839

14. Lefkovits, L., Lefkovits, S., Szilágyi, L.: Brain tumor segmentation with optimized random forest. In: Crimi, A., Menze, B., Maier, O., Reyes, M., Winzeck, S., Handels, H. (eds.) Brainlesion: Glioma, Multiple Sclerosis, Stroke and Traumatic Brain Injuries, vol. 10154, pp. 88–99. Springer, Cham (2016). https://doi.org/10.1007/978-3-319-55524-9_9

15. Long, J., Shelhamer, E., Darrell, T.: Fully convolutional networks for semantic segmentation. In: The IEEE Conference on Computer Vision and Pattern Recognition (CVPR), June 2015

16. Menze, B.H., et al.: The multimodal brain tumor image segmentation benchmark (brats). IEEE Trans. Med. Imaging **34**(10), 1993–2024 (2015). https://doi.org/10.1109/TMI.2014.2377694

17. Myronenko, A.: 3D MRI brain tumor segmentation using autoencoder regularization. In: Crimi, A., Bakas, S., Kuijf, H., Keyvan, F., Reyes, M., van Walsum, T. (eds.) BrainLes 2018. LNCS, vol. 11384, pp. 311–320. Springer, Cham (2019). https://doi.org/10.1007/978-3-030-11726-9_28

18. Pereira, S., Pinto, A., Alves, V., Silva, C.A.: Brain tumor segmentation using convolutional neural networks in MRI images. IEEE Trans. Med. Imaging **35**(5), 1240–1251 (2016). https://doi.org/10.1109/TMI.2016.2538465

19. Phophalia, A., Maji, P.: Multimodal brain tumor segmentation using ensemble of forest method. In: BrainLes@MICCAI (2017)

20. Ronneberger, O., Fischer, P., Brox, T.: U-Net: convolutional networks for biomedical image segmentation. In: Navab, N., Hornegger, J., Wells, W.M., Frangi, A.F. (eds.) MICCAI 2015. LNCS, vol. 9351, pp. 234–241. Springer, Cham (2015). https://doi.org/10.1007/978-3-319-24574-4_28

21. Song, B., Chou, C.R., Chen, X., Huang, A., Liu, M.C.: Anatomy-guided brain tumor segmentation and classification. In: Crimi, A., Menze, B., Maier, O., Reyes, M., Winzeck, S., Handels, H. (eds.) BrainLes 2016. LNCS, vol. 10154, pp. 162–170. Springer, Cham (2016). https://doi.org/10.1007/978-3-319-55524-9_16

22. Tustison, N.J., Avants, B.B., Cook, P.A., Zheng, Y., Egan, A., Yushkevich, P.A., Gee, J.C.: N4itk: improved n3 bias correction. IEEE Trans. Med. Imaging **29**(6), 1310–1320 (2010). https://doi.org/10.1109/TMI.2010.2046908

23. Zhou, Z.H., Feng, J.: Deep forest. National Sci. Rev. **6**(1), 74–86 (10 2018). https://doi.org/10.1093/nsr/nwy108. https://doi.org/10.1093/nsr/nwy108
24. Zikic, D., Glocker, B., Konukoglu, E., Criminisi, A., Demiralp, C., Shotton, J., Thomas, O.M., Das, T., Jena, R., Price, S.J.: Decision forests for tissue-specific segmentation of high-grade gliomas in multi-channel MR. In: Ayache, N., Delingette, H., Golland, P., Mori, K. (eds.) MICCAI 2012. LNCS, vol. 7512, pp. 369–376. Springer, Heidelberg (2012). https://doi.org/10.1007/978-3-642-33454-2_46

Brain Tumor Segmentation and Associated Uncertainty Evaluation Using Multi-sequences MRI Mixture Data Preprocessing

Vladimir Groza[1], Bair Tuchinov[2(✉)], Evgeniya Amelina[2],
Evgeniy Pavlovskiy[2], Nikolay Tolstokulakov[2], Mikhail Amelin[2,3],
Sergey Golushko[2], and Andrey Letyagin[4]

[1] Median Technologies, Valbonne, France
vladimir.groza@mediantechnologies.com
[2] Novosibirsk State University, Novosibirsk, Russia
bairt@nsu.ru, pavlovskiy@post.nsu.ru, n.tolstokulakov@g.nsu.ru
[3] FSBI "Federal Neurosurgical Center", Novosibirsk, Russia
[4] Research Institute of Clinical and Experimental Lymphology, Branch of IC&G SB
RAS, Novosibirsk, Russia
letyaginay@bionet.nsc.ru

Abstract. The brain tumor segmentation is one of the crucial tasks nowadays among other directions and domains where daily clinical workflow requires to put a lot of efforts while studying computer tomography (CT) or structural magnetic resonance imaging (MRI) scans of patients with various pathologies. MRI is the most common method of primary detection, non-invasive diagnostics and a source of recommendations for further treatment of brain diseases. The brain is a complex structure, different areas of which have different functional significance.

In this paper, we extend the previous research work on the robust pre-processing methods which allow to consider all available information from MRI scans by the composition of T1, T1C, T2 and T2-Flair sequences in the unique input. Such approach enriches the input data for the segmentation process and helps to improve the accuracy of the segmentation and associated uncertainty evaluation performance.

Proposed in this paper method also demonstrates strong improvement on the segmentation problem. This conclusion was done with respect to Dice metric, Sensitivity and Specificity compare to identical training/validation procedure based only on any single sequence and regardless of the chosen neural network architecture.

Obtained results demonstrate significant performance improvement while combining three MRI sequences in the 3-channel RGB like image for considered tasks of brain tumor segmentation. In this work we provide the comparison of various gradient descent optimization methods and of the different backbone architectures.

The reported study was funded by RFBR according to the research project No 19-29-01103.

A. Crimi and S. Bakas (Eds.): BrainLes 2020, LNCS 12659, pp. 148–157, 2021.
https://doi.org/10.1007/978-3-030-72087-2_13

Keywords: Medical imaging · Deep learning · Neural network · Segmentation · Brain · MRI

1   Introduction

Segmentation of the different parts of the human body and organs of interest still stays one of the most important problems for the clinical routine as well as for the computer-aided diagnosis systems. Deep learning methods broadly demonstrated their capabilities and various applicability for common medical imaging formats (such as CT/MRI/X-ray), organs localization, and specific problems [22,23].

Recently, several methods and approaches aiming at the optimal usage of different MRI scans [18] sequences were proposed. Usually the goal of these works is to improve the solutions stability, compensate the lack of missing data and to increase the models performance [11–14].

In the previous works [20,21] we proposed and evaluated the data pre-processing method on another more complex brain tumor segmentation problems. This method consists of constructing the RGB-like 3-channel input image by assigning each channel to each of the a) native (T1) and b) post-contrast T1-weighted (T1C), c) T2-weighted (T2), and d) T2 Fluid Attenuated Inversion Recovery (T2-Flair) sequences. In order to obtain a generalized robust solution, we used the dataset consisting of different brain tumors types as well as their high shape and localization variability.

The proposed approach is focused on the search for small areas of contrast. Such methodology is explained by the currently commonly used technique when the per-voxel MR subtraction is used to compare the T1-weighted images before and after contrast injection.

2   Datasets Description

2.1   BraTS Dataset

BraTS 2020 dataset [24–28] consists of training (369 medical cases) and validation (125 medical cases) parts. The data is presented as MRI volumes with 155 slices corresponding to T1, T2, T2-Flair and T1C sequences of high-grade glioblastoma (GBM/HGG) and lower-grade glioma (LGG). These multimodal volume scans are available as NIfTI files. All the imaging datasets have been segmented manually, by one to four raters, following the same annotation protocol, and their annotations were approved by experienced neuro-radiologists. Annotations comprise the GD-enhancing tumor (ET – label 4), the peritumoral edema (ED – label 2), and the necrotic and non-enhancing tumor core (NCR/NET – label 1). The sub-regions considered for evaluation are: 1) the "enhancing tumor" (ET), 2) the "tumor core" (TC), and 3) the "whole tumor" (WT).

2.2 Siberian Brain Tumor Dataset

According to the official BraTS 2020 challenge rules, we also used additional data as it was allowed. This private Siberian Brain Tumor (SBT) dataset consists of 100 brain MRI volume scans in the format of NIfTI files and totally resulting in 16270 2D slices. This limited dataset comprises four different subtypes of brain tumors - meningioma, neurinoma, glioblastoma and astrocytoma. The SBT dataset includes several MRI sequences such as T1, T1C, T2, T2 Flair and diffusion-weighted images (DWI). All separate scans were acquired with different clinical protocols and scanners from Federal Center for Neurosurgery, co-registered to the same anatomical template and interpolated to the same pixel spacing and resolution ($1 \, \mathrm{mm}^3$). This private dataset has been segmented manually by two experienced board-certified neuro-radiologist raters, following the same annotation (labels) protocol and any label misclassification was manually corrected by an expert.

As part of this work we also used this dataset as an external validation cohort and partially enriched the training set with the volumes of appropriate tumor types.

It is important to mention that these two datasets significantly differ from each other due to the represented tumor types: BraTS 2020 dataset consists only of the HGG and LGG, while our private dataset includes four different tumor types.

3   Methods

To perform the segmentation of the brain tumors in head MRI scans, we used the *Encoder-Decoder* type neural networks. We trained three different strong *LinkNet* [30] pipelines for the numerical experiments, namely with the *se-resnext50_swsl*, *se-resnext101* and *resnet34* backbones [31] and corresponding decoders as shown on Fig. 1.

Each block of the decoder consists of the 2D Convolution, Upsampling and another Convolution layer, each followed by the Batch Normalization and ReLU [32] activation function. In the last block of decoder we apply Softmax activation in order to obtain the required probabilities for each sample and get the required label for each pixel after applying the argmax function over the feature maps. The proposed solution also includes a 5-fold cross-validation scheme, so all three networks were trained on 5 different folds with the patient-based split and ensembled in order to obtained final predictions. All data augmentations were applied *on the fly* if required and only to each particular mini-batch without modifying the validation subset.

The training process includes several steps. First, all available data were filtered and prepared for the training - we used only T1C, T1 and T2-Flair sequences as input for our models. We used the combination of these sequences to create the pseudo RGB like images with three channels [19]. We also performed a comparison and analysis of the possible aggregation options and found that

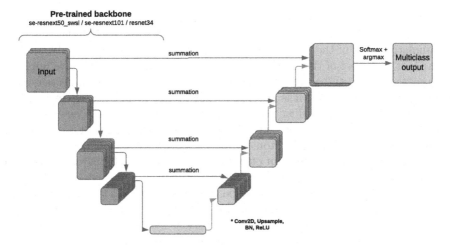

Fig. 1. LinkNet-like network with custom backbones

inclusion of the T2 sequence instead of any of the already used leads to the worseness of results. All NIfTI volumes were processed to get 2D PNG images which were normalized to [0;1] range and padded from original the size of 240×240 pixels to the common size of 256×256 pixels.

After that, the pipeline was trained for 30 epochs using the Binary cross-entropy (BCE) loss, stochastic gradient descent optimizer *Adam* [29] with initial learning rate (LR) of 1e−4 and the *ReduceLROnPlateau* scheduling with the "patience" of 5 epochs and reducing factor of 0.1.

We used the following augmentation for both images and GT references: random 90° rotations, horizontal/vertical flips, ShiftScaleRotate, transpose transformation as well as by random changes in brightness, contrast and gamma. The use of heavy augmentation enhances the models generalization performance and prevents the overfitting problem.

Our solution has two important steps - the inference step to get segmentation prediction and the post-processing operations.

On the inference, we performed identical data preparation step as for the training phase. To improve the solution performance the test time augmentations (TTA) from the D4 transformation group were used (rotations of 90/180/270/360° as well as its mirrored variations). The idea to use such augmentation is explained by the presence of all these transformations during the training phase.

After analyzing the obtained results and associated uncertainty maps for each of the objective classes, we performed an additional forced increment of the probabilities for classes 1 and 4 (NCR/NET and ET respectively) by 10%. Such a simple trick allowed to slightly "shift" the boundaries of required regions and improve the performance on the local validation and leader board.

One more specific post-processing technique was applied in order to improve the segmentation of the ET region. There are a quite well known studies showing that direct subtraction of the T1 from T1C images gives quite an informative picture of the ET region. The main idea of the algorithm is that the contrast-accumulating part of the tumor has a high intensity in the second study and low in the first. The rest of the brain structures on tomograms should have minimal differences in both studies relative signal intensity. Before performing a per-voxel comparison of tomograms, it is necessary to compare the brightness (intensities) of the two sets of images, since the tomograms actual images are very different. The change in brightness (intensity) is associated with the peculiarities of the image formation process in the tomograph.

Original input images were cut off with the obtained WT binary segmentation masks, then the obtained region of interest (ROI) was normalized to the [0; 1] values range and binarized again by some threshold (from 0.1 to 0.9). At the end of these manipulation were assigned all non-zeros pixels with the appropriate ET label (4) and substitute in the final segmentation masks.

One more branch of this work was focused on the estimation of uncertainty associated with the given segmentation masks. These uncertainty maps can be obtained directly from our network outputs after the Softmax layers and before applying the argmax function. One has to combine the feature maps related to the required classes to get the uncertainty maps for the metric objectives which are different from output classes. For example, for the whole tumor (WT) uncertainty it is needed to summarize three feature maps, invert it and scale to the range of [0; 100] where value 0 stays for the most certain pixels and 100 - for the most uncertain one.

The validation phase defines the best model checkpoint, which will be used for the prediction phase and it requires strong attention to details. First, we created the patient-based split to get the pure and proper validation of our models and to avoid any kind of overfitting. Moreover, there was no augmentation used during the validation step.

Taking into account that this competition includes two separate verification as validation and test phases, we also performed the pseudo-labeling method. We took the validation submission details, which demonstrate the quality of the predictions for each particular volume and selected only those that scores more than 0.94 for the WT Dice and have no abnormalities such as low Dice score for another region. This method provided us with 57 additional volumes that we included in the training set. First, all these cases were included only to a single fold, and training phase was repeated. Since the validation subset did not change we were able to correctly estimate the impact of these additional pseudo labeled data. After that, all data were split again into new 5 patient-base folds and all networks were retrained with no changes. Such a method allowed to improve our solution by up to 1.5% in terms of DICE metric depending on the segmentation class.

4   Results

In this work we present the segmentation performance as the main results, including the comparison of different gradient descent optimizers and several most commonly used architectures as well in the Table 1.

Table 1. The part of results Task 1 - Segmentation on the validation dataset.

Number of submit	Dice_ET	Dice_WT	Dice_TC
4_submit	0.72505	0.89408	0.79384
20_submit	0.7268	0.89433	0.8076
28_submit	0.73167	0.89292	0.8175
29_submit	0.73166	0.8933	0.81428

Calculating the mean Dice score with the consideration of the "empty" slices where Dice is assigned equal to 1 is very common approach nevertheless not really fair. So, according to the BraTS challenge rules the mean Dice metric for each particular experiment was calculated only on the slices where ground truth (GT) or segmentation output is present at least for one of the given classes.

In the Table 1 we presented 4 different submits just to highlight the improvement obtained regarding the modification and proposed hypothesis. "4_submit" stands for the common ensemble including three different LinkNet networks described above in the Methods section without any post-processing or additional tricks. "20_submit" presents the solution based on the previous but with the selected model of the previous ensemble achieving the best metric performance on the validation set. The two last given submissions "28" and "29" stand for the improvement obtained with the forced multiplication of the features maps by factor 1.05 and 1.1 respectively. Both of these solutions also includes additional post-processing step that is based on the "a-priori" knowledge that there can be no big edema region without tumor core region, so it was corrected for all cases where automated prediction returned corresponding labels.

In the Fig. 2 and Fig. 3 we present the examples of segmentation prediction on the validation dataset. The 40 and 82 patients, with labels and areas: Yellow - the peritumoral edema, Red - GD-enhancing tumor, Green - the non-enhancing tumor and the necrotic tumor core.

From the main Task 1 - Segmentation, we present the results of complementary research task - evaluation of uncertainty measures in the context of glioma region segmentation in the Table 2.

In the Table 2 we present the performance DICE metrics for the "Uncertainty" task for our best submission on the validation part of the competition. It is possible to mention that all proposed techniques and hypotheses lead to a slight improvement of the results.

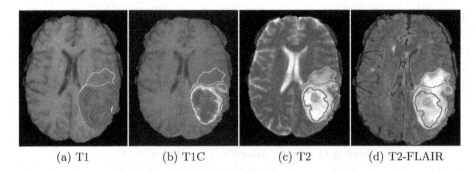

(a) T1   (b) T1C   (c) T2   (d) T2-FLAIR

Fig. 2. The 40 patient with MRI T1, T1C, T2, T2-Flair sequences (Color figure online)

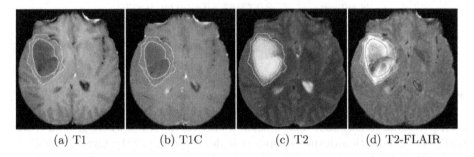

(a) T1   (b) T1C   (c) T2   (d) T2-FLAIR

Fig. 3. The 82 patient with MRI T1, T1C, T2, T2-Flair sequences (Color figure online)

Table 2. The part of results Task 3 - Evaluation of uncertainty measures in segmentation.

Number of submit	Dice_AUC_WT	Dice_AUC_TC	Dice_AUC_ET
4_submit	0.9556	0.9203	0.8650
9_submit	0.9556	0.9342	0.8651
29_submit	0.9568	0.9371	0.8625

Table 3. Selected submission for the test set in Task 1 - Segmentation on the test dataset.

Label	Dice_ET	Dice_WT	Dice_TC	Sens_ET	Sens_WT	Sens_TC	Spec_ET	Spec_WT	Spec_TC	Hsdrf95_ET	Hsdrf95_WT	Hsdrf95_TC
Mean	0.78221	0.87624	0.82762	0.81301	0.86954	0.82639	0.99963	0.99925	0.99967	16.17126	8.71754	18.04194
StdDev	0.20268	0.12893	0.24356	0.22608	0.15108	0.24091	0.00039	0.00081	0.00086	69.48095	40.51499	69.40839
Median	0.83334	0.91081	0.91212	0.88806	0.91817	0.92287	0.99972	0.99955	0.99983	2.0	3.08114	2.44949
25quantile	0.75493	0.86211	0.83333	0.78742	0.83998	0.82525	0.9995	0.99887	0.99968	1.41421	2.05902	1.41421
75quantile	0.88553	0.94124	0.94882	0.93893	0.9556	0.9618	0.99988	0.99977	0.99994	2.73369	5.31363	4.47214

All of our models were equally evaluated on the test set of the BraTS 2020 challenge. Detailed results can be observed below in the Tables 3 and 4 for the Multi-class Segmentation and Uncertainty Estimation tracks respectively according to the official BraTS 2020 challenge metrics. We observe very strong

Table 4. Selected submission for the test set in Task 3 - Evaluation of uncertainty measures in segmentation.

Label	DICE_WT	DICE_TC	DICE_ET	FTP_WT	FTP_TC	FTP_ET	FTN_WT	FTN_TC	FTN_ET	Score_WT	Score_TC	Score_ET
Mean	0.94276	0.91679	0.88106	0.04294	0.06424	0.11291	0.98459	0.98628	0.98643	0.95760	0.93313	0.89085
StdDev	0.05712	0.17286	0.13537	0.03058	0.12061	0.07909	0.00197	0.00119	0.00087	0.03440	0.11002	0.07814
Median	0.95936	0.95980	0.91141	0.03492	0.03587	0.09781	0.98504	0.98660	0.98665	0.96883	0.96952	0.90665
25quantile	0.93906	0.93871	0.86443	0.02496	0.02366	0.06196	0.98362	0.98595	0.98602	0.95185	0.94673	0.86817
75quantile	0.97240	0.97719	0.94625	0.05063	0.05962	0.14429	0.98602	0.98703	0.98706	0.97861	0.97945	0.93487

stability of our results for both considered problems compare to the local and official validation with the constant increase of the metric values.

5 Conclusions

This work presents an efficient approach to combining three MRI sequences in the 3-channel RGB-like image in order to significantly improve the model's performance of the particular applied numerical problem of brain tumor segmentation.

The proposed method also demonstrates significant improvement on the segmentation problem with respect to Dice metrics compare to similar training/validation procedures based on any single sequence regardless of the chosen neural network architecture.

The stability and robustness of this method was confirmed with the various numerical experiments and with very high probability can be extended to many other applied problems.

These particular results are very important from the clinical point of view and can strongly increase the quality of the computer-aided systems where similar solutions are deployed.

One of the possibilities for future research directions is the investigation of the proposed pipeline behavior through the increased private dataset: volume and type of tumors.

Acknowledgment. The reported study was funded by RFBR according to the research project No 19-29-01103.

References

1. Bray, F., Ferlay, J., Soerjomataram, I., Siegel, R.L., Torre, L.A., Jemal, A.: Global cancer statistics 2018: GLOBOCAN estimates of incidence and mortality worldwide for 36 cancers in 185 countries. CA Cancer J. Clin. **68**(6), 394–424 (2018)
2. Bobinski, M., Greco, C.M., Schrot, R.J.: Giant intracranial medullary thyroid carcinoma metastasis presenting as apoplexy. Skull Base **9**(5), 359–362 (2009)
3. Chrastina, J., Novak, Z., Riha, I., et al.: Diagnostic value of brain tumor neuroendoscopic biopsy and correlation with open tumor resection. J. Neurol. Surg. A Cent. Eur. Neurosurg. **75**(2), 110–115 (2012)

4. Kamnitsas, K., et al.: Ensembles of multiple models and architectures for robust brain tumour segmentation. In: Crimi, A., Bakas, S., Kuijf, H., Menze, B., Reyes, M. (eds.) BrainLes 2017. LNCS, vol. 10670, pp. 450–462. Springer, Cham (2018). https://doi.org/10.1007/978-3-319-75238-9_38

5. Li, C., et al.: Multi-parametric and multi-regional histogram analysis of MRI: modality integration reveals imaging phenotypes of glioblastoma. Eur. Radiol. **29**(9), 4718–4729 (2019). https://doi.org/10.1007/s00330-018-5984-z

6. Myronenko, A.: 3D MRI brain tumor segmentation using autoencoder regularization. In: Crimi, A., Bakas, S., Kuijf, H., Keyvan, F., Reyes, M., van Walsum, T. (eds.) BrainLes 2018. LNCS, vol. 11384, pp. 311–320. Springer, Cham (2019). https://doi.org/10.1007/978-3-030-11726-9_28

7. Chaurasia, A., Culurciello, E.: LinkNet: exploiting encoder representations for efficient semantic segmentation, CoRR, vol. abs/1707.03718 (2017)

8. Milletari, et al., : V-Net: Fully convolutional neural networks for volumetric medical image segmentation. In: 2016 4th International Conference on 3D Vision, pp. 565–571. IEEE (2016)

9. Xie, S., Girshick, R., Dollár, P., Tu, Z., He, K.: Aggregated residual transformations for deep neural networks. In: Proceedings of the 30th IEEE Conference on Computer Vision and Pattern Recognition, CVPR 2017, vol. 1, pp. 5987–5995 (2017)

10. Hu, J., Shen, L., Albanie, S., Sun, G., Wu, E.: Squeeze-and-excitation networks, CoRR abs/1709.01507. arXiv:1709.01507 (2017)

11. Ge, C., Gu, I.Y., Store Jakola, A., Yang, J.: Cross-modality augmentation of brain MR images using a novel pairwise generative adversarial network for enhanced glioma classification. In: 2019 IEEE International Conference on Image Processing (ICIP), pp. 559–563 (2019)

12. Havaei, M., Guizard, N., Chapados, N., Bengio, Y.: HeMIS: hetero-modal image segmentation. In: Ourselin, S., Joskowicz, L., Sabuncu, M.R., Unal, G., Wells, W. (eds.) MICCAI 2016. LNCS, vol. 9901, pp. 469–477. Springer, Cham (2016). https://doi.org/10.1007/978-3-319-46723-8_54

13. Varsavsky, T., Eaton-Rosen, Z., Sudre, C.H., Nachev, P., Cardoso, M.J.: PIMMS: permutation invariant multi-modal segmentation, CoRR, vol. abs/1807.06537. arXiv:1807.06537 (2018)

14. Dorent, R., Joutard, S., Modat, M., Ourselin, S., Vercauteren, T.: Hetero-modal variational encoder-decoder for joint modality completion and segmentation. arXiv:1907.11150 (July 2019)

15. Roy, A.G., Navab, N., Wachinger, C.: Recalibrating fully convolutional networks with spatial and channel 'squeeze & excitation' blocks, CoRR abs/1808.08127. arXiv:1808.08127 (2018)

16. Ronneberger, O., Fischer, P., Brox, T.: U-Net: convolutional networks for biomedical image segmentation. In: Navab, N., Hornegger, J., Wells, W.M., Frangi, A.F. (eds.) MICCAI 2015. LNCS, vol. 9351, pp. 234–241. Springer, Cham (2015). https://doi.org/10.1007/978-3-319-24574-4_28

17. Buslaev, A., Parinov, A., Khvedchenya, E., Iglovikov, V., Kalinin, A.: Albumentations: fast and flexible image augmentations. arXiv:1809.06839 (2018)

18. Letyagin, A.Y., et al.: Artificial intelligence for imaging diagnostics in neurosurgery. In: 2019 International Multi-conference on Engineering, Computer and Information Sciences (SIBIRCON), pp. 336–337. IEEE-Institute of Electrical and Electronics Engineers Inc. (2019)

19. Groza, V., et al.: Data preprocessing via multi-sequences MRI mixture to improve brain tumor segmentation. In: Rojas, I., Valenzuela, O., Rojas, F., Herrera, L.J., Ortuño, F. (eds.) IWBBIO 2020. LNCS, vol. 12108, pp. 695–704. Springer, Cham (2020). https://doi.org/10.1007/978-3-030-45385-5_62

20. Tolstokulakov, N., et al.: Data preprocessing via compositions multi-channel mri images to improve brain tumor segmentation. In: IEEE 17th International Symposium on Biomedical Imaging Workshops (ISBI Workshops), Iowa City, IA, USA, vol. 2020, pp. 1–4 (2020). https://doi.org/10.1109/ISBIWorkshops50223.2020.9153416

21. Letyagin, A., et al.: Multi-class brain tumor segmentation via multi-sequences MRI mixture data preprocessing. In: Cognitive Sciences, Genomics and Bioinformatics (CSGB), Novosibirsk, Russia, vol. 2020, pp. 185–189 (2020). https://doi.org/10.1109/CSGB51356.2020.9214645

22. Yan, Q., et al.: COVID-19 chest CT image segmentation - a deep convolutional neural network solution. arXiv:2004.10987 (2020)

23. Groza, V., Kuzin, A.: Pneumothorax segmentation with effective conditioned post-processing in chest x-ray. In: IEEE 17th International Symposium on Biomedical Imaging Workshops (ISBI Workshops), Iowa City, IA, USA, vol. 2020, pp. 1–4 (2020). https://doi.org/10.1109/ISBIWorkshops50223.2020.9153444

24. Menze, B.H., Jakab, A., Bauer, S., Kalpathy-Cramer, J., Farahani, K., Kirby, J., et al.: The multimodal brain tumor image segmentation benchmark (BraTS). IEEE Trans. Med. Imaging **34**(10), 1993–2024 (2015). https://doi.org/10.1109/TMI.2014.2377694

25. Bakas, S., Akbari, H., Sotiras, A., Bilello, M., Rozycki, M., Kirby, J.S., et al.: Advancing the cancer genome atlas glioma MRI collections with expert segmentation labels and radiomic features. Nat. Sci. Data **4**, 170117 (2017). https://doi.org/10.1038/sdata.2017.117

26. Bakas, S., Reyes, M., Jakab, A., Bauer, S., Rempfler, M., Crimi, A., et al.: Identifying the best machine learning algorithms for brain tumor segmentation, progression assessment, and overall survival prediction in the BraTS challenge. arXiv preprint arXiv:1811.02629 (2018)

27. Bakas, S., Akbari, H., Sotiras, A., Bilello, M., Rozycki, M., Kirby, J., et al.: Segmentation labels and radiomic features for the pre-operative scans of the TCGA-GBM collection. The Cancer Imaging Archive (2017). https://doi.org/10.7937/K9/TCIA.2017.KLXWJJ1Q

28. Bakas, S., Akbari, H., Sotiras, A., Bilello, M., Rozycki, M., Kirby, J., et al.: Segmentation labels and radiomic features for the pre-operative scans of the TCGA-LGG collection. The Cancer Imaging Archive (2017). https://doi.org/10.7937/K9/TCIA.2017.GJQ7R0EF

29. Kingma, D.P., Ba, J.: Adam: a method for stochastic optimization. arXiv:1412.6980 (2014)

30. Chaurasia, A., Culurciello, E.: LinkNet: exploiting encoder representations for efficient semantic segmentation. arXiv preprint arXiv:1707.03718 (2017)

31. Hu, J., Shen, L., Sun, G.: Squeeze-and-excitation networks. arXiv preprint arXiv:1709.01507 (2017)

32. Agarap, A.F.: Deep learning using rectified linear units (ReLU). arXiv preprint arXiv:1803.08375 (2018)

A Deep Supervision CNN Network for Brain Tumor Segmentation

Shiqiang Ma[1], Zehua Zhang[1], Jiaqi Ding[1], Xuejian Li[1], Jijun Tang[1,2,3], and Fei Guo[1(✉)]

[1] School of Computer Science and Technology, College of Intelligence and Computing, Tianjin University, Tianjin 300350, China
`fguo@tju.edu.cn`
[2] Department of Computer Science and Engineering, University of South Carolina, Columbia, SC 29208, USA
[3] Key Laboratory of Systems Bioengineering (Ministry of Education), Tianjin University, Tianjin 300072, People's Republic of China

Abstract. The brain tumor segmentation is essential for diagnosis and treatment of brain diseases. However, most of current 3D deep learning technologies require large number of magnetic resonance images (MRIs). In order to make full use of small dataset like BraTS 2020, we propose a deep supervision-based 2D residual U-net for efficient and automatic brain tumor segmentation. In our network, residual blocks are used to alleviate the gradient dispersion caused by excessive depth of network, while multiple deep supervision branches are used as the regularization of the network, they can improve the training stability and enable the encoder to extract richer visual features. The CBICA's IPP's evaluation of the segmentation results verifies the effectiveness of our method. The average Dice of ET, WT and TC are 0.7593, 0.8726 and 0.7879 respectively.

Keywords: Brain tumor segmentation · Deep supervision · U-net · Residual block

1 Introduction

Glioma is the most common primary brain tumor, it has strong aggressiveness and high mortality rate. The average life expectancy of patients with high-grade tumors is usually no more than two years, but finding the tumor on MRI images as early as possible can improve the survival time and survival probability of patients. However, it is time-consuming and inefficient for doctors to manually label brain tumors on such huge number of MRIs, so automatic brain tumor segmentation plays an important role in assisting doctors in diagnosis, surgical planning and evaluation of postoperative recovery effects.

The Multimodal Brain Tumor Segmentation Challenge (BraTS) is currently one of the most authoritative competitions in the field of brain tumor segmentation. It aims to evaluate the performances of the latest methods in this field [1–3].

© Springer Nature Switzerland AG 2021
A. Crimi and S. Bakas (Eds.): BrainLes 2020, LNCS 12659, pp. 158–167, 2021.
https://doi.org/10.1007/978-3-030-72087-2_14

The BraTS 2020 training dataset [4–8] provides 369 multi-institutional routine clinically-acquired multimodal MRI scans of glioblastoma (GBM/HGG) and lower grade glioma (LGG). This dataset is available as NIfTI files (.nii.gz), including MRIs of native (T1), post-contrast T1-weighted (T1Gd), T2-weighted (T2), and T2 Fluid Attenuated Inversion Recovery (T2-FLAIR) volumes (Fig. 1). Our task is to segment each tumor area into enhancing tumor, the peritumoral edema, and the necrotic and non-enhancing tumor core.

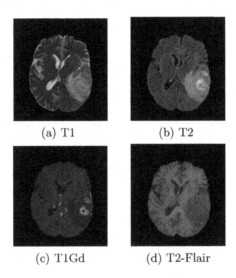

(a) T1   (b) T2

(c) T1Gd   (d) T2-Flair

Fig. 1. Multimodal MRI scans

2   Related Work

With the development of deep learning, the automatic segmentation of brain tumors gets excellent performance. Especially since the emergence of Convolutional Neural Networks (CNN), it greatly boosts the development of brain tumor segmentation. It is very suitable for processing image data because it can obtain local features more efficiently. Thus CNN-based brain tumor segmentation methods emerge in endlessly.

The first to mention is U-Net [9] (see Fig. 2(a)), it is extensively studied in the field of brain tumor segmentation. Its encoder-decoder structure and ship-connection became the basis of many later researches. Isensee *et al.* [10,11] proposed a modified version of U-net for brain tumor segmentation task. It increased the depth of U-net by pre-activation residual blocks, so that achieved excellent results in BraTS 2017 and BraTS 2018. Specifically, the skip-connections ensured that deep feature maps can contain more low-level informations. Zhou *et al.* proposed a segmentation network called U-Net++ [13]. As shown in Fig. 2(b),

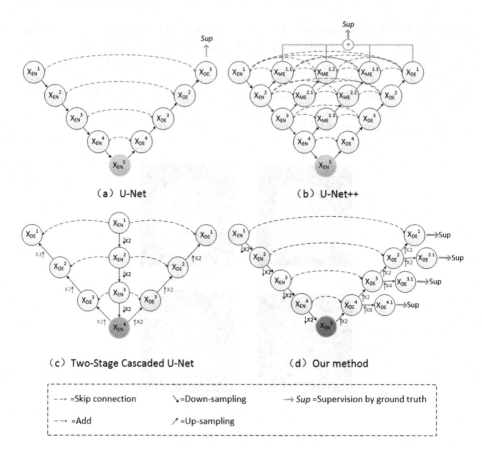

Fig. 2. Comparison of various U-Net. (a) U-Net. (b) U-Net++. (c) Two-Stage Cascaded U-Net. (d) Our proposed network

it was developed as an improved U-net by nested and dense skip-connections. Then multi-scale features were fused through dense connections like DenseNet [14–17]. In addition, deep supervision was used to train multiple sub-networks in this method.

There are also some improvements to the encoder-decoder structure of U-net. Myronenko [18] proposed an asymmetric encoder-decoder architecture, it used multi-task learning [19,20] method to train the network. The network contained two sub-networks, especially one of them used variational U-net structure to reconstruct original input images. The variational U-net can be regarded as regularization of the network to prevent over-fitting caused by small size of training set, also it enabled encoder to extract more features. In addition, Chen et al. [21] proposed MASSL, it used tumor masks obtained by roughly segmentation as the labels of the reconstruction task. This method encouraged model to learn differences between tumor area and background through pseudo-labels mentioned above. Thus it can avoid the network from reconstructing original images, which may lead encoder to pay much attention to original image features.

Furthermore, the first place in BraTS 2019 challenge was a two-stage cascaded U-Net [22] (see Fig. 2(c)). The first stage of this network was an asymmetric encoder-decoder architecture. The prediction results of the first stage were concatenated with original input images, then they were used as the input of the second stage. The second stage network architecture included two branches, and they perform segmentation tasks simultaneously. The purpose of two branches is to help encoder learn richer features. And the difference between the two branches lies in the decoder's up-sampling technologies, which leads to different segmentation results.

Based on the above-mentioned development of the U-net, we propose a new network. As can be seen in Fig. 2(d), we use residual U-net as the main structure. Then we replace the skip-connection of U-net with shortcut. Also, we use different annotations as deep supervision labels to regularize the network, which enable the encoder to obtain more features and correlations between tumor regions.

Almost all segmentation methods use 2D or 3D CNN architecture. 2D CNNs use 2D image slices to train the network, they use less memory but cannot obtain the spatial context information of image slices. 3D CNNs use voxel to train the network, it use enormous memory but can obtain the spatial context information, thus can improve accuracy of segmentation. Due to the small size of training set, we use 2D CNN architecture in this experiment.

3   Methods

In this paper, our network has a main network and three sub-networks. We use U-net with the residual block as the backbone to perform different brain tumor segmentation tasks, i.e.segmentation of GD-enhancing tumor, the peritumoral edema, or the necrotic and non-enhancing tumor core. And use deep supervision method on sub-networks to realize multi-brunch architecture with different labels. We following introduce the main structure and deep supervision of the network respectively.

3.1   Main Network

As can be seen in Fig. 3, the input of network includes four slices, which belong to four modalities respectively, so that the network can obtain richer information. The encoder includes input, four down-sampling layers and 2, 2, 2, 3 residual blocks after each down-sampling layer. Among them, each residual block contains two sub-blocks, which are composed of a layer normalization, a relu activation function and a 3×3 2D convolutional layer. In the down-sampling stage, we use 3×3 2D convolutional layer with strides of 2 for operations. Basically, the input slices first pass through a 2D convolutional layer with relu activation function

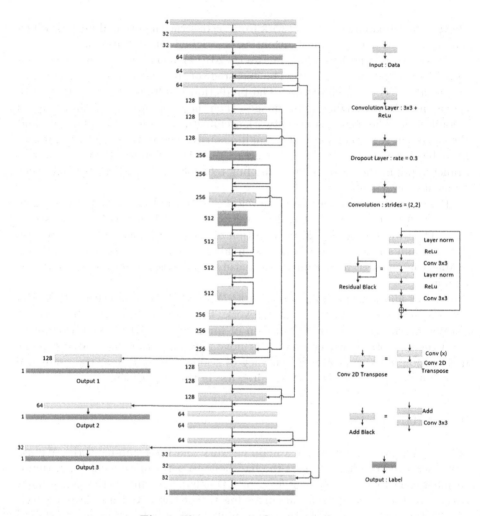

Fig. 3. The structure of our network.

and a dropout layer with dropout rate of 0.3, then enter followed residual blocks and down-sampling layers successively.

The decoder includes transpose blocks, add blocks and residual blocks. The transpose block is used for up-sampling. It includes a 1×1 2D convolution and a 3×3 deconvolution layer with strides of 2. The 1×1 2D convolution is used for dimensionality reduction, and the deconvolution layer is used to restore the size of feature map. The add block includes an add layer and a 3×3 2D convolution layer for the fusion of low-dimensional and high-dimensional features. Every time an up-sampling and add operation is over, a residual block will be used for further feature extraction, and there are 3 residual blocks in decoder. At different stages of decoder, we match different labels for the three sub-networks to obtain different types of segmentation results. In this way, the encoder can learn more diverse features. For the main network, we set the corresponding label for the required type of segmentation, so as to we can obtain the corresponding segmentation result.

3.2 Deep Supervision Method

We use the relevance between three types of labels to reduce the mutual misjudgment between multiple tumor regions, thus we can get better segmentation result of a single label. Therefore, for three different labels, we add three additional sub-networks for segmentation. We elicit a branch before each transpose block as decoder of a sub-network. The decoder of the first sub-network includes three transpose blocks, the decoder of the second sub-network includes two transpose blocks, and the third sub-network has only one transpose block. Multiple networks share a same encoder, it can increase the demand for features of encoder and enable the encoder to obtain richer features.

For different segmentation tasks of the main network, the tasks of three sub-networks are variable. For example, when our main network task is to segment enhancing tumor, the label of the first sub-network is necrotic and non-enhancing tumor core, the label of the second sub-network is peritumoral edema, and the label of the third sub-network is GD-enhancing tumor. In this way, the task of each sub-network depends on the segmentation task of main network. We need to choose different solutions according to different tasks.

3.3 Loss Functions

Due to the imbalance between the tumor area and the background in the brain tumor segmentation task, the target tumor area only occupies a small part of the entire slice. So cross-entropy loss function always pay more attention to background area. To solve this problem, we use the combination of binary cross-entropy and dice loss [23] as our loss function.

Among them, dice loss is used to solve the problem of class imbalance, it is expressed as follows:

$$L_{dice} = 1 - \frac{2 \, | \, y \cap \hat{y} \, |}{| \, y \, | \cup | \, \hat{y} \, |} \tag{1}$$

where y is the ground truth and $\hat{y}$ is the prediction.

And binary cross-entropy loss function is as follows:

$$BCE = -\frac{1}{2} \sum_{i=0}^{n} (y_i \log \hat{y}_i + (1 - y_i) \log(1 - \hat{y}_i)) \tag{2}$$

where n is the number of categories, y_i is the ground truth and $\hat{y}_i$ is the prediction.

Therefore, the total loss of our network is described as follows:

$$L_{total} = L_{dice} + BCE_{weight} * BCE \tag{3}$$

where $BCE_{weights}$ is the weight of BCE in the total loss, which is set to 0.5.

4   Experiments

In this section, we introduce our pre-processing method, post-processing method and some experimental details.

4.1   Pre-processing

The data of each patient contains slices of four modalities. On a axial, each modal has 155 slices with a size of 240×240 pixels. In the process of pre-processing, by dividing by the maximum pixel value of 155 slices, we normalize the pixel values of all slices to 0–1. Then, so as to obtain more training data, we flip the slices and therefore get the same number of data as original.

4.2   Post-processing

In order to reduce the influence of false positive on prediction maps, we identify the relative positions of GD-enhancing tumor, the necrotic and non-enhancing tumor core and peritumoral edema in the segmentation results. Then we delete those pixels that belong to the GD-enhancing tumor but are outside of the peritumoral edema. It can eliminate some theoretically impossible pixels.

4.3   Training Details

We use the Adam optimizer and train for 60 epochs. The initial learning rate is set to $1e^{-4}$. Then it becomes $2e^{-5}$ when reach to the 10th epoch. And when the 20th epoch is reached, the learning rate is reduced to $1e^{-5}$. After 30 epochs, the learning rate remains at $2e^{-6}$.

5   Results

We train our network on BraTS 2020 training set. After data enhancement, we use 520 cases as the training set, 115 cases as the validation set and 115 cases as the test set. Then we adjust the network parameters through the performance of the test set. After that we make predictions on the validation dataset provided by BraTS 2020, and finally we submit them to the online evaluation platform to evaluate the segmentation results. Finally, the average Dice we get on ET, WT, and TC are 0.7040, 0.8794 and 0.7731. And the median of our Dice scores on ET, WT and TC are 0.8350, 0.9101, and 0.8642 respectively. The performances are shown in Table 1.

The results of test set are presented in Table 2. We can see that the results of the test set are slightly higher than of the validation set.

Table 1. Performances on BraTS 2020 validation dataset

Validation dataset	Dice_mean			Dice_median		
	ET	WT	TC	ET	WT	TC
Our method	0.7040	0.8794	0.7731	0.8350	0.9101	0.8642

Table 2. Performances on BraTS 2020 test dataset

Test dataset	Dice_mean			Dice_median		
	ET	WT	TC	ET	WT	TC
Our method	0.7593	0.8726	0.7879	0.8326	0.9062	0.8909

6 Conclusion

In this paper, we propose a CNN network that uses residual blocks and deep supervision method. There are two key points in this paper. Firstly, the shortcut in the residual blocks can effectively alleviate the problem of gradient dispersion, so that the network can capture high-level visual features. Secondly, deep supervision method helps to improve the training stability of deeper networks, and enables the encoder to obtain richer features. Currently, our method is dedicated to 2D segmentation based on a single axis. However, due to the lack of spatial information in 2D data, it cannot effectively segment brain tumors with spatial correlation. So in the future work, we will apply the above method to 3D networks to improve segmentation performance.

References

1. McKinley, R., Rebsamen, M., Meier, R., Wiest, R.: Triplanar ensemble of 3D-to-2D CNNs with label-uncertainty for brain tumor segmentation. In: Crimi, A., Bakas, S. (eds.) BrainLes 2019. LNCS, vol. 11992, pp. 379–387. Springer, Cham (2020). https://doi.org/10.1007/978-3-030-46640-4_36
2. Zhao, Y.-X., Zhang, Y.-M., Liu, C.-L.: Bag of tricks for 3D MRI brain tumor segmentation. In: Crimi, A., Bakas, S. (eds.) BrainLes 2019. LNCS, vol. 11992, pp. 210–220. Springer, Cham (2020). https://doi.org/10.1007/978-3-030-46640-4_20
3. Zhou, C., Chen, S., Ding, C., Tao, D.: Learning contexualand attentive information for brain tumor segmentation. In: Pre-conference Proceedings of the 2018 7th MICCAI BraTS Challenge, pp. 571–578 (2018)
4. Menze, B.H., Jakab, A., Bauer, S., Kalpathy-Cramer, J., Farahani, K., Kirby, J., et al.: The multimodal brain tumor image segmentation benchmark (BRATS). IEEE Trans. Med. Imaging **34**(10), 1993–2024 (2015). https://doi.org/10.1109/TMI.2014.2377694
5. Bakas, S., Akbari, H., Sotiras, A., Bilello, M., Rozycki, M., Kirby, J.S., et al.: Advancing the cancer genome atlas glioma mri collections with expert segmentation labels and radiomic features. Nat. Sci. Data **4**, 170117 (2017). https://doi.org/10.1038/sdata.2017.117

6. Bakas, S., Reyes, M., Jakab, A., Bauer, S., Rempfler, M., Crimi, A., et al.: Identifying the best machine learning algorithms for brain tumor segmentation, progression assessment, and overall survival prediction in the BRATS challenge. arXiv preprint arXiv:1811.02629 (2018)

7. Bakas, S., Akbari, H., Sotiras, A., Bilello, M., Rozycki, M., Kirby, J., et al.: Segmentation labels and radiomic features for the pre-operative scans of the TCGA-GBM collection. The Cancer Imaging Archive (2017). https://doi.org/10.7937/K9/TCIA.2017.KLXWJJ1Q

8. Bakas, S., Akbari, H., Sotiras, A., Bilello, M., Rozycki, M., Kirby, J., et al.: Segmentation labels and radiomic features for the pre-operative scans of the TCGA-LGG collection. The Cancer Imaging Archive (2017). https://doi.org/10.7937/K9/TCIA.2017.GJQ7R0EF

9. Ronneberger, O., Fischer, P., Brox, T.: U-Net: convolutional networks for biomedical image segmentation. In: Navab, N., Hornegger, J., Wells, W.M., Frangi, A.F. (eds.) MICCAI 2015. LNCS, vol. 9351, pp. 234–241. Springer, Cham (2015). https://doi.org/10.1007/978-3-319-24574-4_28

10. Isensee, F., Kickingereder, P., Wick, W., Bendszus, M., Maierhein, K.H.: Brain tumor segmentation and radiomics survival prediction: Contribution to the BRATS 2017 challenge. In: 2017 Proceedings of the 6th MICCAI BraTS Challenge, pp. 100–107 (2017)

11. Isensee, F., Kickingereder, P., Wick, W., Bendszus, M., Maierhein, K.H.: No newnet. In: 2018 Pre-conference Proceedings of the 7th MICCAI BraTS Challenge, pp. 222–231 (2018)

12. He, K., Zhang, X., Ren, S., Sun, J.: Deep residual learning for image recognition. In: 2016 Proceedings of the IEEE Conference on Computer Vision and Pattern Recognition, pp. 770–778 (2016)

13. Zhou, Z.W., Siddiquee, M.M.R., Tajbakhsh, N., Liang, J.M.: UNet++: a nested U-Net architecture for medical image segmentation. In: Deep Learning in Medical Image Anylysis and Multimodal Learning for Clinical Decision Support, pp. 3–11 (2018)

14. Huang, G., Liu, Z., Van Der Maaten, L., Weinberger, K.Q.: Densely connected convolutional networks. In: 2017 Proceedings of the 30th IEEE Conference on Computer Vision and Pattern Recognition, CVPR 2017, pp. 2261–2269 (2017)

15. Zhang, X., Jian, W., Cheng, K.: 3D dense U-nets for brain tumor segmentation. In: 2018 Pre-conference Proceedings of the 7th MICCAI BraTS Challenge, pp. 562–570 (2018)

16. Stawiaski, J.: Leveraging a DenseNet encoder pre-trained on ImageNet for brain tumor segmentation. In: 2018 Pre-conference Proceedings of the 7th MICCAI BraTS Challenge, pp. 438–447 (2018)

17. Mckinley, R., Meier, R., Wiest, R.: Ensembles of densely-connected CNNs with label-uncertainty for brain tumor segmentation. In: 2018 Pre-conference Proceedings 7th MICCAI BraTS Challenge, pp. 322–330 (2018)

18. Myronenko, A.: 3D MRI brain tumor segmentation using autoencoder regularization. In: Crimi, A., Bakas, S., Kuijf, H., Keyvan, F., Reyes, M., van Walsum, T. (eds.) BrainLes 2018. LNCS, vol. 11384, pp. 311–320. Springer, Cham (2019). https://doi.org/10.1007/978-3-030-11726-9_28

19. Weninger, L., Liu, Q., Merhof, D.: Multi-task learning for brain tumor segmentation. In: Crimi, A., Bakas, S. (eds.) BrainLes 2019. LNCS, vol. 11992, pp. 327–337. Springer, Cham (2020). https://doi.org/10.1007/978-3-030-46640-4_31

20. Zhou, C., Ding, C., Lu, Z., Wang, X., Tao, D.: One-pass multi-task convolutional neural networks for efficient brain tumor segmentation. In: Frangi, A.F., Schnabel, J.A., Davatzikos, C., Alberola-López, C., Fichtinger, G. (eds.) MICCAI 2018. LNCS, vol. 11072, pp. 637–645. Springer, Cham (2018). https://doi.org/10.1007/978-3-030-00931-1_73
21. Chen, S., Bortsova, G., García-Uceda Juárez, A., van Tulder, G., de Bruijne, M.: Multi-task attention-based semi-supervised learning for medical image segmentation. In: Shen, D., et al. (eds.) MICCAI 2019. LNCS, vol. 11766, pp. 457–465. Springer, Cham (2019). https://doi.org/10.1007/978-3-030-32248-9_51
22. Jiang, Z., Ding, C., Liu, M., Tao, D.: Two-stage cascaded U-Net: 1st place solution to BraTS challenge 2019 segmentation task. In: Crimi, A., Bakas, S. (eds.) BrainLes 2019. LNCS, vol. 11992, pp. 231–241. Springer, Cham (2020). https://doi.org/10.1007/978-3-030-46640-4_22
23. Milletari, F., Navab, N., Ahmadi, S.-A.: V-net: fully convolutional neural networks for volumetric medical image segmentation. In: 2016 4th International Conference on 3D Vision (3DV). IEEE (2016)

Multi-threshold Attention U-Net (MTAU) Based Model for Multimodal Brain Tumor Segmentation in MRI Scans

Navchetan Awasthi[1,2]([✉]) [ID], Rohit Pardasani[3] [ID], and Swati Gupta[4] [ID]

[1] Massachusetts General Hospital, Boston, USA
[2] Harvard University, Cambridge, USA
[3] General Electric Healthcare, Bangalore, India
[4] Indian Institute of Science, Bangalore, India

Abstract. Gliomas are one of the most frequent brain tumors and are classified into high grade and low grade gliomas. The segmentation of various regions such as tumor core, enhancing tumor etc. plays an important role in determining severity and prognosis. Here, we have developed a multi-threshold model based on attention U-Net for identification of various regions of the tumor in magnetic resonance imaging (MRI). We propose a multi-path segmentation and built three separate models for the different regions of interest. The proposed model achieved mean Dice Coefficient of 0.59, 0.72, and 0.61 for enhancing tumor, whole tumor and tumor core respectively on the training dataset. The same model gave mean Dice Coefficient of 0.57, 0.73, and 0.61 on the validation dataset and 0.59, 0.72, and 0.57 on the test dataset.

Keywords: Attention U-Net · Brain tumor · Segmentation · Gliomas · Multi-threshold

1 Introduction

In adults, glioma is one of the frequent brain tumor originating from glial cells and infiltrating the surrounding tissue [17]. The gliomas are classified into two main categories depending on the severity of gliomas -

- High grade gliomas require immediate treatment and have median survival of two years or less [8,9,11,18].
- Low grade gliomas have life expectancy of several years, hence aggressive treatment is often delayed [8,9,11,18].

N. Awasthi and R. Pardasani—The authors contributed equally to the work.

© Springer Nature Switzerland AG 2021
A. Crimi and S. Bakas (Eds.): BrainLes 2020, LNCS 12659, pp. 168–178, 2021.
https://doi.org/10.1007/978-3-030-72087-2_15

Various neuroimaging protocols are used for evaluating the progression of the disease before and after treatment to determine the success of the treatment strategy as well as any changes in the brain. In clinical settings, these are evaluated on the basis of qualitative criteria or by quantitative measures [13,21].

The automatic analysis of the tumor structures will produce highly accurate and reproducible measurements of the relevant structures, thus these image processing routines are of utmost importance. Since the amount of data available for processing is huge, there is a need of an automatic method for segmenting the various regions of the brain. The analysis will help in improving the treatment planning, improving the diagnosis and follow-up of individual patients for further procedures [6,7,9,17]. The development of the automatic procedures for segmentation is not an easy task as the lesions are defined in terms of intensity changes compared to the surrounding tissues and even the segmentation done by expert radiologist show significant variations because of the partial volume effects, bias field artifacts and intensity gradients between the adjacent structures [17]. The tumor structures also vary across patients in terms of extension, localization, size and thus affecting the use of strong priors for the segmentation of the anatomical structures. The modalities provide different complementary information and thus different features are involved in segmentation [8,17]. The imaging modalities that are used to map tumor-induced tissue changes include T2 and Fluid-Attenuated Inversion Recovery (FLAIR) MRI, post-Gadolinium T1 MRI, perfusion and diffusion MRI, and Magnetic Resonance Spectroscopic Imaging (MRSI), among others [8,9,17].

Previously, many models have been proposed for improving the reconstruction and improving the segmentation such as U-Net, U-Net++ etc. [20,22]. Recently, there has been focus on modified U-Net architectures such as attention-U-Net as well as hybrid U-Net architectures for image reconstruction as well as image super-resolution based enhancement [3–5,20]. Similarly, many architectures have been developed using texture analysis, active contours, random forests as well as probabilistic models for tumor segmentation [14].

Here, we have proposed a Multi-Threshold Attention U-Net (MTAU) [19] based 2D model for segmentation of the multimodal brain tumor images of MRI scans into three different regions (Necrotic (NCR) and the Non-Enhancing Tumor (NET), Enhancing Tumor (ET) and Edema (ED)) by individually training three different models. The three models use identical architecture, thus we reduce the memory requirement (the 2D model being less complex than a 3D one and has lesser number of parameters) at the time of training/inference without increasing the effort of model design. Alternately, in the absence of memory constraint we can use same models in a multi-path architecture format where 2D U-Net models get trained in parallel.

2   Methods

We started by selecting an architecture for 2D Attention U-Net model [19] followed by training this architecture for three different tasks (or regions) and finally ended up with three different models, one for each region viz. NCR+NET, ET and edema. We can also use the same architecture in multi-path format by stacking and training three models in parallel, if there are no constraints on GPU memory. The model architecture is shown in Fig. 1 with the various connections along with actual parameters involved in each layer. This architecture was replicated and trained for each of the three regions as shown in Fig. 2. All three attention U-Nets are independent of each other, hence we can chose a separate threshold for each region. The purpose of threshold in each model is to binarize the output values to 0 and 1. Training separate model for each region gives the flexibility of choosing a separate threshold based on Area Under Curve (AUC) of the respective outputs and thus the model is termed as 'Multi-Threshold Attention U-Net (MTAU)'.

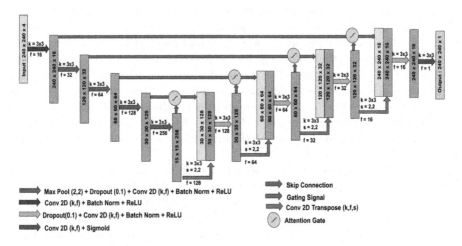

Fig. 1. Architecture of the proposed attention U-Net model utilized in this work. The various size of filters and the corresponding connection are shown in the architecture and the binary cross entropy loss has been used for back-propagation.

2.1 Dataset

Our proposed technique was trained on BraTS 2020 training dataset [8,9]. The training dataset consists of 369 scans while the validation dataset consists of 125 MRI scans of glioblastoma (GBM/HGG) and lower grade glioma (LGG). Also, the dataset has been updated from last year with more routine clinically-acquired 3T multimodal MRI scans and accompanying ground truth labels by expert board-certified neuroradiologists. The various sub -regions considered for the segmentation evaluation are: 1) the "enhancing tumor" (ET), 2) the "tumor core" (TC), and 3) the "whole tumor" (WT). The ET area is shown in hyper-intensity in T1Gd as compared to T1, and also as compared to "healthy" white matter in T1Gd. Here, TC denotes the bulk of the tumor entailing the ET, necrotic (fluid-filled) and the non-enhancing (solid) parts in the tumor. The appearance of the necrotic (NCR) and the non-enhancing tumor core is typically hypo-intense in T1-Gd as compared to T1. The WT describes the full extent of the pathology by entailing the TC and the peritumoral edema (ED), which is typically shown by hyper-intense signal in FLAIR. The provided segmentation for the dataset have values of 1 for NCR and NET, 2 for ED, 4 for ET, and 0 representing everything else [8,9]. An example showing the various regions of interest with the actual segmentation region can be seen in Fig. 3.

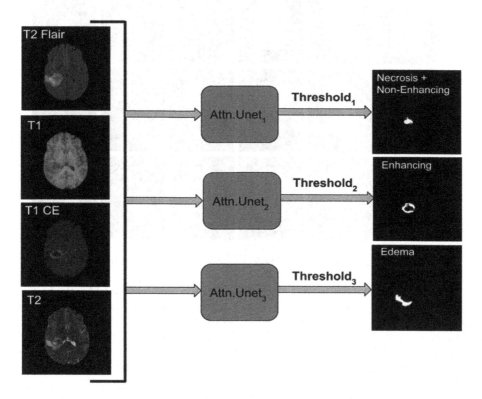

Fig. 2. Diagram of the proposed Multi-Threshold Attention U-Net (MTAU) model utilized in this work. Each of the attention U-Nets are independent of each other and their optimally chosen thresholds are numbered as 1, 2 and 3 respectively.

2.2 Preprocessing

The data is first processed to reduce in-homogeneity by scaling each scan volume matrix (of a particular modality) with maximum value in that matrix. Since our approach is 2D segmentation, we create dataset by extracting slices from the volumes. Each volume matrix is sliced into 2D gray scale images. A total of

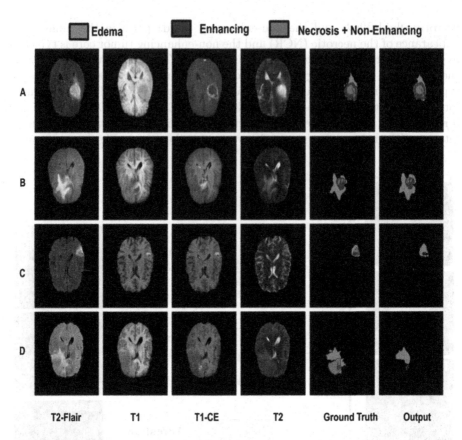

Fig. 3. The segmentation results obtained utilizing the proposed model for the best as well as below average results for the training dataset. Every row represents a different patient data. The columns represent the T2-FLAIR, T1, T1-CE, T2, Ground truth and the Output. The segmentation labels are: Green for Edema, Blue for Enhancing Tumor and Red for (Necrosis + Non-Enhancing). (A) and (B) represents the best segmentation outputs while (C) and (D) represents the below average segmentation obtained using the proposed model. (Color figure online)

57,195 (=369 vols × 155 slices per vol) sets of slices were extracted from the training volumes. Each 2D set comprised of 4 input slices and 3 segmentation maps. The gray scale slices having only zeros in inputs and segmentation maps were removed from the training set. We did this to ensure that results during training do not give a false impression of accuracy while actually model is mapping zeros to zeros. After removal of such slices we were left with 50,899 sets that we split into training, validation, and testing set comprising 40000, 5000, and 5899 sets respectively. The slices of training, validation and test set were taken from different volumes to ensure that model does not over-fit on the given dataset. It is pertinent to mention again that the training, validation and testing set created here were from the 369 training volumes for which ground truth

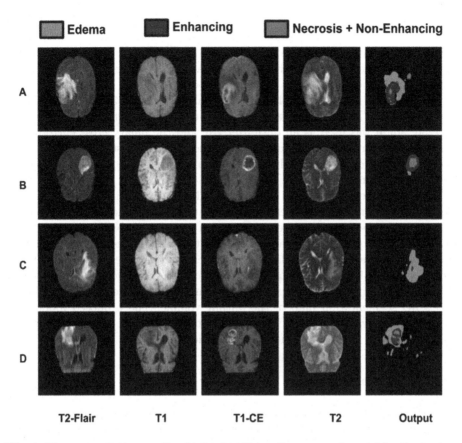

Fig. 4. The segmentation results obtained utilizing the proposed model for the best as well as below average results for the validation dataset. Every row represents a different patient data. The columns represent the T2-FLAIR, T1, T1-CE, T2, Ground truth and the Output. The segmentation labels are: Green for Edema, Blue for Enhancing Tumor and Red for (Necrosis + Non-Enhancing). (A) and (B) represents the best segmentation outputs while (C) and (D) represents the below average segmentation obtained using the proposed model. (Color figure online)

was available. These sets were used for training, validating and testing at our level internally. Apart from this, there is a separate validation set of 125 volumes given by challenge organizers for which ground truth was not provided. The segmentation map on these validation volumes were uploaded on portal to get the performances. The results shown in Table 1 and Table 2 correspond to 369 training volumes and 125 validation volumes as referred by the challenge organizers.

2.3   Model Architecture

The model architecture utilized in this work is shown in Fig. 1. It consists of the U-Net architecture where the attention gates are used to give more focus to the area of segmentation. The total number of parameters in the network are 3,125,013 consisting of 3,121,141 trainable parameters and 3,872 non-trainable parameters. The learning rate was set to be $1e - 5$ while the number of epochs were 50, adam was used as an optimizer [16] and the binary cross entropy loss has been used for back-propagation. The model with the best validation set was saved and used for further analysis. The deep learning model was implemented in Keras [12] using Tensorflow [1] as backend.

As discussed earlier, instead of using a single model we utilized three different models for training as well as testing of the dataset. We made three models with same architecture for each of the segmentation masks (1 for NCR and NET, 2 for ED, 4 for ET) provided in the dataset. Each of the model is trained independently and then the three models are saved for further analysis. These three models are then used for inference and the results are combined to get the final output of the three segmentation maps. The resulting outputs are further combined into a 3D segmentation map of the output volume. The proposed model has lesser complexity, requires lesser memory as compared to a 3D model and can be trained much faster if the models are trained in parallel.

3   Figures of Merit

The reliability of the proposed model or method should be checked by means of parameters that can quantify the accuracy and validity of the test results and the proposed model. Following are commonly used metrics in computer vision utilised in this work.

3.1   Dice Coefficient (DSC)

This parameter is used to calculate similarity between two sample sets. It is also known as Sørensen–Dice index or Dice Similarity Coefficient and calculated as [10]:

$$DSC = \frac{2 * |X \cap Y|}{|X| + |Y|}$$

where X and Y are the two samples/sample sets. $|X|$ and $|Y|$ represent the cardinalities of set X and Y.

3.2   Sensitivity (SN)

This value gives the True positive rates reported by a test. It tells the number of times the test gave a positive result when the sample/person indeed had a condition [2]. It is calculated as:

$$SN = \frac{TP}{TP + FN}$$

where TP are the number of true positives and FN are the number of false negatives.

3.3 Specificity (SP)

This value gives the True negative rates reported by a test. It tells the proportion of samples reported negative by the test which are indeed negative [2]. Specificity is calculated as:

$$SP = \frac{TN}{TN + FN}$$

where TN are the number of true negatives and FN are the number of false negatives.

3.4 Hausdorff Distance (h)

This value is a measure of distance between two sample sets. It is the maximum distance of a set from the nearest point in the other set [15]. Hausdorff Distance is calculated as:

$$h(X,Y) = max_{x \epsilon X} min_{y \epsilon Y} \|x - y\|$$

where X = $\{x_1, x_2, ..., x_n\}$ and Y = $\{ y_1, y_2, ...y_n\}$ are sample sets.

Table 1. Performance of proposed method on BraTS 2020 training dataset for segmentation

Evaluation metrics	Dice			Sensitivity			Specificity			Hausdorff		
	ET	WT	TC	ET	WT	TC	ET	WT	TC	ET	WT	TC
Mean	0.59	0.72	0.61	0.52	0.72	0.63	0.99	0.99	0.99	38.87	20.81	24.22
Std. Dev.	0.30	0.19	0.25	0.31	0.24	0.27	0.00	0.00	0.00	100.75	20.02	28.35
Median	0.70	0.78	0.67	0.57	0.80	0.71	0.99	0.99	0.99	4.33	11.70	14.09
25 Quantile	0.41	0.66	0.46	0.28	0.59	0.47	0.99	0.99	0.99	2.23	7.48	8.17
75 Quantile	0.83	0.86	0.82	0.79	0.93	0.85	0.99	0.99	0.99	12.36	29.51	31.05

Table 2. Performance of proposed method on BraTS 2020 validation dataset for segmentation

Evaluation metrics	Dice			Sensitivity			Specificity			Hausdorff		
	ET	WT	TC	ET	WT	TC	ET	WT	TC	ET	WT	TC
Mean	0.57	0.73	0.61	0.52	0.77	0.62	0.99	0.99	0.99	47.22	24.03	31.53
Std. Dev.	0.33	0.17	0.26	0.34	0.23	0.27	0.00	0.00	0.00	108.70	22.81	41.07
Median	0.71	0.78	0.69	0.60	0.86	0.70	0.99	0.99	0.99	4.12	13.74	15.84
25 Quantile	0.30	0.67	0.38	0.18	0.62	0.47	0.99	0.99	0.99	2.23	7.00	9.69
75 Quantile	0.84	0.84	0.82	0.81	0.95	0.83	0.99	0.99	0.99	15.66	35.76	46.79

Table 3. Performance of proposed method on BraTS 2020 testing dataset for segmentation

Evaluation metrics	Dice			Sensitivity			Specificity			Hausdorff		
	ET	WT	TC	ET	WT	TC	ET	WT	TC	ET	WT	TC
Mean	0.59	0.72	0.57	0.57	0.76	0.60	0.99	0.99	0.99	38.53	22.73	41.82
Std. Dev.	0.29	0.18	0.27	0.30	0.20	0.29	0.00	0.00	0.00	102.68	23.44	78.27
Median	0.70	0.79	0.63	0.57	0.81	0.71	0.99	0.99	0.99	4.30	10.00	16.99
25 Quantile	0.46	0.66	0.47	0.31	0.68	0.40	0.99	0.99	0.99	2.23	5.93	8.12
75 Quantile	0.82	0.85	0.79	0.80	0.93	0.84	0.99	0.99	0.99	10.28	33.56	45.40

4 Results and Discussions

The performance of the proposed model was tested on the training, validation as well as testing dataset. The training dataset consists of 369 scans while the validation dataset consists of 125 MRI scans of glioblastoma (GBM/HGG) and lower grade glioma (LGG). All computations were carried out on a Linux workstation with Intel Xeon Silver 4110 CPU with 2.10 GHz clock speed, having 128 GB RAM and a TITAN RTX GPU with 24 GB memory. The results are obtained using the proposed model for the training, validation as well as testing dataset. Figure 3 shows the segmentation results for training dataset utilized for training the model. Figure 3(A) and (B) represents the best cases while Fig. 3(C) and (D) represents the below average results obtained on the training dataset obtained using the proposed segmentation model. The results obtained for the complete training dataset are provided in Table 1. The dice value obtained for the enhancing tumor, whole tumor and tumor core were found to be 0.59, 0.72, and 0.61 respectively. The values of Std. Dev., Median, 25 Quantile, and 75 Quantile are also provided in Table 1. The values of sensitivity, specificity, and Hausdorff distance are also calculated and are shown in Table 1.

Figure 4 shows the segmentation results for validation dataset utilized for validation the model. Figure 4(A) and (B) represents the best results while Fig. 4(C) and (D) represents the below average results obtained on the validation dataset using the proposed segmentation model. The results obtained for the complete validation dataset are provided in Table 2. The dice value obtained for the enhancing tumor, whole tumor and tumor core were found to be 0.57, 0.73, and 0.61 respectively. The values of Std. Dev., Median, 25 Quantile, and 75 Quantile are also provided in Table 2. The values of sensitivity, specificity, and Hausdorff distance are also calculated and are shown in Table 2.

The results obtained for the complete testing dataset are provided in Table 3. The dice value obtained for the enhancing tumor, whole tumor and tumor core were found to be 0.59, 0.72, and 0.57 respectively. The values of Std. Dev., Median, 25 Quantile, and 75 Quantile are also provided in Table 3. The values of sensitivity, specificity, and Hausdorff distance are also calculated and are shown in Table 3. The training, validation, as well as testing dataset outputs are obtained utilizing the online portal for submission of the results.

5   Conclusions

A multi-threshold model was developed based on attention U-Net for identification of various regions of the tumor in MRI scans. The proposed model offers the advantages of reduced computational complexity, less memory requirements as well as less training time if trained in parallel. The proposed model achieved mean Dice Coefficient of 0.59, 0.72, and 0.61 for enhancing tumor, whole tumor and tumor core respectively on the training dataset. The same model gave mean Dice Coefficient of 0.57, 0.73, and 0.61 on the validation dataset and 0.59, 0.72, and 0.57 respectively on the testing dataset.

References

1. Abadi, M., et al.: Tensorflow: Large-scale machine learning on heterogeneous distributed systems. arXiv preprint arXiv:1603.04467 (2016)
2. Altman, D.G., Bland, J.M.: Diagnostic tests. 1: sensitivity and specificity. BMJ: Br. Med. J. **308**(6943), 1552 (1994)
3. Awasthi, N., Jain, G., Kalva, S.K., Pramanik, M., Yalavarthy, P.K.: Deep neural network based sinogram super-resolution and bandwidth enhancement for limited-data photoacoustic tomography. IEEE Trans. Ultrason. Ferroelectr. Freq. Control **67**(12), 2660–2673 (2020)
4. Awasthi, N., Pardasani, R., Kalva, S.K., Pramanik, M., Yalavarthy, P.K.: Sinogram super-resolution and denoising convolutional neural network (SRCN) for limited data photoacoustic tomography. arXiv preprint arXiv:2001.06434 (2020)
5. Awasthi, N., Prabhakar, K.R., Kalva, S.K., Pramanik, M., Babu, R.V., Yalavarthy, P.K.: PA-Fuse: deep supervised approach for the fusion of photoacoustic images with distinct reconstruction characteristics. Biomed. Opt. Express **10**(5), 2227–2243 (2019)
6. Bakas, S., et al.: Segmentation labels and radiomic features for the pre-operative scans of the TCGA-LGG collection. Cancer Imaging Arch. **286** (2017)
7. Bakas, S., et al.: Segmentation labels and radiomic features for the pre-operative scans of the TCGA-GBM collection. Cancer Imaging Arch. Nat. Sci. Data **4**, 170117 (2017)
8. Bakas, S., et al.: Advancing the cancer genome atlas glioma MRI collections with expert segmentation labels and radiomic features. Sci. Data **4**, 170117 (2017)
9. Bakas, S., et al.: Identifying the best machine learning algorithms for brain tumor segmentation, progression assessment, and overall survival prediction in the brats challenge. ArXiv Preprint ArXiv:1811.02629 (2018)
10. Carass, A., et al.: Evaluating white matter lesion segmentations with refined sørensen-dice analysis. Sci. Rep. **10**(1), 1–19 (2020)
11. Cavenee, W.K., Louis, D.N., Ohgaki, H., Wiestler, O.D.: WHO Classification of Tumours of the Central Nervous System, vol. 1. WHO Regional Office Europe (2007)
12. Chollet, F., et al.: Keras: deep learning library for theano and tensorflow. **7**(8) (2015). https://keras.io/k
13. Eisenhauer, E.A., et al.: New response evaluation criteria in solid tumours: revised recist guideline (version 1.1). Eur. J. Cancer **45**(2), 228–247 (2009)
14. Gordillo, N., Montseny, E., Sobrevilla, P.: State of the art survey on MRI brain tumor segmentation. Magn. Reson.Imaging **31**(8), 1426–1438 (2013)

15. Huttenlocher, D.P., Klanderman, G.A., Rucklidge, W.J.: Comparing images using the hausdorff distance. IEEE Trans. Pattern Anal. Mach. Intell. **15**(9), 850–863 (1993)
16. Kingma, D.P., Ba, J.: Adam: A method for stochastic optimization. ArXiv Preprint ArXiv:1412.6980 (2014)
17. Menze, B.H., et al.: The multimodal brain tumor image segmentation benchmark (brats). IEEE Trans. Med. Imaging **34**(10), 1993–2024 (2014)
18. Ohgaki, H., Kleihues, P.: Population-based studies on incidence, survival rates, and genetic alterations in astrocytic and oligodendroglial gliomas. J. Neuropathol. Exp. Neurol. **64**(6), 479–489 (2005)
19. Oktay, O., et al.: Attention u-net: learning where to look for the pancreas. ArXiv Preprint ArXiv:1804.03999 (2018)
20. Ronneberger, O., Fischer, P., Brox, T.: U-Net: convolutional networks for biomedical image segmentation. In: Navab, N., Hornegger, J., Wells, W.M., Frangi, A.F. (eds.) MICCAI 2015, Part III. LNCS, vol. 9351, pp. 234–241. Springer, Cham (2015). https://doi.org/10.1007/978-3-319-24574-4_28
21. Wen, P.Y., et al.: Updated response assessment criteria for high-grade gliomas: response assessment in neuro-oncology working group. J. Clin. Oncol. **28**(11), 1963–1972 (2010)
22. Zhou, Z., Rahman Siddiquee, M.M., Tajbakhsh, N., Liang, J.: UNet++: a nested U-net architecture for medical image segmentation. In: Stoyanov, D., et al. (eds.) DLMIA/ML-CDS-2018. LNCS, vol. 11045, pp. 3–11. Springer, Cham (2018). https://doi.org/10.1007/978-3-030-00889-5_1

Multi-stage Deep Layer Aggregation for Brain Tumor Segmentation

Carlos A. Silva$^{(\boxtimes)}$, Adriano Pinto, Sérgio Pereira, and Ana Lopes

Center MEMS of University of Minho, Campus of Azurém,
4800-058 Guimarães, Portugal
csilva@dei.uminho.pt

Abstract. Gliomas are among the most aggressive and deadly brain tumors. This paper details the proposed Deep Neural Network architecture for brain tumor segmentation from Magnetic Resonance Images. The architecture consists of a cascade of three Deep Layer Aggregation neural networks, where each stage elaborates the response using the feature maps and the probabilities of the previous stage, and the MRI channels as inputs. The neuroimaging data are part of the publicly available Brain Tumor Segmentation (BraTS) 2020 challenge dataset, where we evaluated our proposal in the BraTS 2020 Validation and Test sets. In the Test set, the experimental results achieved a Dice score of 0.8858, 0.8297 and 0.7900, with an Hausdorff Distance of 5.32 mm, 22.32 mm and 20.44 mm for the whole tumor, core tumor and enhanced tumor, respectively.

Keywords: Brain tumor segmentation · Deep learning · Convolutional Neural Networks · Gaussian filters

1 Introduction

Gliomas present the highest mortality rate and incidence among brain tumors. They can be categorized according to their aggressiveness into two levels: High Grade Gliomas (HGGs) and Low Grade Gliomas (LGGs) [10]. Currently, multi-sequence Magnetic Resonance Imaging (MRI) is the best imaging modality to assess the structure and sub-regions of gliomas, due to their highly heterogeneity in appearance, shape, location, and tissues. MRI allows the volumetric characterization of the lesions, which requires semantic segmentation. However, expert manual segmentation is expensive, susceptible to inter- and intra-rater variability, and time-consuming [4, 10].

The heterogeneous nature of gliomas make automatic segmentation very challenging. So, in the past years, Convolutional Neural Networks (CNNs) that learn complex features directly from the data have achieved the best performances in the Brain Tumor Segmentation (BraTS) challenge [7, 8, 11, 12, 19]. Pereira *et al.* [12] proposed a point-wise segmentation approach using plain CNNs. Despite its success, following approaches [7, 8, 11, 13, 19] were mostly based on more efficient Fully Convolutional Network (FCN) principles [15], especially

© Springer Nature Switzerland AG 2021
A. Crimi and S. Bakas (Eds.): BrainLes 2020, LNCS 12659, pp. 179–188, 2021.
https://doi.org/10.1007/978-3-030-72087-2_16

inspired on the encoder-decoder U-Net architecture [14]. Kamnitsas *et al.* [8] ensembled several different models to tackle the bias-variance trade-off of the models. Myronenko [11] proposed a 3D FCN network for gliomas segmentation, but coupled an extra decoder based on Variational Auto-encoder with the purpose of regularizing the encoder. Zhao *et al.* [19], as the authors define it, proposed a bag of tricks to enhance training and inference. Some of the procedures include careful voxel sampling, multi-size patches, extra unlabeled data for semi-supervised learning, and multi-task learning. Jian *et al.* [7] ranked 1st in BraTS 2019 by developing a two-stage cascade FCN, where the second FCN is able to refine the results from the first. Inspired by [11], they also employ an extra regularization decoder with a simple interpolation scheme.

The U-Net [14] architecture includes long skip connections to merge low and high level features of the same scale. Therefore, these connections are shallow. Instead, the work on Deep Layer Aggregation (DLA) [17] proposes a principled iterative and hierarchical aggregation of layers from all scales. In this way, it enables the model to gather more spatial and semantic information for better segmentation refinement. In this paper, we evaluate the use of DLA blocks for brain tumor segmentation. To further refine the segmentation, we employ a three-stage cascade architecture, where the output of one FCN is directly fed to the next. This is an alternative to directly going deeper by down-sampling layers. The whole model is trained end-to-end.

The remaining of this paper is organized as follows. In Sect. 2, we present the proposed method. The data and experimental setup are described in Sect. 3. Then, in Sect. 4, the experimental results are presented, followed by the main conclusions in Sect. 5.

2   Methods

In this work, we address the problem of brain tumor segmentation using a 2D FCN. We propose a variant of the DLA architecture [17] as the core segmentation architecture. An overview of the proposed architecture is depicted in Fig. 1.

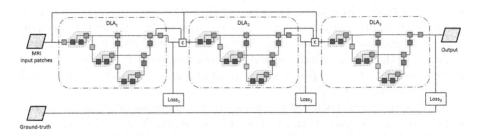

Fig. 1. Overview of the proposed Network. From one stage to another each module considers the MRI input patches alongside the respective feature maps of the last previous layer and the probability maps from the Softmax.

The first stage receives as input multi-sequence MRI images of size 120×120, generating a first rough segmentation of the brain tumor. The subsequent stages of the architecture receive the probability map and also the feature maps from the previous stage. These are combined with the raw input images to provide a more accurate segmentation of the brain tumor. The complete neural network is trained end-to-end, having a loss function for each stage.

2.1 Deep Layer Aggregation

Widely known deep learning-based architectures, e.g. U-Net [14] and FCN [15] consider the information from shallow layers by employing linear skip connections. However, this linear aggregation, *i.e.* the combination of different blocks of a network, restrains the possibility to refine features from shallow stages of the network. Deep layer aggregation (DLA) [17] extends over linear aggregation layers to better fuse across channels and depths (semantic fusion), and across resolutions and scales (spatial fusion). Considering more depth and sharing of features extracted from different stages of the network improves the overall inference.

Deep Layer Aggregation considers two deep aggregation blocks: Iterative Deep Aggregation (IDA) and Hierarchical Deep Aggregation (HDA). IDA focus on spatial fusion, increasing the spatial resolution while simultaneously elaborating over the input. HDA aims for semantic fusion, extending linear skip connections with a tree-based structure that spans the feature hierarchy over different depths. HDA preserves feature channels while, at the same time, combines them with the feature channels of the current depth [17].

Having the capability to better aggregate features from shallow levels of the network and further on refine them, poses as an interesting approach for brain tumor segmentation. With semantic fusion we aim to properly distinguish healthy from glioma tissue, while with spatial fusion we aim to properly locate it and distinguish the different types of glioma tissue.

Down-sampling is usually achieved through pooling or convolutional layers with stride of 2. However, these procedures may lead to aliasing that may impact the details of the segmentation [18]. Instead, we employ Max-pooling with kernel size of 2 and stride of 1, followed by Gaussian filtering with stride 2 and kernel size of 5×5 and a sigma of 1.25. As for the Up-sampling block, we employed Transpose Convolutions with a kernel size of 3×3 and a stride of 2×2, as opposed to typical up-sampling layers.

The DLA architecture is presented in Fig. 2, detailing as well the employed operational block.

2.2 Loss Function

The proposed architecture was trained with three auxiliary losses, one for each level. The equation of the complete loss is:

$$L = 0.3 \times \text{loss}_1 + 0.4 \times \text{loss}_2 + 1.0 \times \text{loss}_3, \tag{1}$$

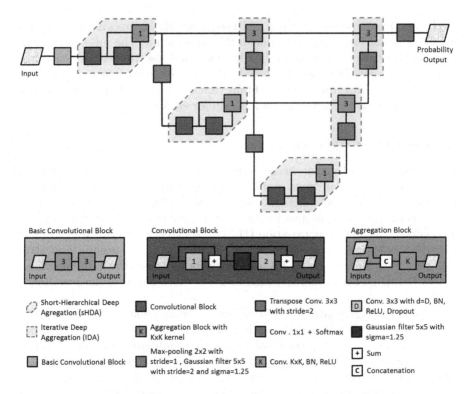

Fig. 2. Deep layer aggregation module, employed at different levels of our proposed architecture.

Each auxiliary loss considers the categorical cross entropy, defined as:

$$\mathcal{L}(y, \hat{y}) = -\sum_i y_i log(\hat{y}_i),\qquad(2)$$

where i indexes the class, $\hat{y}$ the probabilistic prediction and y the true class, when employing one-hot encoding. This loss function was also computed over brain tumorous region [6,19].

3   Experimental Setup

3.1   Data

The proposed methods were evaluated in the BraTS 2020 dataset [1–4,10], which consists of multi-institutional MRI acquisitions with both HGGs and LGGs. Additionally, the acquired MRI sequences follow the clinical practice, hence including 4 sequences: T1, contrast enhanced T1, T2, and FLAIR. The Training set includes 369 subjects with the corresponding manual segmentations. For development, we randomly split the images into training (295 patients), validation (37 patients), and test (37 patients). BRATS 2020 also includes an official Validation set with 125 subjects, where the manual annotations are not publicly available. Therefore, segmentation scores are blindly computed by an online platform.

3.2   Pre-processing and Data Augmentation

The dataset was already pre-processed, namely multi-sequence registration and skull stripping. Hence, we only performed intensity normalization by assuring that each patient brain volume had zero mean and unitary standard deviation, using its own statistics.

The data augmentation used, to further prevent overfitting and improve the generalization capacity of the proposal, encompasses random shifts $(-0.1–0.1)$ and scale $(0.9–1.1)$ to each input channel after patch standardization. Also, we apply bi-dimensional stochastic rotation and cutout [5].

3.3   Settings and Model Training

Due to memory constraints, we employ a 2D axial patch-wise approach, with a patch size of 120×120. The training patches are sampled such that the tumor is partially covered. The model is trained end-to-end by backpropagation. The loss function is optimized using the AdamW [9] optimizer. For regularization purposes, we employ weight decay. It is kept constant at 1×10^{-3} for the first 50 epochs, then it decreases using a cosine annealing to 1×10^{-6}. A similar scheduling is applied to the learning rate; however, it is initialized at 1×10^{-4} and decays till 5×10^{-5}. We also employ spatial dropout [16] with probability of 0.25.

The neural network was trained for 170 epochs. For the choice of the neural network structure and hyperparameters, we used the validation and test sets for selecting the best epoch and to compare different set-ups and choose the final neural network, respectively. These validation and test sets were obtained from the training data.

For training the ensemble we split the training dataset in five folds. For each fold, we split it in two, using half for validation and the other half for testing. The remaining four folds are used for training the neural network. In each fold, we changed the seed used for initializing the weights of the neural network. The initialization of the weights condition the region of the loss where we start the optimization. So by changing the seed, we aim to obtain diverse solutions, which we know that may improve the performance of the ensemble.

For evaluation, we report two metrics: Dice Score, and Hausdorff Distance [3,10]. The proposed methods were implemented using PyTorch, and the experiments were conducted on an Nvidia RTX 2080 TI GPU.

3.4 Post-processing

As post-processing step, we employed morphological filtering to remove small clusters whose volume is less than a threshold. The threshold was chosen based on statistics obtained on the training dataset.

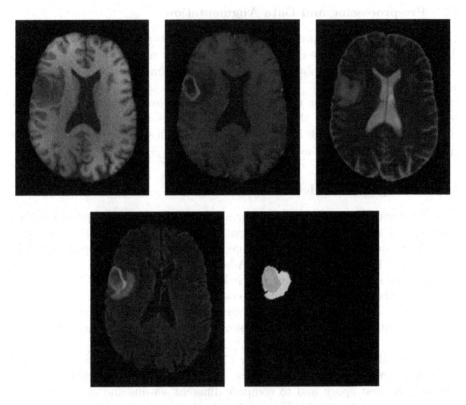

Fig. 3. Qualitative results from patient BraTS20_Validation_002. The enhancing tumor is shown in orange, necrosis in turquoise and edema in green. This subject obtained a mean Dice score of 0.9272, 0.9603 and 0.8939 for whole tumor, core tumor and enhanced tumor, respectively. (Color figure online)

4 Results

We report our results on BRATS 2020. In Table 1, we present the results on BRATS 2020 validation dataset, which were computed using the online platform.

The first entry in the table is DLA-Pool neural network. This uses a 2×2 max-pooling layer with stride of 2 to increase the field of view, which is a common approach. In DLA-GConv, we evaluate the observation by Zhang [18] that max-pooling may cause aliasing, so we employ a max-pooling with kernel size of 2 and stride of 1, followed by a Gaussian filtering with stride 2 and kernel size of 5×5

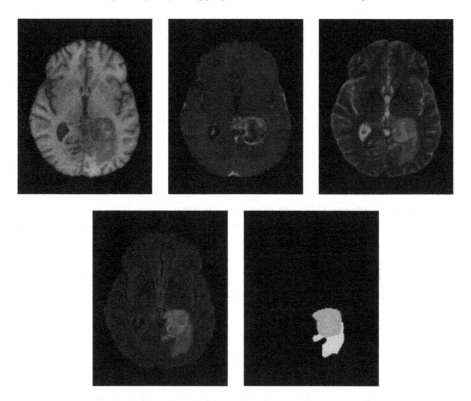

Fig. 4. Qualitative results from patient BraTS20_Validation_017. The enhancing tumor is shown in orange, necrosis in turquoise and edema in green. This subject obtained a mean Dice score of 0.9500, 0.9354 and 0.9446 for whole tumor, core tumor and enhanced tumor, respectively. (Color figure online)

and a sigma of 1.25. The effect of the Gaussian filter is to limit the rapid variation on the input pattern, avoiding aliasing. As can be observed in Table 1, this filtering in the downsampling improved the segmentation in all regions, specially the core and enhancing tumor. In this case, all metrics improved, except for the Hausdorff distance of the core tumor. The improvement was more pronounced in the enhancing tumor. In the last entry, we present the results of our ensemble. As can be observed the ensemble improved all metrics.

In Figs. 3 and 4, we show two example of the segmentation obtained with our method in the BraTS 2020 Validation set using the ensemble.

In Table 2, we present the results obtained in the BraTS 2020 Test set. We observe that there was a large drop in the whole tumor, but the core tumor and enhancing tumor increased when comparing to the results on the BraTS 2020 Validation set. Also, the Haudorff distance improved, except for the core tumor. Comparing the mean Dice with the Dice value in the interquartile, we also note that the mean Dice are closer to the lower interquartile value and in the core tumor it is even lower than the interquartile value. This may indicate that we

Table 1. Dice (DSC) and Hausdorff (HD$_{95}$) metrics of the proposed method on BraTS 2020 validation set. Each metric is represented by the average. WT - whole tumor, TC - tumor core, ET - enhancing tumor core.

	DSC			HD$_{95}$		
	ET	WT	TC	ET	WT	TC
DLA-Pool	0.7246	0.8916	0.7885	33.87	8.06	8.93
DLA-GConv	0.7418	0.8918	0.7945	27.81	7.07	12.99
Ensemble	0.7565	0.9050	0.8062	27.16	4.34	9.39

have some low outliers that are skewing the mean Dice. The identification of these cases may help to spot aspects to improve in our approach, which we leave as future work.

Table 2. Dice (DSC) and Hausdorff (HD$_{95}$) metrics of the proposed method on BraTS 2020 Test set. WT - whole tumor, TC - tumor core, ET - enhancing tumor core.

	DSC			HD$_{95}$		
	ET	WT	TC	ET	WT	TC
Mean	0.7900	0.8858	0.8297	20.44	5.32	22.32
StdDev	0.2278	0.1175	0.2514	79.73	7.60	79.42
Median	0.8514	0.9208	0.9187	1.41	3.00	2.24
25 quantile	0.7698	0.8786	0.8624	1.00	2.00	1.41
75 quantile	0.9181	0.9478	0.9567	2.40	5.00	5.38

5   Conclusion

The segmentation of gliomas in MRI is a challenging task, due to the heterogeneity of the lesion and the imaging modality itself. In this entry to the BraTS 2020 challenge, we propose a multi-cascade FCN for brain tumor segmentation based on the Deep Layer Aggregation principles, which consists in a systematic aggregation of features from the different scales. Also, we investigated the use of Gaussian filters for reducing the aliasing during pooling, which we found to be beneficial.

Our neural network was able to obtain a Dice score of 0.8858, 0.8297 and 0.7900, with an Hausdorff Distance of 5.32 mm, 22.32 mm and 20.44 mm for the whole tumor, core tumor and enhanced tumor, respectively, in the BraTS 2020 Test set.

References

1. Bakas, S., Akbari, H., Sotiras, A., et al.: Segmentation labels for the pre-operative scans of the TCGA-GBM collection. Cancer Imaging Arch. (2017). https://doi.org/10.7937/K9/TCIA.2017.KLXWJJ1Q
2. Bakas, S., et al.: Segmentation labels and radiomic features for the pre-operative scans of the TCGA-LGG collection. Cancer Imaging Arch. **286** (2017). https://doi.org/10.7937/K9/TCIA.2017.GJQ7R0EF
3. Bakas, S., et al.: Advancing the cancer genome atlas glioma MRI collections with expert segmentation labels and radiomic features. Nature Sci. Data **4**, 170117 (2017)
4. Bakas, S., et al.: Identifying the best machine learning algorithms for brain tumor segmentation, progression assessment, and overall survival prediction in the brats challenge. arXiv preprint arXiv:1811.02629 (2018)
5. DeVries, T., Taylor, G.W.: Improved regularization of convolutional neural networks with cutout. arXiv preprint arXiv:1708.04552 (2017)
6. Isensee, F., Kickingereder, P., Wick, W., Bendszus, M., Maier-Hein, K.H.: No new-net. In: Crimi, A., Bakas, S., Kuijf, H., Keyvan, F., Reyes, M., van Walsum, T. (eds.) BrainLes 2018. LNCS, vol. 11384, pp. 234–244. Springer, Cham (2019). https://doi.org/10.1007/978-3-030-11726-9_21
7. Jiang, Z., Ding, C., Liu, M., Tao, D.: Two-stage cascaded U-Net: 1st place solution to BraTS challenge 2019 segmentation task. In: Crimi, A., Bakas, S. (eds.) BrainLes 2019. LNCS, vol. 11992, pp. 231–241. Springer, Cham (2020). https://doi.org/10.1007/978-3-030-46640-4_22
8. Kamnitsas, K., et al.: Ensembles of multiple models and architectures for robust brain tumour segmentation. In: Crimi, A., Bakas, S., Kuijf, H., Menze, B., Reyes, M. (eds.) BrainLes 2017. LNCS, vol. 10670, pp. 450–462. Springer, Cham (2018). https://doi.org/10.1007/978-3-319-75238-9_38
9. Loshchilov, I., Hutter, F.: Decoupled weight decay regularization. arXiv preprint arXiv:1711.05101 (2017)
10. Menze, B.H., et al.: The multimodal brain tumor image segmentation benchmark (BRATS). IEEE Trans. Med. Imaging **34**(10), 1993–2024 (2015)
11. Myronenko, A.: 3D MRI brain tumor segmentation using autoencoder regularization. In: Crimi, A., Bakas, S., Kuijf, H., Keyvan, F., Reyes, M., van Walsum, T. (eds.) BrainLes 2018. LNCS, vol. 11384, pp. 311–320. Springer, Cham (2019). https://doi.org/10.1007/978-3-030-11726-9_28
12. Pereira, S., Pinto, A., Alves, V., Silva, C.A.: Brain tumor segmentation using convolutional neural networks in MRI images. IEEE Trans. Med. Imaging **35**(5), 1240–1251 (2016)
13. Pereira, S., Pinto, A., Amorim, J., Ribeiro, A., Alves, V., Silva, C.A.: Adaptive feature recombination and recalibration for semantic segmentation with fully convolutional networks. IEEE Trans. Med. Imaging **38**(12), 2914–2925 (2019)
14. Ronneberger, O., Fischer, P., Brox, T.: U-Net: convolutional networks for biomedical image segmentation. In: Navab, N., Hornegger, J., Wells, W.M., Frangi, A.F. (eds.) MICCAI 2015. LNCS, vol. 9351, pp. 234–241. Springer, Cham (2015). https://doi.org/10.1007/978-3-319-24574-4_28
15. Shelhamer, E., Long, J., Darrell, T.: Fully convolutional networks for semantic segmentation. IEEE Trans. Pattern Anal. Mach. Intell. **39**(4), 640–651 (2017)
16. Tompson, J., Goroshin, R., Jain, A., LeCun, Y., Bregler, C.: Efficient object localization using convolutional networks. In: Proceedings of the IEEE Conference on Computer Vision and Pattern Recognition, pp. 648–656 (2015)

17. Yu, F., Wang, D., Shelhamer, E., Darrell, T.: Deep layer aggregation. In: Proceedings of the IEEE Conference on Computer Vision and Pattern Recognition, pp. 2403–2412 (2018)
18. Zhang, R.: Making convolutional networks shift-invariant again. arXiv preprint arXiv:1904.11486 (2019)
19. Zhao, Y.-X., Zhang, Y.-M., Liu, C.-L.: Bag of tricks for 3D MRI brain tumor segmentation. In: Crimi, A., Bakas, S. (eds.) BrainLes 2019. LNCS, vol. 11992, pp. 210–220. Springer, Cham (2020). https://doi.org/10.1007/978-3-030-46640-4_20

Glioma Segmentation Using Ensemble of 2D/3D U-Nets and Survival Prediction Using Multiple Features Fusion

Muhammad Junaid Ali[1,2], Muhammad Tahir Akram[1,2], Hira Saleem[1,2], Basit Raza[1,2(✉)], and Ahmad Raza Shahid[1,2]

[1] Medical Imaging and Diagnostics Lab, National Center of Artificial Intelligence (NCAI), Islamabad, Pakistan
[2] Department of Computer Science, COMSATS University, Islamabad, Pakistan
{basit.raza,ahmadrshahid}@comsats.edu.pk

Abstract. Automatic segmentation of gliomas from brain Magnetic Resonance Imaging (MRI) volumes is an essential step for tumor detection. Various 2D Convolutional Neural Network (2D-CNN) and its 3D variant, known as 3D-CNN based architectures, have been proposed in previous studies, which are used to capture contextual information. The 3D models capture depth information, making them an automatic choice for glioma segmentation from 3D MRI images. However, the 2D models can be trained in a relatively shorter time, making their parameter tuning relatively easier. Considering these facts, we tried to propose an ensemble of 2D and 3D models to utilize their respective benefits better. After segmentation, prediction of Overall Survival (OS) time was performed on segmented tumor sub-regions. For this task, multiple radiomic and image-based features were extracted from MRI volumes and segmented sub-regions. In this study, radiomic and image-based features were fused to predict the OS time of patients. Experimental results on BraTS 2020 testing dataset achieved a dice score of 0.79 on Enhancing Tumor (ET), 0.87 on Whole Tumor (WT), and 0.83 on Tumor Core (TC). For OS prediction task, results on BraTS 2020 testing leaderboard achieved an accuracy of 0.57, Mean Square Error (MSE) of 392,963.189, Median SE of 162,006.3, and Spearman R correlation score of -0.084.

Keywords: Brain tumor segmentation · Overall survival prediction · Ensemble · U-Net

1 Introduction

MRI is widely used as an imaging technique as it is a standard non-invasive technique. However, automatically detecting brain tumors from such images is a challenging task because of heterogeneity in dataset and the class imbalance problem. Pixel wise segmentation of tumor sub-regions from MRI images plays a crucial role in patient's treatment as it helps the medical professionals in diagnostic and treatment procedures.

© Springer Nature Switzerland AG 2021
A. Crimi and S. Bakas (Eds.): BrainLes 2020, LNCS 12659, pp. 189–199, 2021.
https://doi.org/10.1007/978-3-030-72087-2_17

The abnormal growth of brain cells is referred to as brain tumor that can be benign or malignant. Most of the malignant tumors occur in the glial cells of the brain and are called gliomas which are very aggressive and deadly. Some forms of the cancer are highly aggressive and such a tumor is classified as high grade. The levels define their degree of aggressiveness. World Health Organization (WHO) [1] categorizes the tumors of level I and II as Low Grade Gliomas (LGG) and that of levels III and IV as High Grade Glioma (HGG).

Early diagnosis saves lives and it requires accurate grading and segmentation of tumor sub-regions with high accuracy. The manual segmentation suffers from inter-rater deviations and it is also very time consuming because of large 3D data. To overcome the varying results and to speed up the diagnosis process, the computer aided segmentation is essential. Various models [12–17] have been proposed over the years in order to improve the accuracy of automated diagnostic systems.

The deep leaning based segmentation methods have gained popularity since the introduction of U-Nets [2] and Fully Convolutional Networks (FCN) [3] models due to their capabilities to provide pixel wise predictions. The U-Net architecture is specifically designed for medical image segmentation and different variants of this architecture have dominated the latest submissions for BraTS [4–8] challenge. The U-Net based models have proven to be accurate for medical image segmentation task especially for 3D brain tumor segmentation as indicated by Ghaffari et al. [9].

The 3D U-Net models face challenges like class imbalance and higher computation cost. The imbalance issue can be tackled with various loss functions but model learning and hyper-tuning of parameters are still major challenges due to high computation costs of these models. On the contrary, the 2D U-Nets have fewer parameters and can be trained quickly thus hyper-tuning of parameters is possible. The 2D U-net underperforms because they do not carry the depth information which is crucial for 3D segmentation. In this research, we demonstrate that the 2D models can achieve the similar or closer performance to 3D models if hyper-tuning has been done efficiently. To overcome the lack of depth information in 2D models, we train it in multiple views like axial, sagittal, and coronal. As 3D and 2D models have their respective advantages and to combine these advantages, we train our model with both 2D net-work by Asra et al. [10] and the lightweight 3D model known as Dilated Multi-Fiber Network (DMFNet) [12]. The main reason for selecting the 3D DMF model is that it has fewer number of parameters which can be trained quickly as compared to other 3D complex models.

The task of survival prediction is normally performed after segmentation of tumors from MRI volumes. In this task, the survival time of patients in days is calculated after the surgery has been done. An accurate survival prognosis is essentially important for treatment planning.

The rest of the paper is organized as follows: in Sect. 2 the proposed methodology is discussed, Sect. 3 discusses the experimental section and in Sect. 4, the results and discussed and finally, the conclusion is discussed in Sect. 5.

2   Proposed Methodology

2.1   Segmentation Task

2D Model. Motivated by the presentation of our previous 2D network [10], which performed better for HGG volumes, we used the same model for this segmentation. The slices were cropped to 224 × 224 in order to design a simpler 2D U-Net model. All of the tumor slices are selected for training but not all of the non-tumor ones are. The model is trained in all three views, which are axial, sagittal, and coronal. The input slices selected for axial, sagittal, and coronal planes are 224 × 224, 152 × 224, and 224 × 152 in size respectively.

3D Model. DMFNet tackles the problem of high computational cost of 3D convolutional layers in brain tumor segmentation task. It combines the group convolution approach with MultiFiber (MF) and dilated convolutions to reduce the number of parameters and to maintain the prediction accuracy. Group convolution reduces the number of parameters by grouping a single connection into 'n' connections called fibers as shown in Fig 2. The flow of information between these fibers is handled by a multiplexer. Multiplexer utilizes two $1 \times 1 \times 1$ convolutions combined with the input channel grouped in 4 channels.

Dilated convolutions are then used to capture the spatial information of brain tumors. Dilation rates of 1, 2, and 3 are used indicating how many pixels are skipped while doing performing convolutions, followed by their weighted sum to capture the most valuable information. The DMFNet receives input MRI patches of size $4 \times 128 \times 128 \times 128$ and operates like an encoder-decoder in which encoder is comprised of DMF units and decoder concatenates the upsampled features with high resolution features. The entire number of parameters of the network are 3.88M which are far fewer than other 3D segmentation networks.

Ensemble of Models. Zhou et al. proposed a dimension fusion UNet known as D-UNet in which they combine the 2D and 3D convolutions in the encoder [11]. Dimension transform block is used to transform 2D network information to 3D and vice versa. Moreover, to manage the number of parameters this dimension transform block is only used in early layers of the network where 3D convolutions are used. However, our proposed strategy combines the predictions of 2D and 3D networks trained separately. We created ensemble of these two models to take advantages of both of them. The predictions of 2D and 3D models are averaged. The overall pipeline for the segmentation task is shown below in Fig. 1.

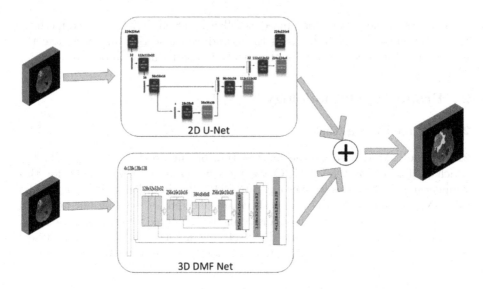

Fig. 1. Ensemble based approach used for segmentation

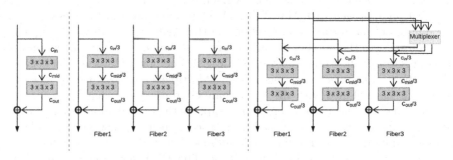

Fig. 2. Group-wise convolution used in DMF-Net

2.2 Survival Prediction Task

The tumorous regions predicted from patient's MRI volume are further used for Overall Survival (OS) time prediction. To predict the OS time, we extracted radiomic and image-based features. Radiomic features are found to be effective in different studies for OS tasks [8,13,14]. They were extracted using pyradiomics library and OpenCV was used for extraction of image-based features. The image based features included volume, volume ratio, surface area, surface area to volume ratio, position of the whole tumor center, position of the enhancing tumor center, relevant location of the whole tumor center to brain center, and relevant location of the enhancing tumor center to brain center.

Algorithm 1. Algorithm for Overall Survival Prediction of Patient's

---

procedure OSPREDICTIONTASK(MRI_Volumes,Segmented_Tumors)
    for Each patient in Patients **do**
        Radiomic_Features=Extract_Radiomic_Features(MRI_Volumes ,
Segmented_Tumors)
        ImageBased_Features=Extract_ImageBasaed_Features(MRI_Volumes,
Segmented_Tumors)
        patients_features.append(ImageBased_Features,
Radiomic_Features,Patient_Age,Patient_ID)

    Selected_Features=RFE_Feature_Selection(patients_features)
    $Predicted\_Survival\_Days$=RandomForestEstimator(Selected_Features)
    return $Predicted\_Survival\_Days$ ▷ Predicted days of patients

---

Radiomic features were extracted by applying the Laplacian of Gaussian (LoG) filters. Texture and intensity based features were extracted from input image and images filtered from LoG filters with sigma values of 1,2 and 3. We extracted 14 shape features (tumor volume, surface area, 2D and 3D diameter etc.), 19 first-order features and 28 Grey Level Co-occurrence Matrix (GLCM) features from segmented tumors. The features were extracted for the tumorous regions comprising of ET, Non-enhancing Edema (NED), TC and WT. Due to the overlapping of tumors, some of their features were common within the feature vector. To select the best performing features, we used Random Forest-Recursive Features Elimination (RF-RFE). We also added the age of the patients as a feature to the feature vector as it contains important information about his/her survival. After selection of important features, Random Forest (RF) regressor with Grid search was used for survival prediction. Figure 3 shows the system model of steps performed for OS prediction. The algorithm of proposed OS prediction approach is shown in Algorithm 1. The scatter-plot indicating the predicted OS prediction values with the actual values on BraTS 2020 training set is shown in Fig. 5.

3 Experiments

3.1 Dataset

BraTS 2020 dataset was used to perform experiments. The overall BraTS dataset contains sets of training, validation and test datasets. The training set consists of data of 369 patients and as compared to previous versions the dataset is not divided into LGG/HGG subsets but combined as a single dataset. The validation set is composed of 125 volumes which are same as for the previous year's challenge. Each patient's data contains MRI volumes of 4 modalities (a) T1 (b) T1-Weighted (c) T2 weighted, and (d) Fluid Attenuated Inversion Recovery (FLAIR).

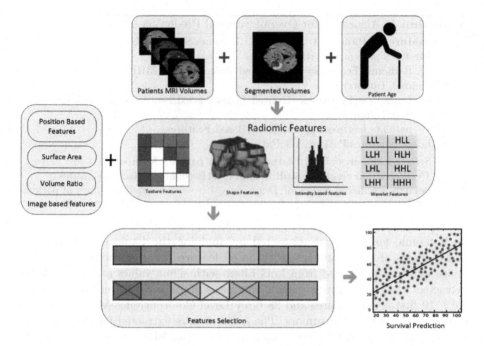

Fig. 3. System Model of whole methodology used for survival prediction

Each modality consists of 155 slices of 240 × 240 pixels. Each patient's segmentation ground truth is given with training set having labels of 1,2 and 4 for the necrotic/non-enhancing tumor (NCR/NET), 2 as peritumoral edema (ED), 4 as enhancing tumor and 0 is everything else including normal brain/background, respectively.

The same set of volumes are used for both the segmentation and the survival prediction tasks. Survival prediction of glioma patients is done after having gone through Gross Total Resection (GTR), where the entire tumor is removed. OS prediction time of patients after GTR is provided with training set with their age.

3.2   Implementation Details

We used Tensor Flow with Keras as top level API for implementing our deep learning based 2D and 3D U-Net models. We executed our models on NVIDIA Titan pascal GPU with 12 GB VRAM. The training data results were computed on local machines whereas the validation data metrics were evaluated using BraTS online leaderboard with team name "COMSATS-MIDL". The learning rate at the beginning of training was set at 1e−05. The number of epochs for 2D models were selected empirically whereas for 3D models, the total number of epochs were set to be 250 with batch size of 2. All the models were trained with all the 4 modalities and each modality was normalized using z-score normalization.

For survival prediction task, we have used pyradiomics and OpenCV for feature extraction and scikit-learn for model training and tuning. The experiments were performed on local computer in jupyter notebook.

4   Results

In order to visually show the performance of the proposed brain tumor segmentation model, we selected a few slices from the training set from all three planes. The results on some slices are shown in the Fig. 4 below.

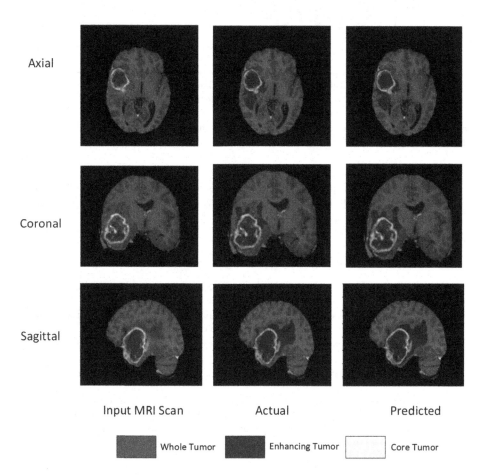

Fig. 4. Comparison of results of 2D U-Net, 3D DMF-Net and Ensemble of 2D U-Net+3D DMF-Net on local validation data split from training set.

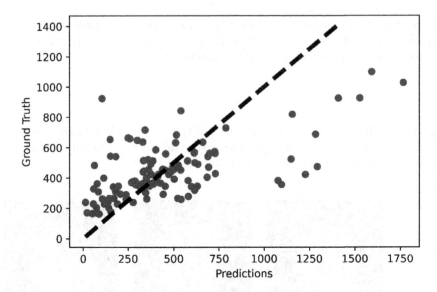

Fig. 5. Scatter plot of regression predictions of actual and predicted overall survival in days

The performance of proposed approach was evaluated on BraTS 2020 dataset validation leaderboard. The results are shown below in Table 1 and segmentation results on testing dataset are shown in Table 2.

Table 1. Results of 3 tumor sub-regions for BraTS 2020 validation dataset

	Dice scores			Sensitivity			Specificity			Hausdorff95		
	ET	WT	TC	ET	WT	TC	ET	WT	TC	ET	WT	TC
Mean	0.748	0.871	0.748	0.751	0.872	0.718	0.998	0.993	0.998	3.929	9.428	10.090
StdDev	0.281	0.128	0.259	0.295	0.148	0.2850	0.003	0.007	0.004	6.796	18.102	14.680
Median	0.843	0.907	0.858	0.852	0.915	0.8310	0.999	0.995	0.999	2.236	3.742	5.8310

Table 2. Results of 3 tumor sub-regions for BraTS 2020 testing dataset

	Dice scores			Sensitivity			Specificity			Hausdorff95		
	ET	WT	TC	ET	WT	TC	ET	WT	TC	ET	WT	TC
Mean	0.79	0.876	0.83	0.862	0.868	0.827	0.999	0.999	0.999	14.699	8.092	21.23
StdDev	0.18	0.118	0.232	0.198	0.135	0.243	0.0047	0	0.0071	63.731	29.621	74.86
Median	0.84	0.910	0.914	0.932	0.906	0.926	0.999	0.990	0.990	2.000	3.239	2.91

Table 3. Results of validation data survival prediction

Accuracy	MSE	medianSE	stdSE	SpearmanR
0.483	105079.4	37004.93	146376	0.134

Table 4. Results of testing data survival prediction

Accuracy	MSE	medianSE	stdSE	SpearmanR
0.579	392963.2	16200.3	920292.8	−0.084

Our proposed solution for segmentation task achieved good performance for enhancing tumor and Core Tumors (CT), however, its performance was below par on whole tumor as shown in Table 1 and Table 2. This indicates that model did not perform well in classifying Edema ('label 2'). This fact is also evident from the median values of the dice scores where dice median value for CT is even higher than the whole tumor. The higher values of sensitivity for enhancing tumor clearly show the model's performance against enhancing tumor where it has missed few instances for enhancing tumor. The median values for sensitivity are also on the higher side for enhancing tumor and CT. The segmentation model perform poorly with respect to Hausdorff distance as it is clear from the higher mean and standard deviation values, but median values are very low. The results of OS prediction on validation and testing set is shown in Table 3 and Table 4 which shown the accuracy and medianSE on testing data is better as compared to validation set.

The 3D DMF-Net captures inter slice information and volumetric information which contains tumor information 2D U-Net captures low level information. Combing the both helped to have volumetric and pixel-wise information which increased the system performance.

5   Conclusion

The proposed segmentation model results clearly show improvements while using the ensemble model. The 3D models provided depth information while 2D models were better hyper-tuned thus resulting in overall performance enhancement. Overall, the proposed solution performance was excellent for more than 50% of the cases and we suspect that the LGG cases are pulling the mean values towards the lower side. Moreover, for survival prediction task the extracted radiomics and image based features trained on random forest regression model achieved better results. Location, size, age and geometric features are found to be most effective features and are helpful in prediction of patient OS time. The results of OS prediction on testing data is quite promising.

Acknowledgments. This work has been supported by the Higher Education Commission under Grant 2(1064) and is carried out at Medical Imaging and Diagnostics (MID) Lab at COMSATS University Islamabad under the umbrella of National Center of Artificial Intelligence (NCAI), Pakistan.

References

1. Louis, D.N., et al.: The 2016 World Health Organization classification of tumors of the central nervous system: a summary. Acta Neuropathologica **131**(6), 803–820 (2016)
2. Ronneberger, O., Fischer, P., Brox, T.: U-Net: convolutional networks for biomedical image segmentation. In: Navab, N., Hornegger, J., Wells, W.M., Frangi, A.F. (eds.) MICCAI 2015. LNCS, vol. 9351, pp. 234–241. Springer, Cham (2015). https://doi.org/10.1007/978-3-319-24574-4_28
3. Long, J., Shelhamer, E., Darrell, T.: Fully convolutional networks for semantic segmentation. In: Proceedings of the IEEE Conference on Computer Vision and Pattern Recognition, pp. 3431–3440 (2015)
4. Menze, B.H., et al.: The multimodal brain tumor image segmentation benchmark (BRATS). IEEE Trans. Med. Imaging **34**(10), 1993–2024 (2014)
5. Bakas, S., et al.: Advancing the cancer genome atlas glioma MRI collections with expert segmentation labels and radiomic features. Sci. Data **4**, 170117 (2017)
6. Bakas, S., Reyes, M., Jakab, A., Bauer, S., Rempfler, M., Crimi, A., et al.: Identifying the best machine learning algorithms for brain tumor segmentation, progression assessment, and overall survival prediction in the BRATS challenge, arXiv preprint arXiv:1811.02629 (2018)
7. Bakas, S., Akbari, H., Sotiras, A., Bilello, M., Rozycki, M., Kirby, J., et al.: Segmentation La-bels and Radiomic features for the Pre-operative Scans of the TCGA-GBM collection. Cancer Imaging Archive (2017). https://doi.org/10.7937/K9/TCIA.2017.KLXWJJ1Q
8. Bakas, S., Akbari, H., Sotiras, A., Bilello, M., Rozycki, M., Kirby, J., et al.: Segmentation labels and radiomic features for the pre-operative scans of the TCGA-LGG collection. Cancer Imaging Archive (2017). https://doi.org/10.7937/K9/TCIA.2017.GJQ7R0EF
9. Ghaffari, M., Sowmya, A., Oliver, R.: Automated brain tumor segmentation using multimodal brain scans: a survey based on models submitted to the BraTS 2012–2018 challenges. IEEE Rev. Biomed. Eng. **13**, 156–168 (2019)
10. Rafi, A., et al.: U-Net based glioblastoma segmentation with patient's overall survival prediction. In: Brito-Loeza, C., Espinosa-Romero, A., Martin-Gonzalez, A., Safi, A. (eds.) ISICS 2020. CCIS, vol. 1187, pp. 22–32. Springer, Cham (2020). https://doi.org/10.1007/978-3-030-43364-2_3
11. Zhou, Y., Huang, W., Dong, P., Xia, Y., Wang, S.: D-UNet: a dimension-fusion U shape network for chronic stroke lesion segmentation. IEEE/ACM Trans. Comput. Biol. Bioinform. **39**(6), 1856–1867 (2019)
12. Chen, C., Liu, X., Ding, M., Zheng, J., Li, J.: 3D dilated multi-fiber network for real-time brain tumor segmentation in MRI. In: Shen, D., Liu, T., Peters, T.M., Staib, L.H., Essert, C., Zhou, S., Yap, P.-T., Khan, A. (eds.) MICCAI 2019. LNCS, vol. 11766, pp. 184–192. Springer, Cham (2019). https://doi.org/10.1007/978-3-030-32248-9_21
13. Isensee, F., Kickingereder, P., Wick, W., Bendszus, M., Maier-Hein, K.H.: Brain Tumor Segmentation and Radiomics Survival Prediction: Contribution to the BRATS 2017 Challenge. In: Crimi, A., Bakas, S., Kuijf, H., Menze, B., Reyes, M. (eds.) BrainLes 2017. LNCS, vol. 10670, pp. 287–297. Springer, Cham (2018). https://doi.org/10.1007/978-3-319-75238-9_25

14. Jiang, Z., Ding, C., Liu, M., Tao, D.: Two-stage cascaded U-Net: 1st place solution to BraTS challenge 2019 segmentation task. In: Crimi, A., Bakas, S. (eds.) BrainLes 2019. LNCS, vol. 11992, pp. 231–241. Springer, Cham (2020). https://doi.org/10.1007/978-3-030-46640-4_22

15. Myronenko, A.: 3D MRI brain tumor segmentation using autoencoder regularization. In: Crimi, A., Bakas, S., Kuijf, H., Keyvan, F., Reyes, M., van Walsum, T. (eds.) BrainLes 2018. LNCS, vol. 11384, pp. 311–320. Springer, Cham (2019). https://doi.org/10.1007/978-3-030-11726-9_28

16. Zhou, C., Ding, C., Lu, Z., Wang, X., Tao, D.: One-pass multi-task convolutional neural networks for efficient brain tumor segmentation. In: Frangi, A.F., Schnabel, J.A., Davatzikos, C., Alberola-López, C., Fichtinger, G. (eds.) MICCAI 2018. LNCS, vol. 11072, pp. 637–645. Springer, Cham (2018). https://doi.org/10.1007/978-3-030-00931-1_73

17. Feng, X., Tustison, N.J., Patel, S.H., Meyer, C.H.: Brain tumor segmentation using an ensemble of 3d u-nets and overall survival prediction using radiomic features. Front. Comput. Neurosci. **14**, 25 (2020)

Generalized Wasserstein Dice Score, Distributionally Robust Deep Learning, and Ranger for Brain Tumor Segmentation: BraTS 2020 Challenge

Lucas Fidon[✉], Sébastien Ourselin, and Tom Vercauteren

School of Biomedical Engineering and Imaging Sciences, King's College London,
London, UK
lucas.fidon@kcl.ac.uk

Abstract. Training a deep neural network is an optimization problem with four main ingredients: the design of the deep neural network, the per-sample loss function, the population loss function, and the optimizer. However, methods developed to compete in recent BraTS challenges tend to focus only on the design of deep neural network architectures, while paying less attention to the three other aspects. In this paper, we experimented with adopting the opposite approach. We stuck to a generic and state-of-the-art 3D U-Net architecture and experimented with a non-standard per-sample loss function, the generalized Wasserstein Dice loss, a non-standard population loss function, corresponding to distributionally robust optimization, and a non-standard optimizer, Ranger. Those variations were selected specifically for the problem of multi-class brain tumor segmentation. The generalized Wasserstein Dice loss is a per-sample loss function that allows taking advantage of the hierarchical structure of the tumor regions labeled in BraTS. Distributionally robust optimization is a generalization of empirical risk minimization that accounts for the presence of underrepresented subdomains in the training dataset. Ranger is a generalization of the widely used Adam optimizer that is more stable with small batch size and noisy labels. We found that each of those variations of the optimization of deep neural networks for brain tumor segmentation leads to improvements in terms of Dice scores and Hausdorff distances. With an ensemble of three deep neural networks trained with various optimization procedures, we achieved promising results on the validation dataset and the testing dataset of the BraTS 2020 challenge. Our ensemble ranked fourth out of 78 for the segmentation task of the BraTS 2020 challenge with mean Dice scores of 88.9, 84.1, and 81.4, and mean Hausdorff distances at 95% of 6.4, 19.4, and 15.8 for the whole tumor, the tumor core, and the enhancing tumor.

Keywords: Brain tumor · Segmentation · BraTS challenge · Dice score · Distributionally robust optimization · Convolutional neural network

© Springer Nature Switzerland AG 2021
A. Crimi and S. Bakas (Eds.): BrainLes 2020, LNCS 12659, pp. 200–214, 2021.
https://doi.org/10.1007/978-3-030-72087-2_18

1   Introduction

Accurate brain tumor segmentation based on MRI is important for diagnosis, surgery planning, follow-up, and radiation therapy [1,2]. However, manual segmentation is time-consuming (1 h per subject for a trained radiologist [26]) and suffers from large inter- and intra-rater variability [26]. Automatic and accurate brain tumor segmentation is thus necessary.

In recent BraTS challenges [3,26], innovations on convolutional neural networks (CNNs) architectures, have led to significant improvement in brain tumor segmentation accuracy [6,7,12,14,20,36]. Recently, the development of nnUNet [16] has shown that a well-tuned 2D U-Net [31] or 3D U-Net [9] can achieve state-of-the-art results for a large set of medical image segmentation problems and datasets, including BraTS. The 2D U-Net and 3D U-Net were among the first convolutional neural network architectures proposed for medical image segmentation. This suggests that the improvement that the design of the deep neural network can bring to brain tumor segmentation is more limited than what was previously thought.

In contrast, little attention has been paid to the design of deep learning optimization methods in deep learning-based pipelines for brain tumor segmentation. We identify three main ingredients other than the design of the deep neural network architecture, in the design of deep learning optimization methods that are illustrated in Fig. 1: 1) The per-sample loss function or simply *loss function* for short (e.g. the Dice loss [27,32]), 2) The population loss function (e.g. the empirical risk) whose minimization is hereby referred as the *optimization problem*. 3) The optimizer (e.g. SGD and Adam [21]), Recent state-of-the-art deep learning pipelines for brain tumor segmentation uses generic choices of those optimization ingredients such as the sum of the Dice loss and the Cross-entropy loss, Stochastic Gradient Descent (SGD), or Adam as an optimizer and empirical risk minimization.

In this paper, we build upon the 3D U-Net [9] architecture-based pipeline of nnUNet [16] and explore alternative loss functions, optimizers, and optimization problems that are specifically designed for the problem of brain tumor segmentation. We propose to use the generalized Wasserstein Dice loss [11] as an alternative per-sample loss function, as discussed in Sect. 2.1, we use distributionally robust optimization [13] as an alternative to empirical risk minimization, as discussed in Sect. 2.2, and we use the Ranger optimizer [23,37] as an alternative optimizer, as discussed in Sect. 2.3.

The generalized Wasserstein Dice loss [11] is a per-sample loss function that was designed specifically for the problem of multi-class brain tumor segmentation. It allows us to take advantage of the hierarchical structure of the tumor regions labeled in BraTS. In contrast to empirical risk minimization, distributionally robust optimization [13] accounts for the presence of underrepresented subdomains in the training dataset. In addition, distributionally robust optimization does **not** require labels about the subdomains in the training dataset, such as the data acquisition centers where the MRI was performed, or whether the patient has high-grade or low-grade gliomas. This makes distributionally robust

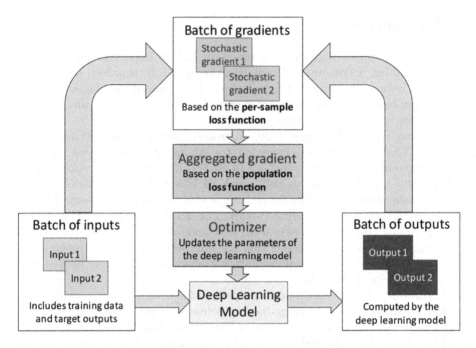

Fig. 1. Diagram of a deep learning optimization pipeline. Deep learning optimization methods are made of four main components: 1) The design of the deep neural network architecture, 2) The **per-sample loss function** (e.g. the Dice loss) that determines the stochastic gradient, 3) The **population loss function** (e.g. the empirical risk) that determines how to merge the stochastic gradients into one aggregated gradient, 4) The **optimizer** (e.g. Adam) that determines how the aggregated gradient is used to update the parameters of the deep neural network at each training iteration. In this work, we explore variants for the per-sample loss function, the population loss function and the optimizer for application to automatic brain tumor segmentation.

optimization easy to apply to the BraTS 2020 dataset in which that information is not available to the participants. Ranger [23,37] is a generalization of the widely used Adam optimizer that is more stable with the small batch sizes and noisy labels encountered in BraTS.

Empirical evaluation of those alternatives on the BraTS 2020 validation dataset suggests that they outperform and are more robust than nnUNet. In addition, our three networks, each one trained with one of the alternative ingredients listed above, appear to be complementary over the three regions of interest in the BraTS challenge: whole tumor, tumor core, and enhancing tumor. The ensemble formed by our three networks outperforms all of the individual networks for all regions of interest and shows promising results compared to our competitors in the BraTS 2020 challenge. **Our ensemble ranked fourth out of 78 at the segmentation task of the BraTS 2020 challenge** after evaluation on the withheld BraTS 2020 testing dataset.

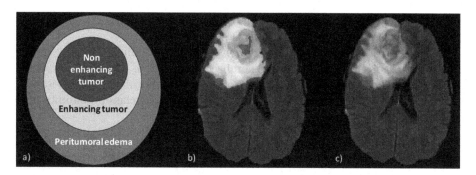

Fig. 2. The brain tumor classes have a hierarchical structure. a) Hierarchy of the regions labeled in the BraTS 2020 dataset. b) Manual segmentation overlaid on the FLAIR image for a case in the BraTS 2020 training dataset. c) FLAIR image.

2 Method: Varying the Three Main Ingredients of the Optimization of Deep Neural Networks

In current state-of-the-art deep learning pipelines for brain tumor segmentation, the *training* of the deep neural network consists in the following optimization problem

$$\boldsymbol{\theta}^{*ERM} := \underset{\boldsymbol{\theta}}{\mathrm{argmin}}\, \frac{1}{n}\sum_{i=1}^{n} \mathcal{L}\left(h(\boldsymbol{x}_i; \boldsymbol{\theta}), \boldsymbol{y}_i\right) \tag{1}$$

where h is a deep neural network with parameters $\boldsymbol{\theta}$, $\mathcal{L}$ is a smooth per-volume loss function, and $\{(\boldsymbol{x}_i, \boldsymbol{y}_i)\}_{i=1}^{n}$ is the training dataset. $\boldsymbol{x}_i$ are the input 3D brain MRI T1, T1-gad, T2, and FLAIR volumes, and $\boldsymbol{y}_i$ are the ground-truth manual segmentations.

Some of the main ingredients of this optimization problem are: 1) The deep neural network architecture for h, 2) The loss function $\mathcal{L}$, 3) The optimization problem (here *empirical risk minimization*, i.e. we minimize the mean of the per-sample loss functions), and 4) The optimizer which is the algorithm that allows finding an approximation of $\boldsymbol{\theta}^*_{ERM}$. In recent years, most of the research effort has been put in the deep neural network architecture. In this work, we set the deep neural network architecture to the 3D U-Net [9] used in nnUNet [16], and explore the three other ingredients.

In this section, we present the per-sample loss function, population loss function, and optimizer that we have used to compete in the BraTS 2020 challenge.

2.1 Changing the Per-Sample Loss Function: The Generalized Wasserstein Dice Loss [11]

The generalized Wasserstein Dice loss [11] is a generalization of the Dice Loss for multi-class segmentation that can take advantage of the hierarchical structure of the set of classes in BraTS. The brain tumor classes hierarchy is illustrated in Fig. 2. Our PyTorch implementation of the generalized Wasserstein Dice loss is publicly available[1].

When the labeling of a voxel is ambiguous or too difficult for the neural network to predict it correctly, the generalized Wasserstein Dice loss is designed to favor mistakes that are semantically more plausible. Formally, the generalized Wasserstein Dice loss between the ground-truth (one-hot) class probability map $\mathbf{p}$ and the predicted class probability map $\hat{\mathbf{p}}$ is defined as [11]

$$\mathcal{L}_{GWDL}(\hat{\mathbf{p}},\mathbf{p}) = \frac{2\sum_{l\neq b}\sum_i \mathbf{p}_{i,l}(1 - W^M(\hat{\mathbf{p}}_i,\mathbf{p}_i))}{2\sum_{l\neq b}[\sum_i p_{i,l}(1 - W^M(\hat{\mathbf{p}}_i,\mathbf{p}_i))] + \sum_i W^M(\hat{\mathbf{p}}_i,\mathbf{p}_i)} \tag{2}$$

where $W^M(\hat{\mathbf{p}}_i,\mathbf{p}_i)$ is the Wasserstein distance between predicted $\hat{\mathbf{p}}_i$ and ground truth $\mathbf{p}_i$ discrete probability distribution at voxel i. $M = (M_{l,l'})_{1\leq l,l'\leq L}$ is a distances matrix between the BraTS 2020 labels, and b is the class number corresponding to the background.

The matrix M informs the generalized Wasserstein Dice loss about the relationships between the classes. For two classes of indices l and l', the smaller the distance $M_{l,l'}$, the less mistaking a voxel of (ground-truth) class l for the class l' is penalized.

The matrix M is a distance matrix. As a result, it is symmetrical with zeros on its diagonal. In addition, by convention, we set the maximal-label distance to 1 that corresponds to the distance between the *background* class and all the other classes. Specifically, we adapted the distances matrix used in [11], by removing the *necrotic core tumor* that has been merged with the *non-enhancing core* since the BraTS 2017 challenge. For the classes indices 0: *background*, 1: *enhancing tumor*, 2: *edema*, 3: *non-enhancing tumor*, this corresponds to the matrix

$$M = \begin{pmatrix} 0 & 1 & 1 & 1 \\ 1 & 0 & 0.6 & 0.5 \\ 1 & 0.6 & 0 & 0.7 \\ 1 & 0.5 & 0.7 & 0 \end{pmatrix} \tag{3}$$

The distances between the classes reflect the hierarchical structure of the tumor regions, as illustrated in Fig. 2. The distances between the tumor classes are all lower than 1 because they have more in common than with the background.

It is worth noting, that since the ground truth segmentation map $\mathbf{p}$ is a one-hot segmentation map, for any voxel i, we have

$$W^M(\hat{\mathbf{p}}_i,\mathbf{p}_i) = \sum_{l=1}^L p_{i,l}\left(\sum_{l'=1}^L M_{l,l'}\hat{p}_{i,l'}\right) \tag{4}$$

Previous Work: Other top performing methods of previous BraTS challenges have proposed to exploit the hierarchical structure of the classes present in BraTS by optimizing directly for the overlapping regions whole tumor, tumor core, and

[1] https://github.com/LucasFidon/GeneralizedWassersteinDiceLoss.

enhancing tumor [18,25,28,36,38]. However, in contrast to those methods, the generalized Wasserstein Dice loss allows optimizing for both the overlapping regions and the non-overlapping regions labeled in the BraTS dataset simultaneously by considering all the inter-class relationships.

2.2 Changing the Optimization Problem: Distributionally Robust Optimization [13]

Distributionally Robust Optimization (DRO) is a generalization of Empirical Risk Minimization (ERM) in which the weights of each training sample are also optimized to automatically reweight the samples with higher loss value [8,13,29,30].

DRO aims at improving the generalization capability of the neural network by explicitly accounting for uncertainty in the training dataset distribution. For example, in the BraTS dataset, we don't know if the different data acquisition centers are equally represented. This can lead the deep neural networks to underperform on the subdomains that are underrepresented in the training dataset. DRO aims at mitigating this problem by encouraging the neural network to perform more consistently on the entire training dataset.

More formally, DRO is defined by the min-max optimization problem [13]

$$\boldsymbol{\theta}_{DRO}^* := \operatorname*{argmin}_{\boldsymbol{\theta}} \max_{\boldsymbol{q} \in \Delta_n} \left(\sum_{i=1}^{n} q_i \, \mathcal{L} \left(h(\boldsymbol{x}_i; \boldsymbol{\theta}), \boldsymbol{y}_i \right) - \frac{1}{\beta} D_{KL} \left(\boldsymbol{q} \, \middle\| \, \frac{1}{n} \mathbf{1} \right) \right) \quad (5)$$

where a new unknown probabilities vector parameter $\boldsymbol{q}$ is introduced, $\frac{1}{n}\mathbf{1}$ denotes the uniform probability vector $\left(\frac{1}{n}, \ldots, \frac{1}{n} \right)$, D_{KL} is the Kullback-Leibler divergence, Δ_n is the unit n-simplex, and $\beta > 0$ is a hyperparameter.

$D_{KL} \left(\boldsymbol{q} \, \middle\| \, \frac{1}{n} \mathbf{1} \right)$ is a regularization term that measures the dissimilarity between $\boldsymbol{q}$ and the uniform probability vector $\frac{1}{n}\mathbf{1}$ that corresponds to assign the same weight $\frac{1}{n}$ to each sample like in ERM. Therefore, this regularization term allows to keep the problem close enough to ERM, and its strength is controlled by β.

Implementation: Recently, it has been shown in [13] that $\boldsymbol{\theta}_{DRO}^*$ can be approximated using any of the optimizers commonly used in deep learning provided the sample volumes are sampled using a *hardness weighted sampling* strategy during training instead of the classic shuffling of the data at each epoch. For more details on how the hardness weighted probabilities vector **q** is approximated online during training while adding negligible computational overhead, we refer the reader to [13, see Algorithm 1].

DRO and Brain Tumor Segmentation: The *hardness weighted sampling* corresponds to a principled hard example mining method and it has been shown to improve the robustness of nnUNet for brain tumor segmentation using the BraTS 2019 dataset [13].

In the BraTS dataset, some cases have no enhancing tumor and the Dice score for this class will be either 0 or 1. As a result, when the mean Dice loss is used as a loss function, those cases with missing enhancing tumor will typically have a higher loss value. This is an example of cases, perceived as *hard examples* with DRO, that have a higher sampling probability in $\mathbf{q}$ during training.

2.3 Changing the Optimizer: Ranger [23,37]

Ranger is an optimizer for training deep neural networks that consists of the combination of two recent contributions in the field of deep learning optimization: the Rectified Adam (RAdam) [23] and the lookahead optimizer [37]. Recently, Ranger has shown promising empirical results for applications in medical image segmentation [33].

RAdam [23] is a modification of the Adam optimizer [21] that aims at reducing the variance of the adaptive learning rate of Adam in the early-stage of training. For more details, we refer the reader to [23, see Algorithm 2].

Lookahead [37] is a generalization of the exponential moving average method that aims at accelerating the convergence of other optimizers for deep neural networks. Lookahead requires to maintain two sets of values for the weights of the deep neural networks: one set of *fast weights* θ, and one set of *slow weights* ϕ. Given a loss function $\mathcal{L}$, an optimizer A (e.g. RAdam), a synchronization period k and a slow weights step size $\alpha > 0$, training a deep neural network with Lookahead is done as follows [37, see Algorithm 1]

$$
\begin{aligned}
&\textbf{for } t = 1, 2, \ldots, T \textbf{ do} && \triangleright \text{ Outer iterations}\\
&\quad \theta_{t,0} \leftarrow \phi_{t-1} && \triangleright \text{ Synchronize weights}\\
&\quad \textbf{for } i = 1, 2, \ldots, k \textbf{ do} && \triangleright \text{ Inner iterations}\\
&\qquad d \sim \mathcal{D} && \triangleright \text{ Sample a batch of training data}\\
&\qquad \theta_{t,i} \leftarrow \theta_{t,i-1} + A(\mathcal{L}, \theta_{t,i-1}, d) && \triangleright \text{ Update the fast weights}\\
&\quad \phi_t \leftarrow \phi_{t-1} + \alpha\left(\theta_{t,k} - \phi_{t-1}\right) && \triangleright \text{ Update the slow weights}\\
&\textbf{return } \phi_T
\end{aligned}
$$

Lookahead can be seen as a wrapper that can be combined with any deep learning optimizer. However, its combination with RAdam has quickly become the most popular. This is the reason why we considered only lookahead in combination with RAdam in our experiments.

It is worth noting that the optimizers used in deep learning also depend on hyperparameters such as the batch size, the patch size, and the learning rate schedule. We did not explore in depth those hyperparameters in this work.

2.4 Deep Neural Networks Ensembling

Deep neural networks ensembling has been used in previous BraTS challenge to average the predictions of different neural network architectures [10,19,24].

In this subsection, we discuss the role of ensembling for segmentation using different deep learning optimization methods.

Different deep learning optimization methods can give similarly good segmentations, but they are likely to have different biases and to make different mistakes. In this case, the ensembling of diverse models can lead to averaging out the inconsistencies due to the choice of the optimization method and improve the segmentation performance and robustness.

Let $\mathbf{x}$ be the random variable corresponding to the input 3D brain MRI T1, T1-gad, T2, and FLAIR volumes, and $\mathbf{y}$ be the random variable corresponding to the ground-truth manual segmentations for cases with a brain tumor. After training, a deep neural network trained for segmentation gives an approximation $P(\mathbf{y}|\mathbf{x};\boldsymbol{\theta}_\mu,\boldsymbol{\mu}) \approx P(\boldsymbol{y}|\boldsymbol{x})$ of the posterior segmentation distribution, where $\boldsymbol{\theta}_\mu$ is the vector of trainable parameters of the network obtained after training, and $\boldsymbol{\mu}$ are the vector of hyperparameters corresponding to the choice of the deep learning optimization method. Assuming that $P(\mathbf{y}|\mathbf{x};\boldsymbol{\theta}_\mu,\boldsymbol{\mu})$ is an unbiased estimator of $P(\mathbf{y}|\mathbf{x})$, and that a set of trained networks corresponding to hyperparameters $\{\boldsymbol{\mu}_1,\ldots,\boldsymbol{\mu}_M\}$ are available, an unbiased *ensembling estimation* of $P(\mathbf{y}|\mathbf{x})$ with reduced variance is given by

$$P(\mathbf{y}|\mathbf{x}) \approx \frac{1}{M}\sum_{m=1}^{M} P(\mathbf{y}|\mathbf{x};\boldsymbol{\theta}_{\mu_m},\boldsymbol{\mu}_m) \qquad (6)$$

3   Experiments and Results

In this section, we first describe the data and the implementation details, and second, we present the models that we compare and analyze their segmentation performance and robustness.

3.1   Data and Implementation Details

Data. The BraTS 2020 dataset[2] has been used for all our experiments. No additional data has been used.

The dataset contains the same four MRI sequences (T1, T1-gad, T2, and FLAIR) for patients with either high-grade Gliomas [4] or low-grade Gliomas [5]. All the cases were manually segmented for peritumoral edema, enhancing tumor, and non-enhancing tumor core using the same labeling protocol [3,6,26]. The training dataset contains 369 cases, the validation dataset contains 125 cases, and the testing dataset contains 166 cases. MRI for training and validation datasets are publicly available, but only the manual segmentations for the training dataset are available. The evaluation on the validation dataset can be done via the BraTS challenge online evaluation platform[3]. The evaluation on the testing dataset was performed only once by the organizers 48 h after they made the testing dataset

---

[2] See http://braintumorsegmentation.org/ for more details.
[3] https://ipp.cbica.upenn.edu/.

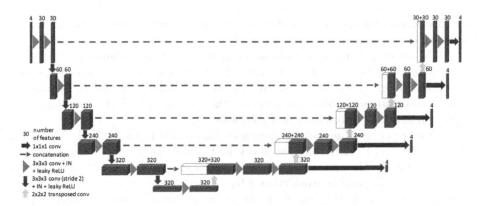

Fig. 3. Illustration of the 3D U-Net [9] architecture used. Blue boxes represent feature maps. IN stands for instance normalization. The design of this 3D U-Net was determined using the heuristics of nnUNet [16]. The main differences between this 3D U-Net and the original 3D U-Net are listed in Sect. 3.1.

available to us. For each case, the four MRI sequences are available after co-registration to the same anatomical template, interpolation to 1mm isotropic resolution, and skull stripping [26].

Convolutional Neural Network Architecture. The same 3D U-Net architecture [9], based on the heuristics of nnUNet [16], was used in all our experiments. The main differences compared to the original 3D U-Net [9] are:

- More levels are used,
- Instance normalization [34] is used instead of batch normalization [15],
- Leaky ReLU is used instead of ReLU (with a negative slope of 0.01),
- Spatial downsampling is performed using convolutions with stride 2 instead of average/max-pooling layers,
- Spatial upsampling is performed using transposed convolutions and the number of features is reduced to match the number of features in the next skip connection before concatenation,
- Deep supervision [22] is used (see the purple $1 \times 1 \times 1$ convolutions in Fig. 3),
- The maximum number of features is capped at 320,
- The initial number of features is 30 instead of 32 (like in nnUNet V1 [17]),
- The number of features is increased only once in the first level.

To help the reader to better appreciate those differences, the 3D U-Net used is illustrated in Fig. 3.

Training Implementation Details. Our code is based on the nnUNet code[4]. By default and when not indicated otherwise, the sum of the Dice loss and the

---

[4] https://github.com/MIC-DKFZ/nnUNet last visited in August 2020.

Cross-entropy loss is used with empirical risk minimization and the SGD with Nesterov momentum optimizer like in [16]. The learning rate is decreased at each epoch t as

$$\lambda_t = \lambda_0 \times \left(1 - \frac{t}{t_{max}}\right)^{0.9}$$

where λ_0 is the initial learning rate and t_{max} is the maximum number of epochs fixed as 1000. The batch size was set to 2 and the input patches were of dimension $128 \times 192 \times 128$. Deep supervision was used as illustrated in Fig. 3. A large number of data augmentation methods are used: random cropping of a patch, random zoom, gamma intensity augmentation, multiplicative brightness, random rotations, random mirroring along all axes, contrast augmentation, additive Gaussian noise, Gaussian blurring, and simulation of low resolution. For more implementation details about nnUNet we refer the interested reader to [16] and the nnUNet GitHub page.

Inference Implementation Details. Following nnUNet inference pipeline [16], we applied test-time data augmentation, as previously studied in [35], using flipping along all three spatial dimensions. When less than 50 voxels, or equivalently 0.05 mL, in the whole volume were predicted as enhancing tumor, we changed their prediction to non-enhancing tumor.

Hardware. GPUs NVIDIA Tesla V100-SXM2 with 16GB of memory were used to train all the deep neural networks. Training each deep neural network took us between 4 and 5 days.

3.2   Models Description

In this paragraph, we describe the different models that are compared in Table 1.

nnUNet [16]. The original nnUNet code with all the default parameters was trained on the BraTS 2020 training set. Specifically to the optimization, this means that the sum of the Dice loss and the Cross-entropy loss, SGD with Nesterov momentum, and empirical risk minimization were used for the *nnUNet* model.

nnUNet + Ranger [23,37]. Exactly the same as for the model *nnUNet* above, except the optimizer was Ranger [23,37] with a learning rate of 3×10^{-3}.

We experimented with different values of the initial learning rate for the Ranger optimizer $\{10^{-3}, 3 \times 10^{-3}, 10^{-2}\}$, and the value of 3×10^{-3} was retained because it performed best on the BraTS 2020 validation dataset.

We also tried Adam [21] and RAdam [23] (without lookahead [37]) optimizers, and we tuned the learning rates for each optimizer using the BraTS 2020 validation dataset and the same values for the initial learning rate as mentioned above. However, we found that Ranger outperformed all the others on the BraTS 2020 validation dataset.

nnUNet + GWDL [11]. Exactly the same as for the model *nnUNet* above, except the per-sample loss function was the sum of the generalized Wasserstein Dice Loss (GWDL) [11] and the Cross-entropy loss. The initial learning rate was not tuned specifically for use with the GWDL, and we used the default value of nnUNet.

nnUNet + DRO [13]. Exactly the same as for the model *nnUNet* above, except that we used distributionally robust optimization using the hardness weighted sampler proposed in [13]. The initial learning rate was not tuned specifically for use of DRO and we used the default value of nnUNet. We choose $\beta = 100$ because it is the value that was found to perform best for brain tumor segmentation in [13].

Ensemble Mean Softmax. This model is obtained by averaging the predicted softmax probabilities of the models *nnUNet + Ranger*, *nnUnet + GWDL* and *nnUNet + DRO*. The model *nnUNet* is not included in the ensemble because the model *nnUNet* performed less well than all the other methods in terms of both Dice scores and Hausdorff distances on the three regions of interest.

3.3   Mean Segmentation Performance

Mean Dice scores and Hausdorff distances for the whole tumor, the core tumor, and the enhancing tumor can be found in Table 1.

In terms of mean Dice scores, *nnUNet + DRO* is the only non-ensembling model to outperform *nnUNet* in all regions of interest. *nnUNet + GWDL* and *nnUNet + Ranger* ouperform *nnUNet* for enhancing tumor and whole tumor. Among the non-ensembling models, *nnUNet + DRO*, *nnUNet + GWDL* and *nnUNet + Ranger* appear as complementary as they all achieve the top mean Dice score for one of the regions of interest. That was the motivation for ensembling those three models.

In terms of mean Hausdorff distances, *nnUNet + DRO*, *nnUNet + GWDL* and *nnUNet + Ranger* outperform *nnUNet* for all regions of interest.

The ensemble outperformed all the other models for all regions in terms of both mean Dice scores and mean Hausdorff distances.

The results of the ensemble on the BraTS 2020 testing dataset are reported in Table 2. It is those results that were used to rank the different competitors. Our ensemble ranked fourth for the segmentation task.

3.4   Robustness Performance

In the summary of the BraTS 2018 challenge, the organizers emphasized the need for more robust automatic brain tumor segmentation algorithms [6]. The authors also suggest using the interquartile range (IQR) of the Dice scores to compare the robustness of the different methods. IQR for the Dice scores for our models can be found in Table 1.

Table 1. Segmentation results on the BraTS 2020 Validation dataset. The evaluation was performed on the BraTS online evaluation platform. The ensemble includes all the single models **except** the original nnUNet (first model). GWDL: Generalized Wasserstein Dice Loss [11], DRO: Distributionally Robust Optimization [13], ET: Enhancing Tumor, WT: Whole Tumor, TC: Tumor Core, Std: Standard deviation, IQR: Interquartile range. Best values restricted to single models are in bold. Best values among all models (including ensemble) are in bold and underlined.

Model	ROI	Dice score (%)				Hausdorff 95% (mm)			
		Mean	Std	Median	IQR	Mean	Std	Median	IQR
nnUNet [16]	ET	74.0	29.9	86.9	15.2	38.9	109.5	2.0	2.2
	WT	90.5	7.3	92.7	6.1	5.2	8.6	3.0	3.3
	TC	83.9	17.0	90.1	13.7	9.4	34.6	3.0	4.6
nnUNet	ET	**77.4**	28.2	87.6	12.1	32.7	101.0	1.7	2.0
+	WT	90.6	7.0	92.8	6.3	4.7	6.5	2.8	2.6
Ranger [23,37]	TC	83.8	18.1	91.3	14.3	9.0	34.5	2.4	4.6
nnUNet	ET	76.7	28.0	87.4	12.6	**29.8**	96.1	2.0	2.0
+	WT	**90.8**	6.6	92.9	5.5	**4.6**	6.7	3.0	2.5
GWDL [11]	TC	83.3	16.0	90.2	15.9	6.9	11.4	3.2	5.3
nnUNet	ET	75.6	28.6	87.5	12.6	32.5	100.9	2.0	2.3
+	WT	90.6	7.0	92.5	5.9	**4.6**	6.1	3.0	3.0
DRO [13]	TC	**84.1**	16.2	90.1	12.5	**6.1**	10.4	3.0	4.0
Ensemble	ET	**<u>77.6</u>**	27.4	87.6	11.1	**<u>26.8</u>**	91.1	1.7	2.0
Mean	WT	**<u>91.0</u>**	6.5	92.9	6.3	**<u>4.4</u>**	6.0	2.8	2.9
Softmax	TC	**<u>84.4</u>**	15.6	90.8	12.4	**<u>5.8</u>**	10.2	2.8	4.3

Table 2. Segmentation results on the BraTS 2020 Testing dataset. The evaluation was performed by the BraTS 2020 organizers. ET: Enhancing Tumor, WT: Whole Tumor, TC: Tumor Core, Std: Standard deviation, IQR: Interquartile range.

Model	ROI	Dice score (%)				Hausdorff 95% (mm)			
		Mean	Std	Median	IQR	Mean	Std	Median	IQR
Ensemble	ET	81.4	19.5	85.9	13.9	15.8	69.5	1.4	1.2
Mean	WT	88.9	11.6	92.4	7.0	6.4	29.0	2.9	3.0
Softmax	TC	84.1	24.5	92.6	9.2	19.4	74.7	2.2	2.6

Ensembling and Distributionally Robust Optimization (DRO) [13] are two methods that have been empirically shown to decrease the IQR for brain tumor segmentation. Among the non-ensembling models, *nnUNet + DRO* is the only one to achieve lower Dice scores IQR than *nnUNet* for all the region of interest. The ensemble achieves the lowest Dice scores IQR for the enhancing tumor and the core tumor regions.

4   Conclusion

In this paper, we experimented with three of the main ingredients of deep learning optimization to compete in the BraTS 2020 challenge.

Our results suggest that the segmentation mean performance and robustness of nnUNet [16] can be improved using distributionally robust optimization [13], the generalized Wasserstein Dice Loss, and the Ranger optimizer [23,37]. Those three features appeared as complementary, and we achieved our top segmentation performance by ensembling three neural networks, each trained using one of them. In future work, we will explore the combination of those three features to train a single deep neural network. Our ensemble ranked fourth out of the 78 participating teams at the segmentation task of the BraTS 2020 challenge after evaluation on the BraTS 2020 testing dataset.

Acknowledgments. This project has received funding from the European Union's Horizon 2020 research and innovation program under the Marie Skłodowska-Curie grant agreement TRABIT No 765148; Wellcome [203148/Z/16/Z; WT101957], EPSRC [NS/A000049/1; NS/A000027/1]. Tom Vercauteren is supported by a Medtronic / RAEng Research Chair [RCSRF1819/7/34]. We would like to thank Luis Carlos Garcias-Peraza-Herrera for helpful discussions and his feedback on a preliminary version of this paper. We also thank the anonymous reviewers for their suggestions.

References

1. Andres, E.A., et al.: Po-1002 pseudo computed tomography generation using 3D deep learning-application to brain radiotherapy. Radiother. Oncol. **133**, S553 (2019)
2. Andres, E.A., et al.: Dosimetry-driven quality measure of brain pseudo computed tomography generated from deep learning for MRI-only radiotherapy treatment planning. Int. J. Radiat. Oncol. Biol. Phys. **108**(3), 813–823 (2020)
3. Bakas, S., et al.: Advancing the cancer genome atlas glioma MRI collections with expert segmentation labels and radiomic features. Sci. Data **4**, 170117 (2017)
4. Bakas, S., et al.: Segmentation labels and radiomic features for the pre-operative scans of the TCGA-GBM collection. The Cancer Imaging Archive (2017). https://doi.org/10.7937/K9/TCIA.2017.KLXWJJ1Q
5. Bakas, S., et al.: Segmentation labels and radiomic features for the pre-operative scans of the TCGA-LGG collection. The Cancer Imaging Archive (2017). https://doi.org/10.7937/K9/TCIA.2017.GJQ7R0EF
6. Bakas, S., et al.: Identifying the best machine learning algorithms for brain tumor segmentation, progression assessment, and overall survival prediction in the brats challenge. arXiv preprint arXiv:1811.02629 (2018)
7. Chandra, S., et al.: Context aware 3D CNNs for brain tumor segmentation. In: Crimi, A., Bakas, S., Kuijf, H., Keyvan, F., Reyes, M., van Walsum, T. (eds.) BrainLes 2018. LNCS, vol. 11384, pp. 299–310. Springer, Cham (2019). https://doi.org/10.1007/978-3-030-11726-9_27
8. Chouzenoux, E., Gérard, H., Pesquet, J.C.: General risk measures for robust machine learning. Found. Data Sci. **1**, 249 (2019)

9. Çiçek, Ö., Abdulkadir, A., Lienkamp, S.S., Brox, T., Ronneberger, O.: 3D U-Net: learning dense volumetric segmentation from sparse annotation. In: Ourselin, S., Joskowicz, L., Sabuncu, M.R., Unal, G., Wells, W. (eds.) MICCAI 2016. LNCS, vol. 9901, pp. 424–432. Springer, Cham (2016). https://doi.org/10.1007/978-3-319-46723-8_49

10. Eaton-Rosen, Z., et al.: Using niftynet to ensemble convolutional neural nets for the brats challenge. Proceedings of the 6th MICCAI-BRATS Challenge, pp. 61–66 (2017)

11. Fidon, L., et al.: Generalised wasserstein dice score for imbalanced multi-class segmentation using holistic convolutional networks. In: Crimi, A., Bakas, S., Kuijf, H., Menze, B., Reyes, M. (eds.) BrainLes 2017. LNCS, vol. 10670, pp. 64–76. Springer, Cham (2018). https://doi.org/10.1007/978-3-319-75238-9_6

12. Fidon, L., et al.: Scalable multimodal convolutional networks for brain tumour segmentation. In: Descoteaux, M., Maier-Hein, L., Franz, A., Jannin, P., Collins, D.L., Duchesne, S. (eds.) MICCAI 2017. LNCS, vol. 10435, pp. 285–293. Springer, Cham (2017). https://doi.org/10.1007/978-3-319-66179-7_33

13. Fidon, L., Ourselin, S., Vercauteren, T.: SGD with hardness weighted sampling for distributionally robust deep learning. arXiv preprint arXiv:2001.02658 (2020)

14. Gibson, E., et al.: NiftyNet: a deep-learning platform for medical imaging. Comput. Methods Programs Biomed. **158**, 113–122 (2018)

15. Ioffe, S., Szegedy, C.: Batch normalization: accelerating deep network training by reducing internal covariate shift. In: Proceedings of Machine Learning Research, vol. 37, pp. 448–456. PMLR, Lille, France (2015). http://proceedings.mlr.press/v37/ioffe15.html

16. Isensee, F., Jäger, P.F., Kohl, S.A., Petersen, J., Maier-Hein, K.H.: Automated design of deep learning methods for biomedical image segmentation. arXiv preprint arXiv:1904.08128 (2020)

17. Isensee, F., Kickingereder, P., Wick, W., Bendszus, M., Maier-Hein, K.H.: No new-net. In: Crimi, A., Bakas, S., Kuijf, H., Keyvan, F., Reyes, M., van Walsum, T. (eds.) BrainLes 2018. LNCS, vol. 11384, pp. 234–244. Springer, Cham (2019). https://doi.org/10.1007/978-3-030-11726-9_21

18. Jiang, Z., Ding, C., Liu, M., Tao, D.: Two-stage cascaded u-net: 1st place solution to brats challenge 2019 segmentation task. In: Crimi, A., Bakas, S. (eds.) BrainLes 2019. LNCS, vol. 11992, pp. 231–241. Springer, Cham (2020). https://doi.org/10.1007/978-3-030-46640-4_22

19. Kamnitsas, K., et al.: Ensembles of multiple models and architectures for robust brain tumour segmentation. In: Crimi, A., Bakas, S., Kuijf, H., Menze, B., Reyes, M. (eds.) BrainLes 2017. LNCS, vol. 10670, pp. 450–462. Springer, Cham (2018). https://doi.org/10.1007/978-3-319-75238-9_38

20. Kamnitsas, K., et al.: DeepMedic for brain tumor segmentation. In: International Workshop on Brainlesion: Glioma, Multiple Sclerosis, Stroke and Traumatic Brain Injuries, pp. 138–149. Springer, Berlin (2016)

21. Kingma, D.P., Ba, J.: Adam: a method for stochastic optimization. arXiv preprint arXiv:1412.6980 (2014)

22. Lee, C.Y., Xie, S., Gallagher, P., Zhang, Z., Tu, Z.: Deeply-supervised nets. In: Artificial Intelligence and Statistics, pp. 562–570 (2015)

23. Liu, L., et al.: On the variance of the adaptive learning rate and beyond. In: 8th International Conference on Learning Representations, ICLR 2020 (2020)

24. Lyksborg, M., Puonti, O., Agn, M., Larsen, R.: An ensemble of 2D convolutional neural networks for tumor segmentation. In: Paulsen, R.R., Pedersen, K.S. (eds.) SCIA 2015. LNCS, vol. 9127, pp. 201–211. Springer, Cham (2015). https://doi.org/10.1007/978-3-319-19665-7_17

25. McKinley, R., Rebsamen, M., Meier, R., Wiest, R.: Triplanar ensemble of 3D-to-2D CNNs with label-uncertainty for brain tumor segmentation. In: Crimi, A., Bakas, S. (eds.) BrainLes 2019. LNCS, vol. 11992, pp. 379–387. Springer, Cham (2020). https://doi.org/10.1007/978-3-030-46640-4_36

26. Menze, B.H., et al.: The multimodal brain tumor image segmentation benchmark (brats). IEEE Trans. Med. Imaging **34**(10), 1993–2024 (2014)

27. Milletari, F., Navab, N., Ahmadi, S.A.: V-net: fully convolutional neural networks for volumetric medical image segmentation. In: 2016 Fourth International Conference on 3D Vision (3DV), pp. 565–571. IEEE (2016)

28. Myronenko, A.: 3D MRI brain tumor segmentation using autoencoder regularization. In: Crimi, A., Bakas, S., Kuijf, H., Keyvan, F., Reyes, M., van Walsum, T. (eds.) BrainLes 2018. LNCS, vol. 11384, pp. 311–320. Springer, Cham (2019). https://doi.org/10.1007/978-3-030-11726-9_28

29. Namkoong, H., Duchi, J.C.: Stochastic gradient methods for distributionally robust optimization with f-divergences. In: Advances in Neural Information Processing Systems, pp. 2208–2216 (2016)

30. Rahimian, H., Mehrotra, S.: Distributionally robust optimization: a review. arXiv preprint arXiv:1908.05659 (2019)

31. Ronneberger, O., Fischer, P., Brox, T.: U-Net: convolutional networks for biomedical image segmentation. In: Navab, N., Hornegger, J., Wells, W.M., Frangi, A.F. (eds.) MICCAI 2015. LNCS, vol. 9351, pp. 234–241. Springer, Cham (2015). https://doi.org/10.1007/978-3-319-24574-4_28

32. Sudre, C.H., Li, W., Vercauteren, T., Ourselin, S., Jorge Cardoso, M.: Generalised dice overlap as a deep learning loss function for highly unbalanced segmentations. In: Cardoso, M.J., et al. (eds.) DLMIA/ML-CDS -2017. LNCS, vol. 10553, pp. 240–248. Springer, Cham (2017). https://doi.org/10.1007/978-3-319-67558-9_28

33. Tilborghs, S., et al.: Comparative study of deep learning methods for the automatic segmentation of lung, lesion and lesion type in CT scans of COVID-19 patients. arXiv preprint arXiv:2007.15546 (2020)

34. Ulyanov, D., Vedaldi, A., Lempitsky, V.: Instance normalization: the missing ingredient for fast stylization. arXiv preprint arXiv:1607.08022 (2016)

35. Wang, G., Li, W., Aertsen, M., Deprest, J., Ourselin, S., Vercauteren, T.: Aleatoric uncertainty estimation with test-time augmentation for medical image segmentation with convolutional neural networks. Neurocomputing **338**, 34–45 (2019)

36. Wang, G., Li, W., Ourselin, S., Vercauteren, T.: Automatic brain tumor segmentation using cascaded anisotropic convolutional neural networks. In: Crimi, A., Bakas, S., Kuijf, H., Menze, B., Reyes, M. (eds.) BrainLes 2017. LNCS, vol. 10670, pp. 178–190. Springer, Cham (2018). https://doi.org/10.1007/978-3-319-75238-9_16

37. Zhang, M., Lucas, J., Ba, J., Hinton, G.E.: Lookahead optimizer: k steps forward, 1 step back. In: Advances in Neural Information Processing Systems, pp. 9593–9604 (2019)

38. Zhao, Y.-X., Zhang, Y.-M., Liu, C.-L.: Bag of tricks for 3D MRI brain tumor segmentation. In: Crimi, A., Bakas, S. (eds.) BrainLes 2019. LNCS, vol. 11992, pp. 210–220. Springer, Cham (2020). https://doi.org/10.1007/978-3-030-46640-4_20

3D Semantic Segmentation of Brain Tumor for Overall Survival Prediction

Rupal R. Agravat[✉][iD] and Mehul S. Raval[iD]

Ahmedabad University, Ahmedabad, Gujarat, India
rupal.agravat@iet.ahduni.edu.in, mehul.raval@ahduni.edu.in

Abstract. Glioma, a malignant brain tumor, requires immediate treatment to improve the survival of patients. The heterogeneous nature of Glioma makes the segmentation difficult, especially for sub-regions like necrosis, enhancing tumor, non-enhancing tumor, and edema. Deep neural networks like full convolution neural networks and an ensemble of fully convolution neural networks are successful for Glioma segmentation. The paper demonstrates the use of a 3D fully convolution neural network with a three-layer encoder-decoder approach. The dense connections within the layer help in diversified feature learning. The network takes 3D patches from T_1, T_2, T_1c, and FLAIR modalities as input. The loss function combines dice loss and focal loss functions. The Dice similarity coefficient for training and validation set is 0.88, 0.83, 0.78 and 0.87, 0.75, 0.76 for the whole tumor, tumor core and enhancing tumor, respectively. The network achieves comparable performance with other state-of-the-art ensemble approaches. The random forest regressor trains on the shape, volumetric, and age features extracted from ground truth for overall survival prediction. The regressor achieves an accuracy of 56.8% and 51.7% on the training and validation sets.

Keywords: Brain tumor segmentation · Deep learning · Dense network · Overall survival · Radiomics features · Random forest regressor · U-net

1 Introduction

Early-stage brain tumor diagnosis can lead to proper treatment planning, which improves patient survival chances. Out of all types of brain tumors, Glioma is one of the most life-threatening brain tumors. It occurs in the glial cells of the brain. Depending on its severity and aggressiveness, Glioma has grades ranging from I to IV. Grade I, II are Low-Grade Glioma (LGG), and grade III and IV are High-Grade Glioma (HGG). It can further be divided into constituent structures like - necrosis, enhancing tumor, non-enhancing tumor, and edema. The core consists of necrosis, enhancing tumor, non-enhancing tumor. In most cases LGG does not contain enhancing tumor, whereas HGG contains necrosis, enhancing, and non-enhancing structures. Edema occurs from infiltrating tumor cells and biological response to the angiogenic and vascular permeability factors

A. Crimi and S. Bakas (Eds.): BrainLes 2020, LNCS 12659, pp. 215–227, 2021.
https://doi.org/10.1007/978-3-030-72087-2_19

released by the spatially adjacent tumor cells [3]. Non-invasive Medical Resonance Imaging (MRI) is the most advisable imaging technique as it captures the
functioning of soft tissue adequately compared to other imaging techniques. MR
images are prone to inhomogeneity introduced by the surrounding magnetic field,
which introduces the artifacts in the captured image. Besides, the appearance of
various brain tissues is different in various modalities. Such issues increase the
time in the study of the image.

The treatment planning is highly dependent on the accurate tumor structure
segmentation, but due to heterogeneous nature of Glioma, the segmentation
task becomes difficult. Furthermore, the human interpretation of the image is
non-reproducible as well as dependent on expertise. It requires computer-aided
MR image interpretation to locate the tumor. Also, even if the initially detected
tumor is completely resected, such patients have poor survival prognosis, as
metastases may still redevelop. It leads to an open question about overall survival
prediction.

Authors in [1] discussed the basic, generative, and discriminative techniques
for brain tumor segmentation. Nowadays, Deep Neural Network (DNN) has
gained more attention for the segmentation of biological images. The Convolution Neural Networks (CNN), like DeepMedic [15], U-net [22], V-Net [20],
SegNet [5], ResNet [12], DenseNet [13] give state-of-the-art results for semantic
segmentation. Out of all these methods, U-net is a widely accepted end-to-end
segmentation architecture for brain tumors. U-net is an encoder-decoder architecture, which reduces the size of feature maps to half and doubles the number
of feature maps at every encoder layer. The process is reversed at every decoder
layer. The skip connections between the peer layers of U-net help in proper
feature reconstruction.

1.1   Literature Review: BraTS 2019

Segmentation. Authors in [14] used the ensemble of twelve encoder-decoder
models, where each model is made up of a cascaded network. The first network
in a model finds the coarse segmentation, which was given as input in the second network, and the input images to predict all the labels. The network losses
combine at different stages for better network parameter tuning. The Dice Similarity Coefficient (DSC) for the validation set is 0.91, 0.87, 0.80 for Whole Tumor
(WT), Tumor Core (TC), and Enhancing Tumor (ET), respectively.

In [28], authors applied various data processing methods, network design
methods, and optimization methods to learn the segmentation labels at every
iteration. The student models combined at the teacher-level model with successive output merging. The loss function is the combination of dice loss and
cross-entropy loss for the networks trained on various input patch sizes. The
method achieved the DSC of 0.91, 0.84, and 0.75 for WT, TC, and ET, respectively.

The approach demonstrated in [18] used thirty Heteroscedastic classification
models to find the variance of all the models for the ensemble. The focal loss

forms the loss function. Various post-processing techniques were applied to fine-tune the network segmentation. The DSC achieved for the approach was 0.91, 0.83, and 0.77 for WT, TC, and ET.

Authors in [23] used fifteen 2D FCNN models working in parallel on axial, coronal and sagittal views of images. The approach used in [11] focused on the attention mechanism applied for the ensemble of four 3D FCNN models. In [16], authors used the ensemble of four 2D FCNN models working on different sets of images based on the size of the tumor.

Survival Prediction. Authors in [2] implemented 2D U-net with dense modules at the encoder part and convolution modules at the decoder part along with focal loss function at training time. The segmentation results fed into Random Forest Regressor(RFR) to predict the Overall Survival (OS) of the patients. The RFR trains on the age, shape, and volumetric features extracted from the ground truth provided with the training dataset. They achieved 58.6% OS accuracy on the validation set.

Authors in [27] used vanilla U-net and U-net with attention blocks to make the ensemble of six models based on various input patches and the presence/absence of attention blocks. The linear regressor trains selected radiomics features along with the relative invasiveness coefficient. The DSC achieved on the validation set was 0.90, 0.83, and 0.79 for WT, TC, and ET, respectively, and the OS accuracy was 59%.

Authors in [10] implemented the ensemble of six models, which are the variation of U-net with different patch sizes, feature maps with several layers in the encoder-decoder architecture. For OS prediction, six features were extracted from the segmentation results to train the linear regression. The DSC achieved on the validation set was 0.91, 0.80, and 0.74 for WT, TC, and ET, and the OS accuracy was 31%.

Authors in [26] used the U-net variation, where the additional branch of prediction uses Variational Encoder. The OS prediction used the volumetric and age features to train ANN with two layers, each with 64 neurons.

Except the approach demonstrated in [2,26], all the other approaches use an ensemble of the segmentation prediction networks. There are certain disadvantages of ensemble approaches: (1) ensemble methods are usually computationally expensive. Therefore, they add learning time, and memory constraints to the problem, (2) use of ensemble methods reduces the model interpretability due to increased complexity and makes it very difficult to understand.

The focus of this paper is to develop a robust 3D fully convolutional neural network (FCNN) for tumor segmentation along with RFR [2] to predict OS of high grade glioma (HGG) patients. The remaining paper is as follows: Sect. 2 focuses on the BraTS 2020 dataset, Sect. 3 demonstrates the proposed methods for tumor segmentation and OS prediction, Sect. 4 provides implementation details, and Sect. 5 discusses the results followed by the conclusion and future work.

2   Dataset

The dataset [8,9,19] contains 293 HGG and 76 LGG pre-operative scans in four Magnetic Resonance Images (MRI) modalities, which are T_1, T_2, T_1c and FLAIR. One to four raters have segmented the images using the same annotation tool to prepare the ground truths. The annotations were approved by experienced neuro-radiologists [6,7]. Annotations have the enhancing tumor (ET label 4), the peritumoral edema (ED label 2), and the necrotic and non-enhancing tumor core (NCR/NET label 1). The T_2, T_1c and FlAIR images of a single scan are co-registered with the anatomical template of T_1 image of the same scan. All the images are interpolated to the same resolution (1 mm × 1 mm × 1 mm), and skull-stripped. Features like age, survival days and resection status for 237 HGG scans are provided separately for OS. The validation and test datasets consist of 125 and 166 MRI scans respectively, with the same pre-processing. The dataset includes age and resection status for all the sets along with survival days for the training set.

3   Proposed Method

3.1   Task 1: Tumor Segmentation

A FCNN provides end-to-end semantic segmentation for the input of the arbitrary size and learns global information related to it. Moreover, the 3D FCNN gathers spatial relationships between the voxels. Our network is an extension of our previous work proposed in [2], where the network had poorly performed on validation set and test set. The network overfitted the training set and could not learn the 3D voxel relationship. The proposed work includes 3D modules with increased network depth. The network uses three-layer encoder-decoder architecture with dense connections between the convolution layers and skip-connections across peer layers. The network is as shown in Fig. 1. The Batch Normalization (BN) and Parametric ReLU (PReLU) activation function follow the convolution layer in each dense module.

Dense connections between the layers in the dense module allows to obtain additional inputs(collective knowledge) from all earlier layers and passes on its feature-maps to all subsequent layers. It allows the gradient to flow to the earlier layers directly, which provides in-depth supervision on preceding layers by the classification layer. Also, dense connections provide diversified features to the layers, which leads to detailed pattern identification capabilities. The layers generate 64, 128 and 256 feature maps and the bottleneck layer generates 512 feature maps. The 1 × 1 × 1 convolution at the end generates a single probability map for multi-class classifications.

Brain tumor segmentation task deals with a highly imbalanced dataset where tumorous slices are less than non-tumorous slices; such an imbalance dataset reduces network accuracy. Two approaches deal with the issue: 1) The patch-based input to the network guarantees that the network does not overlearn the

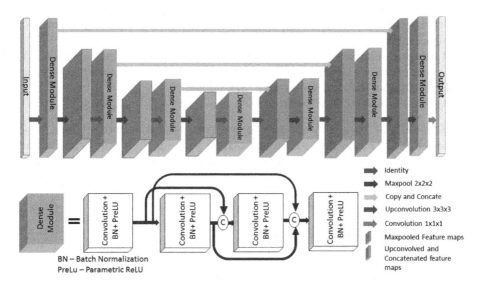

Fig. 1. The proposed 3D encoder-decoder FCNN.

background voxels, 2) the network trains with the combination of the following loss functions.

– Soft Dice Loss: is a measure to find overlap between two regions.

$$Soft\ Dice\ Loss = 1 - \frac{2\sum_{voxels} y_{true} y_{pred}}{\sum_{voxels} y_{pred}^2 + \sum_{voxels} y_{true}^2}$$

y_{true} represents ground truth and y_{pred} represents network output probability. The dice loss function directly considers the predicted probabilities without converting them into binary output. The numerator provides standard correct predictions between input and target, whereas the denominator provides individual separate correct predictions. This ratio normalizes the loss according to the target mask and allows learning even from the minimal spatial representation of the target mask.

– Focal Loss [17]: It is dependent on the network probability p_t. It balances negative and positive samples by tuning weighting parameter α. It also deals with easy and hard examples by tuning the focusing parameter γ.

$$FL(p_t) = -\alpha_t (1 - p_t)^\gamma\ log(p_t) \tag{1}$$

The modulating factor $(1 - p_t)^\gamma$ adjusts the rate at which easy examples are down-weighted.

3.2   Task 2: Overall Survival Prediction

OS prediction deals with predicting the number of days for which patients survive after the tumor is resected and proper post-operative treatment given. We have used the following features to train RFR:

- **Statistical Features**: the amount of edema, amount of necrosis, amount of enhancing tumor, the extent of tumor and proportion of tumor.
- **Radiomic Features** [25] **for necrosis**: Elongation, flatness, minor axis length, primary axis length, 2D diameter row, 2D diameter column, sphericity, surface area, 2D diameter slice, 3D diameter.
- **Age**(available with BraTS dataset).

Necrosis plays a significant role in the treatment of tumors. Gross Total Resection (GTR) of necrosis is comparatively easy vis a vis enhancing tumor. Considering this, shape features of necrosis are extracted using a radiomics package [25]. In addition to these features, whole tumor statistical features from the segmentation results and age are considered to train RFR.

4 Implementation Details

4.1 Pre-processing

Pre-processing boosts network training and improves performance. All four modality images are biased field corrected followed by denoising, and Z-score normalization on individual MR sequence is applied where each sequence was subtracted by its mean from the data and divided by its standard deviation. Data augmentation happens by flipping the patches around the vertical axis.

4.2 Training

Input to the network is patches of size $64 \times 64 \times 64$ from four modalities (T1, T2, T1c, FLAIR). The network trains on the entire training image dataset. The network uses the combination of two loss functions: 1) dice loss function and 2) focal loss function with $\alpha = 1$ and $\gamma = 2$. The network trains for 610 epochs with batch size 1. The batch normalization with batch size 1 does not speed up the network learning but helps during testing, where it uses the statistics of the batch rather than the running statistics. This layer normalization approach [4] normalizes each feature map with its own mean and standard deviation.

The sliding window approach provides the output for each subject. The stride size is reduced to half of the training window size to overcome the boundary voxels' unstable prediction issue. The output of the original patch and flipped patch is predicted and averaged to generate the final output. The prediction of a single image takes around one minute.

4.3 Post-processing

The post-processing includes two methods: 1) The connected component analysis (CCA) removes the tumor with a volume less than thousand voxels, 2) enhancing tumor is formed in the surrounding of the necrosis. and its size cannot be very small in HGG. Such small size enhancing tumor is converted to necrosis. The empirically chosen threshold for the conversion is three hundred.

5   Results

5.1   Segmentation

The achieved DSC, sensitivity, specificity and Hausdorff95 for training set is in
Table 1, and for the validation set in Table 2. The results shows the improvement with post-processing, hence the test set results are generated with post-
processing as shown in Table 3.

Table 1. Various evaluation measures for BraTS 2020 training set.

Evaluation measure	Statistics	Without post-process			With post-process		
		ET	WT	TC	ET	WT	TC
DSC	Mean	0.757	0.881	0.831	0.782	0.882	0.832
	StdDev	0.267	0.115	0.191	0.246	0.116	0.191
	Median	0.863	0.919	0.908	0.872	0.919	0.910
	25quantile	0.751	0.860	0.802	0.779	0.862	0.805
	75quantile	0.906	0.941	0.943	0.912	0.943	0.944
Sensitivity	Mean	0.759	0.846	0.802	0.782	0.844	0.801
	StdDev	0.272	0.156	0.209	0.252	0.158	0.209
	Median	0.858	0.896	0.883	0.864	0.896	0.883
	25quantile	0.748	0.797	0.763	0.762	0.795	0.763
	75quantile	0.921	0.946	0.931	0.925	0.945	0.931
Specificity	Mean	0.999	0.999	0.999	0.999	0.999	0.999
	StdDev	0.000	0.000	0.000	0.000	0.000	0.000
	Median	0.999	0.999	0.999	0.999	0.999	0.999
	25quantile	0.999	0.999	0.999	0.999	0.999	0.999
	75quantile	0.999	0.999	0.999	0.999	0.999	0.999
Hausdorff95	Mean	31.531	06.508	07.275	29.274	06.232	06.999
	StdDev	95.090	08.588	10.946	94.956	08.134	20.507
	Median	01.732	03.606	03.606	01.414	03.464	03.317
	25quantile	01.414	02.236	02.000	01.000	02.236	02.000
	75quantile	04.123	07.071	08.602	03.162	06.939	07.874

The proposed model outperforms some of the ensemble approaches which
is shown in Table 4. Figure 2 shows the successful segmentation of the tumor.
The false positive segmentation voxels are removed in the post-processing. The
network fails to segment the tumor for some HGG images and many LGG images.
One such segmentation failure is shown in Fig. 3. The failure of the network is
observed for: 1) small size of the entire tumor, 2) small size of necrosis, and
3) absence/small size of enhancing tumor. Figure 4 depicts the box plot of the
evaluation metrics, where the red marked cases shows the segmentation failure.

Table 2. Various evaluation measures for BraTS 2020 validation set.

Evaluation measure	Statistics	Without post-process			With post-process		
		ET	WT	TC	ET	WT	TC
DSC	Mean	0.686	0.876	0.725	0.763	0.873	0.753
	StdDev	0.307	0.093	0.284	0.259	0.098	0.263
	Median	0.835	0.914	0.866	0.852	0.908	0.878
	25quantile	0.617	0.863	0.595	0.751	0.856	0.711
	75quantile	0.889	0.932	0.924	0.899	0.935	0.926
Sensitivity	Mean	0.704	0.858	0.674	0.759	0.847	0.713
	StdDev	0.322	0.138	0.312	0.273	0.149	0.288
	Median	0.853	0.901	0.819	0.852	0.897	0.841
	25quantile	0.595	0.826	0.432	0.715	0.809	0.606
	75quantile	0.926	0.952	0.915	0.933	0.954	0.921
Specificity	Mean	0.999	0.999	0.999	0.999	0.999	0.999
	StdDev	0.000	0.000	0.000	0.001	0.000	0.000
	Median	0.999	0.999	0.999	0.999	0.999	0.999
	25quantile	0.999	0.999	0.999	0.998	0.999	0.999
	75quantile	0.999	0.999	0.999	0.999	0.999	0.999
Hausdorff95	Mean	43.635	09.475	14.538	27.704	07.038	10.873
	StdDev	109.143	15.215	38.067	90.918	09.348	33.823
	Median	02.828	04.000	05.099	02.236	03.742	04.690
	25quantile	01.414	02.236	02.236	01.414	02.449	02.236
	75quantile	10.770	07.550	10.724	04.242	06.480	11.045

5.2 OS Prediction

The RFR trains on features extracted from the 213 ground truth images. In the trained RFR, features of network segmented images predict OS days. If the network fails to identify/segment necrosis from the image, then the feature extractor considers the absence of the necrosis and marks all the features except age as zero. OS accuracy on training, validation and test datasets for the images with GTR resection status is in Table 5.

The reduced performance of RFR on validation and test sets shows that it overfits the training dataset. Still, its performance is comparable with other approaches as shown in Table 6.

Table 3. Various evaluation measures for BraTS 2020 test set.

Evaluation measure	Statistics	With post-process		
		ET	WT	TC
DSC	Mean	0.779	0.875	0.815
	StdDev	0.232	0.112	0.250
	Median	0.847	0.910	0.913
	25quantile	0.760	0.855	0.833
	75quantile	0.908	0.935	0.948
Sensitivity	Mean	0.809	0.863	0.817
	StdDev	0.245	0.136	0.244
	Median	0.896	0.910	0.908
	25quantile	0.785	0.823	0.795
	75quantile	0.940	0.947	0.952
Specificity	Mean	0.999	0.999	0.999
	StdDev	0.000	0.001	0.001
	Median	0.999	0.999	0.999
	25quantile	0.999	0.999	0.999
	75quantile	0.999	0.999	0.999
Hausdorff95	Mean	27.078	08.302	21.611
	StdDev	92.554	30.007	74.651
	Median	01.732	03.464	03.162
	25quantile	01.103	02.000	01.799
	75quantile	02.734	06.164	07.858

Table 4. DSC comparison with state-of-art ensemble approaches.

Ref.	Type of network	No. of networks	DSC		
			ET	WT	TC
[11]	3D FCNN	4	0.67	0.87	0.73
[16]	2D FCNN	4	0.68	0.84	0.73
[23]	2D FCNN	15	0.71	0.85	0.71
Proposed	3D FCNN	1	**0.76**	0.87	**0.75**

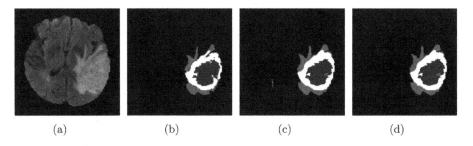

(a)	(b)	(c)	(d)

Fig. 2. Correct segmentation results of the network (a) FLAIR slice (b) Ground truth (c) Segmentation without post-processing (d) Segmentation after post-processing.

Table 5. OS accuracy for BraTS 2020 training, validation and test datasets.

Dataset	Accuracy	MSE	MedianSE	StdSE	SpearmanR
Training	0.568	083165.963	21481.525	0181697.874	0.596
Validation	0.517	116083.477	43974.090	0168176.159	0.217
Test	0.477	382492.357	46612.810	1081670.063	0.333

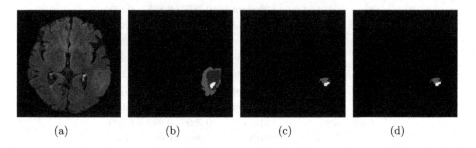

(a)   (b)   (c)   (d)

Fig. 3. Incorrect segmentation results of the network (a) FLAIR slice (b) Ground truth (c) Segmentation without post-processing (d) Segmentation after post-processing.

Table 6. Comparative analysis of OS accuracy on BraTS 2019 validation set.

Ref.	Approach	# Features	Accuracy (%)
[10]	Linear Regression	9	31.0
[21]	Linear Regression	NA	51.5
[26]	Artifical Neural Network	7	45.0
Proposed	RFR	18	**51.7**

According to the study [24], gender plays a vital role in response to tumor treatment. The females respond to the post-operative treatment better compared to males, which improves their life expectancy. The inclusion of the 'gender' feature into the existing feature list can significantly improve OS accuracy.

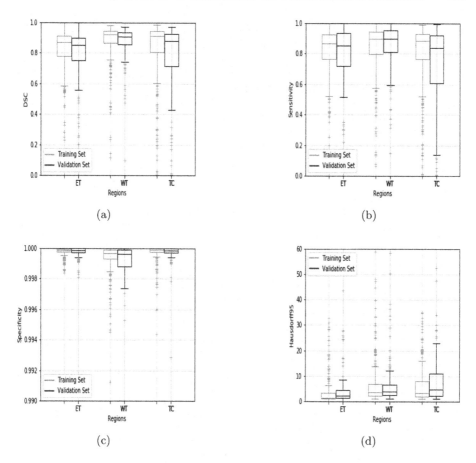

Fig. 4. The box plot (a) DSC (b) Sensitivity (c) Specificity (d) Hausdorff95

6   Conclusion

The proposal uses three-layer deep 3D U-net based encoder-decoder architecture for semantic segmentation. Each encoding and decoding layer module incorporates dense connections which allow the diversified feature learning and gradient propagation to the intial layers. The patch selection and loss function approaches improved the performance of the network. The pre-processing, post-processing and patch selection methods play vital role in the robust performance of the network. The network outperforms some of the state-of-the-art ensemble approaches. The network fails where the size of either the entire tumor or its sub-component (necrosis and enhancing tumor) is comparatively small. The smaller subcomponent size is observed in LGG cases where the network fails significantly. The age, statistical, and necrosis shape features of the ground truth train RFR with five-fold cross-validation for OS prediction. Later, network segmentation for cases with GTR tests RFR for OS prediction. The RFR performs better than other state-of-the-art approaches which use linear regression and ANN.

Acknowledgement. The authors would like to thank NVIDIA Corporation for donating the Quadro K5200 and Quadro P5000 GPU used for this research, Dr. Krutarth Agravat (Medical Officer, Essar Ltd) for clearing our doubts related to medical concepts, Ujjawal Baid for his fruitful discussion in network selection. The authors acknowledge continuous support from Professor Sanjay Chaudhary, Professor N. Padmanabhan, and Professor Manjunath Joshi.

References

1. Agravat, R.R., Raval, M.S.: Deep learning for automated brain tumor segmentation in MRI images. In: Soft Computing Based Medical Image Analysis, pp. 183–201. Elsevier (2018)
2. Agravat, R.R., Raval, M.S.: Brain tumor segmentation and survival prediction. In: Crimi, A., Bakas, S. (eds.) BrainLes 2019. LNCS, vol. 11992, pp. 338–348. Springer, Cham (2020). https://doi.org/10.1007/978-3-030-46640-4_32
3. Akbari, H., et al.: Pattern analysis of dynamic susceptibility contrast-enhanced MR imaging demonstrates peritumoral tissue heterogeneity. Radiology **273**(2), 502–510 (2014)
4. Ba, J.L., Kiros, J.R., Hinton, G.E.: Layer normalization. arXiv preprint arXiv:1607.06450 (2016)
5. Badrinarayanan, V., Kendall, A., Cipolla, R.: Segnet: a deep convolutional encoder-decoder architecture for image segmentation. IEEE Trans. Pattern Anal. Mach. Intell. **39**(12), 2481–2495 (2017)
6. Bakas, S., et al.: Segmentation labels and radiomic features for the pre-operative scans of the TCGA-GBM collection. Cancer Imaging Arch. (2017). https://doi.org/10.7937/K9/TCIA.2017.KLXWJJ1Q
7. Bakas, S., et al.: Segmentation labels and radiomic features for the pre-operative scans of the TCGA-LGG collection. Cancer Imaging Archive (2017). https://doi.org/10.7937/K9/TCIA.2017.GJQ7R0EF
8. Bakas, S., et al.: Advancing the cancer genome atlas glioma MRI collections with expert segmentation labels and radiomic features. Sci. Data **4**, 170117 (2017)
9. Bakas, S., et al.: Identifying the best machine learning algorithms for brain tumor segmentation, progression assessment, and overall survival prediction in the brats challenge. arXiv preprint arXiv:1811.02629 (2018)
10. Feng, X., Dou, Q., Tustison, N., Meyer, C.: Brain tumor segmentation with uncertainty estimation and overall survival prediction. In: Crimi, A., Bakas, S. (eds.) BrainLes 2019. LNCS, vol. 11992, pp. 304–314. Springer, Cham (2020). https://doi.org/10.1007/978-3-030-46640-4_29
11. Guo, X., et al.: Brain tumor segmentation based on attention mechanism and multi-model fusion. In: Crimi, A., Bakas, S. (eds.) BrainLes 2019. LNCS, vol. 11993, pp. 50–60. Springer, Cham (2020). https://doi.org/10.1007/978-3-030-46643-5_5
12. He, K., Zhang, X., Ren, S., Sun, J.: Deep residual learning for image recognition. In: Proceedings of the IEEE Conference on Computer Vision and Pattern Recognition, pp. 770–778 (2016)
13. Iandola, F., Moskewicz, M., Karayev, S., Girshick, R., Darrell, T., Keutzer, K.: Densenet: Implementing efficient convnet descriptor pyramids. arXiv preprint arXiv:1404.1869 (2014)

14. Jiang, Z., Ding, C., Liu, M., Tao, D.: Two-stage cascaded U-net: 1st place solution to BraTS challenge 2019 segmentation task. In: Crimi, A., Bakas, S. (eds.) BrainLes 2019. LNCS, vol. 11992, pp. 231–241. Springer, Cham (2020). https://doi.org/10.1007/978-3-030-46640-4_22

15. Kamnitsas, K., et al.: Efficient multi-scale 3d CNN with fully connected CRF for accurate brain lesion segmentation. Med. Image Anal. **36**, 61–78 (2017)

16. Kotowski, K., Nalepa, J., Dudzik, W.: Detection and segmentation of brain tumors from MRI using U-nets. In: Crimi, A., Bakas, S. (eds.) BrainLes 2019. LNCS, vol. 11993, pp. 179–190. Springer, Cham (2020). https://doi.org/10.1007/978-3-030-46643-5_17

17. Lin, T.Y., Goyal, P., Girshick, R., He, K., Dollár, P.: Focal loss for dense object detection. In: Proceedings of the IEEE International Conference on Computer Vision, pp. 2980–2988 (2017)

18. McKinley, R., Rebsamen, M., Meier, R., Wiest, R.: Triplanar ensemble of 3D-to-2D CNNs with label-uncertainty for brain tumor segmentation. In: Crimi, A., Bakas, S. (eds.) BrainLes 2019. LNCS, vol. 11992, pp. 379–387. Springer, Cham (2020). https://doi.org/10.1007/978-3-030-46640-4_36

19. Menze, B.H., et al.: The multimodal brain tumor image segmentation benchmark (brats). IEEE Trans. Med. Imaging **34**(10), 1993–2024 (2014)

20. Milletari, F., Navab, N., Ahmadi, S.A.: V-net: Fully convolutional neural networks for volumetric medical image segmentation. In: 2016 Fourth International Conference on 3D Vision (3DV), pp. 565–571. IEEE (2016)

21. Pei, L., Vidyaratne, L., Monibor Rahman, M., Shboul, Z.A., Iftekharuddin, K.M.: Multimodal brain tumor segmentation and survival prediction using hybrid machine learning. In: Crimi, A., Bakas, S. (eds.) BrainLes 2019. LNCS, vol. 11993, pp. 73–81. Springer, Cham (2020). https://doi.org/10.1007/978-3-030-46643-5_7

22. Ronneberger, O., Fischer, P., Brox, T.: U-net: convolutional networks for biomedical image segmentation. In: Navab, N., Hornegger, J., Wells, W.M., Frangi, A.F. (eds.) MICCAI 2015. LNCS, vol. 9351, pp. 234–241. Springer, Cham (2015). https://doi.org/10.1007/978-3-319-24574-4_28

23. Starke, S., Eckert, C., Zwanenburg, A., Speidel, S., Löck, S., Leger, S.: An integrative analysis of image segmentation and survival of brain tumour patients. In: Crimi, A., Bakas, S. (eds.) BrainLes 2019. LNCS, vol. 11992, pp. 368–378. Springer, Cham (2020). https://doi.org/10.1007/978-3-030-46640-4_35

24. Sun, T., Plutynski, A., Ward, S., Rubin, J.B.: An integrative view on sex differences in brain tumors. Cell. Mol. Life Sci. **72**(17), 3323–3342 (2015)

25. Van Griethuysen, J.J., et al.: Computational radiomics system to decode the radiographic phenotype. Cancer Res. **77**(21), e104–e107 (2017)

26. Wang, F., Jiang, R., Zheng, L., Meng, C., Biswal, B.: 3D U-net based brain tumor segmentation and survival days prediction. In: Crimi, A., Bakas, S. (eds.) BrainLes 2019. LNCS, vol. 11992, pp. 131–141. Springer, Cham (2020). https://doi.org/10.1007/978-3-030-46640-4_13

27. Wang, S., Dai, C., Mo, Y., Angelini, E., Guo, Y., Bai, W.: Automatic brain tumour segmentation and biophysics-guided survival prediction. arXiv preprint arXiv:1911.08483 (2019)

28. Zhao, Y.-X., Zhang, Y.-M., Liu, C.-L.: Bag of tricks for 3D MRI brain tumor segmentation. In: Crimi, A., Bakas, S. (eds.) BrainLes 2019. LNCS, vol. 11992, pp. 210–220. Springer, Cham (2020). https://doi.org/10.1007/978-3-030-46640-4_20

Segmentation, Survival Prediction, and Uncertainty Estimation of Gliomas from Multimodal 3D MRI Using Selective Kernel Networks

Jay Patel[1,2]([✉]), Ken Chang[1,2], Katharina Hoebel[1,2], Mishka Gidwani[1],
Nishanth Arun[1], Sharut Gupta[1], Mehak Aggarwal[1], Praveer Singh[1],
Bruce R. Rosen[1], Elizabeth R. Gerstner[3], and Jayashree Kalpathy-Cramer[1]

[1] Athinoula A. Martinos Center for Biomedical Imaging, Massachusetts General
Hospital, Boston, MA, USA
{jpatel38,jkalpathy-cramer}@mgh.harvard.edu
[2] Massachusetts Institute of Technology, Cambridge, MA, USA
[3] Stephen E. and Catherine Pappas Center for Neuro-Oncology,
Massachusetts General Hospital, Boston, MA, USA
https://qtim-lab.github.io/

Abstract. Segmentation of gliomas into distinct sub-regions can help
guide clinicians in tasks such as surgical planning, prognosis, and treatment response assessment. Manual delineation is time-consuming and
prone to inter-rater variability. In this work, we propose a deep learning based automatic segmentation method that takes T1-pre, T1-post,
T2, and FLAIR MRI as input and outputs a segmentation map of the
sub-regions of interest (enhancing tumor (ET), whole tumor (WT), and
tumor core (TC)). Our U-Net based architecture incorporates a modified selective kernel block to enable the network to adjust its receptive
field via an attention mechanism, enabling more robust segmentation of
gliomas of all appearances, shapes, and scales. Using this approach on
the official BraTS 2020 testing set, we obtain Dice scores of .822, .889,
and .834, and Hausdorff distances (95%) of 11.588, 4.812, and 21.984
for ET, WT, and TC, respectively. For prediction of overall survival, we
extract deep features from the bottleneck layer of this network and train
a Cox Proportional Hazards model, obtaining .495 accuracy. For uncertainty prediction, we achieve AUCs of .850, .914, and .854 for ET, WT,
and TC, respectively, which earned us third place for this task.

Keywords: CNN · Segmentation · U-Net · Brain tumor

1 Introduction

Gliomas are the most common type of primary brain malignancy in adults, and
can present with variable prognoses depending on anatomic location, histology,
and molecular characteristics [12]. Imaging via Magnetic Resonance Imaging

© Springer Nature Switzerland AG 2021
A. Crimi and S. Bakas (Eds.): BrainLes 2020, LNCS 12659, pp. 228–240, 2021.
https://doi.org/10.1007/978-3-030-72087-2_20

(MRI) is an important step in surgical planning, treatment response assessment, and guidance of patient management [11]. Features derived from imaging have been shown to be useful in the assessment of prognosis and molecular markers without the need of an invasive biopsy [9,10,20]. However, the extraction of robust quantitative imaging features is reliant upon accurate segmentations of the three main sub-regions of a glioma: enhancing tumor (ET), tumor core (TC), and whole tumor (WT). Manual delineation of these sub-region boundaries is challenging due to the heterogenous appearance that these sub-regions can take on, and this may be further complicated by motion artifacts, field inhomogeneities, and differences in imaging protocols both within and across medical institutions. Overall, this leads manual segmentation to have non-negligible amounts of human error as well as incur significant amounts of intra- and inter-rater variability [23].

In recent years, computer automated segmentation methods have become popular as they can produce accurate and reproducible results many orders of magnitude faster than can be accomplished manually. To encourage the development of automated state-of-the-art segmentation models, the Multimodal Brain Tumor Segmentation Challenge (BraTS) provides participants with MRIs of high-grade (HGG) and low-grade (LGG) gliomas with associated ground truth labels derived from experts [2–5,24].

Looking back at previous BraTS challenges, we noted that variations of 3D U-Nets [28] and other convolutional networks have dominated the leaderboards. Placing second in 2018, Isensee et al. [17] showed that a generic U-Net with only minor modifications could achieve highly competitive performance if using optimal hyperparameter settings and training procedures. Placing first that year, Myronenko [26] utilized an asymmetrical residual U-Net, wherein most of the trainable parameters of the model resided in the encoder. Furthermore, in contrast to the standard U-Net framework which uses 4 downsampling operations in the encoder, he applied only 3 downsampling operations in order to preserve more spatial context. Similar approaches were seen in the 2019 competition. Second place was awarded to Zhao et al. [32], who utilized a U-Net with dense blocks along with many optimization strategies, including variable patch/batch size training, heuristic sampling, and semi-supervised learning. Jiang et al. [18] placed first that year using a two-stage cascaded asymmetrical residual U-Net (very similar to what Myronenko had used the previous year), where the second stage of their cascade was used to refine the coarse segmentation maps generated by the first stage. Moreover, we note that test-time augmentations and model ensembling is now used by all top performing teams. Inspired by these successes and building on our approach for the 2017 BraTS competition [8], we propose a symmetrical residual U-Net with modified selective kernel blocks [21], putting specific emphasis on the network design and optimization strategy.

2 Methods

In the following sections, we present the network architecture and training procedures used in our approach. All code was implemented in DeepNeuro [7] with Tensorflow 2.2 backend [1].

2.1   Preprocessing

The BraTS 2020 challenge organizers provided all competitors with MR imaging and associated ground truth labels for 369 patients. To mitigate variability between patients stemming from differences in imaging protocols, challenge organizers provided data that was already co-registered to the same anatomical template, resampled to 1 mm isotropic resolution, and skull-stripped. To further compensate for intensity inhomogeneity in the dataset, we apply N4 bias correction [29] and normalize each modality of each patient independently to have zero mean, unit variance (based on non-zero voxels only). Finally, all volumes are tightly cropped to remove empty voxels outside the brain region.

2.2   Network Architecture

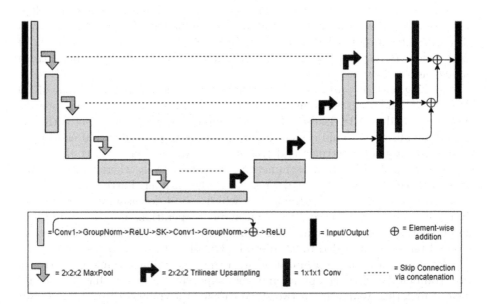

Fig. 1. Schematic of our 3D U-Net architecture. The input to the network is a four channel 128 × 128 × 128 patch. Each layer of the network contains one residual bottleneck block with modified selective kernel convolution used in lieu of a standard 3 × 3 × 3 convolution. The network outputs 3 channels corresponding to each of the tumor sub-regions, with deep supervision used in the decoder to help propagate gradients deep into the network.

Guided by the approaches of top teams from previous BraTS challenges, we utilize a symmetrical 3D U-Net architecture as the backbone for our model. Various architectural modifications to the baseline U-Net have been proposed over the years to create a model that best captures the intrinsic heterogeneity present in gliomas. More specifically, we note that this (histologic and molecular) heterogeneity affects a tumor's radiographic appearance, leading to vast inter-patient differences in scale and shape of the tumor sub-regions (ET, TC, and WT). In this work, we hypothesize that networks utilizing standard convolutions are less capable of generating accurate annotations when presented with the task of segmenting highly heterogeneous gliomas. To this end, we propose to inject modified selective kernel (SK) blocks [21] into our U-Net, such that the network can automatically adjust its receptive field via an attention mechanism to better incorporate spatial information extracted at different scales.

Our network architecture is shown in Fig. 1. We devise a 5 level U-Net which takes four MR modalities (T1 without contrast (T1-pre), T1 with contrast (T1-post), T2, and Fluid Attenuation Inversion Recovery (FLAIR)) as input and outputs the 3 tumor sub-regions of interest. To ensure our model learns both adequate anatomic context and rich representations under the constraints of GPU memory, we use 32 filters in the $3 \times 3 \times 3$ convolutions in the first layer, and double this number of filters as we go deeper into the network. Feature map downsampling and upsampling is accomplished through max pooling and trilin-ear interpolation, respectively, which also helps reduce the memory footprint of the network (when compared to using strided convolution and deconvolution, respectively). Group Normalization [31] (with group size of 16) in lieu of Batch Normalization [16] is used in order to accommodate the smaller batch size nec-essary to train a large patch 3D model. And to encourage faster convergence and ensure that deeper layers of the decoder are learning semantically useful features, we employ deep supervision [19,32] by integrating segmentation outputs from different levels of the network.

Each layer in the network is composed of one residual bottleneck block [14], where we replace the central $3 \times 3 \times 3$ convolution with our modified SK block, which is shown in Fig. 2. Our SK blocks can be decomposed into three distinct operations: 1) Split, 2) Channel Selection, 3) Spatial Attention.

Split: Incoming feature maps are split into two branches. The first branch per-forms a convolutional operation using a standard $3 \times 3 \times 3$ kernel, while the second branch uses a $3 \times 3 \times 3$ kernel with dilation rate of 2. Group Normal-ization and ReLU [27] are subsequently applied to each branch.

Channel Selection: To allow the network to adaptively adjust the receptive field of this block, we apply a channel attention mechanism. The ReLU activated outputs of each branch are summed and then distilled via a global max and average pooling operation. These two feature vectors are passed through a shared multi-layer perceptron (MLP), which has one hidden layer of dimension 8 times smaller than the input/output. The output of the MLP is activated via sigmoid to generate an attention vector. The learned channel attention coefficients α_c are then used to weight the importance of the feature maps coming from standard F_s and dilated F_d convolutions as follows: $F_c^{SK} = \alpha_c F_s + (1 - \alpha_c) F_d$.

Spatial Attention: To focus the network on more salient parts of the image, we apply a simple spatial attention mechanism to F_c^{SK}, similar to the the method from the convolutional block attention module [30]. We aggregate

channel information via a max and average pooling over the channel dimension. These 2 vectors are concatenated, convolved with a $5 \times 5 \times 5$ kernel, and then activated with a sigmoid to output an attention vector. The learned spatial attention coefficients α_s are then used to weight the incoming feature maps as follows:
$$F_s^{SK} = \alpha_s F_c^{SK}$$

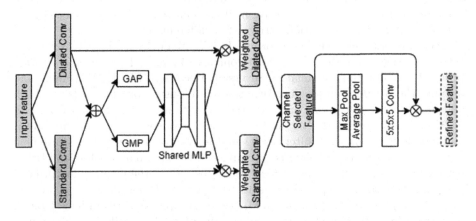

Fig. 2. Modified SK-block used in every layer of our U-Net. The input features are passed through both a standard and dilation rate 2 convolution (with group normalization and ReLU applied). Spatial information is aggregated via a global average and global max pooling operation. The resultant descriptors are passed through a shared MLP, which produces a channel attention map. After channel selection, we apply average and max pooling over the channel dimension, and convolve this two channel vector to produce a spatial attention map.

2.3   Loss Function

As our goal is to maximize the Dice Similarity Coefficient (DSC) [25] between the predicted label maps and the ground truth, we use the following implementation of multi-class soft dice loss:

$$DSC = 1 - \frac{1}{|C|} \sum_{c \in C} \frac{2 * \sum p_c * q_c + \epsilon}{\sum p_c + \sum q_c + \epsilon} \tag{1}$$

where C is the set of classes to predict (ET, TC, and WT), p is the ground truth, q is the prediction, and ϵ is used to prevent floating point instability when the magnitude of the denominator is small (set to 1).

We complement dice loss with cross-entropy, which we believe enables more refined segmentation outputs. To handle the large class imbalance present in this segmentation task, we apply a boundary-reweighting term to the cross-entropy loss. This reweighting map is created as follows. First, all ground truth label maps are binarized and converted into edge images. Next, the euclidean distance transform of these edge images is computed to produce a raw distance map, where the value at each voxel quantifies how far it is from the boundary of a sub-region. This map is subsequently inverted and rescaled such that voxels on the boundary are weighted 6 times more than voxels far away from the boundary. This ensures that easily classified voxels (such as those containing normal tissue distal to the tumor) are given less importance than the difficult to classify voxels (both foreground and background) near the peripheries of the glioma. This boundary-weighted cross-entropy (CE_{BW}) loss is implemented as follows:

$$CE_{BW} = -\sum_{c \in C} BW_c * p_c * log(q_c) \qquad (2)$$

where BW_c is the boundary-reweighting map for the class c. The total loss for our network is the unweighted sum of DSC and CE_{BW}.

2.4 Optimization

We train our network on patches of size $128 \times 128 \times 128$ voxels with batch size 1. As patches of this size cover most of the brain region already, we sample patches at random during training, seeing no performance change from using special sampling heuristics. Training is done using the Adam optimizer with decoupled weight decay [22] and we progressively decrease the learning rate via the following cosine decay schedule:

$$\eta_t = \eta_{min} + 0.5(\eta_{max} - \eta_{min})(1 + cos(\pi T_{curr}/T) \qquad (3)$$

where η_{max} is our initial learning rate (set to $1 * 10^{-4}$), η_{min} is our final learning rate (set to $1 * 10^{-7}$), T_{curr} is the current iteration counter, and T is the total number of iterations to train for (set to 330 epochs).

To mitigate overfitting, we apply weight decay of $2 * 10^{-5}$ to all convolutional kernel parameters, leaving biases and scales un-regularized. Furthermore, we apply real-time data augmentation during the training process. Specifically, we utilize random axis mirror flips (for all 3 axes), isotropic scaling (.75 to 1.25), rotations ($-15°$ to $15°$) around all three axes, and per-channel gamma corrections (.75 to 1.25), all with probability 0.5.

Training a single model using the entire labeled dataset of 369 patients took around 20 h on a NVIDIA Tesla V100 32 GB GPU. No additional data (labeled or unlabeled) was utilized.

2.5 Inference

At test time, we pass the entire image into the network to be segmented in one pass, as opposed to using a sliding window of patch size $128 \times 128 \times 128$. We find that this is both more efficient and also leads to better segmentation quality, presumably due to less edge effects stemming from the use of zero-padded convolutions. We apply simple test-time augmentation by averaging the results from 8 mirror axis-flipped versions of the input volumes. To further boost performance, we average the results from an ensemble of 5 models (all trained from scratch).

2.6   Postprocessing

Many low grade glioma patients present with no ET component, and the presence of even a single false positive ET voxel in our predicted label maps results in a DSC of 0. To reduce the impact of this severe penalty, we replace all ET voxels with TC if the number of predicted voxels is less than 50 voxels. Furthermore, to reduce the number of false positives of any class, we remove any connected component that is less than 10 voxels in size.

2.7   Overall Survival Prediction

For prediction of overall survival (OS), we utilize a Cox Proportional Hazards Model trained on deep features extracted from our segmentation network. More specifically, we pass an entire image into our network and extract 2048 features from the end of the bottleneck layer. This process is repeated for all 8 mirror axis-flipped versions of the input, and for all 5 individual models in our ensemble, producing 40 versions of these 2048 features. To combine these 40 versions together, the features are global average pooled across the batch and spatial dimensions. Training a model using this entire feature set would result in severe overfitting, thus necessitating dimensionalty reduction via principal component analysis (PCA). In addition to these PCA transformed deep features, we also include age at diagnosis and the predicted volumetric size of the three sub-regions in our model.

As only patients with gross total resection (GTR) are evaluated in this task, we choose to train our Cox model on a set of 118 patients (excluding those that may have survival info but lacked GTR). We empirically determined that a model with 10 principal components maximized performance on the validation set.

2.8   Uncertainty Estimation

We use our five network ensemble for estimation of voxel-wise segmentation uncertainty. As mentioned above, we pass mirror axis-flipped inputs through all models in the ensemble, resulting in 40 predictions per sub-region. These predictions are combined by directly averaging the model logits, denoted as l_x. A voxel with high predictive uncertainty will have $|l_x| \approx 0$, whereas a voxel with high predictive certainty will have $|l_x| \gg 5$. To explicitly quantify uncertainty in the range 0 (maximally certain) to 100 (maximally uncertain), we use the following formula [15]:

$$\text{uncertainty}_x = \begin{cases} 200 * \sigma(l_x), & \text{if } 0 \leq \sigma(l_x) < 0.5 \\ 200 * (1 - \sigma(l_x)), & \text{otherwise} \end{cases} \tag{4}$$

where the $\sigma(\cdot)$ function converts the ensembled logits to probabilities via sigmoid.

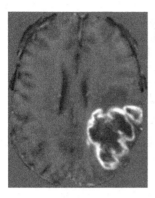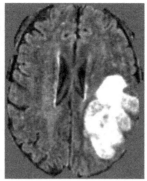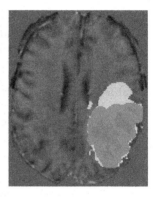

Fig. 3. The left panel shows a T1-post image, the center panel shows a FLAIR image, and the right panel shows the predicted segmentation output overlaid on the T1-post for this validation dataset patient. ET is labeled in blue, TC is the union of the green and blue labels, and WT is the union of the yellow, green, and blue labels. For this patient, we achieved a DSC of 0.949, 0.978, and 0.969 for ET, TC, and WT, respectively. (Color figure online)

3   Results

3.1   Validation Set

We report the DSC and Hausdorff distance (HD95) of our ensemble of five models on the official BraTS 2020 validation dataset (125 cases), as generated by the online evaluation platform. The results for an ensemble trained with just dice loss and an ensemble trained with joint dice/cross-entropy loss can be seen in Table 1. Adding boundary-reweighted cross-entropy significantly improved validation set HD95 for all sub-regions, and produced minor improvements in DSC as well (with the exception of TC).

To assess the effect of the ensemble, we generated the DSC for each of the five models independently. Surprisingly, we note that the ensemble only marginally helps the DSC for ET and WT (average improvement of around 0.5% to 0.75%). Conversely, the ensemble provides a sizable improvement of more than 1.25% to TC. A qualitative example of the output of our ensembled model on a validation set example is shown in Fig. 3.

Table 1. Dice scores and Hausdorff distances (95%) on BraTS 2020 validation data (125 cases).

	Dice			HD95		
Validation	ET	WT	TC	ET	WT	TC
Dice	0.783	0.898	0.853	23.960	6.170	6.685
Dice/Cross-Entropy	0.793	0.901	0.851	15.709	4.884	5.857

236   J. Patel et al.

For prediction of OS, we evaluate our Cox model on a subset of the validation dataset consisting of 29 patients with confirmed GTR. On this small subset, we report classification accuracy of .655. Full results can be viewed in Table 2.

Table 2. Accuracy of OS prediction on BraTS 2020 validation data (29 cases).

Validation	Accuracy	MSE	medianSE	stdSE	SpearmanR
Cox model	0.655	152467	39601	196569	0.479

A well calibrated uncertainty map should show that the network is confident in its correct predictions and unconfident in its incorrect predictions. We assess the fidelity of our uncertainty estimates both qualitatively and quantitatively. An example of our voxel-wise uncertainty maps on a validation set patient (using both our dice loss and dice/cross-entropy trained models) is shown in Fig. 4. Importantly, we note that even though neither model perfectly segmented the whole tumor region for this patient, only the dice/cross-entropy model showed significant uncertainty in the region that it incorrectly predicted as normal. Quantitative metrics (shown in Table 3) were extracted from the online evaluation platform, and are further evidence that our dice/cross-entropy model produces better calibrated uncertainty maps than does a model trained with dice loss alone.

Table 3. AUC for sub-regions on BraTS 2020 validation data (125 cases).

	AUC		
Validation	ET	WT	TC
Dice	0.794	0.914	0.875
Dice/Cross-Entropy	0.815	0.927	0.874

3.2 Testing Set

The official BraTS 2020 testing dataset consisted of 166 cases, and participants were only allowed one submission to the system (for which we submitted results from our joint dice/cross-entropy trained ensemble). Metrics for segmentation accuracy, prediction of OS, and uncertainty estimation are provided in Tables 4, 5, 6 below, respectively.

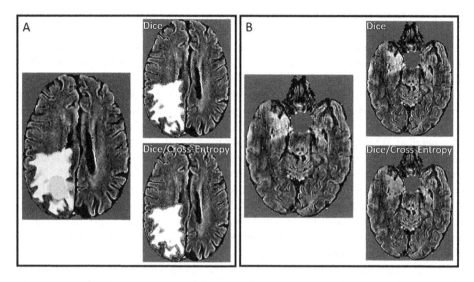

Fig. 4. Voxel-wise uncertainty for WT segmentation overlaid on FLAIR imaging (yellow represents low uncertainty and red represents high uncertainty). Panel A: Both of our models correctly identify the lesion, and thus produce uncertainty estimates that are relatively low (and mainly around the periphery of the lesion). Panel B: Neither model segments this portion of edema. We see that our model trained with only dice loss is highly over-confident in its incorrect prediction, whereas our dice/cross-entropy model produces non-zero uncertainty in that region. (Color figure online)

Table 4. Dice scores and Hausdorff distances (95%) on BraTS 2020 testing data (166 cases).

	Dice			HD95		
Testing	ET	WT	TC	ET	WT	TC
Dice/Cross-Entropy	0.822	0.889	0.834	11.586	4.812	21.984

Table 5. Accuracy of OS prediction on BraTS 2020 testing data (107 cases).

Testing	Accuracy	MSE	medianSE	stdSE	SpearmanR
Cox model	0.495	493194	89401	1250167	0.349

Table 6. AUC for sub-regions on BraTS 2020 testing data (166 cases).

	AUC		
Testing	ET	WT	TC
Dice/Cross-entropy	0.850	0.914	0.854

4   Discussion and Conclusion

In this work, we propose a glioma segmentation network from multimodal MRI in which we utilize selective kernel units to automatically adjust the receptive field of the network to better match the scale of the input tumor. While we settled upon a symmetrical network structure with residual bottleneck blocks, we did experiment with other structures. First, we compared the asymmetrical encoder-decoder structure that has won BraTS two years in a row against a simpler symmetrical structure. Our limited experiments did not seem to indicate any performance difference between the two approaches. Second, we tested whether the addition of residual connections aided optimization of the network. While simply adding residual connections did not improve performance, we noted faster convergence of the model, presumably due to better propagation of gradients throughout the network. There was however a performance boost to switching out standard residual blocks with bottleneck blocks, which we attribute to an increase in network depth and capacity.

While we explored a diversity of approaches in this work, we were not able to test all potentially beneficial modifications. Thus, future work will focus on identifying architectural changes that most improve generalizability of our model. For instance, we plan on thoroughly testing the use of spatial dropout layers, identifying where in the network these dropout layers should be placed and with what probability these dropout layers should use. We also intend to test the efficacy of more intricate test-time augmentation methods, such as scaling and rotating. Moreover, we plan to explore the use of generative adversarial networks as a means of data augmentation during training [6]. Lastly, the BraTS competition focuses on segmentation of pre-operative brain tumors, while another important application of MRI is post-operative evaluation. This presents additional challenges due to the presence of a surgical cavity as well as post-operative, radiation, and chemotherapeutic mediated changes to tumor biology [13]. Future work can extend the segmentation approach proposed in this work to that context.

References

1. Abadi, M., et al.: TensorFlow: a system for large-scale machine learning. In: 12th USENIX Symposium on Operating Systems Design and Implementation (OSDI 2016), pp. 265–283 (2016). https://www.usenix.org/system/files/conference/osdi16/osdi16-abadi.pdf
2. Bakas, S., et al.: Advancing the cancer genome atlas glioma MRI collections with expert segmentation labels and radiomic features. Sci. Data **4**, 170117 (2017). https://doi.org/10.1038/sdata.2017.117
3. Bakas, S., et al.: Segmentation labels and radiomic features for the pre-operative scans of the TCGA-LGG collection [Data Set]. Cancer Imaging Arch. (2017). https://doi.org/10.7937/K9/TCIA.2017.GJQ7R0EF
4. Bakas, S., et al.: Segmentation labels for the pre-operative scans of the TCGA-GBM collection [Data set]. Cancer Imaging Arch. (2017). https://doi.org/10.7937/K9/TCIA.2017.KLXWJJ1Q

5. Bakas, S., et al.: Identifying the Best Machine Learning Algorithms for Brain Tumor Segmentation, Progression Assessment, and Overall Survival Prediction in the BRATS Challenge (2018)

6. Beers, A., et al.: High-resolution medical image synthesis using progressively grown generative adversarial networks (2018)

7. Beers, A., et al.: DeepNeuro: an open-source deep learning toolbox for neuroimaging. Neuroinformatics (2020). https://doi.org/10.1007/s12021-020-09477-5

8. Beers, A., Chang, K., Brown, J., Gerstner, E., Rosen, B., Kalpathy-Cramer, J.: Sequential neural networks for biologically informed glioma segmentation. In: Angelini, E.D., Landman, B.A. (eds.) Medical Imaging 2018: Image Processing, vol. 10574, pp. 807–816. International Society for Optics and Photonics, SPIE (2018). https://doi.org/10.1117/12.2293941

9. Beig, N., et al.: Radiogenomic analysis of hypoxia pathway is predictive of overall survival in Glioblastoma. Sci. Rep. **8**(1), 7 (2018). https://doi.org/10.1038/s41598-017-18310-0

10. Chang, K., et al.: Residual convolutional neural network for the determination of IDH status in low- and high-grade gliomas from MR imaging. Clin. Cancer Res. **24**(5), 1073–1081 (2018). https://doi.org/10.1158/1078-0432.CCR-17-2236, https://clincancerres.aacrjournals.org/content/24/5/1073

11. Chang, K., et al.: Automatic assessment of glioma burden: a deep learning algorithm for fully automated volumetric and bidimensional measurement. Neuro-Oncology **21**(11), 1412–1422 (2019). https://doi.org/10.1093/neuonc/noz106

12. Eckel-Passow, J.E., et al.: Glioma groups based on 1p/19q, IDH, and TERT promoter mutations in tumors. New Engl. J. Med. **372**(26), 2499–2508 (2015). https://doi.org/10.1056/NEJMoa1407279, pMID: 26061753

13. Gerstner, E.R., et al.: Bevacizumab reduces permeability and concurrent temozolomide delivery in a subset of patients with recurrent glioblastoma. Clin. Cancer Res. **26**(1), 206–212 (2020). https://doi.org/10.1158/1078-0432.CCR-19-1739, https://clincancerres.aacrjournals.org/content/26/1/206

14. He, K., Zhang, X., Ren, S., Sun, J.: Deep residual learning for image recognition. In: 2016 IEEE Conference on Computer Vision and Pattern Recognition (CVPR), pp. 770–778 (2016)

15. Hendrycks, D., Gimpel, K.: A baseline for detecting misclassified and out-of-distribution examples in neural networks (2018)

16. Ioffe, S., Szegedy, C.: Batch Normalization: Accelerating Deep Network Training by Reducing Internal Covariate Shift (2015)

17. Isensee, F., Kickingereder, P., Wick, W., Bendszus, M., Maier-Hein, K.H.: No new-net. CoRR abs/1809.10483 (2018), http://arxiv.org/abs/1809.10483

18. Jiang, Z., Ding, C., Liu, M., Tao, D.: Two-stage cascaded u-net: 1st place solution to brats challenge 2019 segmentation task. In: Crimi, A., Bakas, S. (eds.) Brainlesion: Glioma, Multiple Sclerosis, Stroke and Traumatic Brain Injuries, pp. 231–241. Springer International Publishing, Cham (2020). https://doi.org/10.1007/978-3-030-11726-9

19. Kayalibay, B., Jensen, G., van der Smagt, P.: CNN-based segmentation of medical imaging data (2017)

20. Lao, J., et al.: A deep learning-based radiomics model for prediction of survival in glioblastoma multiforme. Sci. Rep. **7**(1), 10353 (2017). https://doi.org/10.1038/s41598-017-10649-8

21. Li, X., Wang, W., Hu, X., Yang, J.: Selective Kernel Networks (2019)

22. Loshchilov, I., Hutter, F.: Decoupled Weight Decay Regularization (2017)

240   J. Patel et al.

23. Mazzara, G.P., Velthuizen, R.P., Pearlman, J.L., Greenberg, H.M., Wagner, H.: Brain tumor target volume determination for radiation treatment planning through automated MRI segmentation. Int. J. Radiat. Oncol. Biol. Phys. **59**(1), 300–312 (2004). https://doi.org/10.1016/j.ijrobp.2004.01.026
24. Menze, B.H., et al.: The Multimodal Brain Tumor Image Segmentation Benchmark (BRATS) 34(10), 1993–2024 (2015). https://doi.org/10.1109/TMI.2014.2377694
25. Milletari, F., Navab, N., Ahmadi, S.A.: V-Net: Fully Convolutional Neural Networks for Volumetric Medical Image Segmentation (2016)
26. Myronenko, A.: 3D MRI brain tumor segmentation using autoencoder regularization (2018)
27. Nair, V., Hinton, G.: Rectified linear units improve restricted Boltzmann machines. In: Proceedings of ICML, vol. 27, pp. 807–814 (2010)
28. Ronneberger, O., Fischer, P., Brox, T.: U-Net: Convolutional Networks for Biomedical Image Segmentation. CoRR abs/1505.0 (2015), http://arxiv.org/abs/1505.04597
29. Tustison, N.J., et al.: N4ITK: improved N3 bias correction. IEEE Trans. Med. Imaging **29**(6), 1310–1320 (2010). https://doi.org/10.1109/TMI.2010.2046908, https://www.ncbi.nlm.nih.gov/pubmed/20378467
30. Woo, S., Park, J., Lee, J.Y., Kweon, I.S.: CBAM: Convolutional Block Attention Module (2018)
31. Wu, Y., He, K.: Group Normalization (2018)
32. Zhao, Y., Zhang, Y.M., Liu, C.L.: Bag of Tricks for 3D MRI Brain Tumor Segmentation, pp. 210–220, May 2020. https://doi.org/10.1007/978-3-030-46640-4_20

3D Brain Tumor Segmentation and Survival Prediction Using Ensembles of Convolutional Neural Networks

S. Rosas González, I. Zemmoura, and C. Tauber[✉]

UMR U1253 iBrain, Université de Tours, Inserm, Tours, France
sarahi.rosasgonzalez@etu.univ-tours.fr

Abstract. Convolutional Neural Networks (CNNs) are the state of the art in many medical image applications, including brain tumor segmentation. However, no successful studies using CNNs have been reported for survival prediction in glioma patients. In this work, we present two different solutions: tumor segmentation and the other for survival prediction. We proposed using an ensemble of asymmetric U-Net like architectures to improve segmentation results in the enhancing tumor region and the use of a DenseNet model for survival prognosis. We quantitatively compare deep learning with classical regression and classification models based on radiomics features and growth tumor models features for survival prediction on the BraTS 2020 database, and we provide an insight into the limitations of these models to accurately predict survival. Our method's current performance on the BraTS 2020 test set is dice scores of 0.80, 0.87, and 0.80 for enhancing tumor, whole tumor, and tumor core, respectively, with an overall dice of 0.82. For the survival prediction task, we got a 0.57 accuracy. In addition, we proposed a voxel-wise uncertainty estimation of our segmentation method that can be used effectively to improve brain tumor segmentation.

Keywords: Tumor segmentation · Survival prediction · Multi-input · Uncertainty estimation · 2.5D convolutions · BraTS

1 Introduction

Glioma is the most frequent primary brain tumor that originates from glial cells [1]. The primary treatment is surgical resection, followed by radiation therapy and chemotherapy. Most tumors of the central nervous system derived from astrocytes. Astrocytomas are classified by both grade and type, established by the WHO according to their molecular properties. Those belonging to grades I astrocytomas are considered benign and slow-growing. Grade II astrocytomas are more likely to recur or progress over time. Making surgical excision with histopathological evaluation is very important. On the contrary, grades III and IV (high grade) have a poor prognosis and are lethal; patients have approximately one-year survival [2].

Imaging analysis of these tumors is challenging, as their shape, extent, and location can differ substantially. Since 2012, the BraTS challenge is held annually to evaluate the

© Springer Nature Switzerland AG 2021
A. Crimi and S. Bakas (Eds.): BrainLes 2020, LNCS 12659, pp. 241–254, 2021.
https://doi.org/10.1007/978-3-030-72087-2_21

state of the art segmentation algorithms by making an extensive preoperative multimodal MRI dataset [3–7].

Recently, Ronneberger et al. [8] proposed a model called U-Net, which is a Fully Convolutional Network (FCN) encoder/decoder architecture. Encoder module consists of multiple connected convolution layers that aim to gradually reduce the spatial dimension of feature maps and capture more high-level semantic features learned to be highly efficient at discriminating between classes. The decoder module uses upsampling layers to recover the spatial dimension and object representation.

The use of an ensemble of models that average the output probabilities of independent models to generate uncertainty estimation for each voxel in the image has been implemented in several recent works as a strategy to improve the generalization power and segmentation performance of their methods [9–14]. The reason is that independently trained models are diverse and biased due to different initialization weights, different choice of hyper-parameters for optimization and regularization and that they are trained in different subsets of the dataset.

Since 2017, an overall survival (OS) prediction task is included. Participants are called to extract imaging/radiomic features and analyze them through machine learning algorithms. In this task, patients must be classified as short-survivors (<10 months), mid-survivors (10–15 months), and long-survivors (>15 months). Intensity, morphologic, histogram-based, and textural features and spatial information, and glioma diffusion properties extracted from glioma growth models can also be considered.

To explore the correlation between medical image features and underlying biological characteristics, radiomics has been proposed. Radiomics refers to a process that extracts high-throughput quantitative features from radiographic images and builds predictive models relating image features to clinical outcomes. The basic idea behind radiomics is that intra-tumor imaging heterogeneity could be the expression of genetic heterogeneity and that information can be extracted from the image [15].

Despite many participants each year, after three years, the accuracy of survival prediction has not increased significantly compared to improvements in the segmentation task. The winners of the first place of 2017 [16], 2018 [17], and 2019 [18] competitions have achieved accuracy values of 0.58, 0.62, and 0.58, respectively. Shboul et al. used radiomic features combined with a Random Forest Regressor (RFR), Feng et al. used geometric features combined with a linear model, and Feng et al. relied on age, statistical, and necrosis shape features to train an RFR.

The developers of top-performing methods of last year followed similar strategies of combining either hand-selected or radiomic features extraction with a supervised machine learning algorithm, mainly based on RFR. A possible explanation for this may be the relatively small dataset and the number of classes to learn. Classical regression techniques typically require learning fewer parameters compared to CNN and perform better with sparse data. Using the age as the only feature with a linear regressor has also achieved competitive results, winning third place in BraTS challenge 2018 [19] and achieving an accuracy of 0.56 on the test dataset. Another top-performant submission that used only two features and a support vector machine linear regression model achieved second place in the BraTS challenge 2019 [20]. The authors chose age and quantitative estimation of tumor invasiveness based on volume measures of tumor subregions.

No successful studies using CNNs have been reported. [21] reported promising results using a small CNN; however, the study concludes that radiomics features are better suited, as features extracted from deep learning networks seemed unstable. In the latest BraTS summary [4], it is stated that deep learning models performed poorly on this open-access data.

In this work, we propose an ensemble of seven models based on U-Net architecture, with two different architectures based on 3D and 2.5D input strategies for segmenting the tumor and a 3D DenseNet architecture to predict survival. We quantitatively compare deep learning with classical regression models based on radiomics features and growth tumor model features for survival prediction, and we provide an insight into the limitations of these models to accurately predict survival. In addition, we provide a voxel-wise uncertainty estimation of our segmentation method.

2 Method

Our proposed method is divided into two workflows; for segmentation and survival prediction, respectively, the preprocessing steps are the same in both workflows. We first describe our dataset and preprocessing and then the segmentation and survival prediction CNN models. Finally, we present our method for the uncertainty estimation task.

2.1 Dataset

Our model was validated in the BraTS 2020 dataset, consisting of preoperative MRI images of glioma patients. Data is divided into three sets by the organizers of the BraTS challenge: training, validation, and test dataset; the first one is the only one that is provided with expert manual segmentation and the grading information of the disease. The training dataset contains images of 369 patients, of which 293 are HGG, and 76 are LGG. However, from these images, we only considered a subset of 118 images for the survival prediction task since only patients provided with survival information and a reported gross total resection status (GTR) must be taken into account. We validate our survival prediction model on 125 patients from the validation dataset. Each patient's MRI images contain four modalities T2-weighted (T2), Fluid-Attenuated Inversion Recovery (FLAIR), T1-weighted (T1), and T1 with gadolinium-enhancing contrast (T1ce). All the subjects in the training dataset are provided with ground truth labels for three tumor tissues: enhancing tumor (ET - label 4), the peritumoral edema (ED - label 2), and the necrotic and non-enhancing tumor core (NCR/NET - label 1). Following tumor subregions were defined: whole tumor (WT) region includes the union of the three tumor tissues ED, ET, and NCR/NET (i.e., label $= 1 \cup 2 \cup 4$), the tumor core (TC) region is the union of the ET and NCR/NET (i.e., label $= 1 \cup 4$) and enhancing tumor (ET) (i.e., label $= 4$).

2.2 Preprocessing and Post-processing

For the preprocessing, we chose to normalize the MRI images first, dividing each modality by the maximum value and then using the z-scores due to its simplicity and good qualitative performance.

We implemented the post-processing of the prediction maps of our proposed model to reduce the number of false positives and enhance tumor detection. A threshold value representing the minimum size of the enhancing tumor region was defined, as suggested in [11], and the label of all the voxels of the enhancing tumor region was replaced with one of the necrosis regions when the total number of predicted ET voxels was lower than the threshold. The threshold value was estimated to optimize the overall performance in this region in the validation dataset. In addition, as proposed by [12], if our model detects no tumor core, we assumed that the detected whole tumor region corresponded to the tumor core, and we relabeled the region as tumor core.

2.3 Segmentation Pipeline

We propose an ensemble of seven models from two kinds of U-Net like models based on 3D and 2.5D convolutions. The 3D model uses concatenated data into a modified 3D U-Net architecture. In contrast, the 2.5D model is based on a multi-input strategy to extract low-level features from each modality independently and on a new 2.5D Multi-View-Inception block that aims at merging features coming from different views of a 3D image and aggregating multi-scale features.

The main component of our network is a modified 3D U-Net architecture of five levels (Fig. 1B). Traditional 3D U-Net architecture has symmetrical decoder and encoder paths. The first path is associated with the extraction of semantic information to make local predictions, while the second section is related to the recovery of the global structure. We propose an Asymmetric U-Net (AU-Net), wider in the encoding path than in the decoding path, to extract more semantic features. To achieve this, we add twice more convolutional blocks in the encoding path than in the decoding section.

In the first implementation of the AU-Net, 3D concatenated MRI sequences are used as input (Fig. 1A'), which we call 3D AU-Net.

In the second implementation of the AU-Net, a different input strategy is proposed. A multi-input module (Fig. 1A) has been developed to maximize learning from independent features associated with each imaging modality before merging in the encoding/decoding architecture, avoiding loss of specific information provided by each modality.

Additionally, we propose that each of these four pathways contains what we define as a 2.5D Multi-View Inception module (Fig. 1C) that allows the extraction of features in the different orthogonal views of the 3D image: axial, coronal and sagittal planes and different scales, merging them all in each forward pass, saving training time and memory consumption. The design of the 2.5D Multi-View Inception module is inspired by the inception module of GoogLeNet [22]. Unlike the inception module, we use 2.5D anisotropic filters instead of 3D or 2D isotropic filters, and we add all the resulted feature maps instead of stacking them. The two U-Net architectures are summarized in Fig. 1.

We propose an ensemble of seven models, four from 3D AU-Net and three from 2.5D AU-Net; all models were trained with different initialization weights and different subsets of the training dataset.

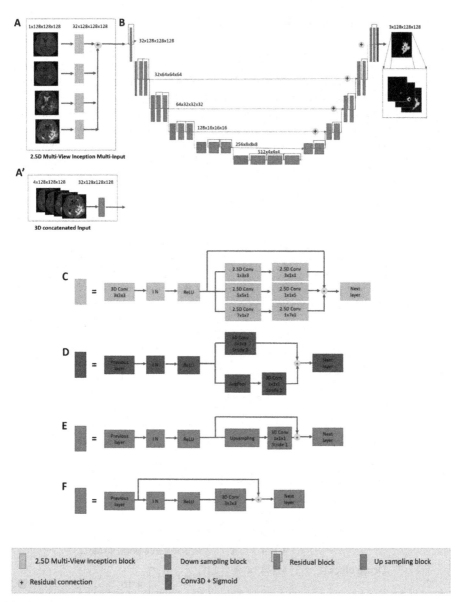

Fig. 1. Proposed two network architectures for the AE AU-Net. 2.5D Multi-View Inception Multi-Input U-net. A Multi-Input module consists of four 2.5D Multi-View Inception blocks, joint with a modified 3D U-Net architecture using residual blocks, instance normalization (IN) instead of Batch normalization, transpose convolutions for upsampling and additive skip connections between encoding and decoding sections.

2.4 Survival Prediction Pipeline

CNN model

For this task, we compared a 3D DenseNet CNN model with classical machine learning methods using handcrafted radiomics features. We first described the CNN model and after the approach based on handcrafted features.

We implemented a 3D DenseNet model, a modified version of the 2D model designed by [23]. We replaced all 2D convolution filters with 3D convolutions. We replaced all Batch normalization operations with Instance normalization operation (IN), which has provided us better results in the segmentation task and is optimal when working with reduced batch size. The network details are illustrated in Table 1.

Table 1. 3D DenseNet architecture.

Layers	Output size	3D DenseNet
Convolution	$64 \times 64 \times 64$	$7 \times 7 \times 7$ conv, stride 2
Pooling	$32 \times 32 \times 32$	$3 \times 3 \times 3$ max pool, stride 2
Dense block 1	$32 \times 32 \times 32$	$\begin{bmatrix} 1 \times 1 \times 1 \\ 3 \times 3 \times 3 \end{bmatrix} \times 6$
Transition layer	$32 \times 32 \times 32$	$1 \times 1 \times 1$ conv
	$16 \times 16 \times 16$	$2 \times 2 \times 2$ avg pool, stride 2
Dense block 2	$16 \times 16 \times 16$	$\begin{bmatrix} 1 \times 1 \times 1 \\ 3 \times 3 \times 3 \end{bmatrix} \times 12$
Transition layer 2	$16 \times 16 \times 16$	$1 \times 1 \times 1$ conv
	$8 \times 8 \times 8$	$2 \times 2 \times 2$ avg pool, stride 2
Dense block 3	$8 \times 8 \times 8$	$\begin{bmatrix} 1 \times 1 \times 1 \\ 3 \times 3 \times 3 \end{bmatrix} \times 24$
Transition layer 3	$8 \times 8 \times 8$	$1 \times 1 \times 1$ conv
	$4 \times 4 \times 4$	$2 \times 2 \times 2$ avg pool, stride 2
Dense block 4	$4 \times 4 \times 4$	$\begin{bmatrix} 1 \times 1 \times 1 \\ 3 \times 3 \times 3 \end{bmatrix} \times 16$
Classification layer	$1 \times 1 \times 1$	$7 \times 7 \times 7$ global avg pool
		3 fully-connected, softmax

For classification, overall survival time was discretized into three survival categories: short survival (<10 months), intermediate survival (10–15 months), and long survival (>15 months), since the prediction is required in days, we assigned a scalar value to each class that corresponds with the median number of survival days of each group (i.e. 150, 363 and 615 days, for short, intermediate and long survival respectively).

Classical survival prediction.

Features extraction: We used our proposed segmentation model described in Sect. 2.3 to segment the different subregions on the training dataset. For each patient, we extracted radiomics features from the whole brain and five subregions from four MRI modalities. The feature extraction subregions include whole tumor (WT), tumor core (TC), peritumoral edema (ED), necrosis (NCR), and enhancing tumor (ET). A total of $2,568 = 107 \times 4 \times 6$ handcrafted features were extracted. All handcrafted features were extracted using Python package PyRadiomics [23].

Feature selection: To reduce the risk of overfitting, we reduced the number of features following the workflow in Fig. 2 through a recursive feature elimination (RFE) scheme with a random forest regressor in a five-fold cross-validation loop to find the optimal number of features. The feature selection procedure was performed with the scikit learn [24] and scipy [25] packages on the whole training data.

Biophysics features: Two additional features were estimated from the extracted radiomics. The first one, the relative invasiveness coefficient (RIC), has been shown to be a predictor of the benefit of gross total resection [26]. The RIC is defined as the extended ratio between the hypoxic tumor core and the infiltration front according to the profile of tumor diffusion. This metric, combined with patients' age were effective to predict survival in the past edition of BraTS challenge 2019 whose model achieved second place in the competition [20]. The second feature calculated was the relative necrosis coefficient (RNC); we defined this value as the relative amount of necrotic tumor within the tumor core. We estimate this value as the ratio between the volume corresponding to the necrosis region and the volume of the tumor core region. We made a model based on each of these features combined with age.

Age is the only available clinical factor and significantly correlated with the survival prognosis (Pearson $r = -0.419$, $p < 1e - 5$) on the training set. We constructed a linear regression model using age as the only predictor.

Model selection: A support vector regression (SVR) model was found as the best model to predict survival using handcrafted features.

2.5 Training

We used a five-fold cross-validation scheme for the training of all our models, and then we got five trained models from each architecture design. We then selected the best seven models to build our ensemble. We used the same hyper-parameters in all our models. We use a soft dice loss function to cope with class imbalances and some data augmentation to prevent overfitting. We implemented random crops and random axis mirror flips for all three axes on the fly during training with a probability of 0.5.

The networks for segmentation were trained for 100 epochs using patches of $128 \times 128 \times 128$ and Adam as optimizer with an initial learning rate of 0.0001 and batch size of 1. The learning rate was decreased by a factor of 5 if no improvement is seen within 30 epochs.

The models for survival prediction were trained for five epochs using patches of $128 \times 128 \times 128$ and Adam as optimizer with an initial learning rate of 0.00005 and batch

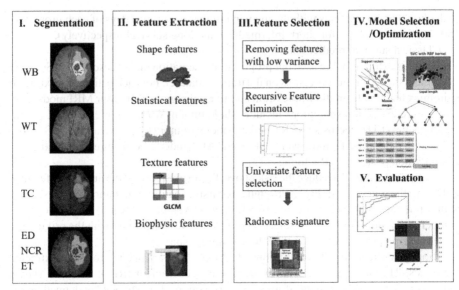

Fig. 2. The workflow for feature-based models. WB: Whole Brain, WT: Whole Tumor, TC: tumor core, ED: peritumoral edema, NCR: Necrosis and ET: Enhancing Tumor.

size of 2. The learning rate was set small, as well as the number of epochs to avoid overfitting. We used a categorical cross-entropy loss function to classify between three classes. All experiments were conducted on a workstation Intel-i7 2.20 GHz CPU, 48G RAM, and an NVIDIA Titan Xp 12 GB GPU.

2.6   Uncertainty Estimation

A new task evaluated this year by the organizers of the challenge is the uncertainty estimation in the segmentation task. This task consists of rewarding methods with predictions that are confident when correct and uncertain when incorrect. To participate in this task, we provide, in addition to our segmentation results, three uncertainty maps associated with the resulting labels at every voxel. The uncertainty maps are associated with the three tumor sub-regions: ET, TC, and WT.

To measure uncertainty, we averaged the model's output probabilities for each label in each voxel to obtain a new probability from the ensemble. Since our model makes a binary classification of each voxel, the highest uncertainty corresponds with a probability of 0.5. Then we used the normalized entropy (Eq. 1) to get an uncertainty measure of the prediction for each voxel:

$$H = -\sum_{c \in C} \frac{p_c \log(p_c)}{\log(|C|)} \in [0, 1], \tag{1}$$

where p_c is the sigmoid output average probability of class c and C is the set of classes, ($C = \{0, 1\}$ in our case). We multiply values by 100 to be between 0 and 100.

3 Experiments and Results

3.1 Segmentation Performance

We experiment with two variants of U-Net, using two different input strategies. The segmentation performance on BraTS 2020 validation dataset is reported in Table 2 in terms of Dice score and Hausdorf distance. Even when the performance of 2.5D AE U-Net is not as good as 3D AE U-Net, adding both architectures improves the overall performance of the ensemble.

Table 2. Ablation study on BraTS validation data (125 cases). Ensemble results from 5-fold cross-validation are reported for 3D, 2.5D, and AE AU-Net models. Reported metrics were computed by the online Brats evaluation platform. WT – whole tumor, TC – tumor core and ET – enhancing tumor.

	Dice			Hausdorff 95 (mm)		
	WT	TC	ET	WT	TC	ET
3D AU-Net	0.897 ± 0.070	0.805 ± 0.186	0.764 ± 0.284	8.20 ± 9.21	8.40 ± 12.09	24.9 ± 10.54
2.5D AU-Net	0.902 ± 0.066	0.814 ± 0.189	0.753 ± 0.295	7.07 ± 12.6	7.84 ± 9.13	30.48 ± 9.92
AE AU-Net	$\mathbf{0.902 \pm 0.066}$	$\mathbf{0.815 \pm 0.161}$	$\mathbf{0.773 \pm 0.257}$	$\mathbf{6.16 \pm 11.2}$	$\mathbf{7.55 \pm 11.4}$	$\mathbf{21.8 \pm 79}$

In Table 3, we can see that our proposed model for segmentation AE AU-Net got overall similar results in the validation and test datasets. We obtained significantly higher results in the ET region and lowered in the WT region, which possibly means that there is an overfitting problem in this region. It is also possible that our post-processing method improves ET and TC segmentation while it impairs the detection of the WT region.

Table 3. Model generalization. AE AU-Net model was evaluated in the validation dataset (125 cases) and test dataset (166 cases).

	Dice			Hausdorff 95 (mm)		
	WT	TC	ET	WT	TC	ET
AE AU-Net (Validation)	$\mathbf{0.902 \pm 0.066}$	$\mathbf{0.815 \pm 0.161}$	0.773 ± 0.257	$\mathbf{6.16 \pm 11.2}$	$\mathbf{7.55 \pm 11.4}$	21.8 ± 79
AE AU-Net (Test)	0.870 ± 0.140	0.805 ± 0.263	$\mathbf{0.795 \pm 0.200}$	7.43 ± 12.6	23.71 ± 79.5	$\mathbf{16.4 \pm 69}$

3.2 Survival Prediction

We experimented using different network architectures well known and designed to classify semantic images to classify patients within three groups. We also experimented

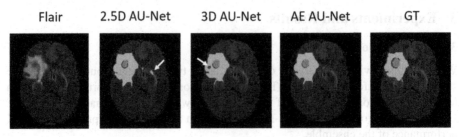

Flair   2.5D AU-Net   3D AU-Net   AE AU-Net   GT

Fig. 3. Example of segmentation result on an image from the BRATS 2020 training dataset. Segmentation is made by two different U-Net architectures and an ensemble of both. Edema (green), necrosis and non-enhancing (red), and enhancing (yellow) (Color figure online).

using four different numbers of layers (21,121,169 and 201) in DenseNet architecture and two different numbers of layers in ResNet (50 and 101). We got better results using a DenseNet architecture with 121 layers. We tested adding one more channel to the input of this model to include the segmentation map obtained from the segmentation model, but no improvement adding this information was obtained. The comparison of the performance of these models is shown in Table 4.

Table 4. 3D models comparison. Ensembles from 5-fold cross-validation. Results on BraTS2020 validation data (29 cases). The online evaluation platform computed metrics. * 3D DenseNet 121 is the submitted method (using a single model).

	Accuracy	MSE	medianSE	stdSE	SpearmanR
3D DenseNet21	0.31	172446	54756	250232	−0.05
*3D DenseNet 121	**0.552**	**87581**	**51529**	**141983**	**0.442**
3D DenseNet 121_5chan	0.448	111063	34225	201790	0.153
3D DenseNet 169	0.276	165349	51076	250235	−0.091
3D DenseNet 201	0.31	155404	46656	252254	−0.042
3D ResNet50	0.448	101673	34225	188017	-----
3D ResNet101	0.31	160284	54756	245504	0
3D IncepResNetv2	0.379	147511	54756	243306	−0.069

Note that Spearman R is not reported for 3D ResNet50 architecture; this is because the model predicted all patients as long survivors.

In Table 5, we show that similar performance was obtained with biophysics models (Invasiveness and Necrosis ratio) and with our proposed CNN model in terms of accuracy. However, a higher Spearman R coefficient is obtained when using a Deep-learning approach; we also got a smaller error.

In Table 6, we show that similar performance was obtained in the validation and test sets for the survival prediction in terms of accuracy, which means that our model

Table 5. Comparing different methods in the BraTS2020 validation dataset (29 cases). The online evaluation platform computed metrics. Baseline is a linear regression model using age as the only predictor. * This is the submitted method.

	Accuracy	MSE	medianSE	stdSE	SpearmanR
Baseline	0.448	89049	33124	121838	0.265
Radiomics	0.414	113965	**29584**	201225	0.17
Invasiveness	**0.552**	94446	40804	142604	0.25
Necrosis ratio	**0.552**	92798	40804	**139411**	0.274
*DenseNet121	**0.552**	**87581**	51529	141983	**0.442**

generalized well. However, our model got significantly lower performance in terms of MSE and Spearman R coefficient.

Table 6. Survival model generalization. DenseNet 121 model evaluated on validation dataset (29 cases) and in test dataset (107 cases).

	Accuracy	MSE	medianSE	stdSE	SpearmanR
DenseNet121 (Validation)	0.552	**87581**	**51529**	**141983**	**0.442**
DenseNet121 (Test)	**0.57**	402332	73441	1132462	0.109

3.3 Evaluation of Uncertainty Measures in Segmentation

In Table 7, we evaluate our proposed model for segmentation, AE AU-Net, in terms of mean Dice Area Under the Curve (Dice AUC) and Filtered True Positives (FTP) ratio scores after filtering out the uncertain voxels obtained with our proposed method for uncertainty estimation.

Table 7. Effect of uncertainty thresholding on Dice AUC (higher is better) and FTP ratio (lower is better) scores. Measures on Brats validation data (125 cases) and test data (166 cases).

	Dice AUC			FTP ratio		
	WT	TC	ET	WT	TC	ET
AE AU-Net (Validation)	**0.940**	0.840	0.797	**0.083**	0.217	0.161
AE AU-Net (Test)	0.909	**0.843**	**0.828**	0.089	**0.146**	**0.150**

We show that filtering out the uncertain voxels improves the segmentation results significantly in all regions (compare results in Table 3 and Table 7), which means that our method for estimating uncertainties is effective. We can also see that our method generalized well to the test dataset.

4   Discussion

The use of an ensemble of models to improve the performance of independently trained models has been efficient in previous challenges. Our results confirm this. We further demonstrated that adding a more complex input into a U-Net architecture can improve segmentation results on enhancing tumor region, which was demonstrated by an ablation study quantitative and qualitatively (See Table 2 and Fig. 3). The use of an ensemble of models not only improved the segmentation results and the generalization power of the method, but it also allowed the measurement of epistemic uncertainty by measuring output variations between the model predictions. Filtering out uncertain voxels demonstrated to improve the segmentation results in all regions.

The use of CNNs has been reported inadequate to predict survival in small datasets [21] mainly because CNNs require to learn many parameters. Previous works proposed networks containing a short number of hidden layers to reduce the number of parameters and the risk of overfitting. In this work, we explored different sizes of networks, ranging from 21 layers to more than 200 layers. We got the best performance with a DenseNet model using 121 total layers. Even when we experimented with smaller networks, we did not observe an improvement in reducing the number of layers. We found high variability in performance with the different architectures evaluated and between slightly different architectures and or when trained with slightly different parameters, making the optimization process challenging to assess. All CNNs showed overfitting behavior after a few epochs independently of the number of layers of the model, which confirm previous statements regarding the need for more training examples for the model to learn.

Comparing between CNN models and handcrafted feature-based models, we found that models relying on biophysics features and age got promising results, over performing many CNNs architectures, which are more complex models with many parameters using only two parameters. We got comparatively lower performance using radiomics features. This can be due to a large number of features initially extracted since we extracted more features from more regions than in other related works [27] this increases the difficulty in selecting the more relevant features by increasing the risk of selecting non-informative features by chance. Then, further studies are required to evaluate the robustness and reproducibility of the feature selection process.

Regarding all survival models, most models predicted only between two classes: long or short survivors, only few models were able to predict mid survivors, which means that distinguishing between three classes is more difficult than separate between two groups and that more patients in the group of mid survivors are required. Using loss functions to favor mid survivor's class predictions might help to alleviate this problem.

5   Conclusions

We have proposed an end-to-end FCN architecture for preoperative MRI tumor segmentation and survival prediction. For the tumor segmentation task, we have experimented with combining two U-Net like architectures in an ensemble (AE AU-Net) to enhance tumor detection. For survival prediction, we obtained better performance using a 3D DenseNet model than with different network architectures and handcrafted based models in our experimental setup. In addition, we showed that voxel-wise uncertainty estimation using an ensemble of models is an effective method that can be used to improve brain tumor segmentation results.

References

1. Ostrom, Q.T., et al.: CBTRUS statistical report: primary brain and central nervous system tumors diagnosed in the United States in 2007–2011. Neuro-oncology **16**(suppl. 4), iv1–63 (2014). https://doi.org/10.1093/neuonc/nou223.
2. Ellor, S.V., Pagano-Young, T.A., Avgeropoulos, N.G.: Glioblastoma: background, standard treatment paradigms, and supportive care considerations. J. Law Med. Ethics **42**(2), 171–182 (2014). https://doi.org/10.1111/jlme.12133
3. Bakas, S., et al.: Advancing The Cancer Genome Atlas glioma MRI collections with expert segmentation labels and radiomic features. Sci Data **4**, 170117 (2017). https://doi.org/10.1038/sdata.2017.117
4. Bakas, S., et al.: Identifying the Best Machine Learning Algorithms for Brain Tumor Segmentation, Progression Assessment, and Overall Survival Prediction in the BRATS Challenge. arXiv:1811.02629 [cs, stat], April 2019 (2020). https://arxiv.org/abs/1811.02629. Accessed 22 April
5. Bakas, S., et al.: Segmentation Labels for the Pre-operative Scans of the TCGA-GBM collection. The Cancer Imaging Archive (2017)
6. Bakas, S., et al.: Segmentation Labels for the Pre-operative Scans of the TCGA-LGG collection. The Cancer Imaging Archive (2017)
7. Menze, B.H., et al.: The multimodal brain tumor image segmentation benchmark (BRATS). IEEE Trans. Med. Imaging **34**(10), 1993–2024 (2015). https://doi.org/10.1109/TMI.2014.2377694
8. Ronneberger, O., Fischer, P., Brox, T.: U-Net: Convolutional Networks for Biomedical Image Segmentation, arXiv:1505.04597 [cs], May 2015, https://arxiv.org/abs/1505.04597. Accessed 22 Apr 2020
9. Kamnitsas, K., et al.: Ensembles of multiple models and architectures for robust brain tumour segmentation. In: Crimi, A., Bakas, S., Kuijf, H., Menze, B., Reyes, M. (eds.) BrainLes 2017. LNCS, vol. 10670, pp. 450–462. Springer, Cham (2018). https://doi.org/10.1007/978-3-319-75238-9_38
10. Myronenko, A.: 3D MRI brain tumor segmentation using autoencoder regularization. In: Crimi, A., Bakas, S., Kuijf, H., Keyvan, F., Reyes, M., van Walsum, T. (eds.) BrainLes 2018. LNCS, vol. 11384, pp. 311–320. Springer, Cham (2019). https://doi.org/10.1007/978-3-030-11726-9_28
11. Isensee, F., Kickingereder, P., Wick, W., Bendszus, M., Maier-Hein, K.H.: No New-Net. In: Crimi, A., Bakas, S., Kuijf, H., Keyvan, F., Reyes, M., van Walsum, T. (eds.) BrainLes 2018. LNCS, vol. 11384, pp. 234–244. Springer, Cham (2019). https://doi.org/10.1007/978-3-030-11726-9_21
12. McKinley, R., Rebsamen, M., Meier, R., Wiest, R.: Triplanar Ensemble of 3D-to-2D CNNs with label-uncertainty for brain tumor segmentation. In: Crimi, A., Bakas, S. (eds.) BrainLes 2019. LNCS, vol. 11992, pp. 379–387. Springer, Cham (2020). https://doi.org/10.1007/978-3-030-46640-4_36

13. Zhao, Y.-X., Zhang, Y.-M., Liu, C.-L.: Bag of tricks for 3D MRI brain tumor segmentation. In: Crimi, A., Bakas, S. (eds.) BrainLes 2019. LNCS, vol. 11992, pp. 210–220. Springer, Cham (2020). https://doi.org/10.1007/978-3-030-46640-4_20

14. Jiang, Z., Ding, C., Liu, M., Tao, D.: Two-stage cascaded U-Net: 1st place solution to BraTS challenge 2019 segmentation task. In: Crimi, A., Bakas, S. (eds.) BrainLes 2019. LNCS, vol. 11992, pp. 231–241. Springer, Cham (2020). https://doi.org/10.1007/978-3-030-46640-4_22

15. Lambin, P., et al.: Radiomics: Extracting more information from medical images using advanced feature analysis. Eur. J. Cancer 48(4), 441–446 (2012). https://doi.org/10.1016/j.ejca.2011.11.036

16. Shboul, Z.A., Vidyaratne, L., Alam, M., Iftekharuddin, K.M.: Glioblastoma and survival prediction. In: Crimi, A., Bakas, S., Kuijf, H., Menze, B., Reyes, M. (eds.) BrainLes 2017. LNCS, vol. 10670, pp. 358–368. Springer, Cham (2018). https://doi.org/10.1007/978-3-319-75238-9_31

17. Feng, X., Dou, Q., Tustison, N., Meyer, C.: Brain tumor segmentation with uncertainty estimation and overall survival prediction. In: Crimi, A., Bakas, S. (eds.) BrainLes 2019. LNCS, vol. 11992, pp. 304–314. Springer, Cham (2020). https://doi.org/10.1007/978-3-030-46640-4_29

18. Agravat, R.R., Raval, M.S.: Brain tumor segmentation and survival prediction. In: Crimi, A., Bakas, S. (eds.) BrainLes 2019. LNCS, vol. 11992, pp. 338–348. Springer, Cham (2020). https://doi.org/10.1007/978-3-030-46640-4_32

19. Weninger, L., Haarburger, C., Merhof, D.: Robustness of radiomics for survival prediction of brain tumor patients depending on resection status. Front. Comput. Neurosci. 13 (2019). https://doi.org/10.3389/fncom.2019.00073.

20. Wang, S., Dai, C., Mo, Y., Angelini, E., Guo, Y., Bai, W.: Automatic brain tumour segmentation and biophysics-guided survival prediction. In: Crimi, A., Bakas, S. (eds.) BrainLes 2019. LNCS, vol. 11993, pp. 61–72. Springer, Cham (2020). https://doi.org/10.1007/978-3-030-46643-5_6

21. Suter, Y., et al.: Deep learning versus classical regression for brain tumor patient survival prediction. arXiv:1811.04907 [cs], November 2018. https://arxiv.org/abs/1811.04907. Accessed 22 July 2020

22. Szegedy, C., Ioffe, S., Vanhoucke, V., Alemi, A.A.: Inception-v4, inception-ResNet and the impact of residual connections on learning. Presented at the Thirty-First AAAI Conference on Artificial Intelligence, February 2017. https://www.aaai.org/ocs/index.php/AAAI/AAAI17/paper/view/14806. Accessed 22 Apr 2020

23. Huang, G., Liu, Z., van der Maaten, L., Weinberger, K.Q.: Densely Connected Convolutional Networks, January 2018, arXiv:1608.06993 [cs]. https://arxiv.org/abs/1608.06993. Accessed 21 July 2020

24. Pedregosa, F., et al.: Scikit-learn: Machine Learning in Python. Journal of Machine Learning Research 12, 2825–2830 (2011)

25. Virtanen, P., et al.: SciPy 1.0: fundamental algorithms for scientific computing in python. Nat. Methods 17, 261–272 (2020). https://doi.org/10.1038/s41592-019-0686-2

26. Baldock, A.L., et al.: Patient-specific metrics of invasiveness reveal significant prognostic benefit of resection in a predictable subset of gliomas. PLoS ONE 9(10), e99057 (2014). https://doi.org/10.1371/journal.pone.0099057

27. Lao, J., et al.: A deep learning-based radiomics model for prediction of survival in glioblastoma multiforme. Sci. Rep. 7, September 2017. https://doi.org/10.1038/s41598-017-10649-8.

Brain Tumour Segmentation Using Probabilistic U-Net

Chinmay Savadikar$^{(\boxtimes)}$, Rahul Kulhalli, and Bhushan Garware

Persistent Systems Ltd., Pune, India
{chinmay_savadikar,rahul_kulhalli,bhushan_garware}@persistent.com

Abstract. We describe our approach towards the segmentation task of the BRATS 2020 challenge. We use the Probabilistic UNet to explore the effect of sampling different segmentation maps, which may be useful to experts when the opinions of different experts vary. We use 2D segmentation models and approach the problem in a slice-by-slice manner. To explore the possibility of designing robust models, we use self attention in the UNet, and the prior and posterior networks, and explore the effect of varying the number of attention blocks on the quality of the segmentation. Our model achieves Dice scores of 0.81898 on Whole Tumour, 0.71681 on Tumour Core, and 0.68893 on Enhancing Tumour on the Validation data, and 0.7988 on Whole Tumour, 0.7771 on Tumour Core, and 0.7249 on Enhancing Tumour on the Testing data. Our code is available at https://github.com/rahulkulhalli/BRATS2020.

Keywords: Brain Tumour Segmentation · UNet · Probabilistic UNet · Deep learning

1 Introduction

Although Deep Learning has achieved significant progress in medical imaging tasks, recent research has pushed towards interpretable models and uncertainty quantification of Deep Learning models for human-in-loop systems. This work explores the possible use of uncertainty and attention mechanisms as principles for designing and training robust models, and describes our approach for the Brain Tumour Segmentation Task. Several measures of uncertainty, based on Monte Carlo Dropout [5], have been proposed for quantifying uncertainty in Deep Learning models [10]. Nair et al. [12] use multiple uncertainty measures to evaluate filtering of segmentation samples based on the model uncertainty. Apart from these pixel–wise uncertainty methods, Probabilistic UNet [9] takes a different approach, wherein a neural network learns to sample different possible segmentation maps. This technique may be useful when the opinion of multiple experts differs on segmentation. We use the Probabilistic UNet framework to explore the effect of sampling various segmentation maps.

C. Savadikar and R. Kulhalli–Equal Contribution

© Springer Nature Switzerland AG 2021
A. Crimi and S. Bakas (Eds.): BrainLes 2020, LNCS 12659, pp. 255–264, 2021.
https://doi.org/10.1007/978-3-030-72087-2_22

2   Data

We use the data provided for the BRATS 2020 Segmentation task [1–4,11]. The data is in the form of multimodal glioblastoma (HGG) and lower grade glioma (LGG) structural MRI volumes, with each volume being a (240 × 240 × 155) tensor encoded in the NIfTI format. Each patient has four associated MRI volumes – Native (T1), Post-contrast T1-weighted (T1Gd), T2-weighted (T2), and T2 Fluid Attenuated Inversion Recovery (T2-FLAIR) volumes. We stratify our train-test split with respect to the frequency of the HGG and LGG tumors. We use 295 patients as our training dataset and 74 patients as our hold-out validation dataset. The segmentation labels contain values of 1 for necrotic (NCR) and non-enhancing (NET) tumour core, 2 for peritumoral edema (ED), 4 for enhancing tumour (ET), and 0 for everything else. The segmentation task requires the model to segment 3 classes of the segmentation masks: Enhancing Tumour, Tumour Core (consisting of ET, NCR and NET), and Whole Tumour (consisting of ED and TC).

2.1   Preprocessing

In order to reduce the complexity of data, and to fit the networks in the GPU memory, we convert the data into 2D slices by slicing along the coronal plane (since this gives slices with aspect ratio of 1). During training, we use all the 4 modalities of the data–T1, T2, T1ce and Flair–stacked as the channel dimension of the network. We take 128 × 128 centrals crops of each slice. We normalize the data such that the values in each slice map to range 0 to 1 using the following function:

$$A_{ijk} = \begin{cases} \frac{A_{ijk}-min(A_k)}{max(A_k)-min(A_k)} & if\, A_{ijk} > 0 \\ 0 & otherwise \end{cases} \tag{1}$$

3   Architecture

We use a Probabilistic UNet with self–attention blocks to obtain the segmentation maps. A separate mask is predicted for each of the 3 classes. The following sections describe the architecture in detail.

3.1   Probabilistic UNet

Figure 1 shows the architecture used for the submission. UNet [13] has achieved the best results on medical image segmentation tasks and has become a de-facto model. The Probabilistic UNet [9] uses 4 parts:

UNet: The UNet extracts segmentation features from the input volume.

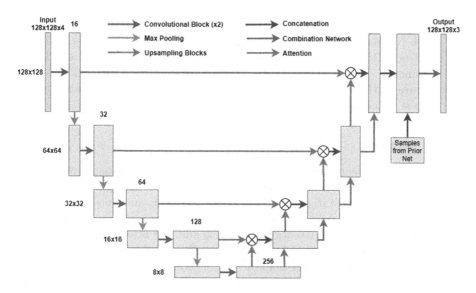

Fig. 1. The architecture of the Probabilistic U-Net used for the final submission. The Prior and Posterior networks use the same architecture as that of the downsampling branch of the U-Net. The convolutional block consists of a sequence of Convolutional, Batch Normalization and ReLU layers. The upsampling block consists of a deconvolutional layer with 2 × 2 filters followed by 2 convolutional blocks. The Combination Network is composed of 2 convolutional blocks without Batch Normalization. All convolutional layers are composed of 3 × 3 filters.

Prior Network: By sampling from a prior distribution P, it is possible to obtain variable segmentation maps. The prior distribution is assumed to be a multivariate, axis–aligned Gaussian distribution with mean $\mu_{prior}(X;\omega)$ and variance $\sigma_{prior}(X;\omega)$.

$$z \sim P(z|X) = N(\mu_{prior}(X;\omega), \sigma_{prior}(X;\omega)) \qquad (2)$$

where X denotes the input and ω denotes the parameters of the Prior Network.

Combination Network: Combines the UNet features and the samples drawn from the Prior distribution to generate the segmentation map.

Posterior Network: The prior network is trained to find an appropriate representation of a variant of a segmentation map by using a posterior distribution Q. Q is assumed to be an axis–aligned Gaussian with mean $\mu_{posterior}(X,Y;\theta)$ and variance $\sigma_{posterior}(X,Y;\theta)$. Y denotes the output, i.e. the segmentation map and θ denotes the parameters of the Posterior Network.

The UNet downsamples the input from dimensions 128 × 128 to 8 × 8 in 4 levels. Each level has 2 convolutional blocks, which is a sequence of a convolutional layer, a Batch Normalization layer [7], and ReLU. Similarly, we use 4 levels of downsampling for the prior and posterior networks. We use prior and posterior networks with latent dimension of size 6. To preserve the mean and variance the of samples drawn from the prior, the combination network does not use Batch Normalization. The UNet consists of 26,70,556 parameters, the prior net consists of 15,08,466 parameters, and the combination network consists of 531 parameters. The posterior net is not used at test time and has the same architecture as that of the prior net (15,08,898 parameters since the segmentation map is also given as an input).

3.2 Attention

We incorporate attention into our architecture to facilitate stronger representational learning and analyse the network's capability to efficiently localise objects. We focus on a specific variant of Attention mechanisms and borrow inspiration from AG (Attention Gates) [14]. This specific attention backbone adds minimal computational overhead (Attention Gates use additive attention as opposed to the conventional Dot Product Attention, as well as perform computations in a low dimensional space with the help of 1×1 convolutions) whilst reducing false positives. Attention Gates compute an attention coefficient $\alpha \in [0, 1]$ using a Gating Signal (a global encoding of the input representation) and the input feature vector obtained from the previous layer. The coefficient tensor is then multiplied element-wise with the previous layer's output. This induces a strong prior in the subsequent layers of the network as the output of the AG block will only contain activations that are relevant to the task, while suppressing the activations that are irrelevant. We hypothesize that this significantly reduces downstream learning complexity. We incorporate these AG blocks at each stage of the UNet. Moreover, we use multi-block attention at levels where the dimensions are 32×32 and 16×6 for the prior and posterior networks.

4 Training Details

We train the models to minimize the soft Dice loss between the predicted and the ground–truth segmentation maps, and the KL Divergence between the prior and the posterior distributions.

$$L = L_{Dice}(y, \hat{y}) + L_{KL}(Q||P) \tag{3}$$

The Dice loss is given by

$$L_{Dice}(y, \hat{y}) = \sum_c (1 - \frac{2 * \sum_{i,j} y * \hat{y} + 1}{\sum_{i,j} y + \sum_{i,j} \hat{y} + 1}) \tag{4}$$

where i and j denote the indices of the height and width respectively, C denotes the number of classes (3, for Whole Tumour, Tumour Core and Enhancing Tumour), and 1 is the smoothing factor used for regularization.

The models are initialized using the He normal initialization [6]. The models are trained for 100 epochs using the Adam optimizer [8], with an initial learning learning rate of 10^{-3} which is decayed by 20% every 3 epochs.

5   Visualization and Analysis

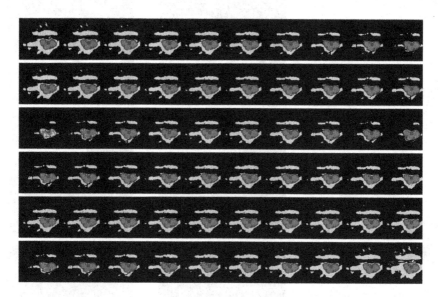

Fig. 2. Visualization of the effect of samples drawn from the prior network. Green denotes NCR and NET, Yellow denotes ED, and Red denotes ET. (Color figure online)

5.1   Probabilistic UNet

By varying one dimension of the latent vector and keeping the others constant, the effect of that dimension on the output can be observed. Figure 2 shows the visualization of the 6 dimensions. Dimensions 1, 3 and 6 encode the thickness of Whole Tumour. Dimension 3 encodes the thickness of the Enhancing Tumour. However, none of the dimensions seem to explicitly encode the Tumour Core.

5.2   Attention Maps

As has been previously documented, the attention blocks in the deeper layers encode the overall location of the tumour (blocks of spatial dimensions 16 and 32). Figure 5 shows the attention maps at all the spatial levels of the UNet. By analyzing the attention maps, one may be able to tune the number of attention blocks required at each stage, and may aid interpretability by giving an insight into the features deemed important at each stage. Following is an analysis of what the attention maps may encode.

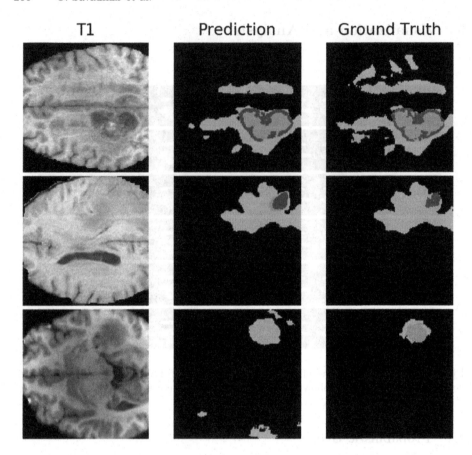

Fig. 3. Comparison of predictions with ground truth. Each row corresponds to a slice from different patients from the hold–out validation set. The top row corresponds to BraTS20_Training_069 slice 100, the middle row corresponds to BraTS20_Training_183 slice 90, and the third row corresponds to BraTS20_Training_268 slice 65. Green denotes NCR and NET, Yellow denotes ED, and Red denotes ET. (Color figure online)

Attention Blocks of Dimensions 64: Block 1 seems to encode the NCR and NET, and the ED regions of the tumour. Blocks 2 seems to encode the Whole Tumour segmentation map, which is reflected in the output of the model. Finally, Block 3 seems to encode the rough location of the tumour.

Attention Blocks of Dimensions 128: Block 1 seems to encode the ED region of the tumour. Blocks 2 seems to encode Tumour Core, and Finally, Block 3 seems to encode the precise boundary of the NCR, NET and ED regions, excluding the Enhancing Tumour.

Effect of the Number of Attention Blocks: Upon increasing the number of attention blocks in each stage on the UNet, an increase in the qualitative performance

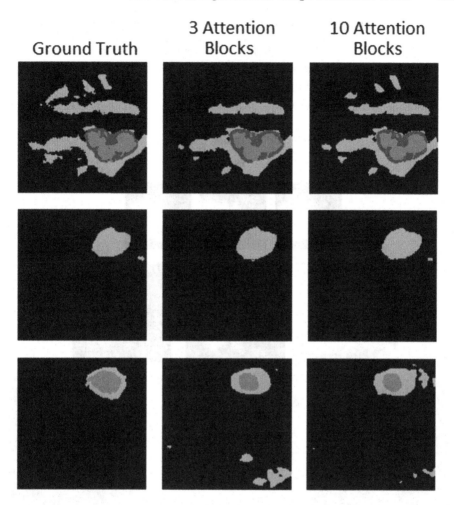

Fig. 4. Visualization of some samples of comparison of predictions using a UNet with 3 attention blocks and 10 attention blocks with the ground truth, while keeping all the other parameters constant. Each row corresponds to a slice from different patients from the hold–out validation set. The top row corresponds to BraTS20_Training_069 slice 100, the middle row corresponds to BraTS20_Training_183 slice 103, and the third row corresponds to BraTS20_Training_268 slice 60. Green denotes NCR and NET, Yellow denotes ED, and Red denotes ET. The UNet with 10 attention blocks is better at segmenting small parts of the tumour. (Color figure online)

of the segmentation maps can be observed. Figure 4 shows the effect of increasing the number of attention blocks from 3 to 10 while keeping all the other parameters constant. A qualitative improvement can be seen for the smaller regions of the tumour. A comparison of quantitative performance is left for future work.

Table 1. Results of the method on the validation set.

Class	Dice score	Hausdorff distance	Sensitivity	Specificity
Whole tumour	0.81898	41.52379	0.84613	0.9981
Tumour core	0.71681	26.2753	0.69957	0.99942
Enhancing tumour	0.68893	36.88595	0.69339	0.99965

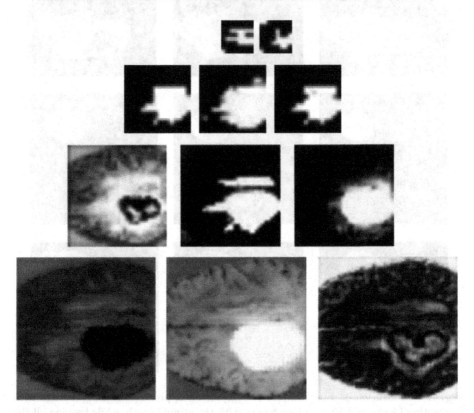

Fig. 5. Starting from top, the three attention blocks at each of the spatial levels of the UNet (16 × 16, 32 × 32, 64 × 64, 128 × 128) for slice 100 of BraTS20_Training_069.

Table 2. Results of the method on the testing set.

Class	Dice score	Hausdorff distance	Sensitivity	Specificity
Whole tumour	0.7988	39.5342	0.8719	0.9979
Tumour core	0.7771	28.4712	0.7813	0.9996
Enhancing tumour	0.7249	29.8357	0.7513	0.9996

6 Results

Our model achieves Dice scores of 0.81898 on Whole Tumour, 0.71681 on Tumour Core, and 0.68893 on Enhancing Tumour on the Validation data, and 0.7988 on Whole Tumour, 0.7771 on Tumour Core, and 0.7249 on Enhancing Tumour on the Testing data. Even though we use 2 dimensional models, we achieve decent results on the validation set. Table 1 shows the results of the approach on the validation set, and Table 2 shows the results of the approach on the testing set. Figure 3 shows the qualitative comparison of the predicted segmentation maps and the ground truth segmentation maps on different slices of 3 different patients from the hold–out validation set.

7 Discussion and Future Scope

From Fig. 3, it can be seen that the models are limited in their representational capacity for segmentation. Our models are 2 dimensional, i.e. they operate on a slice–by–slice fashion. This may inherently limit the performance. Future directions may include increasing the capacity of the models and using 3D models instead of 2D.

8 Conclusion

We experiment with the Probabilistic UNet in conjunction with Gated Attention and find that the results seem to be promising. We see that the attention mechanism plays a role in the quality of the segmentation of the tumours, with the network with more number of attention blocks showing better quality of segmentation maps. As discussed in the Future Scope, extending our hypothesis to the 3D domain may lead to better results.

References

1. Bakas, S., et al.: Segmentation labels and radiomic features for the pre-operative scans of the TCGA-LGG collection. Cancer Imaging Arch. **286** (2017). https:// doi.org/10.7937/K9/TCIA.2017.KLXWJJ1Q
2. Bakas, S., et al.: Segmentation labels and radiomic features for the pre-operative scans of the TCGA-LGG collection. Cancer Imaging Arch. **286** (2017). https:// doi.org/10.7937/K9/TCIA.2017.GJQ7R0EF
3. Bakas, S., et al.: Advancing the cancer genome atlas glioma MRI collections with expert segmentation labels and radiomic features. Scientific data **4**, 170117 (2017)
4. Bakas, S., et al.: Identifying the best machine learning algorithms for brain tumor segmentation, progression assessment, and overall survival prediction in the brats challenge. arXiv preprint arXiv:1811.02629 (2018)
5. Gal, Y., Ghahramani, Z.: Dropout as a Bayesian approximation: representing model uncertainty in deep learning. In: International Conference on Machine Learning, pp. 1050–1059 (2016)

6. He, K., Zhang, X., Ren, S., Sun, J.: Delving deep into rectifiers: surpassing human-level performance on ImageNet classification. In: Proceedings of the IEEE International Conference on Computer Vision, pp. 1026–1034 (2015)

7. Ioffe, S., Szegedy, C.: Batch normalization: accelerating deep network training by reducing internal covariate shift. In: Proceedings of the 32nd International Conference on International Conference on Machine Learning, vol. 37, pp. 448–456. ICML 2015, JMLR.org (2015)

8. Kingma, D.P., Ba, J.: Adam: a method for stochastic optimization. arXiv preprint arXiv:1412.6980 (2014)

9. Kohl, S., et al.: A probabilistic u-net for segmentation of ambiguous images. In: Advances in Neural Information Processing Systems, pp. 6965–6975 (2018)

10. Leibig, C., Allken, V., Ayhan, M.S., Berens, P., Wahl, S.: Leveraging uncertainty information from deep neural networks for disease detection. Sci. Rep. **7**(1), 1–14 (2017)

11. Menze, B.H., et al.: The multimodal brain tumor image segmentation benchmark (brats). IEEE Trans. Med. Imaging **34**(10), 1993–2024 (2014)

12. Nair, T., Precup, D., Arnold, D.L., Arbel, T.: Exploring uncertainty measures in deep networks for multiple sclerosis lesion detection and segmentation. Med. Image Anal. **59**, 101557 (2020). https://doi.org/10.1016/j.media.2019.101557, http://www.sciencedirect.com/science/article/pii/S1361841519300994

13. Ronneberger, O., Fischer, P., Brox, T.: U-net: convolutional networks for biomedical image segmentation. In: Navab, N., Hornegger, J., Wells, W.M., Frangi, A.F. (eds.) Medical Image Computing and Computer-Assisted Intervention - MICCAI 2015, pp. 234–241. Springer International Publishing, Cham (2015). https://doi.org/10.1007/978-3-319-24574-4_28

14. Schlemper, J., et al.: Attention gated networks: learning to leverage salient regions in medical images. Med. Image Anal. **53**, 197–207 (2019)

Segmenting Brain Tumors from MRI Using Cascaded 3D U-Nets

Krzysztof Kotowski[1], Szymon Adamski[1], Wojciech Malara[1],
Bartosz Machura[1], Lukasz Zarudzki[3], and Jakub Nalepa[1,2(✉)]

[1] Future Processing, Gliwice, Poland
{kkotowski,sadamski,wmalara,bmachura,jnalepa}@future-processing.com
[2] Silesian University of Technology, Gliwice, Poland
jnalepa@ieee.org
[3] Maria Sklodowska-Curie Memorial Cancer Center and Institute of Oncology,
Gliwice, Poland

Abstract. In this paper, we exploit a cascaded 3D U-Net architecture to perform detection and segmentation of brain tumors (low- and high-grade gliomas) from multi-modal magnetic resonance scans. First, we detect tumors in a binary-classification setting, and they later undergo multi-class segmentation. To provide high-quality generalization, we investigate several regularization techniques that help improve the segmentation performance obtained for the unseen scans, and benefit from the expert knowledge of a senior radiologist captured in a form of several post-processing routines. Our preliminary experiments elaborated over the BraTS'20 validation set revealed that our approach delivers high-quality tumor delineation.

Keywords: Brain tumor · Segmentation · Deep learning · U-Net

1 Introduction

Brain tumor segmentation from multi-modal magnetic resonance (MR) scans is an important step in oncology care. Accurate delineation of tumorous tissue is pivotal for further diagnosis, prognosis and treatment, and it can directly affect the treatment pathway. Hence, ensuring the reproducibility, robustness, e.g., against different scanner types, and quality of an automated segmentation process are critical to design personalized patient care and reliable patient monitoring. The state-of-the-art brain tumor segmentation algorithms are commonly divided into the *atlas-based*, *unsupervised*, *supervised*, and *hybrid* approaches. In the *atlas-based* techniques, manually segmented images (referred to as *atlases*) are used to infer the segmentation of unseen scans [24] by warping and applying other (non-rigid) registration. Since such methods solely base on the reference sets, capturing representative examples is critical to make them generalize well. This can be a fairly user-dependent and cumbersome task in practice[5].

Unsupervised approaches elaborate intrinsic characteristics of the *unlabeled* data [7] with the use of clustering [12,25,27], Gaussian modeling [26], and

© Springer Nature Switzerland AG 2021
A. Crimi and S. Bakas (Eds.): BrainLes 2020, LNCS 12659, pp. 265–277, 2021.
https://doi.org/10.1007/978-3-030-72087-2_23

other techniques [22]. Once the labeled data is available, we can utilize the *supervised* techniques. Such supervised learners include, among others, decision forests [9,31], conditional random fields [28], support vector machines [15], extremely randomized trees [23], and supervised self-organizing maps [21].

We have been witnessing breakthroughs delivered by deep learning in a variety of fields, and brain tumor detection and segmentation are not the exceptions here [4,10,14,17,18]. Such deep learning-powered segmentation systems are built upon various networks architectures and approaches, and they encompass holistically nested neural nets [30], ensembles of deep neural nets [13], U-Net-based architectures [11,20], autoencoder networks, convolutional encoder-decoder approaches [6], and more [8], very often coupled with a battery of regularization techniques [19]. Finally, the *hybrid* algorithms combine and benefit from advantages of various methods belonging to different categories [29].

We propose a cascaded U-Net-based architecture (Sect. 3) operating over multi-modal MR scans, in which brain tumors are delineated in the first step, and they are later segmented into the enhancing tumor, peritumoral edema, and necrotic and non-enhancing tumor core. In both stages, we employ 3D U-Nets that process 3D patches to better capture the nearest context of the lesion within an input scan. To enhance the generalization abilities of our approach, we utilize and experimentally verify various regularizers, alongside the post-processing routines that capture the expert knowledge of a senior radiologist. The experiments performed over the newest release of the BraTS dataset showed that our architecture delivers accurate multi-class segmentation (Sect. 4).

2   Data

The newest release of the Brain Tumor Segmentation (BraTS'20) dataset [1–4] includes MRI data of 369 patients with diagnosed gliomas—293 high-grade glioblastomas (HGG), and 76 low-grade gliomas (LGG). Each study was manually annotated by one to four experienced and trained readers. The data comes in four co-registered modalities: native pre-contrast (T1), post-contrast T1-weighted (T1Gd), T2-weighted (T2), and T2 Fluid Attenuated Inversion Recovery (T2-FLAIR). All pixels are labeled, and the following classes are considered: healthy tissue, Gd-enhancing tumor (ET), peritumoral edema (ED), the necrotic and non-enhancing tumor core (NCR/NET) [16].

The data was acquired with different protocols and scanners, and at 19 institutions. The studies were interpolated to the same shape ($240 \times 240 \times 155$, hence there are 155 images of 240×240 size, with voxel size of $1\,mm^3$), and they were skull-stripped. Finally, there are 125 patients in the validation set V (see examples in Fig. 1), for which the manual annotations are not provided.

T1	T1Gd	T2	T2-FLAIR

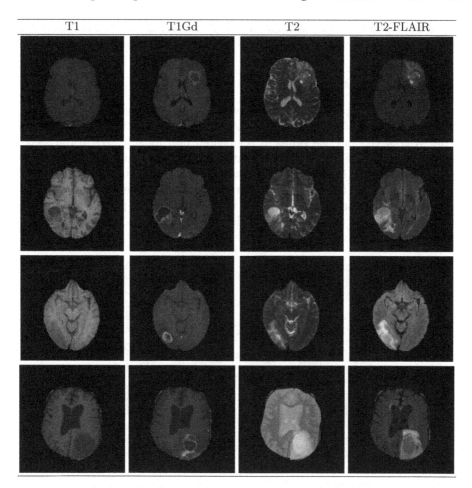

Fig. 1. Example scans of four patients (separate rows) included in the validation set.

3   Methods

3.1   Data Standardization

The data was acquired with different clinical protocols, various scanners, and at different institutions, therefore the pixel intensity distribution across the scans may vary. Thus, we standardize the input volumes of each modality. Specifically, we employ a modified version of the Z-score normalization:

$$z = \frac{p_i - m}{\text{IQR}}, \tag{1}$$

where p_i is the i-th input volume (a single modality), m denotes the median of all pixels *within* the corresponding volume, and IQR is their interquartile range.

We exploit median instead of mean, and IQR instead of the standard deviation
to make the standardization more robust against possible outlying pixel values.

3.2 Our U-Net-Based Architecture

In this proposed deep network architecture, we employ a cascaded processing in
which the tumor is detected in the first stage (hence, we perform the delineation
of the WT class), and the tumor is segmented into its subparts in the second
stage. Note that the latter stage operates *only* on the parts of the scan that
contain the detected lesions, thus the remaining part of the brain scan is pruned.
In both steps, we use the very same underlying U-Net-based architecture in which
the modalities are analyzed separately in the contracting part of the network
(Fig. 2). Finally, the features are fused in the expanding U-Net path.

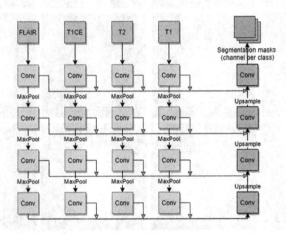

Fig. 2. In our multipath approach, each modality is processed in a separate path in
the contracting part of the network. Note that the details of the architecture, alongside
the number and sizes of the corresponding kernels, are presented in Fig. 3.

To benefit from the information available in the nearest neighboring images
in an input scan, we include the additional slices along Z-axis, i.e., the one slice
below and one above the target slice, to the input patch that is processed by the
network. Then, we utilize 3D convolutional kernels of size $3 \times 3 \times 3$ (Fig. 3). The
output of the network is a 2D probability map for the target slice.

3.3 Post-processing

After the detection process, the U-Net response is binarized using a threshold
τ. This set of binary masks becomes an input to the multi-class segmentation
engine—note that we discard frames in which a lesion was not detected. The
segmentation engine produces a $240 \times 240 \times 3$ sigmoid activation map, where

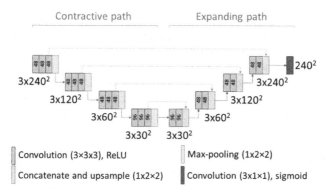

Fig. 3. To exploit the information available in the neighboring slices, we employ 3D convolutional kernels of size $3 \times 3 \times 3$. We present the dimensionality of the tensor within the network, alongside the number of corresponding filters within the architecture.

the last dimension represents the number of classes. The activation is finally passed through a softmax operation, which performs the classification.

To further exploit the "expert knowledge"[1], we employ additional post-processing steps (in the following order):

1. All voxels that are annotated as "healthy" and lay inside a closed tumor volume after detection are relabeled to "tumorous".
2. Small volumes (here, less than 500 mm^3) are removed after multi-class segmentation (to remove false positives which are often manifested as such small volumes). Hence, we assume that the minimal lesion's volume is 500 mm^3.
3. If our model annotated more than 99% voxels as ED in the largest lesion, we convert these voxels to NCR/NET.
4. If there exist voxels surrounded by the tumor core in the axial plane, and are labeled as ED, such voxels are relabeled to NCR/NET.

3.4 Regularization Strategies

To help our detection and segmentation generalize well over the unseen data, we employed and experimentally validated the following regularization strategies:

- **Ensembling.** The training dataset was split into five non-overlapping folds (see Sect. 4.2 for details). Here, we train separate base models, for both detection and segmentation, and then ensemble them by either averaging their probability maps and thresholding the resulting map (with threshold τ), or by employing the majority voting.
- **Data augmentation.** For all models, we exploit *test-time data augmentation* via flipping an input patch horizontally. Additionally, we investigated the

---

[1] These post-processing steps were co-designed in co-operation with a senior radiologist with 12 years of experience.

impact of *training-time data augmentation* on the abilities of our models. It included a chain of the following augmentation steps (run in a random order), each executed with 50% probability (therefore, for one original training patch we elaborate a single augmented example): horizontal flip, scaling (with a random scale $\mathcal{S} \in [0.9, 1.1]$, separately for both X and Y axes), rotation ($\alpha \in [-10°, 10°]$), and linear voxel-wise contrast (the voxel value v becomes $v = 127 + \alpha \cdot (v - 127)$; $\alpha \in [0.9, 1.1]$).

4   Experiments

4.1   Experimental Setup

The DNN models were implemented using `Python3` with the `Keras` library over CUDA 10.0 and CuDNN 7.6.5. The experiments were run on a machine equipped with an Intel i7-6850K (15 MB Cache, 3.80 GHz) CPU with 32 GB RAM and NVIDIA GTX Titan X GPU with 12 GB VRAM.

4.2   Training Process

The metric for training was the DICE score for both stages (detection and segmentation), hence we have the soft DICE loss $\mathcal{L} = 1 - \text{DICE}(P, GT)$, where

$$\text{DICE}(P, GT) = \frac{2\,|P \cdot GT|}{|P|^2 + |GT|^2}, \tag{2}$$

and P denotes the prediction, GT is the ground truth, and $\cdot$ denotes the element-wise multiplication. For the multi-class segmentation training, we take the soft DICE averaged across all classes. The DICE score is collectively calculated for the full batch of size 6 for both detection and segmentation. The optimizer was Nadam (Adam with Nesterov momentum) with the initial learning rate of 10^{-4}, and $\beta_1 = 0.9$, $\beta_2 = 0.999$. The training ran until DICE over V did not increase by at least 0.001 in 7 epochs for detection, and in 5 epochs for segmentation.

The training dataset was split into five non-overlapping stratified folds (each base model is trained over four folds in the training set, and one fold is used for validation during the training process; see their characteristics in Table 1):

- **Detection (WT)**—we stratify the dataset with respect to the distribution of the size of the WT class examples. The fold sizes were equal to 74, 73, 75, 74, and 73 patients.
- **Segmentation (NET, ED, ET)**—we stratify the dataset with respect to the distribution of the size of the NET, ED, ET, and WT class examples. The fold sizes were equal to 72, 77, 72, 73, and 75 patients.

Table 1. Characteristics of the folds (average volume of the corresponding tissue in cm^3) used for training our detection and segmentation parts of the cascaded architecture.

Fold	Detection		Segmentation				
	Brain vol.	WT vol.	Brain vol.	WT vol.	NET vol.	ED vol.	ET vol.
Fold 1	1435.80	96.83	1451.83	94.53	21.09	52.31	21.14
Fold 2	1450.37	97.87	1457.54	104.08	22.77	62.05	19.26
Fold 3	1455.54	96.89	1446.58	96.27	20.60	56.60	19.07
Fold 4	1447.58	101.14	1440.99	102.21	24.15	57.31	20.74
Fold 5	1419.94	105.09	1412.82	100.26	21.92	59.99	18.35

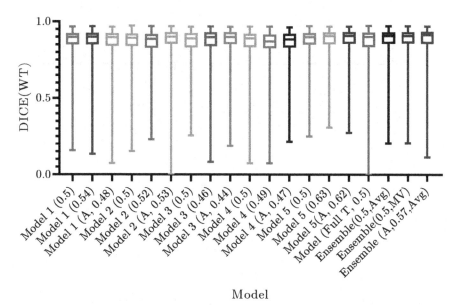

Fig. 4. The DICE values obtained over the BraTS'20 validation set. In parentheses, we report the threshold used for binarization (0.5 vs. τ obtained over our internal validation set extracted from the original training set), the A variants used data augmentation. For the ensembles, we use either averaging (Avg) or majority voting (MV).

4.3 The Results

In this section, we gather the results obtained over the BraTS'20 validation set (as returned by the validation server)[2]. Although the differences in DICE for WT (Fig. 4) are not statistically important in most cases (Kruskal-Wallis test with the Dunn's multiple comparisons tests at $p < 0.05$; only Model(0.49) and Ensemble(A, 0.57) gave statistically different segmentations), the investigated

---

[2] Our team name is `FutureHealthcare`.

variants led to improved Hausdorff distances for WT (Fig. 5). It is also worth noting that a model trained over the entire training set (in which we do not benefit from ensembling) leads to fairly unstable results with large deviations in DICE—see Model (Full T, 0.5) in Fig. 4.

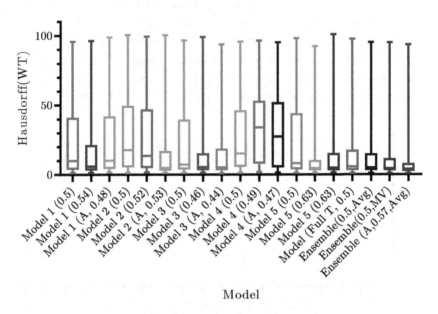

Fig. 5. The Hausdorff distance obtained over the BraTS'20 validation set. In parentheses, we report the threshold used for binarization (0.5 vs. τ obtained over our internal validation set extracted from the original training set), the A variants used data augmentation. For the ensembles, we use either averaging (Avg) or majority voting (MV).

For the whole process (detection and further segmentation), we exploited the ensembles with augmentation—the results obtained over the unlabeled validation data are reported in Table 2. In this table, we gather the quality metrics obtained for the models with and without the post-processing steps that benefit from the expert knowledge (see Sect. 3.3). The results, backed up with statistical tests showed that utilizing such additional routines help significantly improve the DICE scores for all classes (ET, WT, and TC). To this end, the model that exploits these post-processing steps became our final pipeline that was utilized over the unseen BraTS'20 test data—the results obtained over this dataset are reported in Table 3. We can appreciate that it can generalize well over the unseen scans, and delivers high-quality delineation of brain tumors. Investigating the average and median values of DICE and Hausdorff distances indicates that there exist outlying low-quality segmentations (especially of the enhancing-tumor and tumor-core areas)—see the examples of high- and low-quality delineations rendered in Fig. 6. Tackling such cases is currently our research focus, and we work on introducing additional regularization techniques that could potentially help improve the performance in scans of varying quality.

Table 2. Segmentation performance quantified by DICE, sensitivity, specificity, and Hausdorff (95%) distance over the BraTS'20 validation set obtained using our methods—without and with post-processing steps that exploit the expert knowledge (as returned by the validation server). The scores (average μ, standard deviation s, median m, and the 25 and 75 quantile) are presented for the whole tumor (WT), tumor core (TC), and enhancing tumor (ET), and the best scores are boldfaced. The differences between the models with and without the post-processing steps are statistically important ($p < 0.01$; Wilcoxon tests) for all DICE scores: ET, WT, and TC.

	DICE ET	DICE WT	DICE TC	Sens. ET	Sens. WT	Sens. TC	Spec. WT	Spec. ET	Spec. TC.	Haus. ET	Haus. WT	Haus. TC
Without "expert knowledge" in post processing												
μ	0.68530	0.87120	0.74525	0.68190	0.86591	0.70842	**0.99971**	0.99905	**0.99969**	47.45212	11.39316	21.89342
s	0.31474	**0.12650**	0.26815	0.32784	**0.14907**	0.28976	0.00039	0.00105	**0.00040**	112.71078	18.27181	58.58751
m	0.83937	0.91444	0.86610	0.82454	**0.91836**	0.82225	**0.99984**	0.99934	**0.99984**	2.82843	4.12311	6.70820
25q	0.61786	0.85683	0.69756	0.59544	0.84418	0.57997	**0.99963**	0.99872	**0.99962**	1.41421	2.44949	**2.23607**
75q	0.89081	0.93690	0.92384	0.90238	0.95777	0.92886	**0.99996**	0.99975	**0.99993**	14.00713	8.25829	13.89244
With "expert knowledge" in post processing												
μ	**0.69377**	**0.87285**	**0.79121**	**0.68988**	**0.86614**	**0.77520**	**0.99971**	0.99908	0.99956	43.95107	10.48891	11.72002
s	**0.31028**	0.12695	**0.19117**	**0.32350**	0.15004	**0.21679**	0.00038	0.00103	0.00065	108.92972	17.34182	18.33488
m	**0.84176**	**0.91557**	**0.87378**	**0.82714**	**0.91836**	**0.86765**	**0.99984**	0.99940	0.99980	2.82843	4.00000	5.38516
25q	**0.62331**	**0.86179**	**0.71584**	**0.60049**	**0.84695**	**0.66444**	**0.99963**	0.99878	0.99950	1.41421	2.23607	2.23607
75q	**0.89360**	**0.94153**	**0.92374**	**0.90546**	**0.95789**	**0.93516**	**0.99996**	0.99975	0.99992	11.53256	8.30662	12.42174

Table 3. Segmentation performance quantified by DICE, sensitivity, specificity, and Hausdorff (95%) distance over the BraTS'20 test set obtained using our algorithm. The scores (average μ, standard deviation s, median m, and the 25 and 75 quantile) are presented for the whole tumor (WT), tumor core (TC), and enhancing tumor (ET).

	DICE ET	DICE WT	DICE TC	Sens. ET	Sens. WT	Sens. TC	Spec. ET	Spec. WT	Spec. TC	Haus. ET	Haus. WT	Haus. TC
μ	0.73623	0.85213	0.79042	0.77186	0.87297	0.80920	0.99962	0.99878	0.99950	32.28306	12.18863	26.71511
s	0.26635	0.15586	0.27019	0.28547	0.14768	0.25756	0.00042	0.00140	0.00101	96.10351	17.49966	79.51748
m	0.83545	0.90468	0.89410	0.89585	0.92208	0.91432	0.99974	0.99926	0.99979	2.00000	4.12311	3.67360
25q	0.70874	0.83935	0.79860	0.77227	0.84470	0.79424	0.99945	0.99843	0.99951	1.41421	2.23607	2.00000
75q	0.89489	0.93542	0.94339	0.93703	0.95855	0.95489	0.99992	0.99969	0.99992	3.57019	11.27970	8.49559

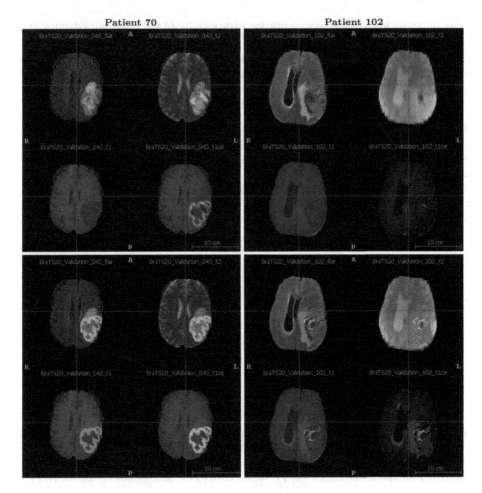

Fig. 6. Examples high- and low-quality segmentations of the patients (patients 40 and 102, respectively) from the validation set (green—ED, yellow—ET, red—NCR/NET). The DICE values amounted to 0.93644 (ET), 0.95381 (WT), and 0.96914 (TC) for patient 40, and to 0.46792 (ET), 0.70347 (WT), and 0.44076 (TC) for patient 102. (Color figure online)

5   Conclusion

In this paper, we exploited cascaded 3D U-Nets for brain tumor detection and segmentation. Our experiments showed that the models deliver accurate delineation and work end-to-end without any user intervention. Additionally, we enhanced our deep learning-powered pipeline with the additional post-processing steps that utilize the "expert knowledge" captured by a senior radiologist.

We currently focus on improving the segmentation quality of the proposed models by applying additional regularization techniques, alongside image analysis algorithms that can be coupled with the deep learning approach. Such techniques might be beneficial especially near the tumor boundaries, and could help improve the Hausdorff distance between automatic and manual segmentations.

Acknowledgments. JN was supported by the Silesian University of Technology funds through the grant for maintaining and developing research potential, and by the Rector's Research and Development Grant (02/080/RGJ20/0003).

This paper is in memory of Dr. Grzegorz Nalepa, an extraordinary scientist and pediatric hematologist/oncologist at Riley Hospital for Children, Indianapolis, USA, who helped countless patients and their families through some of the most challenging moments of their lives.

References

1. Bakas, S., et al.: Advancing the cancer genome atlas glioma MRI collections with expert segmentation labels and radiomic features. Nat. Sci. Data, **4**, 1–13 (2017). https://doi.org/10.1038/sdata.2017.117
2. Bakas, S., et al.: Segmentation labels and radiomic features for the pre-operative scans of the TCGA-GBM collection. Cancer Imaging Arch. (2017). https://doi.org/10.7937/K9/TCIA.2017.KLXWJJ1Q
3. Bakas, S., et al.: Segmentation labels and radiomic features for the pre-operative scans of the TCGA-LGG collection. Cancer Imaging Arch. (2017). https://doi.org/10.7937/K9/TCIA.2017.GJQ7R0EF
4. Bakas, S., et al.: Identifying the best machine learning algorithms for brain tumor segmentation, progression assessment, and overall survival prediction in the BRATS challenge. CoRR abs/1811.02629 (2018)
5. Bauer, S., Seiler, C., Bardyn, T., Buechler, P., Reyes, M.: Atlas-based segmentation of brain tumor images using a Markov Random Field-based tumor growth model and non-rigid registration. In: Proceedings of IEEE EMBC, pp. 4080–4083 (2010)
6. Bontempi, D., Benini, S., Signoroni, A., Svanera, M., Muckli, L.: Cerebrum: a fast and fully-volumetric convolutional encoder-decoder for weakly-supervised segmentation of brain structures from out-of-the-scanner MRI. Medical Image Analy. **62**, 101688 (2020)
7. Chander, A., Chatterjee, A., Siarry, P.: A new social and momentum component adaptive PSO algorithm for image segmentation. Expert Syst. Appl. **38**(5), 4998–5004 (2011)
8. Estienne, T., et al.: Deep learning-based concurrent brain registration and tumor segmentation. Front. Comput. Neurosci. **14**, 17 (2020)
9. Geremia, E., Clatz, O., Menze, B.H., Konukoglu, E., Criminisi, A., Ayache, N.: Spatial decision forests for MS lesion segmentation in multi-channel magnetic resonance images. NeuroImage **57**(2), 378–390 (2011)
10. Ghafoorian, M., et al.: Transfer learning for domain adaptation in MRI: application in brain lesion segmentation. In: Descoteaux, M., Maier-Hein, L., Franz, A., Jannin, P., Collins, D.L., Duchesne, S. (eds.) MICCAI 2017. LNCS, vol. 10435, pp. 516–524. Springer, Cham (2017). https://doi.org/10.1007/978-3-319-66179-7_59
11. Isensee, F., Kickingereder, P., Wick, W., Bendszus, M., Maier-Hein, K.H.: No new-net. In: Crimi, A., Bakas, S., Kuijf, H., Keyvan, F., Reyes, M., van Walsum, T. (eds.) Brainlesion: Glioma, Multiple Sclerosis, Stroke and Traumatic Brain Injuries, pp. 234–244. Springer International Publishing, Cham (2019)

12. Ji, S., Wei, B., Yu, Z., Yang, G., Yin, Y.: A new multistage medical segmentation method based on superpixel and fuzzy clustering. Comput. Math. Meth. Med. 2014, 747549:1–747549:13 (2014)
13. Kamnitsas, K., et al.: Ensembles of multiple models and architectures for robust brain tumour segmentation. In: Crimi, A., Bakas, S., Kuijf, H., Menze, B., Reyes, M. (eds.) BrainLes 2017. LNCS, vol. 10670, pp. 450–462. Springer, Cham (2018). https://doi.org/10.1007/978-3-319-75238-9_38
14. Korfiatis, P., Kline, T.L., Erickson, B.J.: Automated segmentation of hyperintense regions in FLAIR MRI using deep learning. Tomography: J. Imaging Res. 2(4), 334–340 (2016)
15. Ladgham, A., Torkhani, G., Sakly, A., Mtibaa, A.: Modified support vector machines for MR brain images recognition. In: Proceedings CoDIT, pp. 032–035 (2013)
16. Menze, B.H., et al.: The multimodal brain tumor image segmentation benchmark (BRATS). IEEE Trans. Med. Imaging 34(10), 1993–2024 (2015)
17. Moeskops, P., Viergever, M.A., Mendrik, A.M., de Vries, L.S., Benders, M.J.N.L., Isgum, I.: Automatic segmentation of MR brain images with a convolutional neural network. IEEE Trans. Med. Imaging 35(5), 1252–1261 (2016)
18. Myronenko, A.: 3D MRI brain tumor segmentation using autoencoder regularization. In: Crimi, A., Bakas, S., Kuijf, H., Keyvan, F., Reyes, M., van Walsum, T. (eds.) Brainlesion: Glioma, Multiple Sclerosis, Stroke and Traumatic Brain Injuries, pp. 311–320. Springer International Publishing, Cham (2019)
19. Nalepa, J., Marcinkiewicz, M., Kawulok, M.: Data augmentation for brain-tumor segmentation: a review. Front. Comput. Neurosci. 13, 83 (2019)
20. Nalepa, J., et al.: Fully-automated deep learning-powered system for DCE-MRI analysis of brain tumors. Artif. Intell. Med. 102, 101769 (2020)
21. Ortiz, A., Gorriz, J.M., Ramirez, J., Salas-Gonzalez, D.: MRI brain image segmentation with supervised SOM and probability-based clustering method. In: Ferrández, J.M., Álvarez Sánchez, J.R., de la Paz, F., Toledo, F.J. (eds.) IWINAC 2011. LNCS, vol. 6687, pp. 49–58. Springer, Heidelberg (2011). https://doi.org/10.1007/978-3-642-21326-7_6
22. Ouchicha, C., Ammor, O., Meknassi, M.: Unsupervised brain tumor segmentation from magnetic resonance images. In: Proc. IEEE WINCOM, pp. 1–5 (2019)
23. Pinto, A., Pereira, S., Correia, H., Oliveira, J., Rasteiro, D.M.L.D., Silva, C.A.: Brain tumour segmentation based on extremely randomized forest with high-level features. In: Proceedings of IEEE EMBC, pp. 3037–3040 (2015)
24. Pipitone, J., et al.: Multi-atlas segmentation of the whole hippocampus and subfields using multiple automatically generated templates. NeuroImage 101, 494–512 (2014)
25. Saha, S., Bandyopadhyay, S.: MRI brain image segmentation by fuzzy symmetry based genetic clustering technique. In: Proceedings of IEEE CEC, pp. 4417–4424 (2007)
26. Simi, V., Joseph, J.: Segmentation of glioblastoma multiforme from MR images - a comprehensive review. Egypt. J. Radiol. Nucl. Med. 46(4), 1105–1110 (2015)
27. Verma, N., Cowperthwaite, M.C., Markey, M.K.: Superpixels in brain MR image analysis. In: Proceedings of IEEE EMBC, pp. 1077–1080 (2013)
28. Wu, W., Chen, A.Y.C., Zhao, L., Corso, J.J.: Brain tumor detection and segmentation in a CRF (conditional random fields) framework with pixel-pairwise affinity and superpixel-level features. Int. J. Comput. Assist. Radiol. Surg. 9(2), 241–253 (2013). https://doi.org/10.1007/s11548-013-0922-7

29. Zhao, X., Wu, Y., Song, G., Li, Z., Zhang, Y., Fan, Y.: A deep learning model integrating FCNNs and CRFs for brain tumor segmentation. CoRR abs/1702.04528 (2017)
30. Zhuge, Y., et al.: Brain tumor segmentation using holistically nested neural networks in MRI images. Med. Phys. 1–10 (2017)
31. Zikic, D., et al.: Decision Forests for Tissue-Specific Segmentation of High-Grade Gliomas in Multi-channel MR. In: Ayache, N., Delingette, H., Golland, P., Mori, K. (eds.) MICCAI 2012. LNCS, vol. 7512, pp. 369–376. Springer, Heidelberg (2012). https://doi.org/10.1007/978-3-642-33454-2_46

A Deep Supervised U-Attention Net for Pixel-Wise Brain Tumor Segmentation

Jia Hua Xu$^{(\boxtimes)}$, Wai Po Kevin Teng, Xiong Jun Wang,
and Andreas Nürnberger

Data and Knowledge Engineering Group, Faculty of Computer Science,
Otto von Guericke University Magdeburg, Magdeburg, Germany
{jiahua.xu,andreas.nurnberger}@ovgu.de,
{wai.teng,xiongjun.wang}@st.ovgu.de

Abstract. Glioblastoma (GBM) is one of the leading causes of cancer death. The imaging diagnostics are critical for all phases in the treatment of brain tumor. However, manually-checked output by a radiologist has several limitations such as tedious annotation, time consuming and subjective biases, which influence the outcome of a brain tumor affected region. Therefore, the development of an automatic segmentation framework has attracted lots of attention from both clinical and academic researchers. Recently, most state-of-the-art algorithms are derived from deep learning methodologies such as the U-net, attention network. In this paper, we propose a deep supervised U-Attention Net framework for pixel-wise brain tumor segmentation, which combines the U-net, Attention network and a deep supervised multistage layer. Subsequently, we are able to achieve a low resolution and high resolution feature representations even for small tumor regions. Preliminary results of our method on training data have mean dice coefficients of about 0.75, 0.88, and 0.80; on the other hand, validation data achieve a mean dice coefficient of 0.67, 0.86, and 0.70, for enhancing tumor (ET), whole tumor (WT), and tumor core (TC) respectively.

Keywords: Brain tumor · Attention network · U-Net · BraTS2020 · Multistage features

1 Introduction

Glioblastoma (GBM) is the most common and aggressive type of malignant brain tumor in adults [8]. The Public Health England estimates the median survival period as 6 months from diagnosis without treatment [5]; however, timely diagnosis and reasonable treatment can help prolong the survival time and increase the survival rate of the patients. Therefore, diagnosis assessment can be a crucial part to enhance the quality of the patients' lives [15]. Imaging techniques such as magnetic resonance imaging (MRI) scans, which allows the complexity and the heterogeneity of the tumor lesion with a better-visualized performance than a

© Springer Nature Switzerland AG 2021
A. Crimi and S. Bakas (Eds.): BrainLes 2020, LNCS 12659, pp. 278–289, 2021.
https://doi.org/10.1007/978-3-030-72087-2_24

CT scan, is more routinely used for the purpose of visualizing the tumors in medical practice [14]. Although it sounds trivial, the imaging diagnostics are critical for all phases in the treatment of a brain tumor. However, manually identifying the brain tumor region by a radiologist has several drawbacks such as laborious annotation, time consuming and subjective uncertainties, so it indicates that brain tumor segmentation are susceptible to human errors. With the advancement of contemporary technologies, the development of an accurate automatic segmentation framework is viewed as the holy grail among domain experts as well as academic researchers. Emerging machine learning methods, especially deep learning, induced more accurate and reliable solutions to increase clinical workflow efficiency, whilst support decision making [16].

Though attractive, an automatic and reliable segmentation procedure is one of the most challenging tasks for the diagnostic of the brain tumor. Convolutional neural network (CNN) was proposed to solve various computer vision tasks, proving its capability and accuracy without compromising performance. CNN has successfully been the de facto state of the art performers in many applications related to images [19]. On the other hand, the availability of a unique dataset of MRI scans of low- and high-grade glioma patients with repetitive manual tumor delineations by several human experts are scarce. BraTS challenge dataset is based on ample multi-institutional routine clinically-acquired pre-operative multi-modal MRI scans of glioblastoma (GBM/HGG) and lower-grade glioma (LGG) [1,2] (see Fig. 1). As compared to the previous dataset, the BraTS2020 dataset has more routine clinically-acquired 3T multi-modal MRI scans, with accompanying ground truth labels by expert board-certified neurologists [4]. The main task of the BraTS2020 challenge is to develop an automatic method and produce segmentation labels of the different glioma sub-regions with the usage of provided clinically-acquired training data [2].

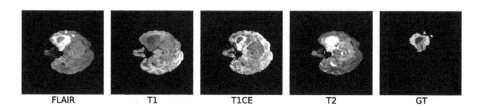

Fig. 1. Multimodal imaging with Flair, t1, t1ce, t2 and annotated ground truth from BraTS2020 training dataset *(from left to right)*. The segmentation are combined to generate the final labels of the tumor sub-regions, each color depicts *Peritumoral Edema(yellow)*, *Necrotic* and *Non-enhancing Tumor Core(blue)*, *Necrotic* and *GD-Enhancing Tumor(green)* and *Background(black)*, the main task of this challenge is to develop a method to segment labels of the different glioma sub-regions. (Color figure online)

The various methods of automatic image segmentation have been proposed by researchers all over the world [4]. As early solution of image segmentation task,

CNN and U-net are the most popular framework structures [17]. For example, Chen et al. [6] proposed an auto-context version of the VoxResNet by combining low-level features, implicit shape information, and high-level context together, which achieved the first place for the 2018 BraTS challenge. On the other hand, Feng et al. [7] developed a 3D U-Net with adaptations in the training and testing strategies, network structures, and model parameters for brain tumor segmentation. Lee et al. [11] proposed a patch-wise U-Net architecture. In this method, the model is used to overcome the drawbacks of conventional U-net with more retention of local information. Subsequently, attention network was another highlighted framework in recent research papers. For example, Oktay et al. [16] proposed a novel attention gate to focus on target structures of varying shapes and sizes. Models trained with attention gate implicitly learn to suppress irrelevant regions in an input image while highlighting salient features. Noori et al. [15] has designed a low-parameter network based on 2D U-Net in which employs an attention mechanism. This technique prevented redundancy for the model by weighting each of the channels adaptively.

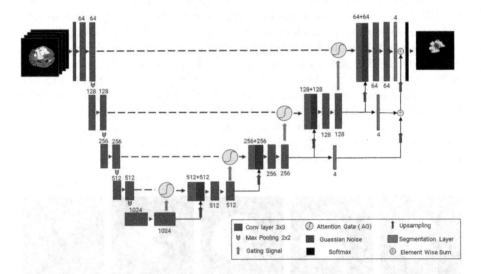

Fig. 2. Proposed Network architecture. Our architecture is inspired by [9,15,16]. An attention mechanism based 2D U-net integrated with multistage segmentation layer features through deep supervision as proposed network structure. The context pathway aggregates high level information in the right part. The low level features was filtered by a attention gate before concatenation with the low high features, the segmentation layer combines the multistage feature maps by a element wise summation for final network output as well as propagation.

In this paper, inspired by [9,10], we proposed a deep supervised U-Attention Net framework for multi-label pixel-wise brain tumor segmentation (Fig. 2). In addition to structure and parameters modification for U-net, we implemented an attention gate prior to concatenating features from skip connection.

Follows, a multistage segmentation layer was added to summarize features element-wise during the upsampling path for network output. Generally, this framework includes three main parts: to speed up computation with equivalent performance of the network in terms of memory, the U-net was chosen as a backbone network structure that learns the high and low-level features, as well as, superposition of multiscale feature maps during upsampling path to enhance gradient signal. However, it has been considered a challenge to reduce false-positive prediction for small objects that show large shape morphology. Conventionally, concatenate operation was implemented with skip-connection directly passing the information from the encoder to the decoder at the same level. Concatenation of these different level feature maps without emphasizing important features brings about redundancies that hinder low-level feature extractions. This may promote errors during the model prediction, leading to wrong segmentation of tumors in the pixel space. The detail of the structure was introduced in the following.

2 Method

2.1 Network Architecture

Our proposed backbone network is based on an optimized U-Net architecture [17]. The U-Net structure approach allows the network to reintegrate the low-level and high-level features throughout the upsampling and downsampling pathway. The input images are injected with gaussian noise of standard deviation of 0.01 to the MRI images before passing through the model, whilst to prevent overfitting from the model. Each block in the downsampling pathway has two convolutional layers with a kernel size of 3 × 3, and the max pooling layer with kernel size 2 and stride 2. The activation function is the 'ReLU' (rectified linear unit).

The numbers of filters in five blocks of downsampling pathway are 64, 128, 256, 512 and 1024 which is consistent with upsampling pathway. These blocks in the downsampling pathway shape will eventually downsample 240 × 240 input image into a 15 × 15 size feature maps. Consequently, the same parameters have been applied in the upsampling pathway except an extra convolutional layer, the network output of each deconvolutional layers was concatenated with the output of the attention gate signal from the same level downsampling pathway. In the end, the feature map (15 × 15) was restored to the original image size(240 × 240).

Meanwhile, grid-based attention gate [16] was introduced before concatenate operation as a filter to suppress low-level feature extractions in irrelevant regions. Attention gated signal was obtained by element wise multiplication between the input signal and grid-based attention coefficient, which learns to focus on wanted region during training phase. The attention network performance could learn to attend specific regions on a pixel level as shown in the architecture in Fig. 3.

During the upsampling process, we would like to keep more feature information from the lower upsampling layer, so a convolutional layer after a deconvolution block was added, which maps the deconvolutional block output(256 × 60 × 60)

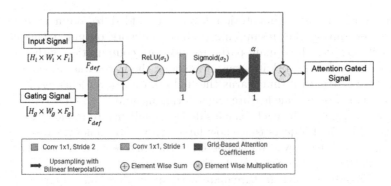

Fig. 3. Attention gate network architecture

to a $(4 \times 60 \times 60)$ feature map in the third lowest upsampling layer. This feature map was element-wisely summed with the output of the second lowest upsampling convolutional layer $(4 \times 120 \times 120)$. In the last stage, the convolutional layer combines the second lowest upsampling convolutional layer to generate the final output $(4 \times 240 \times 240)$ which is then fed into softmax function for multilabel segmentation.

Furthermore, during the upsampling pathway, the segmentation layers enhanced the output of the previous convolutional layer feature maps created at a different stage. Element-wise summation of feature maps from different levels directly pass local details found in the third lowest resolution segmentation map and the second-lowest resolution segmentation map when upsampling, and summed with the final convolutional layer output. The method was inspired in [10,12], this multistage layer forwards the lowest resolution gradient signals to the network output and back-propagate the signals to the whole upsampling network, which allows the model to localize concurrent brain tumor regions precisely.

Though above, the proposed model in this paper implemented a deep supervised U-Attention Net for pixel-wise brain tumor segmentation, which combines the U-net, grid-based attention network and a deep supervised multistage segmentation layer. In this way, we could achieve both the low resolution and high-resolution tumor regions feature representations.

2.2   Evaluation Metrics

To evaluate the model performance of our proposed method, Dice Score and Hausdorff Distance(HD95) was applied for Enhanced Tumor(ET), Whole Tumor(WT), and Tumor Core(TC) regions. The range of Dice coefficient will be from 0 to 1. For two sets A and B define the Dice coefficient:

$$Dice = \frac{2(|A \cap B|)}{|A| + |B|} \tag{1}$$

The Hausdorff Distance (HD) measures how far two subsets of a metric space are from each other and is defined as the longest distance between a point set A and the most adjacent point of set B :

$$HD(A, B) = max\{\sup_{a \in A} \inf_{b \in B} \ d(a, b), \sup_{a \in A} \inf_{b \in B} \ d(b, a)\} \tag{2}$$

where $d(a, b)$ is the Euclidean Distance between a $\in$ A and b $\in$ B. Taking into inconsistent predictions problems, 95th percentile is used, which in the paper, abbreviated as HD95.

2.3 Loss Function

The choice of the loss function plays a vital role in determining the proposed model performance as same as the network structure [18]. Conventional loss function for image segmentation task, such as cross entropy, would yield the inclination of the model to learn the label with the highest count and this effect is apparent in the state of heavy class imbalance. To tackle class imbalance problem induced by the nature of the provided label, in this paper, we adapted a loss functions with regularized term implemented by [15]. This loss function combines a Generalized Dice Loss (GDL) [18] together with a Cross-Entropy loss, which adaptively weights the classes to tackle class imbalance whilst accelerate the convergence, respectively.

$$L_{allloss} = L_{GDL}(G, P) + \lambda \times L_{CE}(G, P) \tag{3}$$

Here the λ is empirically set to 1.25. The adaptive weight of the weighted coefficient is derived as the fraction of the total number of labels as the denominator for each class respectively. This would hamper the tendency to learn labels with higher counts while encouraging the learning process of the model on labels with relatively low counts. Where the weighted coefficient is inversely proportional with the number of label count.

$$L_{ce}(G, P) = w \times L_{ce}(G, P) \tag{4}$$

$$w = \frac{1}{\sum_{i=0}^{G} n_i} \tag{5}$$

Generalized Dice Loss calculates the intersection of union(IOU) of the segmented output with respect to the given labels for each class respectively.

3   Experiments

3.1   Dataset Description

All BraTS multimodal scans per subject provide with a T1 weighted, a post-contrast T1-weighted, a T2-weighted and a FLAIR MRI, which was collected with different clinical protocols and various scanners from multiple (n=19) institutions [3]. The imaging from BraTS2020 dataset have been segmented manually by one to four raters, following the same annotation protocol, and their annotations were approved by experienced neuro-radiologists [4,13]. Each tumor was segmented into GD-enhancing tumor (label4), the peritumoral edema (label2), and the necrotic and non-enhancing tumor core (label1) with dimensions of $240 \times 240 \times 155$. The provided data has been preprocessed such as co-registered to the same anatomical template, interpolated to the same resolution ($1\,mm^3$) and skull-stripped.

3.2   Data Pre-processing

As opposed to conventional RGB images or gray scale images adapt pixel value within the range of $[0, 255]$, MRI intensity values varies throughout different modalities. It is crucial for MRI intensity values to be standardized so that the distribution of the pixel values would be compatible with normal images while feeding it into the model. In our work, we normalize each modality of each patient respectively by first removing the outliers of the image intensities by purging top and bottom 1% intensities. Follows, the image is subtracted by the mean and dividing it with the standard deviation, which normalized the image to compatible value. Each imaging modalities receive the same normalization treatment. Each imaging modalities for an individual patient are stacked together as the modalities are treated as color channels, where an input image of a patient would output a dimension of [S,W,H,M] where *S: Number of slices, W: Image width, H: Image height, M: Number of modalities.* Subsequently, MRI slices that does not consisted brain region are remove due to redundancy as well as to leverage the computation power of the proposed network.

3.3   Data Augmentation

Data augmentation is an approach to prevent overfitting of the model by acquire more diversity of training set via applying feasible transformations such as image rotation. The common solution in medical imaging application is to flip the image up-down or left-right. In this paper, each image has a 50% chance of flipping, specifically, left-right and up-down. In this case, two additional images per slice were generated and zipped into the training dataset. The corresponding labels had been processed following the same protocol as training dataset.

3.4   Label Distribution

The brain tumor label *[0,1,2,4]* represents the *[background, Non-enhancing Tumor Core, Peritumoral Edema, GD-Enhancing Tumor]* in pixel level for this challenge, during the training process, a one-hot vector was implemented as a categorized mapping for label 1 to [0,1,0,0], which also is consistent with the softmax function. For the validation data result submission, the one-hot vector was re-projected to the original label via an argmax function so that the

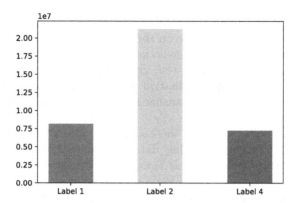

Fig. 4. Distribution of labels for training dataset with *(blue)*: Necrotic and Non-enhancing Tumor Core, *(yellow)*: Peritumoral Edema, *(green)*: GD-Enhancing Tumor, respectively (Color figure online)

evaluation system could recognize the predicted label. Although trivial, we also count all the numbers of label and visualize the label numbers for each class in the training dataset (Fig. 4). This provides a more general view of the label distribution and indicates class imbalance between the labels, specifically label 2 which is the dominant label. It is worth noting that the background labels (label 0) is excluded in the bar chart.

3.5   Training Procedure

The proposed model has been trained on a workstation with a NVIDIA GPU gtx-1080ti, with 11GB VRAM. Before feeding into the training, the input instance shape of training dataset was set at $240 \times 240 \times 4$ with a batch size 24. Adam optimizer was applied and learning rate was set to 10^{-5}. To reduce the over-fitting issues, we fix the dropout value 0.3 and a normal distribution kernel was initiated for layer weights. The max epoch was set 30 because of the time and memory limitations, for each cycle of the training, the time-cost is approximately 28 h for the training process. There are 369 subjects in the training data and 125 subjects in the validation dataset from BraTS2020 challenge. Total parameter is 35,563,044 for updating when training the model.

4   Results

We submitted the predicted original data and augmented data labels for training, validation and test dataset to the online evaluation system. The preparation for all model output scenarios(training, validation and testing) are the same. However, test results were not available from the output resulting from the online evaluation result. Consequently, we were only able to display the results for training and validation dataset. The results were as shown in Table 1 for training

data and Table 2 for validation data. The Dice coefficient and HD95 were used
to evaluate the model performance on three labels: ET, WT, CT, respectively.
To better understand the model prediction performance, the ground truth and
prediction label for single subject were visualized in Fig. 5 for original training
dataset and Fig. 6 for original validation data, to be noted, for the validation
data, only prediction label was visualized since the ground truth is not made
available for the time being.

Preliminary results of our method have mean dice coefficients of
$[0.81, 0.92, 0.86]$ on original training data and $[0.65, 0.86, 0.66]$ on validation
dataset for enhancing tumor, whole tumor, and tumor core respectively. The
result showed a clear over-fitting issue occurs on training dataset perdition
labels from online evaluation system, which led a low performance on the unseen
data. After a data augmentation was performed, the augmented training dataset
has mean dice coefficients of $[0.75, 0.88, 0.80]$ and $[0.67, 0.86, 0.70]$ on validation
dataset for enhancing tumor, whole tumor, and tumor core respectively. The
model performance based on the augmented training data gained a $3 \sim 4\%$
higher dice score than the previous model on the validation dataset, which proves
that the data augmentation was effective method to enhance the model performance; in addition, the median dice coefficient is observed higher than the mean
value, which shows the predicted dice value is not normally distributed. We
noticed that enhanced tumor label was not predicted correctly as expected,
mainly because the proposed model is not sensitive enough to detect small
objects and cannot be segmented from neighbors.

The similar results are observed for the measure of HD95, the value of HD95
on augmentation data model $[40.60, 7.94, 15.75]$ are smaller than the original
data model $[46.40, 11.01, 16.03]$ on validation data for enhancing tumor, whole
tumor, and tumor core respectively.

Table 1. Dice and HD95 for BraTS 2020 training dataset, two groups: No Augmentation and With Augmentation

	Label	Dice			Hausdorff95		
		ET	WT	TC	ET	WT	TC
No Augmentation	Mean	0.81793	0.91904	0.86358	23.49635	5.10187	4.29509
	StdDev	0.23004	0.05521	0.19895	86,11617	11.11984	7.84377
	Median	0.89225	0.93622	0.9372	1.41421	2.23607	1.73205
	25quantile	0.82142	0.90353	0.88291	1.0	1.73205	1.41421
	75quantile	0.93159	0.95362	0.95967	2.0	3.74166	3.0
With Augmentation	Mean	0.75983	0.8854	0.80669	24.72246	6.01101	10.15846
	StdDev	0.25022	0.08867	0.17964	85.92247	7.67428	11.41269
	Median	0.85605	0.91036	0.8652	2.0	3.60555	5.91608
	25quantile	0.73996	0.86622	0.77794	1.41421	2.82843	3.16228
	75quantile	0.90012	0.93592	0.91625	3.74166	5.91608	12.04160

Table 2. Dice and HD95 for BraTS 2020 validation dataset, two groups: No Augmentation and With Augmentation

	Label	Dice			Hausdorff95		
		ET	WT	TC	ET	WT	TC
No Augmentation	Mean	0.64974	0.85767	0.66157	46.40101	11.01570	16.03994
	StdDev	0.3058	0.13076	0.31598	113.01136	18.03175	47.18426
	Median	0.79353	0.89633	0.83136	3.16228	4.58258	6.08276
	25quantile	0.5166	0.84108	0.4773	2.0	3.0	3.16228
	75quantile	0.8673	0.92781	0.9053	12.20656	9.43398	13.78405
With Augmentation	Mean	0.67357	0.86084	0.70421	40.60792	7.94162	15.75085
	StdDev	0.30806	0.12601	0.25173	109.01934	10.05485	35.83261
	Median	0.8083	0.8899	0.79739	3.0	4.24264	8.94427
	25quantile	0.58241	0.84655	0.62058	1.73205	3.0	4.24264
	75quantile	0.88178	0.92575	0.88494	8.06226	8.54400	15.0

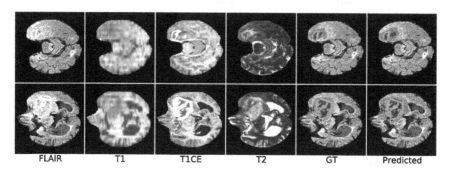

Fig. 5. Segmentation results for BraTS 2020 training dataset. From left to right: FLAIR, T1, T1CE, T2, ground truth(GT), Predicted; Colors: Necrotic and Non-enhancing Tumor Core(Blue), Peritumoral Edema(Yellow), GD-Enhancing Tumor(Green). (Color figure online)

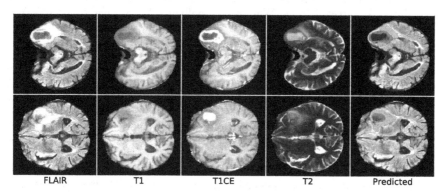

Fig. 6. Segmentation results for BraTS 2020 validation dataset. From left to right: FLAIR, T1, T1CE, T2, Predicted; Colors: Necrotic and Non-enhancing Tumor Core(Blue), Peritumoral Edema(Yellow), GD-Enhancing Tumor(Green). (Color figure online)

5   Conclusion

We have presented a deep supervised U-Attention Net framework for pixel-wise brain tumor segmentation, which can focus on brain tumor low resolution and high resolution feature representations with an attention mechanism and a multistage segmentation layer. The result of mean Dice coefficient is [0.67, 0.86, 0.70] on validation dataset, the ET(enhanced tumor) shows a lower recognition rate compared with other two labels as consistent with previous challenge, the most possibly reason would be ET was presented as small object in the label and very sensitive for the network architecture. Considering the median performance on validation data, we believe that our proposed method is competitive and promising compared with others. The test data will be evaluated once it is available. In the future work, we will focus on the small objects segmentation ability to reduce the false negative error, optimize network structure and fine-tune the parameters to enhance the model performance.

Acknowledgement. Jia Hua Xu and Wai Po Kevin Teng contribute equally to this paper.

References

1. Bakas, S., et al.: Segmentation labels and radiomic features for the pre-operative scans of the TCGA-LGG collection. Cancer Imaging Arch. **286** (2017)
2. Bakas, S., et al.: Segmentation labels and radiomic features for the pre-operative scans of the TCGA-GBM collection. Cancer Imaging Arch. Nat. Sci. Data **4**, 170117 (2017)
3. Bakas, S., et al.: Advancing the cancer genome atlas glioma MRI collections with expert segmentation labels and radiomic features. Sci. Data **4**, 170117 (2017)
4. Bakas, S., et al.: Identifying the best machine learning algorithms for brain tumor segmentation, progression assessment, and overall survival prediction in the BRATS challenge. CoRR abs/1811.02629 (2018). http://arxiv.org/abs/1811.02629
5. Brodbelt, A., Greenberg, D., Winters, T., Williams, M., Vernon, S., Collins, V.P.: Glioblastoma in England: 2007–2011. Eur. J. Cancer **51**(4), 533–542 (2015). https://doi.org/10.1016/j.ejca.2014.12.014. https://pubmed.ncbi.nlm.nih.gov/256 61102/
6. Chen, H., Dou, Q., Yu, L., Qin, J., Heng, P.A.: Voxresnet: deep voxelwise residual networks for brain segmentation from 3D MR images. NeuroImage **170**, 446–455 (2018). https://doi.org/10.1016/j.neuroimage.2017.04.041
7. Feng, X., Tustison, N.J., Patel, S.H., Meyer, C.H.: Brain tumor segmentation using an ensemble of 3D U-nets and overall survival prediction using radiomic features. Front. Comput. Neurosci. **14**, 25 (2020). https://doi.org/10.3389/fncom.2020.00025
8. Hanif, F., Muzaffar, K., Perveen, K., Malhi, S.M., Simjee, S.U.: Glioblastoma multiforme: a review of its epidemiology and pathogenesis through clinical presentation and treatment. Asian Pac. J. Cancer Prev. APJCP **18**(1), 3–9 (2017). https://doi.org/10.22034/APJCP.2017.18.1.3
9. Isensee, F., Kickingereder, P., Wick, W., Bendszus, M., Maier-Hein, K.H.: Brain tumor segmentation and radiomics survival prediction: Contribution to the BRATS 2017 challenge. CoRR abs/1802.10508 (2018). http://arxiv.org/abs/1802.10508

10. Kayalibay, B., Jensen, G., van der Smagt, P.: CNN-based segmentation of medical imaging data. CoRR abs/1701.03056 (2017). http://arxiv.org/abs/1701.03056
11. Lee, B., Yamanakkanavar, N., Choi, J.Y.: Automatic segmentation of brain MRI using a novel patch-wise U-net deep architecture. PloS one **15**(8), e0236493 (2020). https://doi.org/10.1371/journal.pone.0236493
12. Long, J., Shelhamer, E., Darrell, T.: Fully convolutional networks for semantic segmentation. CoRR abs/1411.4038 (2014). http://arxiv.org/abs/1411.4038
13. Menze, B.H., Jakab, A., Bauer, E.A.: The multimodal brain tumor image segmentation benchmark (brats). IEEE Trans. Med. Imaging **34**(10), 1993–2024 (2015). https://doi.org/10.1109/TMI.2014.2377694
14. Nelson, S.J., Cha, S.: Imaging glioblastoma multiforme. Cancer J. (Sudbury, Mass.) **9**(2), 134–145 (2003). https://doi.org/10.1097/00130404-200303000-00009
15. Noori, M., Bahri, A., Mohammadi, K.: Attention-guided version of 2D UNet for automatic brain tumor segmentation. In: 2019 9th International Conference on Computer and Knowledge Engineering (ICCKE). pp. 269–275. IEEE, 24–25 October 2019. https://doi.org/10.1109/ICCKE48569.2019.8964956
16. Oktay, O., et al.: Attention U-Net: Learning where to look for the pancreas. https://arxiv.org/pdf/1804.03999
17. Ronneberger, O., Fischer, P., Brox, T.: U-net: Convolutional networks for biomedical image segmentation. CoRR abs/1505.04597 (2015). http://arxiv.org/abs/1505.04597
18. Sudre, C.H., Li, W., Vercauteren, T., Ourselin, S., Cardoso, M.J.: Generalised dice overlap as a deep learning loss function for highly unbalanced segmentations. CoRR abs/1707.03237 (2017). http://arxiv.org/abs/1707.03237
19. Yamashita, R., Nishio, M., Do, R.K.G., Togashi, K.: Convolutional neural networks: an overview and application in radiology. Insights into Imaging **9**(4), 611–629 (2018). https://doi.org/10.1007/s13244-018-0639-9

A Two-Stage Atrous Convolution Neural Network for Brain Tumor Segmentation and Survival Prediction

Radu Miron[1,2], Ramona Albert[1,2], and Mihaela Breaban[1,2(✉)]

[1] SenticLab, Iasi, Romania
[2] Faculty of Computer Science, "Alexandru Ioan Cuza" University of Iasi,
Iasi, Romania
pmihaela@info.uaic.ro

Abstract. Glioma is a type of heterogeneous tumor originating in the brain, characterized by the coexistence of multiple subregions with different phenotypic characteristics, which further determine heterogeneous profiles, likely to respond variably to treatment. Identifying spatial variations of gliomas is necessary for targeted therapy. The current paper proposes a neural network composed of heterogeneous building blocks to identify the different histologic sub-regions of gliomas in multiparametric MRIs and further extracts radiomic features to estimate a patient's prognosis. The model is evaluated on the BraTS 2020 dataset.

1 Introduction

Gliomas are tumors that arise from the glial cells in the brain, usually in the cerebral hemispheres. They are characterized by the presence of various heterogeneous histologic sub-regions which are depicted by varying intensity profile within multi-parametric MRIs [1]. Characterizing accurately gliomas by the different histologic sub-regions is of critical importance for preoperative planning and risk assessment. This is an area where the advancements in artificial neural networks can bring important contributions by automatically segmenting the heterogeneous regions of the tumor.

The current paper proposes an artificial neural network made of heterogeneous building blocks for brain tumor segmentation and experimentally evaluates it in the context of the BraTS 2020 competition [2–6]. The aim is to accurately identifying three regions of the tumor denoted as *tumor core* (TC) - which comprises GD-enhancing tumor, necrotic and non-enhancing tumor core, *enhancing tumor* structures (ET) and *whole tumor* (WT) - which comprises peritumoral edema and the structures from TC. Based on the segmentation results we further extract radiomic features which are used for survival prediction.

The paper is structured as follows. Section 2 describes the data provided in the competition and the main tasks in order to better understand the decisions made at the architectural level within our approach. Section 3 details the method we use for segmentation and Sect. 4 shortly presents the approach we employ for

A. Crimi and S. Bakas (Eds.): BrainLes 2020, LNCS 12659, pp. 290–299, 2021.
https://doi.org/10.1007/978-3-030-72087-2_25

survival prediction; experimental results for the two tasks are reported at the end of each section. Section 5 concludes the paper.

2   Data

The data comprises volumetric images in the form of NIfTI files corresponding to brain MRI scans. A number of four different modalities are available as follows: 1) native (T1) 2) post-contrast T1-weighted (T1Gd), 3) T2-weighted (T2), and 4) T2 Fluid Attenuated Inversion Recovery (FLAIR) volume. The training data records 369 cases; for each case, in addition to the four different images/files, a mask is provided (also in the form of a NIfTI file) specifying the presence of the three regions of interest for the tumor. The validation set consists of 125 cases (with the four modalities) where masks have to be inferred. The test set consists of 166 cases.

From the 369 cases in the training data, 118 contain additional information such as age, the resection status (known as GTR, i.e. Gross Total Resection) and survival days number. Based on this reduced number of cases, a model for survival estimation is built. The validation dataset in this case is a subset of the segmentation validation set, consisting of images from 29 patients with a GTR status and age information. The test set consists of 107 cases.

3   The Segmentation Algorithm

Inspired by the last two years winners [7,8], we designed a variant of a two-stage cascaded asymmetrical U-Net, with even more inspiration from DeepLabv3+ [9,10] model for the segmentation task. We add further regularization by introducing decoders that are aware of the contour of segmentation area. The contours are made using erode operations for 3 iterations on the segmentation mask with a 3×3 kernel for all three classes (WT, TC, ET). We erode the mask and then subtract this dilation from the original mask, obtaining the contour of the WT, TC, ET area respectively.

The overall flow of our algorithm is detailed in Fig. 1.

3.1   Brief Description of the Model

At first, a concatenation on the channel dimension of the 4 brain MRI acquisitions is input to the neural network. The input has fixed size: $4 \times 128 \times 240 \times 240$. We chose to use such large dimensions in order to preserve as much global information as possible. The prediction is a 3-channel segmentation map, each channel representing the voxels scores for being either WT, TC or ET respectively. Due to limitation of memory we use a two branch decoder, one with interpolation operation and one with deconvolution on the first stage of the cascaded model. For the second stage, the model will take as input the initial volumes concatenated with the segmentation maps obtained by the first stage. As a regularization

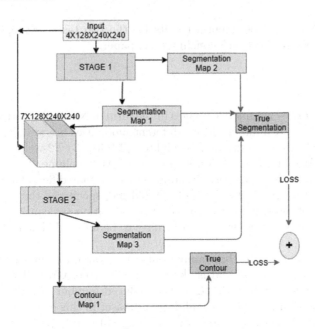

Fig. 1. Flow of the neural network.

step, we add two different convolutions at the end of the second stage, one for the segmentation and the other one for the contour of the segmentation area. The model was trained in an end-to-end fashion.

3.2 Details of the First Stage

We adopt the strategy of using an asymmetric U-Net architecture, as shorter decoders may better keep the information throughout upsampling operations. The difference we bring in comparison with the previous winners is that we use an even shorter decoder. Instead of upsampling the resulted feature maps each time by 2, we upsample first by a scale of 4 and then by a scale of 2. Residual additive connections are added when necessary. We also modify the structure of the encoder, adding atrous convolutions with various dilation rates for extracting features at different resolution. Inspired by the architecture of the DeepLabv3+ model, we also use an additional Spatial Pooling module for similar reasons of using the dilated convolutions in the encoder. The Spatial Pooling module is composed of 5 operations: 3D convolution with kernel size 1, 3 dilated 3D convolutions at different rates and one global average pooling. The five obtained feature maps are concatenated and fed into the decoder. The motivation of using such architecture is due to the fact that we noticed during training that a conventional U-Net [11] performs badly when facing volumes with a small number of voxels on a certain glioma area. The first stage has two branches for segmentation: one using interpolation operations for upsampling and the other one using transposed convolutions. We apply sigmoid activation on top of the output of the last covolutional layer of the decoder, on the channel axis, to obtain segmentation scores for the voxels. In this stage, the input is of size $4 \times 128 \times 240 \times 240$ and the output is a tuple of two tensors of size $3 \times 128 \times 240 \times 240$ - one obtained through upsampling and the other one through transposed convolutions. The output contains 3 binary segmentation maps concatenated on the first channel, each representing the predicted segmentation for WT, TC, ET. This stage is shallower than the other one of the model, each layer having half of the numbers of filters of the second stage. Figure 2 illustrates the architecture of the neural network used in stage 1.

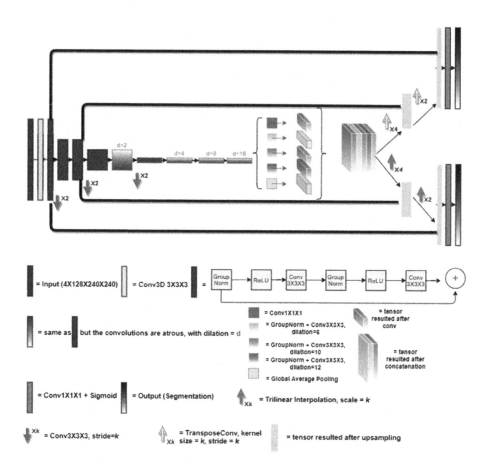

Fig. 2. The first stage of the cascaded model. The black, thick lines coupling parts of the encoder and decoder represent residual additive connections. (Color figure online)

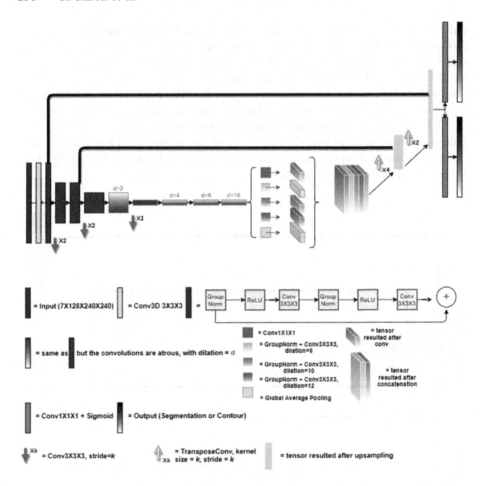

Fig. 3. The second stage of the cascaded model. The black, thick lines coupling parts of the encoder and decoder represent residual additive connections. (Color figure online)

3.3 Details of the Second Stage

The second stage uses as input a 7-channel volume: the 4 acquisitions adding the segmentation maps resulted in the previous step. As mentioned before, the second stage is wider than the first stage. The principles are kept the same with the difference that we no longer use two branches for segmentation. Using dilate and erode operations we construct the contour of the segmentation and use these contours as a helping final branch of the second stage model. Figure 3 illustrates the architecture of the second stage. An example of the contour used in this stage for the whole tumor is given in Fig. 4.

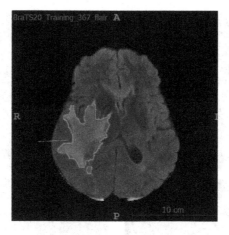

Fig. 4. Contour of a whole tumor

3.4 Preprocessing

Before feeding the data to the neural network, we bring all volumes to mean 0 and standard deviation 1 (taking into account only the non-zero voxels). We also use random flipping on axis, with a probability of 0.5, intensity scaling and intensity shifting.

3.5 Training Details

We used a polynomial learning rate policy $\alpha = \alpha_0 \left(1 - \frac{e}{N_e}\right)^{0.9}$, with $\alpha_0 = 1e - 4$, where e represents the epoch counter and N_e is the total number of epochs. We also used linear warm-up for the first 5 epochs. Batch size used is 1. Since our input volume is quite large and the model is large as well, we use large model support from pytorch framework, provided through anaconda installation through ibm channel[1] in order to make the model fit into our 11 GB GPU. We use soft dice loss as loss function for the training process. The total number of epochs was 250. Due to the large input and intermediate tensors and with the switching time between GPU and CPU, the duration of one epoch of training is 2 h.

Both the code and the trained model are accessible on github[2].

3.6 Inference

During inference, we use 6 flips on a volume on different axes to make the predictions. We use an ensemble of the models saved during the last 10 epochs of training, resulting in a total of 60 predictions per one volume. We add all the

---

[1] https://github.com/IBM/pytorch-large-model-support.
[2] https://github.com/maduriron/BraTS2020.

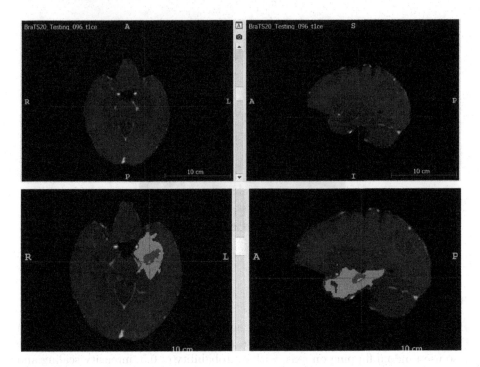

Fig. 5. Example where the model confuses a small vessel of blood for ET (yellow mask). (Color figure online)

segmentation maps (probabilities) resulted and divide by 60. The final binary mask is decided upon threshold values for WT(0.7), TC(0.6), ET(0.7). We tried with several thresholding values, ranging in [0.45, 0.5, 0.55, 0.6, 0.65, 0.75] noticing only small variations in the final scores. In the end we chose the ones with the highest scores on the validation set. If there are less voxels than an empirically derived threshold on ET, we consider those voxels as part of TC. During our experiments we found that the value of 150 voxels was enough to correct the segmentation maps with falsely predicted ET voxels. Without this kind of thresholding, the presence of false positives was observed on the validation set which resulted in scores equal to 0 on ETs in cases where a small number of voxels were detected by our algorithm as ET. These false positives can be explained by confusing the presence of small vessels with ET, situation illustrated in Fig. 5. We also found helpful to move the ET predicted voxels with scores between 0.6 and 0.7 from ET to TC.

3.7 Results

The scores obtained on the training set, the validation set and the test set are reported in Table 1.

Table 1. Results of the segmentation algorithm

Dataset	Dice - ET	Dice - WT	Dice - TC
Training set	0.86348	0.94528	0.93866
Validation set	0.7889	0.90405	0.8482
Test set	0.82135	0.87632	0.83897

Consistent with other results reported by the participants in the competition, the worst score is obtained on the enhancing tumor segment, which seems to still suffer often from false positives in our case.

4 Overall Survival Prediction

4.1 Feature Extraction

The segmentation labels from the first task are used for the estimation of the overall survival days. Radiomic features [12] are extracted, which mainly describe the three-dimensional size and shape of the region of interest and the indexes of the center of mass. For each tumor sub-region we added the ratios of the volume to the size of the whole brain and the size of the whole tumor. A new binary feature was created, entitled 'High Risk': it is set if the tumor is in a lobe with higher probability of risk or if it is in the left hemisphere. We consider a higher risk a center of mass located between 128 and 150 slides on the sagittal plane or between 60 and 150 slides on the coronal plane or between 0 and 41 slides on the transversal plane. The age from the CSV file proved to be an important feature.

The problem tackled is numerical estimation (regression), and we chose to tackle it with Extra Trees [13]; it mainly fits a number of randomized decision trees on various sub-samples of the dataset and uses averaging to improve the predictive accuracy and control over-fitting.

Table 2 illustrates the list of features used during train, ordered by their importance, as reported by the ExtraTrees algorithm.

4.2 Results

Tables 3 and 5 report the results obtained with ExtraTrees on the training and test data. Since the size of the training set is small, the method is prone to overfiting. So far, the best results have been obtained when limiting the depth of the trees to only 7 levels (Table 4). Therefore, we chose to apply this model on the test set, where it achieved the results presented in Table 5.

Table 2. Feature ranking within ExtraTrees

Feature	Importance rate
Age	0.1319
High risk	0.0706
Shape sphericity	0.0502
Center of mass	0.0466
Maximum 2D diameter slice	0.0359
%ED + %ET	0.0356
Maximum 3D diameter	0.0354
%ET	0.0335
%WT	0.0316
%ED	0.0304
%TC	0.0294

Table 3. ExtraTrees regressor, 100 estimators, training dataset

Max depth	Accuracy	MSE	medianSE	stdSE	SpearmanR
6	0.551	71306.459	14072.219	171779.794	0.558
7	0.72	59522.223	4701.321	169803.067	0.665
8	0.695	60195.04	5146.843	169446.718	0.658

Table 4. ExtraTrees regressor, 100 estimators, validation dataset

Max depth	Accuracy	MSE	medianSE	stdSE	SpearmanR
6	0.517	97777.08	36397.01	148209.1	0.249
7	0.414	87744.14	37636	118896.9	0.321
8	0.414	92449.38	32041	127965	0.313

Table 5. ExtraTrees regressor, 100 estimators, test dataset

Max depth	Accuracy	MSE	medianSE	stdSE	SpearmanR
7	0.533	410097.261	54766.909	1243753.011	0.416

5   Conclusion

The two-stage architecture of our algorithm, based on asymmetric U-Net enhanced with dilated convolutions and regularization shows promising results in the segmentation task, being able to detect with high accuracy the spatial variations in gliomas. Regarding the survival task, the model is built on top of the segmentation task. Using ExtraTrees on radiomic features, our method showed to be very sensitive to small changes in the parameters and this is mainly due to the small size of the training set.

References

1. Fathi Kazerooni, A., et al.: Characterization of active and infiltrative tumorous subregions from normal tissue in brain gliomas using multiparametric MRI. J. Magn. Reson. Imaging **48**(4), 938–950 (2018)
2. Menze, B.H., et al.: The multimodal brain tumor image segmentation benchmark (brats). IEEE Trans. Med. Imaging **34**(10), 1993–2024 (2014)
3. Bakas, S., et al.: Advancing the cancer genome atlas glioma MRI collections with expert segmentation labels and radiomic features. Sci. Data **4**, 170117 (2017)
4. Bakas, S., et al.: Identifying the best machine learning algorithms for brain tumor segmentation, progression assessment, and overall survival prediction in the brats challenge. arXiv preprint arXiv:1811.02629 (2018)
5. Bakas, S., et al.: Segmentation labels and radiomic features for the pre-operative scans of the TCGA-LGG collection. Cancer Imaging Arch.
6. Bakas, S., et al.: Segmentation labels and radiomic features for the pre-operative scans of the TCGA-GBM collection. Cancer Imaging Arch.
7. Myronenko, A.: 3D MRI brain tumor segmentation using autoencoder regularization. In: Crimi, A., Bakas, S., Kuijf, H., Keyvan, F., Reyes, M., van Walsum, T. (eds.) BrainLes 2018, Part II. LNCS, vol. 11384, pp. 311–320. Springer, Cham (2019). https://doi.org/10.1007/978-3-030-11726-9_28
8. Jiang, Z., Ding, C., Liu, M., Tao, D.: Two-stage cascaded u-net: 1st place solution to BraTS challenge 2019 segmentation task. In: Crimi, A., Bakas, S. (eds.) BrainLes 2019, Part I. LNCS, vol. 11992, pp. 231–241. Springer, Cham (2020). https://doi.org/10.1007/978-3-030-46640-4_22
9. Chen, L., Papandreou, G., Kokkinos, I., Murphy, K., Yuille, A.L.: DeepLab: semantic image segmentation with deep convolutional nets, atrous convolution, and fully connected CRFs. IEEE Trans. Pattern Anal. Mach. Intell. **40**(4), 834–848 (2018)
10. Chen, L.C., Zhu, Y., Papandreou, G., Schroff, F., Adam, H.: Encoder-decoder with atrous separable convolution for semantic image segmentation. CoRR, abs/1802.02611 (2018)
11. Ronneberger, O., Fischer, P., Brox, T.: U-Net: convolutional networks for biomedical image segmentation. In: Navab, N., Hornegger, J., Wells, W.M., Frangi, A.F. (eds.) MICCAI 2015, Part III. LNCS, vol. 9351, pp. 234–241. Springer, Cham (2015). https://doi.org/10.1007/978-3-319-24574-4_28
12. van Griethuysen, J., et al.: Computational radiomics system to decode the radiographic phenotype. Cancer Res. **77**, e104–e107 (2017)
13. Geurts, P., Ernst, D., Wehenkel, L.: Extremely randomized trees. Mach. Learn. **63**(1), 3–42 (2006)

TwoPath U-Net for Automatic Brain Tumor Segmentation from Multimodal MRI Data

Keerati Kaewrak[1](✉), John Soraghan[1](✉), Gaetano Di Caterina[1](✉), and Derek Grose[2]

[1] Centre for Signal and Image Processing, Department of Electronic and Electrical Engineering, University of Strathclyde, Glasgow, UK
{keerati.kaewrak,j.soraghan,gaetano.di-caterina}@strath.ac.uk
[2] Beatson West of Scotland Cancer Centre, Glasgow, UK
Derek.Grose@ggc.scot.nhs.uk

Abstract. A novel encoder-decoder deep learning network called TwoPath U-Net for multi-class automatic brain tumor segmentation task is presented. The network uses cascaded local and global feature extraction paths in the down-sampling path of the network which allows the network to learn different aspects of both the low-level feature and high-level features. The proposed network architecture using a full image and patches input technique was used on the BraTS2020 training dataset. We tested the network performance using the BraTS2019 validation dataset and obtained the mean dice score of 0.76, 0.64, and 0.58 and the Hausdorff distance 95% of 25.05, 32.83, and 37.57 for the whole tumor, tumor core and enhancing tumor regions.

Keywords: Brain tumor · Deep learning · Segmentation

1 Introduction

Glioma is a type of brain tumor that abnormally grows from glial cells of the brain and it is the most common type of brain tumor that is found in both adults and children. Early and accurate diagnosis are important keys to patients' survival rate [1]. Because the tumor appearances vary from patient to patient, brain tumor segmentation represents a particular challenging image segmentation problem. MRI is currently the imaging modality of choice for brain tumor assessment because of its superior soft-tissue contrast. The brain tumor segmentation (BraTS) challenge has been publishing the BraTS datasets that are widely used for the brain tumor segmentation study for almost a decade [2]. The dataset provided multimodal MRI scans which were acquired from different MRI sequence setting. The multimodal dataset consists of a T1-weighted (T1) scan, a post-contrast T1-weighted (T1 + Gd) scan, T2-weighted (T2) scan and a T2 Fluid Attenuated Inversion Recovery (FLAIR) scan. Since the tumor appears differently on each MRI modality, it is important to use multimodal MRI for brain tumor segmentation [3].

Deep learning based models have been reported and proven to be the *state-of-the-art* method for brain tumor segmentation task in the previous BraTS challenges [2, 4]. Convolutional neural networks using small kernels to extract the essential features from the

© Springer Nature Switzerland AG 2021
A. Crimi and S. Bakas (Eds.): BrainLes 2020, LNCS 12659, pp. 300–309, 2021.
https://doi.org/10.1007/978-3-030-72087-2_26

multimodal MRI data and the technique of increasing the depth of the networks were presented in [5]. Two-pathways CNNs [6] included a cascaded local and global feature extraction paths using different size of the kernels. These local and global feature extractions introduced the idea of giving the network different aspects of the input data to learn. However, these conventional convolutional neural network use fully connected layers as a tumor classification and required post-processing to construct the prediction probability maps of tumor segmentation. U-Net [7] was originally introduced for biomedical image segmentation. The encoder-decoder network architecture that proposed skip connection or concatenation of feature maps from contraction or down-sampling path to the corresponding up-sampling result features in the same block level to provided more precise localization information back into the dense feature maps levels. The up-sampling path of the U-Net restores the dense feature maps back to the original input data dimension with the prediction of the segmentation mask and background of the images. This up-sampling process eliminates the fully connected layers in the conventional convolutional neural network. Furthermore, U-Net is able to give a precise segmentation results using only a few hundreds of annotated training data. Hence, U-Net has become the powerful baseline method for medical imaging segmentation problems including automatic brain tumor segmentation [8–10]. The data augmentation is often used to increase the number of training data [7–9, 11]. However, the previous U-Net based models [9, 10] produced binary segmentation of tumor mask and background of the images. To obtain all tumor sub-regions, the process involves the network training for each sub-regions segmentation separately and then post-processing was applied to obtain all tumor structures segmentation.

In this paper, we present a novel encoder-decoder network architecture that is based on U-Net for multi-class brain tumor segmentation. We replaced the contraction or down-sampling paths of the U-Net with the local and global feature extraction paths inspired by cascaded Two-pathways CNNs and applied the random flipping along axis for data augmentation. We also implemented the proposed network using different input strategies by comparing the segmentation results of full-size images input training and patches input training approach.

The remainder of the paper is organized as follows. Data pre-processing and methodology are presented in Sect. 2. Experimental results are reported in Sect. 3. Finally, the discussion is presented in Sect. 4.

2   Methods

2.1   Data Pre-processing and Augmentation

BraTS2020 training dataset consists of multimodal MRI scans from 369 patients that come from 293 high-grade glioma (HGG) and 76 low-grade glioma (LGG) patients. Each patient has 4 MRI scans; T1-weighted (T1), T1-weighted with gadolinium enhancing contrast (T1 + Gd), T2-weighted (T2) and FLAIR image volumes. Each type of the MRI scan obtained by using different MRI sequence setting acquisition. These images were resampled and interpolate into $1 \times 1 \times 1$ m^3 with the size of $240 \times 240 \times 155$. The data set provides segmentation ground truth annotated by expert neuro-radiologists [2, 12–14]. Annotation label comprises of label 1 for the necrotic and non-enhancing

tumor (NCR/NET), label 2 for the peritumoral edema (ED), and label 4 for the enhancing
tumor (ET). Figure 1 illustrates original FLAIR images and overlaid of annotated ground
truth labels. Red, yellow and blue contours are NCR/NET, ED and ET tumor regions,
respectively.

Data normalization was performed on each MRI scan by subtracting the mean of
each MRI scan and dividing by its standard deviation. We also applied random flipping
along the axis to increase training data as shown in Fig. 1. We split data into the ratio of
0.9:0.1 for cross-validation during the training phase.

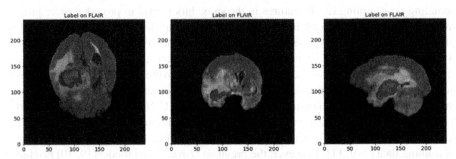

Fig. 1. An example of original image (left) and random flipping results (middle and right) from
training dataset with overlaid ground truth labels; NCR/NET (red), ED (yellow), and ET (blue).
(Color figure online)

2.2 Network Architecture

Figure 2 shows an illustration of the proposed TwoPath U-Net, which is the encoder-
decoder model based on U-Net [7]. It comprises 3 down-sampling blocks, a bottleneck
layer and 3 up-sampling blocks with the multimodal MRI input data. Two feature extraction paths which are 3×3 convolution layers with ReLU and 12×12 follows by $9 \times
9$ convolution layers with ReLU are used as local and global feature extractions paths.
These local and global feature extraction paths capture low-level essential features from
the input. The concatenated local and global features provide multi-perspective of the
essential context from the input to the model. Then, 2×2 max-pooling operator is performed to halve the size of the feature maps dimension and these features maps become
the input of the next down-sampling block. The feature extractions are repeated until
the feature maps increase from 4 to 512 dense features. Then two repeated 3×3 convolution layers followed by ReLU are performed in the bottleneck block of the network
architecture.

For up-sampling blocks, a 3×3 convolution with bilinear interpolation with stride
of 2 is used to double images resolution in both dimensions and to halve the feature
maps, followed by 3×3 convolution with ReLU. The result of the up-sampling is then
concatenated with the corresponding features from the down-sampling side of the same
block level as shown in Fig. 2. These concatenations or skip connections provide the
higher resolution features with local and global localization contextual information to the

up-sampling process. The process repeated until the feature resolution increase back to original resolution with 64 feature maps. Finally, 1×1 convolution layer [15] of length $1 \times 1 \times 64$ with softmax function is employed to produce the output of segmentation mask with 4 classes prediction of all tumor regions.

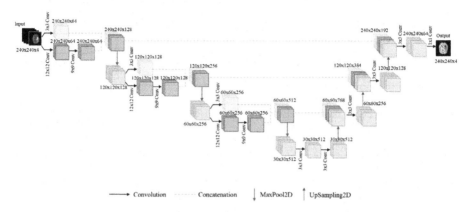

Fig. 2. TwoPath U-Net network architecture.

2.3 Experiments

We performed two experiments using a full-image training and patches training approach. For full-size image training, input image of the size $240 \times 240 \times 4$ were fed through 3 blocks of the proposed networks as shown in Fig. 2. At the end of the down-sampling path, the resolution of the input data were halved in both dimension from 240×240 to 30×30 and the feature maps were increase from 4 to 512. For up-sampling path, the dense feature maps resolution from bottleneck of the network were doubled in each block back to original image resolution and final output was the prediction of 4 classes tumor regions segmentation.

Since the majority of the segmentation results are the background of the images. We wanted to eliminate the background of the images that do not contain the brain and the tumor extents during the network training process. Hence, we performed another experiment using patches training. We cropped the original image in to the size of $160 \times 160 \times 4$ and then trained the network using cropped input data in the same manner as the full-size images training. The summary of the operations and its output dimension of the patches training approach are illustrated in Table 1.

2.4 Network Training

An Adam optimizer [16] was used with a learning rate of 0.00001, $\beta_1 = 0.9$, $\beta_2 = 0.99$, all weights were initialized using a normal distribution with mean of 0 and standard deviation of 0.01, and al biases were initialized as 0. We used training batch size of 16

Table 1. Model summary for patches segmentation training approach.

Layer	Operations	Output Shape
Input	-	(160,160,4)
Down Sampling 1	12x12 Conv2D+ReLU, 9x9 Conv2D+ReLU, 3x3 Conv2D+ReLU — Concatenation	(80,80,128)
Pool 1	Max Pooling 2D	(80,80,128)
Down Sampling 2	12x12 Conv2D+ReLU, 9x9 Conv2D+ReLU, 3x3 Conv2D+ReLU — Concatenation	(80,80,256)
Pool 2	Max Pooling 2D	(40,40,256)
Down Sampling 3	12x12 Conv2D+ReLU, 9x9 Conv2D+ReLU, 3x3 Conv2D+ReLU — Concatenation	(40,40,512)
Pool 3	Max Pooling 2D	(20,20,512)
Bottleneck	Dropout, 3x3 Conv2D+ReLU, 3x3 Conv2D+ReLU, Dropout	(20,20,512)
Up Sampling 1	UpSampling stride of 2, 3x3 Conv2D+ReLU	(40,40,256)
Skip Connection 1	Up Sampling 1 / Down Sampling 3 — Concatenation	(40,40,768)
Up Sampling 2	UpSampling stride of 2, 3x3 Conv2D+ReLU	(80,80,128)
Skip Connection 2	Up Sampling 2 / Down Sampling 2 — Concatenation	(160,160,384)
Up Sampling 3	UpSampling stride of 2, 3x3 Conv2D+ReLU	(160,160,64)
Skip Connection 3	Up Sampling 3 / Down Sampling 1 — Concatenation	(160,160,192)
Prediction	1x1 Convulution+Softmax	(160,160,4)

and trained the network for 50 epochs for each segmentation experiment. Categorical cross-entropy for multiclass segmentation was used as a loss function and can be defined as

$$Loss = -\sum_{n=1}^{N} y_c \log p_c \qquad (1)$$

where N is the number of classes which are 4 including the background for this study. y_c is the ground truth of the nth class and p_c is the prediction from softmax function. We implemented the proposed network using Tensorflow and Keras library on PC equipped with NVIDIA GeForce GTX 1070 GPU, an Intel® Core™ i5–8400 CPU 2.80 GHz processor, 16 GB of RAM.

3   Results

3.1   Evaluation Metrics

We reported the Dice Similarity Coefficient which gives the similarity between predicted tumor regions segmentation and ground truth by comparing the overlapped area and can be defined as

$$DSC = \frac{2TP}{FP + 2TP + FN} \tag{2}$$

where TP, FP and FN denote the number of true positive, false positive and false negative counts, respectively.

We also reported the model performance using Hausdorff Distance measure. Given two finite point sets $A = \{a_1, ..., a_p\}$ and $B = \{b_1, ..., b_q\}$, the Hausdorff Distance can be defined as

$$H(A, B) = \max(h(A, B), h(B, A)) \tag{3}$$

$$h(A, B) = \max_{a \in A} \left(\min_{b \in B}(d(a, b)) \right) \tag{4}$$

$$h(B, A) = \max_{b \in B} \left(\min_{a \in A}(d(b, a)) \right) \tag{5}$$

where $d(a,b)$ is the Euclidean distance between a and b. $h(A,B)$ is called the directed Hausdorff distance from A to B, which identifies the point $a \in A$ that is farthest from any point of B and measures the distance from a to its nearest neighbor in B. This means that $h(A, B)$ first looks for the nearest point in B for every point in A, and then the largest of these values are taken as the distance, which is the most mismatched point of A. Hausdorff distance $H(A, B)$ is the maximum of $h(A, B)$ and $h(B, A)$. Hence, it is able to measure the degree of mismatch between two sets from the distance of the point of A that is farthest from any point of B, and vice versa [17].

3.2   Preliminary Results

We used BraTS2019 validation dataset which provided multimodal MRI scans of 125 patients as unknown testing data for our proposed network. The BraTS2019 validation dataset does not provide ground truth. Hence, we uploaded the prediction masks to CBICA image processing portal for BraTS 2019 validation segmentation task and obtained the evaluation results. We evaluated the proposed network performance on 3 segmentation tasks; segmentation of whole tumor region (WT) which is the union of all annotated tumor labels (ED + NCR/NET + ET), segmentation of the gross tumor core (TC) which is the union of annotated label 1 and 4 (NCR/NET + ET), and segmentation of annotated label 4 (ET) [4].

Table 2 shows that the proposed network with full-images training obtained the segmentation accuracy of mean dice score of 0.57, 0.75 and 0.64, and Hausdorff distance (95th percentile) of 37.56, 25.05 and 32.83 for enhancing tumor, whole tumor and tumor core segmentation tasks. Table 3 shows that the proposed network with middle cropped images training obtained the segmentation accuracy of mean dice score of 0.55, 0.73 and 0.62, and Hausdorff distance (95th percentile) of 39.82, 64.43 and 49.75 for enhancing tumor, whole tumor and tumor core segmentation tasks. Table 4 shows that the proposed network with overlapping cropped images training obtained the segmentation accuracy of mean dice score of 0.50, 0.68 and 0.56, and Hausdorff distance (95th percentile) of 53.10, 66.26 and 59.42 for enhancing tumor, whole tumor and tumor core segmentation tasks. We can also see from Tables 2, 3 and 4 that for 75% quantile, our proposed network has the potential of reaching >80% dice score accuracy for whole tumor and tumor core and >75% dice score accuracy for enhancing tumor regions for all experimental approaches. These results means that with further improvement training strategy, the proposed method could give better performance across whole dataset.

Finally, we compared our experimental results from the proposed network to the segmentation results using the original U-Net in Table 5. We can see that the segmentation results using TwoPath U-Net with full-images training approach achieved the highest dice score for all tumor segmentation tasks. TwoPath U-Net with full-images training approach also obtained lowest Hausdorff distance (95th percentile) which means it obtained the lowest degree of mismatch between the ground truth and the prediction results. An example of the comparison results between the proposed method and ground truth segmentation of all tumor regions from the training dataset is shown in Fig. 3.

Table 2. Segmentation results from full-image training.

Metrics/tumor regions	Dice score			Hausdorff distance 95%		
	ET	WT	TC	ET	WT	TC
Mean	0.57758	0.7586	0.6414	37.56539	25.04765	32.82901
Median	0.73927	0.80849	0.7432	3.52914	43.00697	9.89949
25% quantile	0.30451	0.70338	0.48014	64.24719	70.00714	69.31666
75% quantile	0.82914	0.87162	0.86007	37.56539	25.04765	32.82901

Table 3. Segmentation results from patches training.

Metrics/tumor regions	Dice score			Hausdorff distance 95%		
	ET	WT	TC	ET	WT	TC
Mean	0.55216	0.73045	0.61785	39.82449	64.42918	49.75158
Median	0.69698	0.77228	0.71616	25.11967	61.22908	52.33546
25% quantile	0.32106	0.67741	0.44093	4.50848	52.02259	18.62794
75% quantile	0.80972	0.84034	0.84906	72.59666	79.0231	76.84693

Table 4. Segmentation results from overlapping patches training.

Metrics/tumor regions	Dice score			Hausdorff distance 95%		
	ET	WT	TC	ET	WT	TC
Mean	0.50895	0.67751	0.56053	53.09798	66.26139	59.41933
Median	0.61275	0.73703	0.64553	55.17246	63.97656	60.3034
25% quantile	0.2444	0.58653	0.37928	21.54167	52.20153	40.2368
75% quantile	0.75477	0.80855	0.80325	79.04524	77.13365	79.45817

Table 5. The comparison of segmentation results using different training methods on BraTS 2019 validation dataset.

Metrics/tumor regions	Dice score			Hausdorff distance 95%		
	ET	WT	TC	ET	WT	TC
Original U-Net	0.37982	0.66885	0.46319	58.26046	64.16464	60.10336
TwoPath U-Net Full-images	**0.57758**	**0.7586**	**0.6414**	**37.56539**	**25.04765**	**32.82901**
TwoPath U-Net middle crop	0.55216	0.73045	0.61785	39.82449	64.42918	49.75158
TwoPath U-Net overlapping crop	0.50895	0.67751	0.56053	53.09798	66.26139	59.41933

3.3 BraTS 2020 Challenge

We submitted the BraTS 2020 segmentation task using TwoPath U-Net on BraTS 2020 testing dataset. The dataset contains multimodal MRI volumes of 166 patients. The segmentation results validated by the challenge are shown in Table 6. We obtained mean dice score of 0.72, 0.66, and 0.64 for the whole tumor, tumor core, and enhancing tumor segmentations. The results from BraTS 2020 Challenge has a similar profile to the preliminary results. At 50% and 75% quantile of dataset, the proposed method achieved over 70% and 80% mean dice score accuracy for all tumor regions segmentation. With an improvement of hyperparameters tuning, the proposed method could perform better across the entire dataset.

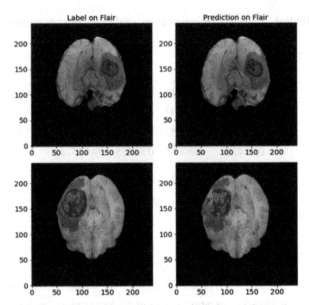

Fig. 3. Comparison between ground truth (left) and prediction (right) segmentation results on original FLAIR images from training dataset with overlaid labels NCR/NET (red), ED (yellow), and ET (blue) regions. (Color figure online)

Table 6. Segmentation results from BraTS 2020 testing dataset.

Metrics/tumor regions	Dice score			Hausdorff distance 95%		
	ET	WT	TC	ET	WT	TC
Mean	0.63615	0.71819	0.65697	57.982	56.15457	61.4418
Median	0.74136	0.79081	0.78923	18.40394	57.05636	49.08382
25% quantile	0.60252	0.64168	0.55418	4.12311	38.90559	15.141
75% quantile	0.81023	0.85834	0.85119	73.6825	74.62648	76.6368

4   Conclusion

In conclusion, we developed a novel network architecture for fully automatic multi-class brain tumor regions segmentation. The proposed network consists of cascaded local and global feature extraction paths to improve segmentation accuracy. We implemented the proposed network architecture using different input data strategies on the BraTS2020 training dataset and tested the network performance using the BraTS2019 validation dataset as unknown testing data. The proposed network gives better segmentation accuracy than the original U-Net and obtained the mean dice score of 0.76, 0.64 and 0.58 on the validation data and 0.72, 0.66, and 0.64 on the testing data for the whole tumor, tumor core and enhancing tumor regions.

References

1. Louis, D.N., et al.: The 2016 world health organization classification of tumors of the central nervous system: a summary. Acta Neuropathol. **131**(6), 803–820 (2016). https://doi.org/10. 1007/s00401-016-1545-1
2. Bakas, S., et al.: Identifying the best machine learning algorithms for brain tumor segmentation, progression assessment, and overall survival prediction in the BRATS challenge. arXiv Prepr. arXiv:1811.02629 (2018)
3. Corso, J.J., Sharon, E., Dube, S., El-Saden, S., Sinha, U., Yuille, A.: Efficient multilevel brain tumor segmentation with integrated bayesian model classification. IEEE Trans. Med. Imaging. **27**, 629–640 (2008). https://doi.org/10.1109/TMI.2007.912817
4. Menze, B.H., et al.: The multimodal brain tumor image segmentation benchmark (BRATS). IEEE Trans. Med. Imaging. **34**, 1993–2024 (2015). https://doi.org/10.1109/TMI.2014.237 7694
5. Pereira, S., Pinto, A., Alves, V., Silva, C.A.: Brain tumor segmentation using convolutional neural networks in MRI images. IEEE Trans. Med. Imaging. **35**, 1240–1251 (2016). https:// doi.org/10.1109/TMI.2016.2538465
6. Havaei, M., Dutil, F., Pal, C., Larochelle, H., Jodoin, P.-M.: A convolutional neural network approach to brain tumor segmentation. In: Crimi, A., Menze, B., Maier, O., Reyes, M., Handels, H. (eds.) Brainlesion: Glioma, Multiple Sclerosis, Stroke and Traumatic Brain Injuries. BrainLes 2015. LNCS, vol. 9556, pp. 195–208. Springer, Cham (2016). https://doi.org/10. 1007/978-3-319-30858-6_17
7. Navab, N., Hornegger, J., Wells, W.M., Frangi, A.F. (eds.): MICCAI 2015. LNCS, vol. 9349. Springer, Cham (2015). https://doi.org/10.1007/978-3-319-24553-9
8. Gu, Z., et al.: CE-Net: context encoder network for 2D medical image segmentation. IEEE Trans. Med. Imaging. **38**, 2281–2292 (2019). https://doi.org/10.1109/TMI.2019.2903562
9. Dong, H., Yang, G., Liu, F., Mo, Y., Guo, Y.: Automatic brain tumor detection and segmentation using u-net based fully convolutional networks. In: Valdés Hernández, M., González-Castro, V. (eds.) Medical Image Understanding and Analysis. MIUA 2017. CCIS, vol. 723, pp. 506–517. Springer, Cham (2017). https://doi.org/10.1007/978-3-319-60964-5_44
10. Kaewrak, K., Soraghan, J., Caterina, G.D., Grose, D.: Modified U-Net for automatic brain tumor regions segmentation. In: 2019 27th European Signal Processing Conference (EUSIPCO), pp. 1–5 (2019)
11. Jiang, Z., Ding, C., Liu, M., Tao, D.: Two-stage cascaded u-net: 1st place solution to BraTS challenge 2019 segmentation task. In: Crimi, A., Bakas, S. (eds.) Brainlesion: Glioma, Multiple Sclerosis, Stroke and Traumatic Brain Injuries. BrainLes 2019. LNCS, vol. 11992, pp. 231–241. Springer, Cham (2020). https://doi.org/10.1007/978-3-030-46640-4_22
12. Bakas, S., et al.: Advancing The Cancer Genome Atlas glioma MRI collections with expert segmentation labels and radiomic features. Sci. Data. **4**, 170117 (2017). https://doi.org/10. 1038/sdata.2017.117
13. Bakas, S., Akbari, H., Sotiras, A.: Segmentation labels for the pre-operative scans of the TCGA-GBM collection. The Cancer Imaging Archive (2017)
14. Bakas, S., et al.: Segmentation labels and radiomic features for the pre-operative scans of the TCGA-LGG collection. Cancer Imaging Arch. 286 (2017)
15. Lin, M., Chen, Q., Yan, S.: Network in network. arXiv Prepr. arXiv:1312.4400 (2013)
16. Kingma, D.P., Ba, J.: Adam: A method for stochastic optimization. arXiv Prepr. arXiv:1412. 6980 (2014)
17. Sim, K.S., Nia, M.E., Tso, C.P., Kho, T.K.: Brain Ventricle Detection Using Hausdorff Distance. In: Tran, Q.N., Arabnia Bioinformatics, and Systems Biology, H.R.B.T.-E.T. in A. and I. for C.B. (eds.) Emerging Trends in Computer Science and Applied Computing. pp. 523–531. Morgan Kaufmann, Boston (2016)

Brain Tumor Segmentation and Survival Prediction Using Automatic Hard Mining in 3D CNN Architecture

Vikas Kumar Anand[1], Sanjeev Grampurohit[1], Pranav Aurangabadkar[1],
Avinash Kori[1], Mahendra Khened[1], Raghavendra S. Bhat[2],
and Ganapathy Krishnamurthi[1]

[1] Indian Institute of Technology Madras, Chennai 600036, India
gankrish@iitm.ac.in
[2] Intel Technology India Pvt. Ltd., Bengaluru, India

Abstract. We utilize 3-D fully convolutional neural networks (CNN) to segment gliomas and its constituents from multimodal Magnetic Resonance Images (MRI). The architecture uses dense connectivity patterns to reduce the number of weights and residual connection and is initialized with weights obtained from training this model with BraTS 2018 dataset. Hard mining is done during training to train for the difficult cases of segmentation tasks by increasing the dice similarity coefficient (DSC) threshold to choose the hard cases as epoch increases. On the BraTS2020 validation data ($n = 125$), this architecture achieved a tumor core, whole tumor, and active tumor dice of 0.744, 0.876, 0.714, respectively. On the test dataset, we get an increment in DSC of tumor core and active tumor by approximately 7%. In terms of DSC, our network performances on the BraTS 2020 test data are 0.775, 0.815, and 0.85 for enhancing tumor, tumor core, and whole tumor, respectively. Overall survival of a subject is determined using conventional machine learning from rediomics features obtained using generated segmentation mask. Our approach has achieved 0.448 and 0.452 as the accuracy on the validation and test dataset.

Keywords: Gliomas · MRI · 3D CNN · Segmentation · Hard mining · Overall survival

1 Introduction

A brain tumor is an abnormal mass of tissue that can be malignant or benign. Furthermore, based on risk, a malignant tumor can be classified into two categories, High-Grade Glioma (HGG) and Low-Grade Glioma (LGG). MR imaging is the most commonly used imaging solution to detect the tumor location, size, and morphology. Different modalities of MR imaging enhances separate components of a brain tumor. The Enhancing Tumor (ET) appears as a hyperintense region in the T1Gd image with respect to T1- weighted image and T1Gd image of

A. Crimi and S. Bakas (Eds.): BrainLes 2020, LNCS 12659, pp. 310–319, 2021.
https://doi.org/10.1007/978-3-030-72087-2_27

healthy white matter. Typically resection is performed on the Tumor Core (TC) region. The necrotic region (NCR), non-enhancing region (NET), and ET constitutes the TC. The NCR and NET tumor core appear as hypointense areas in T1Gd with respect to T1. The TC and peritumoral edema (ED) constitutes the WT and describes the disease's full extent. The WT appears as a hyper-intense area in FLAIR. Delineation of the tumor and its component, also called segmentation of tumor region, on several modalities is the first step towards diagnosis. Radiologists carry out this process in a clinical setup, which is time-consuming, and manual segmentation becomes cumbersome with an increase in patients' numbers. Therefore, automated techniques are required to perform segmentation tasks and reduce the radiologist effort. The diffused boundary of the tumor and partial volume effect in the MRI further enhance the challenge in the segmentation of the different regions of the tumor on several MR imaging modalities. In recent years, Deep Learning methods, especially Convolutional Neural Networks (CNN), have achieved the state of the art results in the segmentation of different tumor components from a different sequence of MR images [9,17]. Typically, due to the volumetric nature of medical images, organs are being imaged as 3-D entities, and subsequently, we utilize the nature of 3D CNN based architectures for segmentation task.

In this manuscript, we have used patch-based 3D encoder-decoder architecture to segmentation brain tumors from MR volumes. We have also used conditional random field and 3D connected component analysis for post-processing of the segmentation maps.

2 Related Work

BraTS 2018 winner, Myronenko et al. [13], has proposed 3D encoder-decoder architecture with variational autoencoder as a regularization for a large encoder. He has used a non-cuboid patch of a fairly bigger size to train the network with a batch size of 1. Instead of using softmax on several classes or several networks for a different class, all three nested tumor sub-regions are being taken as output after sigmoid. The ensemble of the different networks has given the best result. Isenee et al. [7] has used basic U-Net [18] with minor modifications. They secure second place in BraTS 2018 challenge by care-full training with data augmentation during training and testing time. For training, a 128^3 patch has been used with a batch size of 2. Due to the small batch size instance, normalization has been used. They found an increase in performance using Leaky ReLU instead of the ReLU activation function. BraTS 2019 winner Jiang et al. [8] have used two-stage cascaded U-Net architecture to segmentation brain tumors. Segmentation maps obtained in the first stage are being fed to the second stage along with inputs of the first stage. They have also used two decoders in the second stage to get two different segmentation maps. The loss function incorporates all losses that occur due to these segmentations. Data augmentation during training and testing has further improved performance. Data sampling, random patch-size training as a data processing method, semi-supervised learning, architecture development, and fusion of results as a model devising methods and warming-up learning and multi-task learning optimizing processes have used as different tricks by [20] for 3D brain tumor segmentation. Bag of tricks for segmentation has secured second place in BraTS 2019 challenge.

This work utilizes a single 3D encoder-decoder architecture for segmentation for different components of a brain tumor. We have used a smaller patch size and hard mining to train our model. Smaller patch size gives us leverage to deploy our model on a smaller GPU, and the hard mining step finds the hard example during training for weighting the loss function. We have not used additional training data and used only data that are provided by challenge organizers.

3   Materials and Methods

A 3D fully convolutional neural network (3DFCNN) [11] is devised to segment brain tumors and its constituents ET, NER, NET, and ED, from multi-parametric MR volume. This network is used to achieve semantics segmentation task. Each pixel or volex, which is fed to the network, is assigned with a class label by model. This network has dense connectivity patterns that enhance the flow of information and gradients through the model. This enables us to make a deep network tractable. The predictions produced by the model are smoothened by using Conditional random fields followed by class wise 3D connected component analysis. Post-processing techniques help in decreasing the number of false positives in final segmentation maps.

3.1   Data

BraTS 2020 challenge dataset [1–4,12] has been utilized to train the network architecture which is discussed in this manuscript. The training dataset comprises 396 subjects (number of HGG case = 320 and LGG cases = 76). Each subject has 4 MR sequences, namely FLAIR, T2, T1, T1Gd, and segmentation maps, annotated by an expert on each sequence. Each volume is skull-stripped, rescaled to the same resolution (1 mm × 1 mm × 1 mm), and co-registered to the common anatomical template. The BraTS 2020 challenge organizer has issued 125 cases and 166 cases to validate and test the algorithm, respectively. Features such as age, survival days, and resection status are provided separately for the training, validation, and testing phases for 237, 29, and 166 HGG scans.

Data Pre-processing. Each volume is normalized to have zero mean and unit standard deviation as a part of pre-processing.

$$img = (img - mean(img))/std(img)$$

img = Only brain region of the volume
$mean(img)$ = mean of a volume
$std(img)$ = standard deviation of a volume.

3.2   Task1: Brain Tumor Segmentation

3D Fully Convolutional Encoder-Decoder Architecture: The fully convolutional network is used for semantic segmentation task. The input to the network is 64^3 sized cubes. The network predicts the respective class of voxels in the input cube. Each input to the network has to follow two paths, an encoding, and a decoding path. The encoding architecture of the network comprises Dense blocks along with Transition Down blocks. A series of convolution layers followed by ReLU [14] & each convolutional layer receives input from all the preceding convolutional layers that make the Dense block. This connectivity pattern leads to the explosion of many feature maps with the network's depth. To overcome the explosion in parameters, set the number of output feature maps per convolutional layer to a small value ($k = 4$). The spatial dimension of the feature maps is reduced by utilizing the Transition down blocks in the network. The decoding or the up-sampling pathway in the network consists of the Dense blocks and Transition Up blocks. Transposed convolution layers are utilized to up sample feature maps in the Transition Up blocks. In the decoding section, the dense blocks take features from the encoding part in concatenation with the up-sampled features as input. The network architecture for semantic segmentation is depicted in Fig. 1.

Patch Extraction: Each patch is of size 64^3. These are extracted from the brain. Many patches are extracted from less frequent classes such as necrosis compared to more frequent classes. This scheme of patch extraction helps in reducing the class imbalance between different classes. The 3DFCNN accepts an input of size 64^3 and predicts the respective class of the input voxels. There are 77 layers in the network architecture. The effective reuse of the model's features is ensured by utilizing the dense connections among various convolutional layers. The dense connections among layers increase the number of computations, which is subdued by fixing the number of output feature maps per convolutional layer to 4.

Training: The dataset is split into training, validation, and testing in the ratio 70: 20: 10 using stratified sampling based on tumor grade. The network is trained on 205 HGG volumes and 53 LGG volumes. The same ratio of HGG and LGG volumes has been maintained during the validation and testing of the network on held-out data. To further address the issue of class imbalance in the network, the network parameters are trained by minimizing weighted cross-entropy. The weight associated with each class is equivalent to the ratio of the median of the class frequency to the frequency of the class of interest [6]. The number of samples per batch is set at 4, while the learning rate is initialized to 0.0001 and decayed by a factor of 10% every-time the validation loss plateaued.

Hard Mining: Our network performed poorly on the hard examples. We have resolved this issue by hard mining such cases and fine-tuned the trained network with these hard mined cases [10]. We implement a threshold-based selection of hard examples. This threshold is obtained using DSC. If a subject has a DSC, which is less than a threshold DSC then this subject is considered a hard example. We choose all such hard examples for a particular set threshold, and

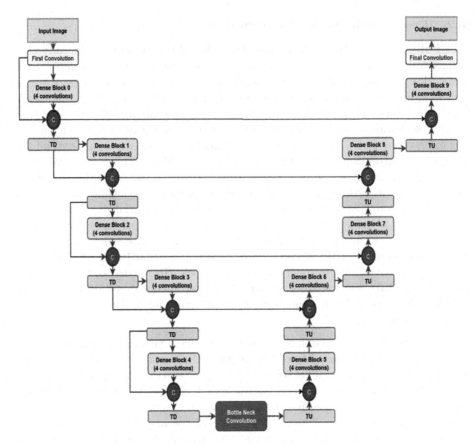

Fig. 1. 3DFCNN used for segmentation of Brain Tumor and its constituents. TD: Transition Down block; C: Concatenation block; TU: Transition Up block

we fine-tune our model with these cases. We have chosen two threshold values to fine-tune our model.

3.3   Task2: Overall Survival Prediction

For training, we have extracted radiomic features using ground truth. There are five types of rediomic features have been extracted for this purpose. These features are first-order radiomic features, which comprises 19 features (mean, median, entropy, etc.); second-order features are Gray Level Co-Occurrence Matrix (GLCM), Gray Level Run Length Matrix (GLRLM), Gray Level Dependence Matrix (GLDM), Gray Level Size Zone Matrix (GLSZM), and Neighboring Gray Tone Difference Matrix (NGTDM), these are altogether 75 different features and 2D and 3D Shape features consists 26 features. We have used pyradicomis [19] to extract all radiomic features. Using a different combination of segmentation maps, we have extracted 1022 different features. We have assigned an importance value to each feature by using a forest of trees [5,16]. Thirty-two most important features out of 1022 features are being used to train the Random Forest Regressor (RFR). The pipeline of overall survival prediction is illustrated in Fig. 2.

Fig. 2. Pipeline used for prediction of overall survival of a patient.

4   Results

The algorithm is implemented in PyTorch [15]. The network is trained on NVIDIA GeForce RTX 2080 Ti GPU with Intel Xeon(R) CPU E5-2650 v3 @ 2.30 GHz × 20, 32 GB RAM CPU. We have not used any additional data set for training our network. We have taken $64 \times 64 \times 64$ sized non-overlapping patches for training. Four patches from different parametric images such as Flair, T1ce, T2, and T1 images are concatenated as the input of the network. We have taken the same size of overlapping patches during inference that reduce the edge effect in segmentation maps.

We have reported our network performance on validation and test data sets. The proportion of HGG and LGG in these data sets is not known to us. We have uploaded our segmentation and OS prediction result for validation and test data sets on the BraTS 2020 server. DSC, Sensitivity, Specificity, and Hausdorff Distance have been used as metrics for the network performance evaluation.

4.1   Segmentation Task:

Performance on Our Network the BraTS 2020 Validation Data: The model is trained using dice loss to generate segmentation maps. Figure 3 shows different parametric images and the segmentation map obtained on a validation data

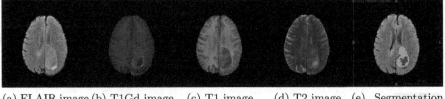

(a) FLAIR image (b) T1Gd image (c) T1 image (d) T2 image (e) Segmentation maps

Fig. 3. Segmentation results on validation: Green, Yellow and Red regions represent edema, enhancing tumor and necrosis respectively. (Color figure online)

On the BraTS validation data (n = 125), the performance of the network is listed in Table 1.

Table 1. Different metrics for all component of tumor on the validation data set

	DSC			Sensitivity			Specificity			Hausdroff distance		
	ET	WT	TC	ET	WT	TC	ET	WT	TC	ET	WT	TC
Mean	0.71	0.88	0.74	0.74	0.92	0.74	0.99	0.99	0.99	38.31	6.88	32.00
StdDev	0.31	0.13	0.29	0.33	0.14	0.31	0.0005	0.001	0.0005	105.32	12.67	90.55
Median	0.85	0.90	0.89	0.89	0.96	0.89	0.99	0.99	0.99	2.23	3.61	4.24

Performance of Our Network on the BraTS 2020 Test Data: Figure 4 depicts the different MR images and segmentation maps generated by our network on test data set. Table 2 summarizes all performance metrics obtained on test data set (n = 166).

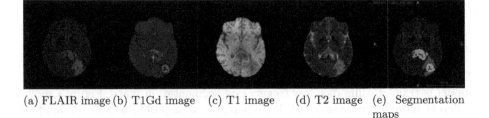

(a) FLAIR image (b) T1Gd image (c) T1 image (d) T2 image (e) Segmentation maps

Fig. 4. Segmentation results on test data: Green, Yellow and Red regions represent edema, enhancing tumor and necrosis respectively. (Color figure online)

Table 2. Different Metrics for all component of Tumor on the Test data set

	DSC			Sensitivity			Specificity			Hausdroff distance		
	ET	WT	TC	ET	WT	TC	ET	WT	TC	ET	WT	TC
Mean	0.776	0.8507	0.815	0.833	0.923	0.838	0.999	0.999	0.999	19.11	8.07	21.276
StdDev	0.223	0.164	0.255	0.249	0.153	0.257	0.0004	0.0016	0.0007	74.87	12.21	74.66
Median	0.836	0.902	0.907	0.926	0.968	0.940	0.999	0.999	0.999	2	4.12	3

4.2 OS Prediction Task:

Figure 2 explains the overall flowchart of the experiment used to find the survival day. Feature extractor module is used to drive all rediomics features using all 4 type sequenced images and corresponding ground truth with several combinations. Before obtaining the importance of a feature for this task, we have standardized the training and validation feature matrix. Feature importance is obtained using a forest of trees. RFR has been used as a regressor for this task. Table 3 comprises the different metrics obtained during training and validation.

Prediction of OS Using Training and Validation Data: During validation phase, clinical information of 29 cases are provided to find the OS of case. Table 3 contains all performance metrics for OS prediction.

Table 3. Different metrics obtained on training and validation data to evaluate the survival of a patient.

	Accuracy	MSE	medianSE	stdSE	SpearmanR
Training	0.59	44611.096	13704.672	109337.932	0.775
Validation	0.448	110677.443	22874.178	142423.687	0.169

Prediction of over All Survival Using Test Data : For testing phase, clinical information is available for 166 cases. Table 4 is the summary of performance metrics of our algorithms for survival prediction.

Table 4. Different metrics obtained on test data to evaluate the survival of a patient.

	Accuracy	MSE	medianSE	stdSE	SpearmanR
Test	0.452	4122630758	67136.258	1142775.390	−0.014

5   Conclusion

This manuscript deals with 2 out of 3 problems posed by the challenge organizer. We have illustrated the use of a fully convolutional neural network for brain tumor segmentation. Overall survival prediction is calculated using generated segmentation maps with conventional machine learning algorithms. We have used the single network from our previous participation in the BraTS 2018 challenge, described in [11]. This year, we have introduced the hard mining steps during training the network. Our model has achieved DSC of 0.71 and 0.77 on enhancing tumors, 0.88 and 0.85 on the whole tumor, and 0.74 and 0.81 on the tumor core for validation and test data. We have observed that the hard mining step improves DSC for tumor core by 9%, DSC for ET by 8%, and the whole tumor by 2%. Hard mining step makes it easy to learn hard examples by the network during training.

References

1. Bakas, S., et al.: Segmentation labels and radiomic features for the pre-operative scans of the TCGA-GBM collection. The Cancer Imaging Archive, vol. 286 (2017)
2. Bakas, S., et al.: Segmentation labels and radiomic features for the pre-operative scans of the TCGA-LGG collection. The Cancer Imaging Archive, vol. 286 (2017)
3. Bakas, S., et al.: Advancing the cancer genome atlas glioma MRI collections with expert segmentation labels and radiomic features. Sci. Data **4**, 170117 (2017)
4. Bakas, S., et al.: Identifying the best machine learning algorithms for brain tumor segmentation, progression assessment, and overall survival prediction in the brats challenge. arXiv preprint arXiv:1811.02629 (2018)
5. Buitinck, L., et al.: API design for machine learning software: experiences from the scikit-learn project. In: ECML PKDD Workshop: Languages for Data Mining and Machine Learning, pp. 108–122 (2013)
6. Eigen, D., Fergus, R.: Predicting depth, surface normals and semantic labels with a common multi-scale convolutional architecture. In: Proceedings of the IEEE International Conference on Computer Vision, pp. 2650–2658 (2015)
7. Isensee, F., Kickingereder, P., Wick, W., Bendszus, M., Maier-Hein, K.H.: No new-net. In: Crimi, A., Bakas, S., Kuijf, H., Keyvan, F., Reyes, M., van Walsum, T. (eds.) BrainLes 2018. LNCS, vol. 11384, pp. 234–244. Springer, Cham (2019). https://doi.org/10.1007/978-3-030-11726-9_21
8. Jiang, Z., Ding, C., Liu, M., Tao, D.: Two-stage cascaded U-Net: 1st place solution to BraTS challenge 2019 segmentation task. In: Crimi, A., Bakas, S. (eds.) BrainLes 2019. LNCS, vol. 11992, pp. 231–241. Springer, Cham (2020). https://doi.org/10.1007/978-3-030-46640-4_22
9. Kamnitsas, K., et al.: Efficient multi-scale 3D CNN with fully connected CRF for accurate brain lesion segmentation. Med. Image Anal. **36**, 61–78 (2017)
10. Khened, M., Kori, A., Rajkumar, H., Srinivasan, B., Krishnamurthi, G.: A generalized deep learning framework for whole-slide image segmentation and analysis. arXiv preprint arXiv:2001.00258 (2020)

11. Kori, A., Soni, M., Pranjal, B., Khened, M., Alex, V., Krishnamurthi, G.: Ensemble of fully convolutional neural network for brain tumor segmentation from magnetic resonance images. In: Crimi, A., Bakas, S., Kuijf, H., Keyvan, F., Reyes, M., van Walsum, T. (eds.) BrainLes 2018. LNCS, vol. 11384, pp. 485–496. Springer, Cham (2019). https://doi.org/10.1007/978-3-030-11726-9_43

12. Menze, B.H., et al.: The multimodal brain tumor image segmentation benchmark (BRATS). IEEE Trans. Med. Imaging **34**(10), 1993–2024 (2014)

13. Myronenko, A.: 3D MRI brain tumor segmentation using autoencoder regularization. In: Crimi, A., Bakas, S., Kuijf, H., Keyvan, F., Reyes, M., van Walsum, T. (eds.) BrainLes 2018. LNCS, vol. 11384, pp. 311–320. Springer, Cham (2019). https://doi.org/10.1007/978-3-030-11726-9_28

14. Nair, V., Hinton, G.E.: Rectified linear units improve restricted Boltzmann machines. In: ICML (2010)

15. Paszke, A., et al.: PyTorch: an imperative style, high-performance deep learning library. In: Wallach, H., Larochelle, H., Beygelzimer, A., d'Alché-Buc, F., Fox, E., Garnett, R. (eds.) Advances in Neural Information Processing Systems, vol. 32, pp. 8024–8035. Curran Associates, Inc. (2019). http://papers.neurips.cc/paper/9015-pytorch-an-imperative-style-high-performance-deep-learning-library.pdf

16. Pedregosa, F., et al.: Scikit-learn: machine learning in Python. J. Mach. Learn. Res. **12**, 2825–2830 (2011)

17. Pereira, S., Pinto, A., Alves, V., Silva, C.A.: Brain tumor segmentation using convolutional neural networks in MRI images. IEEE Trans. Med. Imaging **35**(5), 1240–1251 (2016)

18. Ronneberger, O., Fischer, P., Brox, T.: U-Net: convolutional networks for biomedical image segmentation. In: Navab, N., Hornegger, J., Wells, W.M., Frangi, A.F. (eds.) MICCAI 2015. LNCS, vol. 9351, pp. 234–241. Springer, Cham (2015). https://doi.org/10.1007/978-3-319-24574-4_28

19. Van Griethuysen, J.J., et al.: Computational radiomics system to decode the radiographic phenotype. Cancer Res. **77**(21), e104–e107 (2017)

20. Zhao, Y.-X., Zhang, Y.-M., Liu, C.-L.: Bag of tricks for 3D MRI brain tumor segmentation. In: Crimi, A., Bakas, S. (eds.) BrainLes 2019. LNCS, vol. 11992, pp. 210–220. Springer, Cham (2020). https://doi.org/10.1007/978-3-030-46640-4_20

Some New Tricks for Deep Glioma Segmentation

Chase Duncan[1]([✉]), Francis Roxas[1], Neel Jani[1], Jane Maksimovic[2], Matthew Bramlet[2], Brad Sutton[1], and Sanmi Koyejo[1]

[1] University of Illinois Urbana-Champaign, Champaign, USA
{cddunca2,roxas2,bsutton,sanmi}@illinois.edu
[2] OSF HealthCare, Peoria, USA

Abstract. This manuscript outlines the design of methods, and initial progress on automatic detection of glioma from MRI images using deep neural networks, all applied and evaluated for the 2020 Brain Tumor Segmentation (BraTS) Challenge. Our approach builds on existing work using U-net architectures, and evaluates a variety deep learning techniques including model averaging and adaptive learning rates.

Keywords: Glioma segmentation · Uncertainty in glioma segmentation · Deep learning · Learning rates

1 Introduction

The automatic detection of glioma using MRI images is a promising application area for deep learning [15]. Deep learning – the convolutional neural network in particular – has been profitably brought to bear on related tasks in computer vision [14] such as semantic segmentation of images [22].

The need for delineating a glioma from a standard clinical brain MR study extends beyond just identifying the presence of disease. There are significant clinical needs for accurate segmentation to enable quantitative monitoring of the progression of the disease and to enable surgical planning applications. Some advanced treatment options, such as gamma knife or proton therapy, require precise delineation of active tumor tissues. In addition to clinical needs, automated segmentation enables the capture and recreation of disease for training purposes either physically (through 3D printing) or virtually (through VR-based educational software). With the advancement of VR and augmented reality viewing hardware, the need to rapidly create 3D models of tissues from medical images will rapidly increase in the next several years. Unfortunately, many of these workflows are currently limited by time-consuming manual tracing of tumor regions by a trained viewer. The development of methods for fully automated, quick, and accurate segmentation will usher in a rapid transformation of the use of medical imaging data and provide clinicians a better understanding of the patient's 3D anatomy to improve treatment outcomes.

© Springer Nature Switzerland AG 2021
A. Crimi and S. Bakas (Eds.): BrainLes 2020, LNCS 12659, pp. 320–330, 2021.
https://doi.org/10.1007/978-3-030-72087-2_28

A common issue a machine learning practitioner encounters when using a neural network to model a problem is a lack of data. The trade-off for the expressiveness of neural networks, achieved by the millions of model parameters that must be estimated, is that they require an enormous amount of data to train. This issue is acute when using neural models on problems where the data distribution is made up of biomedical images. Many tasks in the area of ML for biomedical image processing require exacting effort on behalf of doctor's. This makes acquiring labeled data prohibitively expensive and thus very scarce. There are privacy issues that must be addressed before data is released, another trammel requiring time and effort.

Finally, biomedical images are just a small subset of the set of all images in the world, many of which can be freely accessed on the Internet. ImageNet [5], a dataset which is widely used by the computer vision community, has over 14 million examples, for instance. Whereas the BraTS 2020 training dataset [17] has just 369 examples which, in the area, is a treasure trove of data. On the other hand, there is an inverse relationship between the dimensionalities of the two datasets. Although the source images in ImageNet range in dimensionality, it is very common practice to use subsampled 256×256 images. BraTS examples are $4 \times 240 \times 240 \times 155$. Which is to say, for the problem of glioma segmentation using deep learning, there are *orders of magnitude less data* with which to train. Yet, those models need *orders of magnitude more parameters* to model the conditional distribution (i.e., the probability score) of the labels given the data.

Glioma segmentation is unique from other forms of biomedical image segmentation because of the variance in the structure in glioma. Glioma is an infiltrative disease which makes boundary detection challenging. This means that sophisticated forms of data augmentation [28] may not work out of the box, see Fig. 1.

The scale of the data makes it so that common heuristics for setting hyperparameters do not apply. Our work led us to study the relationship between batch size and learning rates in glioma segmentation. We found that our model performed poorly using the standard learning rate schedule for glioma segmentation but following common rules of thumb in deep learning did not point in a useful direction. Indeed, we only found a learning rate schedule that improved the training on accident. Our experimental results suggest that even incremental increases in the batch size may require a completely different learning rate schedule than used in the literature. We find larger initial learning rates with smaller gradients work well for small batches.

2   Related Work

UNet was first introduced in [21] and has since become popular template architecture for glioma segmentation using deep learning as in [16,19,29], among others. Likewise, our architecture is an adaptation of the UNet. It is often useful to consider the UNet modularly in terms of the *encoder*. Which learns a mapping from the input MRI images to a low dimensional feature space, and the *decoder* which learns to map elements of that feature space to a segmentation mask.

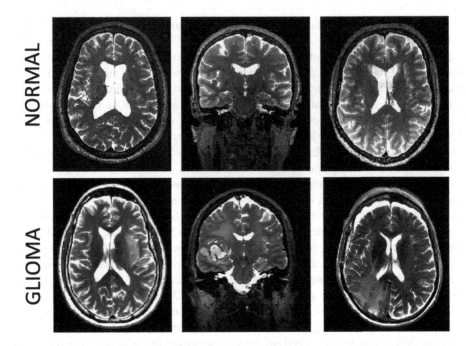

Fig. 1. Top row: T2-weighted MRI images of healthy brains from three different subjects. Bottom row: T2-weighted MRI images of brains from three patients with glioma.

Previous work [19] dealt with the lack of data by regularizing in the latent space, the output of the encoder. They do this using a VAE [11] architecture. The VAE is constructed using the encoder of the segmentation model. The model is optimized in the usual way and the KL-divergence loss and reconstruction loss are added to the segmentation loss. This causes the VAE loss to propagate through the segmentation encoder.

Other prior work [16] use a two-stage cascade architecture. The first stage labels the whole tumor which is then input to the second stage which segments the enhancing tumor and tumor core labels.

In [29], the authors employ data sampling, semi-supervision and ensembling, averaging the predictions of multiple, separately-trained models. Averaging predictions was shown by [13] to useful and computationally cheap method for modeling uncertainty. Additionally, [8] show that simply averaging models can lead to better generalization. Note the difference between model averaging and ensembling. In the first instance we average over the parameters, in the second case we average over the predictions.

3   Methods

The BraTS datasets are already bias-field corrected, coregistered, resampled, and brain extracted. There are four contrasts available for each subject, representing the output of a standard brain MR exam, a T1-weighted precontrast image (T1), a T1-weighted post-contrast image (T1ce), a T2-weighted image (T2), and a T2-weighted fluid attenuated inversion recovery (FLAIR) image.

3.1   Data Augmentation and Preprocessing

We separately center and scale to unit variance each mode – T1, T2, T1ce, FLAIR – of each patient in the dataset. As in,

$$X'_{mp} = \frac{X_{mp} - \mu_{X_{mp}}}{\sigma_{X_{mp}}}, \tag{1}$$

where X_{mp} is the mth mode of MRI data for patient p.

Additionally, we perform an additive shift which is uniformly sampled from $(-0.1, 0.1)$ and then a multiplicative scaling sampled uniformly from $(0.9, 1.1)$. Finally, we mirror the set of MRI images along a random axis. These are popular transformations [16,19,28], but in our experiments we found them to have little impact on the performance of the model, see Table 1. Nevertheless, we chose to include this data augmentation in our data preprocessing pipeline.

3.2   Architecture

Our architecture is a modified UNet [21], the specifications of which are given in Table 2 The network has an initialization layer that is a standard convolutional layer. That is, the feature space is expanded to 32 and the activations are passed along.

Table 1. Comparison of baseline with and without data augmentation, i.e. shifting, scaling and mirroring. In our experiments these operations had little impact on the relevant performance metrics.

Model	Dice			Hausdorf		
	Enhancing t.	Whole tumor	Tumor core	Enhancing t.	Whole tumor	Tumor core
Base + aug	0.66	0.87	0.79	46.34	10.66	10.53
	0.66	0.87	0.79	44.07	11.52	10.71

After the initial layer, the network is assembled from preactivated ResNet blocks [6]. Pre-activation blocks begin with the activation function and end with a convolutional layer. This is done so that a skip connection can be made directly from the input to the output. Doing this means that any information about the data can be propagated without transformation to the deepest layers of the

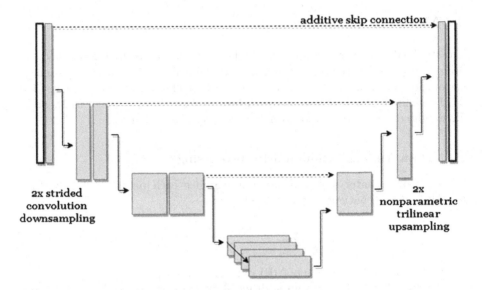

Fig. 2. Basic structure of the network. Our work boiled to searching for appropriate hyperparamters for the best generalization.

network Fig. 3 A single dropout [25] layer directly after the initialization layer is used for regularization (Fig. 2).

Upsampling in the decoder is performed using trilinear interpolation. We also experimented with deconvolution [27] for upsampling. In contrast to [9] our experiments found deconvolution to be slower to train and to perform worse according to the performance metrics.

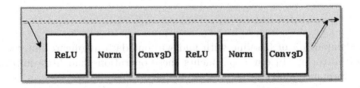

Fig. 3. This is done so that a skip connection can be made directly from the input to the output. Doing this means that any information about the data can be propagated without transformation to the deepest layers of the network.

We use GroupNorm [26] for normalization. This is an adaptation of Batch-Norm, normalization technique in [7], to be used in situations where large batches are prohibitive to use.

The final output is $C \times H \times W \times D$ tensor, where H, W, D are the dimensions of the input volume and C corresponds to the classification labels, of real numbers. These *logits* are the passed through a softmax layer which compresses each value of the tensor to a value, p, between 0 and 1. We take p_{ijk}, the ijkth output of the softmax layer, to be the probability that the ijkth voxel of the input is one of *enhancing tumor, whole tumor,* and *tumor core,* respectively. That is, the multi-class classification problem is recast as a joint classification of 3 separate binary classes.

3.3 Loss and Optimization

The most common cost function for training models for binary classification is the cross entropy loss. This is given by,

$$CE(\theta) = -y_i \log(p) - (1 - y_i) \log(1 - p).$$

This rewards a prediction p for being high with $y_i = 1$ and low when $y_i = 0$. Cross-entropy loss is tried and true and simple to understand but it has an unfortunate downside for glioma segmentation. Binary cross-entropy can perform poorly in a setting where the classes are not balanced. This is the case with glioma segmentation. Most of the brain is not tumor. Some address this issue by using a weighting scheme [21]. This solution uses some prior knowledge to weight the classes according to their counts.

Another approach was introduced by [18] as part of their VNet, the 'soft' Dice loss function. We use this loss function to train the network and it is given by,

$$L_{Dice}(P, T; \theta) = 1 - 2 \frac{\sum_{i,j,k} p_{ijk}(\theta) t_{ijk}}{\sum_{i,j,k} p_{ijk}(\theta) p_{ijk}(\theta) + \sum_{i,j,k} t_{ijk} t_{ijk}}. \tag{2}$$

where $p_{ijk} \in P$ and $t_{ijk} \in T$ are the ijkth elements of the tensor of predictions and the tensor of targets, respectively. We compute this seperately for each of the 3 output volumes, i.e. according to each class. This loss function allows us to optimize according to the performance metric, the Dice coefficient, which is not differentiable.

We use the stochastic optimization algorithm Adam [11] which has previously been shown to have good results when using the UNet representation for this problem [19], et al.

A very common learning rate schedule in previous BraTS competitions has been

$$l_{N_e}(e) = a_0 (1 - \frac{e}{N_e})^\beta, \tag{3}$$

which is a function of the epoch, parameterized by N_e, the maximum epoch; β, commonly set to 0.9; and a_0, the initial learning rate, often $1e - 4$.

In our experiments we do not find good results with this schedule. Instead we use a constant step learning rate

$$l_{N_e}(e) = a_0 - eC, \tag{4}$$

where C is the step constant and a_0 is the initial learning rate. We initialize $a_0 = 2e - 4$ and use $C = 1e - 7$. Therefore we find a larger starting learning rate and less steep gradient for the schedule to perform the best. Weight decay [12] of $1e - 5$ is used.

We use a batch size of 5. We suspect the batch size is the reason the larger learning rate performs better than the multiplicative learning rate schedule. There is a nuanced relationship between learning rate and batch size [23,24], which is an active area of research. Intuitively, as the batch size increases so too should the learning rate. A larger batch size implies a better estimate of the $\nabla L_{Dice}(P, T; \theta)$. This inspires confidence to take less tentative steps in the estimated direction of steepest descent.

3.4 Model Ensembling for Uncertainty Estimation

Results in [13] show that model ensembling can lead to better predictive uncertainty estimates. Here we use a uniform mixture model

$$p(y|\mathbf{x}) = M^{-1} \sum_{p_{\theta_i}} p_{\theta_i}(y|\mathbf{x}) \tag{5}$$

for prediction. In practice we train M separate models and take mean model output to be the final prediction.

4 Experimental Setup and Results

We train our model on an Nvidia Quadro RTX 8000 with 48G of memory. The model architecture is implemented in PyTorch [20]. Table 2 describes the architecture of the system.

Tables 4 and 6 show our results on the validation data for segmentation and uncertainty respectively.

Table 2. Model architecture. Each ResNet layer is pre-activated with two convolutional layers and group normalization using 8 groups. Every convolutional layer has zero padding. Upsampling is performed using trilinear interpolation. Each upsampling layer has a scale factor of 2. The final layer is sigmoid which maps the logits to the unit interval.

	Channels	Filter size	Stride	Notes
3D convolution	32	3 × 3 × 3	1	Initialization, encoder
Resnet	32	3 × 3 × 3	1	
3D convolution	3 × 3 × 3	64	2	Downsample by 2 and double the number of filters
Resnet	64	3 × 3 × 3	1	
Resnet	64	3 × 3 × 3	1	
3D convolution	128	3 × 3 × 3	2	Downsample by 2 and double the number of filters
Resnet	128	3 × 3 × 3	1	
Resnet	128	3 × 3 × 3	1	
3D convolution	256	3 × 3 × 3	2	Downsample by 2 and double the number of filters
Resnet	256	3 × 3 × 3	1	
Resnet	256	3 × 3 × 3	1	
Resnet	256	3 × 3 × 3	1	
Resnet	256	3 × 3 × 3	1	
Trilinear upsampling				
3D convolution	128	1 × 1 × 1	1	Halve the number of features
Resnet	128	3 × 3 × 3	1	
3D convolution	64	1 × 1 × 1	1	Halve the number of features, begin decoder
Trilinear upsampling				
Resnet	64	3 × 3 × 3	1	
3D convolution	32	1 × 1 × 1	1	Halve the number of features
Trilinear upsampling				
Resnet	32	3 × 3 × 3	1	
3D convolution	3	1 × 1 × 1	1	Final compression of features, output are logits
Sigmoid				

Table 3. Comparison of the two learning rate schedules on the BraTS 2020 segmentation task validation data. The multiplicative learning rate is Eq. 3. The additive learning rate is Eq. 4.

Model	Dice		
	Enhancing tumor	Whole tumor	Tumor core
Multiplicative additive	0.66	0.87	0.79
	0.68	**0.88**	**0.81**
	Hausdorf		
	Enhancing tumor	Whole tumor	Tumor core
Multiplicative additive	46.34	10.66	10.53
	37.45	**6.71**	**7.07**

Our results on the segmentation task test set are given in Table 5. As expected we see a performance drop on the test set from the validation set Table 3. Our results on the uncertainty task test set are given in Table 7.

Table 4. Model performance on the BraTS 2020 validation set for the segmentation task.

Dice		
Enhancing tumor	Whole tumor	Tumor core
0.66	0.87	0.79
Hausdorf		
Enhancing tumor	Whole tumor	Tumor core
46.34	10.66	10.53

Table 5. Model performance on the BraTS 2020 test set for the segmentation task.

Dice		
Enhancing tumor	Whole tumor	Tumor core
0.75	0.84	0.77
Hausdorf		
Enhancing tumor	Whole tumor	Tumor core
27.41	9.79	26.63
46.34	10.66	10.53

Table 6. Model performance on BraTS 2020 validation set for the uncertainty task.

Dice		
Whole tumor AUC	Tumor core AUC	Enhancing tumor AUC
0.94	0.86	0.85
FTN Ratio		
Whole tumor AUC	Tumor core AUC	Enhancing tumor AUC
0.03	0.52	0.06
FTP Ratio		
Whole tumor AUC	Tumor core AUC	Enhancing tumor AUC
0.98	0.98	0.98

Table 7. Model performance on BraTS 2020 test set for the uncertainty task.

Dice		
Whole tumor AUC	Tumor core AUC	Enhancing tumor AUC
FTP Ratio		
Whole tumor AUC	Tumor core AUC	Enhancing tumor AUC
0.03	0.57	0.05
FTN Ratio		
Whole tumor AUC	Tumor core AUC	Enhancing tumor AUC
0.98	0.99	0.62

References

1. Bakas, S., et al.: Advancing the cancer genome atlas glioma MRI collections with expert segmentation labels and radiomic features. Sci. Data 4, 1–13 (2017). https://doi.org/10.1038/sdata.2017.117

2. Bakas, S., et al.: Segmentation labels and radiomic features for the pre-operative scans of the TCGA-GBM collection (2017). https://doi.org/10.7937/K9/TCIA.2017.KLXWJJ1Q

3. Bakas, S., et al.: Segmentation labels and radiomic features for the pre-operative scans of the TCGA-LGG collection (2017). https://doi.org/10.7937/K9/TCIA.2017.GJQ7R0EF

4. Bakas, S., et al.: Identifying the best machine learning algorithms for brain tumor segmentation, progression assessment, and overall survival prediction in the brats challenge (2019)

5. Deng, J., Dong, W., Socher, R., Li, L.J., Li, K., Fei-Fei, L.: ImageNet: a large-scale hierarchical image database. In: CVPR09 (2009)

6. He, K., Zhang, X., Ren, S., Sun, J.: Identity mappings in deep residual networks. In: Leibe, B., Matas, J., Sebe, N., Welling, M. (eds.) ECCV 2016, Part IV. LNCS, vol. 9908, pp. 630–645. Springer, Cham (2016). https://doi.org/10.1007/978-3-319-46493-0_38

7. Ioffe, S., Szegedy, C.: Batch normalization: accelerating deep network training by reducing internal covariate shift. In: Bach, F., Blei, D. (eds.) Proceedings of the 32nd International Conference on Machine Learning. Proceedings of Machine Learning Research, vol. 37, pp. 448–456. PMLR, Lille, France (Jul 2015). http://proceedings.mlr.press/v37/ioffe15.html

8. Izmailov, P., Podoprikhin, D., Garipov, T., Vetrov, D., Wilson, A.G.: Averaging weights leads to wider optima and better generalization (2018). http://arxiv.org/abs/1803.05407, cite arxiv:1803.05407Comment: Appears at the Conference on Uncertainty in Artificial Intelligence (UAI) (2018)

9. Jiang, Z., Ding, C., Liu, M., Tao, D.: Two-stage cascaded u-net: 1st place solution to BraTS challenge 2019 segmentation task. In: Crimi, A., Bakas, S. (eds.) BrainLes 2019, Part I. LNCS, vol. 11992, pp. 231–241. Springer, Cham (2020). https://doi.org/10.1007/978-3-030-46640-4_22

10. Kingma, D., Ba, J.: Adam: a method for stochastic optimization. In: International Conference on Learning Representations (2014)

11. Kingma, D.P., Welling, M.: Auto-encoding variational bayes. In: Bengio, Y., LeCun, Y. (eds.) ICLR (2014). http://dblp.uni-trier.de/db/conf/iclr/iclr2014.html#KingmaW13

12. Krogh, A., Hertz, J.A.: A simple weight decay can improve generalization. In: Proceedings of the 4th International Conference on Neural Information Processing Systems, pp. 950–957. NIPS 1991, Morgan Kaufmann Publishers Inc., San Francisco (1991)

13. Lakshminarayanan, B., Pritzel, A., Blundell, C.: Simple and scalable predictive uncertainty estimation using deep ensembles. In: Guyon, I., et al. (eds.) Advances in Neural Information Processing Systems 30, pp. 6402–6413. Curran Associates, Inc. (2017). http://papers.nips.cc/paper/7219-simple-and-scalable-predictive-uncertainty-estimation-using-deep-ensembles.pdf

14. Lecun, Y., Bottou, L., Bengio, Y., Haffner, P.: Gradient-based learning applied to document recognition. Proc. IEEE 86(11), 2278–2324 (1998)

15. LeCun, Y., Bengio, Y., Hinton, G.: Deep learning. Nature **521**, 436–44 (2015). https://doi.org/10.1038/nature14539
16. Ma, J., Yang, X.: Automatic brain tumor segmentation by exploring the multi-modality complementary information and cascaded 3D lightweight CNNs. In: Crimi, A., Bakas, S., Kuijf, H., Keyvan, F., Reyes, M., van Walsum, T. (eds.) Brain-Les 2018, Part II. LNCS, vol. 11384, pp. 25–36. Springer, Cham (2019). https://doi.org/10.1007/978-3-030-11726-9_3
17. Menze, B.H., et al.: The multimodal brain tumor image segmentation benchmark (brats). IEEE Trans. Med. Imaging **34**(10), 1993–2024 (2015). https://doi.org/10.1109/TMI.2014.2377694. https://www.ncbi.nlm.nih.gov/pubmed/25494501
18. Milletari, F., Navab, N., Ahmadi, S.A.: V-net: fully convolutional neural networks for volumetric medical image segmentation, pp. 565–571 (2016). https://doi.org/10.1109/3DV.2016.79
19. Myronenko, A.: 3D MRI brain tumor segmentation using autoencoder regularization. In: Crimi, A., Bakas, S., Kuijf, H., Keyvan, F., Reyes, M., van Walsum, T. (eds.) BrainLes 2018, Part II. LNCS, vol. 11384, pp. 311–320. Springer, Cham (2019). https://doi.org/10.1007/978-3-030-11726-9_28
20. Paszke, A., et al.: Pytorch: an imperative style, high-performance deep learning library. In: Wallach, H., Larochelle, H., Beygelzimer, A., d' Alché-Buc, F., Fox, E., Garnett, R. (eds.) Advances in Neural Information Processing Systems 32, pp. 8024–8035. Curran Associates, Inc. (2019). http://papers.neurips.cc/paper/9015-pytorch-an-imperative-style-high-performance-deep-learning-library.pdf
21. Ronneberger, O., Fischer, P., Brox, T.: U-net: Convolutional networks for biomedical image segmentation. CoRR abs/1505.04597 (2015). http://arxiv.org/abs/1505.04597
22. Shelhamer, E., Long, J., Darrell, T.: Fully convolutional networks for semantic segmentation. IEEE Trans. Pattern Anal. Mach. Intell. **39**(4), 640–651 (2017). https://doi.org/10.1109/TPAMI.2016.2572683
23. Smith, L.N.: Cyclical learning rates for training neural networks. In: 2017 IEEE Winter Conference on Applications of Computer Vision (WACV), pp. 464–472 (2017)
24. Smith, S., Kindermans, P.J., Ying, C., Le, Q.V.: Don't decay the learning rate, increase the batch size (2018). https://openreview.net/pdf?id=B1Yy1BxCZ
25. Srivastava, N., Hinton, G., Krizhevsky, A., Sutskever, I., Salakhutdinov, R.: Dropout: a simple way to prevent neural networks from overfitting. J. Mach. Learn. Res. **15**(56), 1929–1958 (2014). http://jmlr.org/papers/v15/srivastava14a.html
26. Wu, Y., He, K.: Group normalization. In: Ferrari, V., Hebert, M., Sminchisescu, C., Weiss, Y. (eds.) ECCV 2018, Part XIII. LNCS, vol. 11217, pp. 3–19. Springer, Cham (2018). https://doi.org/10.1007/978-3-030-01261-8_1
27. Zeiler, M.D., Taylor, G.W., Fergus, R.: Adaptive deconvolutional networks for mid and high level feature learning. In: 2011 International Conference on Computer Vision, pp. 2018–2025 (2011)
28. Zhao, A., Balakrishnan, G., Durand, F., Guttag, J., Dalca, A.: Data augmentation using learned transformations for one-shot medical image segmentation. pp. 8535–8545 (2019). https://doi.org/10.1109/CVPR.2019.00874
29. Zhao, Y.-X., Zhang, Y.-M., Liu, C.-L.: Bag of tricks for 3D MRI brain tumor segmentation. In: Crimi, A., Bakas, S. (eds.) BrainLes 2019, Part I. LNCS, vol. 11992, pp. 210–220. Springer, Cham (2020). https://doi.org/10.1007/978-3-030-46640-4_20

PieceNet: A Redundant UNet Ensemble

Vikas L. Bommineni$^{(\boxtimes)}$

University of Pennsylvania, Philadelphia, PA 19104, USA
vikas.bommineni@pennmedicine.upenn.edu

Abstract. Segmentation of gliomas is essential to aid clinical diagnosis and treatment; however, imaging artifacts and heterogeneous shape complicate this task. In the last few years, researchers have shown the effectiveness of 3D UNets on this problem. They have found success using 3D patches to predict the class label for the center voxel; however, even a single patch-based UNet may miss representations that another UNet could learn. To circumvent this issue, I developed PieceNet, a deep learning model using a novel ensemble of patch-based 3D UNets. In particular, I used uncorrected modalities to train a standard 3D UNet for all label classes as well as one 3D UNet for each individual label class. Initial results indicate this 4-network ensemble is potentially a superior technique to a traditional patch-based 3D UNet on uncorrected images; however, further work needs to be done to allow for more competitive enhancing tumor segmentation. Moreover, I developed a linear probability model using radiomic and non-imaging features that predicts post-surgery survival.

Keywords: UNet · Multilabel · Patches

1 Introduction

Segmentation of brain tumors is challenging due to their highly heterogeneous appearance and shape, which may be further complicated by imaging artifacts. Convolutional neural networks have become the most promising technique for brain tumor segmentation. In particular, modifications of the 3D UNet variation proposed by Çiçek et al. [1] have been used to great success in this task. A UNet is a widely used convolutional network structure that consists of a contracting path to capture context and a symmetric expanding path that enables precise localization. In the context of this challenge, it is commonly used on extracted 3D patches (for memory purposes) to provide class labels for all input voxels when padding is used. Researchers nominally refer to this specific class of models as either a 'patch-based' UNet or 'tile-based' UNet [2].

Still, it's difficult to segment particular sub-regions of tumors by using a single patch-based UNet. The main reason is the low contrast in image intensity and missed subregion connections by a specific UNet design. For example, the whole tumor is easy to segment, but traditional UNets have underperformed on

© Springer Nature Switzerland AG 2021
A. Crimi and S. Bakas (Eds.): BrainLes 2020, LNCS 12659, pp. 331–341, 2021.
https://doi.org/10.1007/978-3-030-72087-2_29

enhancing tumor segmentations in this challenge. To counteract this underperformance on particular subregions, Wang et al. [3] created a way to decompose this multi-class segmentation task into a sequence of three binary segmentation problems based on subregion hierarchy; in optimizing each subregion separately, they found this minimized false positives. Moreover, Feng et al. [4] found that an ensemble of multiple UNets can generally improve the segmentation accuracy as individual models may make different errors. Through a voting schema, the final number of errors can be reduced.

Using elements from both of these two papers, I propose a novel ensemble strategy to segment gliomas. I trained one 'inclusive' UNet with all 3 label classes (using all modalities). In addition, I trained one 'specialized' 3D UNet for each specific part of the label map (again, using all modalities). For example, one specialized UNet is trained with just the enhancing tumor part of the label map, as shown in Fig. 1. During the validation and testing phases, I adopted a sliding approach and a voting procedure to predict class labels. Using this segmentation output and non-imaging features, I developed a linear regression model.

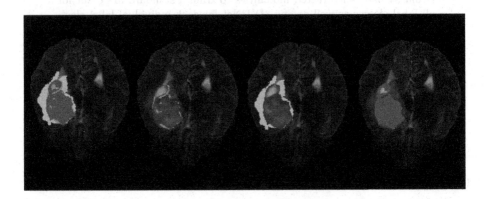

Fig. 1. Label maps on same T2 modality from a randomly selected training subject. From the left, first image shows all 3 non-zero label classes. Second image is label 1: the necrotic and non-enhancing tumor core. Third image is label 2: the peritumoral edema. The fourth image is label 4: the GD-enhancing tumor. Each of the seperated labels were fed into their own UNet.

1.1   Data

The datasets used in the Multimodal Brain Tumor Segmentation Challenge ('BraTS') consist of pre-operative multimodal MRI scans of glioblastoma and lower grade glioma as well as survival data. [5–9]. The data were acquired with different clinical protocols and various scanners from multiple (n = 19) institutions. The image modalities and label map were provided in NIfTI format ('.nii.gz'). In the training set, each of the 369 subjects has FLAIR, T1, T2, and

T1ce modalities as well as a label map with 3 non-zero labels (leftmost of Fig. 1) and overall survival data. In the validation set, age, resection status, and the same modalities were given for 125 subjects.

2 Methods

2.1 Modality Normalization

Researchers have found that bias correction and denoising algorithms negligibly improved UNet ensemble performance [4]. So, I opted not to use these preprocessing steps to save computational time. However, since these images differ in voxel intensity, it was necessary to normalize all modalities before training any of the networks. I normalized each modality of each subject independently by subtracting the mean and dividing by the standard deviation of the cropped brain region.

2.2 Label Map Separation

I extracted individual label maps from the provided segmentation map using NiBabel. This generated three new NIfTI files, one for each non-zero label class. A visual example is shown in Fig. 1.

2.3 Patch Extraction

Due to limited GPU and time constraints, I extracted patches with size 128 by 128 by 128 from each subject; this patch size is the largest I could fit into my GPU. The network is trained on these patches. The patch center is determined based on local voxel intensities post-normalization with foreground voxels weighted much more heavily than background voxels; the exact weighting was decided empirically with a relative weighting of 4 to 1; other weighting schemes considered but ultimately producing poorer results included 1 to 1, 2 to 1, 2 to 1.5, and 3 to 1.

2.4 Network Structure and Training

In the context of the BraTS challenge, patch-based 3D UNets have long been used [2]. The basic idea behind PieceNet is that UNets trained on binary label maps are more likely to achieve higher segmentation accuracy as there are fewer components to simultaneously optimize. For the tumor segmentation task, I trained 4 networks. Each network was asynchronously trained with randomly sampled patches of size 128 by 128 by 128 voxels and batch size 1. The first network was an 'inclusive' 3D UNet, which took in all label classes as input. Each of the other networks used almost identical architecture but were modified to permit only a binary label map; I refer to these UNets as label-class 'specific' UNets.

I used an identical 3D UNet architecture for all 4 networks; the only difference was their inputs. For encoding, I used a VGG style network with two consecutive 3D convolution layers (with kernel size 3). These layers are followed by batch norm layers and a leaky ReLU activation function (constant alpha of .2). Corresponding symmetric decoding blocks were then used. Features were finally concatenated to the deconvoluted outputs. The extracted segmentation map of the input patch was then expanded to the ground truth label(s) (which vary depending on the network).

Training was performed on an NVIDIA TITAN Xp GPU with 12 GB memory. 600 epochs were used. I refer to an epoch as an interation over all 369 images. The model was constructed in Tensorflow with the Adam optimizer and a cross-entropy loss function. A constant learning rate of .001 was used (i.e. no decay). Total training time was 55 h for each model.

In total, I ended up with four trained networks: 3 label-class specific UNets and 1 label-class inclusive UNet.

2.5   Ensemble Construction

I used a 'voting' scheme to determine the labels of each voxel label under an ensemble model. The probability output of all classes from each model was obtained and the final probability was calculated via weighted averaging. The weights were determined based on the average label distribution across all the training set images. The non-zero class with the highest probability at each voxel was selected as the final segmentation label for that particular voxel.

2.6   Volume Prediction

A sliding approach is used to combine the output from each patch input. A stride size of 16 by 16 by 16 was used on each 128 by 128 by 128 patch, and the output probability was averaged for each overlapping stride. The downside to this method is that voxels on the image's edge will have fewer predictions to average; this risk can be mitigated by using increasingly smaller stride sizes, although the tradeoff between marginal returns on performance and exponentially increasing GPU use makes such a tradeoff unreasonable.

2.7   Survival Prediction

With the provided information, it's impossible to decisively determine the survival days post-glioma surgery for a random group of patients; patients likely die for a number of reasons unrelated to the tumor or due to side effects in other regions of the patient's body. For example, older patients with comorbidities may die from complications of glioma such as pneumonia. As such, this portion of the competition is geared to minimize errors and not necessarily attempt to perfectly guess a specific patient's survival time. As such, I chose to construct a linear regression model using the surface area, volume, and spatial location of a patient's tumor along with the patient's age and resection status. To determine spatial location, I roughly split each patient's brain into the 8 following different lobes: (L/R) Frontal Lobe, (L/R) Parietal Lobe, (L/R) Temporal Lobe, and (L/R) Occipital Lobe. The spatial location of a tumor was assigned based on the maximum overlap between the tumor and any single lobe region. This linear regression model prevents overfitting and explosive error compounding.

Processing of Survival Data. One of the subjects in the training set was still alive and was indicated as such. I removed this subject from the provided CSV file and trained the model on the remaining data.

3   Results

3.1   Tumor Segmentation

All 369 provided training subjects were used in the training process, and all 125 provided validation subjects were used in the validation process. The dice scores ('Dice'), sensitivities and specificities, Hausdorff distances ('H95') of the enhanced tumor (ET), whole tumor (WT) and tumor core (TC) were automatically calculated after submitting to the processing portal. As shown below in Table 1, we were able to compare the mean performances of a label-class inclusive 3D UNet ('IUNet') to that of an ensemble between inclusive and all specific UNets (SUNet):

Table 1. Mean performance metrics for validation subjects

Model	DiceET	DiceWT	DiceTC	H95ET	H95WT	H95TC
IUNet	0.68116	0.87326	0.76591	43.29379	9.7209	16.23283
AllSUNets + IUNet	0.71841	0.88382	0.78765	30.76704	4.83357	9.25773

As one can see, the segmentations' metrics were helped by an ensemble between the specific UNets and the inclusive UNet; this is a commonly observed feature of ensemble models and was expected to happen. Adjusting the voting scheme weights may enable even better performance metrics for these segmentations, but further exploration would be required.

Testing Segmentation Statistics. As shown below in Table 2, the model performed about the same on the testing data set as on the validation data set; however, ET tumor segmentation saw a significant improvement when comparing the testing segmentations to the validation segmentations (in terms of both Hausdorff distance and Dice score).

Table 2. Mean performance metrics for testing subjects

Model	DiceET	DiceWT	DiceTC	H95ET	H95WT	H95TC
AllSUNets + IUNet	0.75671	0.87213	0.78131	23.55325	6.56378	29.36651

Validation Segmentation Metrics' Distributions. By looking at the distributions of the segmentations' metrics, one can understand my model's biases. Figures 3, 4, 5, 6, 7 and 8 show the Dice Scores and Hausdorff distances for each label class of the validation subjects. The segmentations were generated by the AllSUNets + IUNet ensemble. In certain cases, note the complete failure of the ensemble to segment the enhancing tumor (Fig. 3). Upon qualitative inspection with FSLeyes, the features of these tumors' cores deviated significantly from the norm (Fig. 2). A more diverse training set would enable segmentation of these tricky cores. Future researchers should focus on augmentation of the provided data set with more spatially diverse tumor cores.

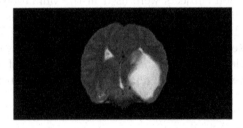

Fig. 2. Example of a deviant tumor on T2 modality from validation subject 082. In contrast to the T2 image from Fig. 1, this tumor presents as a large homogenous region of oedema without necrosis or enhancement. It lacks the subtly defined areas that most gliomas in the training set had.

3.2   Survival Prediction

Validation. All training subjects with survival data were used in the training process. On the complete validation set, accuracy was .379, mean squared error (mean SE) was 93859.54, median SE was 67348.26, and std SE was 102092.41.

Testing. On the testing set, accuracy was .589, mean SE was 380178.978, median SE was 48286.239, and std SE was 1145879.517. For two of the testing subjects, my model outputted negative survival days; in these cases, I replaced these negative values with the average of all other predicted survival days. Failure of the segmentation model seems to be the causal factor here, as the segmentation volumes of these two subjects were anomalous (below total average of non-zero labeled voxels across all subjects).

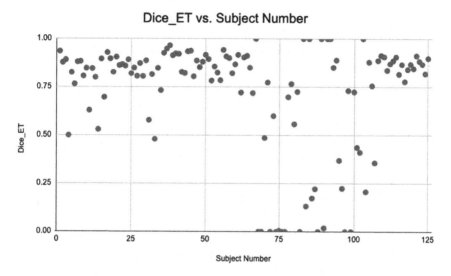

Fig. 3. Scatterplot of Dice Scores for enhancing tumor segmentations for all 125 validation subjects. Subject number is the same as in the provided dataset.

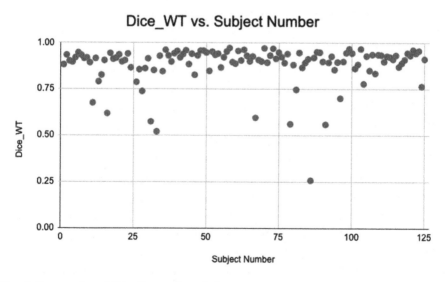

Fig. 4. Scatterplot of Dice Scores for whole tumor segmentations for all 125 validation subjects. Subject number is the same as in the provided dataset.

338   V. L. Bommineni

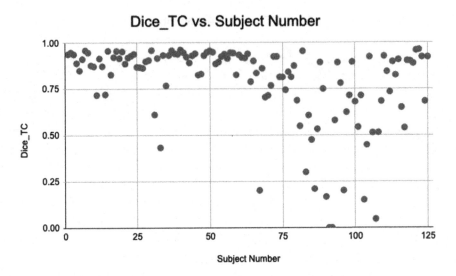

Fig. 5. Scatterplot of Dice Scores for tumor core segmentations for all 125 validation subjects. Subject number is the same as in the provided dataset.

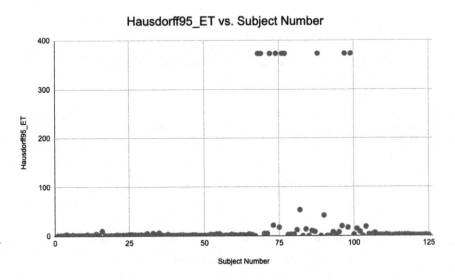

Fig. 6. Scatterplot of Hausdorff distances for enhancing tumor segmentations for all 125 validation subjects. Subject number is the same as in the provided dataset.

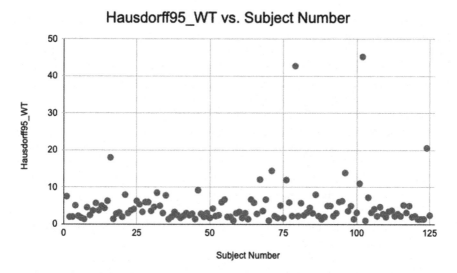

Fig. 7. Scatterplot of Hausdorff distances for whole tumor segmentations for all 125 validation subjects. Subject number is the same as in the provided dataset.

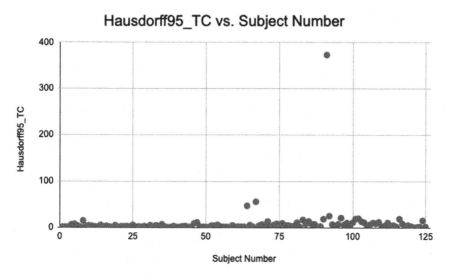

Fig. 8. Scatterplot of Hausdorff distances for tumor core segmentations for all 125 validation subjects. Subject number is the same as in the provided dataset.

4 Conclusion

In this paper, I developed a segmentation method for multimodal MRI scans with a glioma. The ensemble consisted of an overall network trained on all labels classes and of a network trained for each individual label class; predicted labels were determined via a simple voting scheme. Out of the three classes, the model still performed relatively poorly on segmenting the enhancing tumor; I attribute this to a lack of training image diversity. Future work should involve post-processing (i.e. conditional random field [10]) as well as augmentation of current data with diverse glioma presentations.

For the survival task, I played it safe and used a linear regression model to minimize the test errors. The model perfomed relatively well for this particular task, placing second in the 2020 BraTS competition. In the future, improvements to the segmentation may improve the survival prediction. Moreover, as Feng et al. [4] note, further exploration of imaging features is expected to improve the predictive power of the model; however, further exploration should be balanced while minimizing error levels.

In sum, we developed a novel ensemble of 3D U-Nets for brain tumor segmentation. A linear regression model was also developed for the survival prediction task.

References

1. Çiçek, Ö., Abdulkadir, A., Lienkamp, S.S., Brox, T., Ronneberger, O.: 3D U-Net: Learning Dense Volumetric Segmentation from Sparse Annotation. arXiv:1606.06650 [cs]. (2016)
2. F. Isensee, P. Kickingereder, W. Wick, M. Bendszus, K. H. Maier-Hein, "No New-Net", arXiv preprint arXiv:1809.10483 (2019)
3. Wang, G., Li, W., Ourselin, S., Vercauteren, T.: Automatic brain tumor segmentation using cascaded anisotropic convolutional neural networks. In: Crimi, A., Bakas, S., Kuijf, H., Menze, B., Reyes, M. (eds.) BrainLes 2017. LNCS, vol. 10670, pp. 178–190. Springer, Cham (2018). https://doi.org/10.1007/978-3-319-75238-9_16
4. Feng, X., et al.: Brain Tumor Segmentation Using an Ensemble of 3D U-Nets and Overall Survival Prediction Using Radiomic Features. Frontiers Comput. Neurosci. **14** 25 (2020). https://doi.org/10.3389/fncom.2020.00025
5. Menze, B.H., Jakab, A., Bauer, S., Kalpathy-Cramer, J., Farahani, K., Kirby, J., et al.: The multimodal brain tumor image segmentation benchmark (BRATS). IEEE Trans. Med. Imaging **34**(10), 1993–2024 (2015). https://doi.org/10.1109/TMI.2014.2377694
6. Bakas, S., Akbari, H., Sotiras, A., Bilello, M., Rozycki, M., Kirby, J.S., et al.: Advancing the cancer genome atlas glioma MRI collections with expert segmentation labels and radiomic features. Nature Sci. Data **4**, 170117 (2017). https://doi.org/10.1038/sdata.2017.117
7. Bakas, S., Reyes, M., Jakab, A., Bauer, S., Rempfler, M., Crimi, A., et al.: Identifying the Best Machine Learning Algorithms for Brain Tumor Segmentation, Progression Assessment, and Overall Survival Prediction in the BRATS Challenge, arXiv preprint arXiv:1811.02629 (2018)

8. Bakas, S., Akbari, H., Sotiras, A., Bilello, M., Rozycki, M., Kirby, J., et al.: Segmentation labels and radiomic features for the pre-operative scans of the TCGA-GBM collection. Cancer Imaging Arch. (2017). https://doi.org/10.7937/K9/TCIA.2017.KLXWJJ1Q

9. Bakas, S., Akbari, H., Sotiras, A., Bilello, M., Rozycki, M., Kirby, J., et al.: Segmentation labels and radiomic features for the pre-operative scans of the TCGA-LGG collection. Cancer Imaging Arch. (2017). https://doi.org/10.7937/K9/TCIA.2017.GJQ7R0EF

10. Kamnitsas, K., et al.: Efficient multi-scale 3D CNN with fully connected CRF for accurate brain lesion segmentation. Med. Image Anal. **36**, 61–78 (2017)

Cerberus: A Multi-headed Network for Brain Tumor Segmentation

Laura Daza$^{(\boxtimes)}$, Catalina Gómez, and Pablo Arbeláez

Universidad de los Andes, Bogotá, Colombia
{la.daza10,c.gomez10,pa.arbelaez}@uniandes.edu.co

Abstract. The automated analysis of medical images requires robust and accurate algorithms that address the inherent challenges of identifying heterogeneous anatomical and pathological structures, such as brain tumors, in large volumetric images. In this paper, we present Cerberus, a single lightweight convolutional neural network model for the segmentation of fine-grained brain tumor regions in multichannel MRIs. Cerberus has an encoder-decoder architecture that takes advantage of a shared encoding phase to learn common representations for these regions and, then, uses specialized decoders to produce detailed segmentations. Cerberus learns to combine the weights learned for each category to produce a final multi-label segmentation. We evaluate our approach on the official test set of the Brain Tumor Segmentation Challenge 2020, and we obtain dice scores of 0.807 for enhancing tumor, 0.867 for whole tumor and 0.826 for tumor core.

Keywords: Semantic segmentation · Brain tumor · MRI

1 Introduction

The use of Magnetic Resonance Imaging (MRI) for detection, treatment planning and monitoring of brain tumors has spurred interest in automatic segmentation of these structures. However, this task comprises many challenges, including the difficulty in annotating tumors with irregular shapes and appearances in large diagnostic images acquired through different protocols and scanners. Consequently, the availability of datasets for this task is highly restricted. These limitations call for automated algorithms that are robust to class imbalance and reduced training sets.

The recent success of Deep Neural Networks for segmentation tasks in natural images has promoted the development of specialized approaches for processing volumetric medical data. One notable example is U-Net [1], an encoder-decoder architecture that is used as foundation in most methods for biomedical segmentation. For instance, a recent variation of the 3D U-Net, dubbed No New-Net [12], achieved the second place on the segmentation task of the Brain Tumor Segmentation (BraTS) challenge [2–5,14]. This method uses the standard U-Net with an extensive training procedure that includes cross-validation of five

© Springer Nature Switzerland AG 2021
A. Crimi and S. Bakas (Eds.): BrainLes 2020, LNCS 12659, pp. 342–351, 2021.
https://doi.org/10.1007/978-3-030-72087-2_30

models aided with additional annotated data from the same task in the Medical Segmentation Decathlon challenge [15].

In the field of brain tumor segmentation, a common approach is the use of a cascade of networks that first segment the coarsest structures, and then use those results as input for the following networks [13,16]. This approach specializes each network to a specific target, allowing the detection of objects with varying characteristics. However, cascaded networks necessarily require training more than one architecture, which significantly increases the number of parameters and, in most cases, hinders end-to-end training of the system. Deep Cascaded Attention Network (DCAN) [18] attempts to address these issues by sharing a low-level feature extractor followed by three independent encoder-decoder branches specialized for each brain region. The cascaded component is an attention mechanism that inputs the features of the branch for the coarsest categories to the fine-grained ones. Another recent method [9] replaces complex cascades with a multitask learning approach that employs two decoders at different scales for the coarse and fine categories. In contrast, we present a single model with a robust backbone to extract rich features that can be used to obtained specialized segmentations.

A second family of techniques focuses on reducing the computational cost inherent to processing 3D data. Among the best performing methods in the BraTS Challenge 2018 is DMFNet [7], a 3D encoder-decoder architecture that uses dilated multi-fiber units [8] to limit the number of parameters and FLOPS. In [6], Reversible Layers [10] are introduced to the No-New-Net architecture as an alternative to reduce memory consumption. Instead of storing all the activations in the forward pass, they are re-calculated during the backward pass using the next layer's activations. This strategy allows to process complete volumes rather than patches, but the performance gains are small and the training time is increased by 50% with respect to the non-reversible equivalent. Another method that does not require image patching is introduced in [11], in which the method alleviates memory consumption by swapping data from GPU to CPU memory inside the forward propagation.

In this paper, we propose Cerberus, a single lightweight network to address the task of brain tumor segmentation. Figure 1 shows an overview of our model. We hypothesize that learning representations that are common to all the categories, followed by specialized modules to recover brain tumor regions, allows our model to exploit both the shared coarse characteristics of neurological pathologies, and the unique features that are inherent to each structure type. Cerberus leverages a shared encoder to learn a common representation space with sufficient expressive capabilities for identifying heterogeneous brain tumors. Our method also learns how to combine the parameters learned for the specialized tasks to solve the more challenging multi-label segmentation task. With less than 4M of parameters, our model trains faster and can be trained on a single GPU. We demonstrate the competitive performance of Cerberus in the official validation sets of the BraTS Challenge 2018 and 2020. We will make our code publicly available in order to ensure reproducibility and encourage further research.

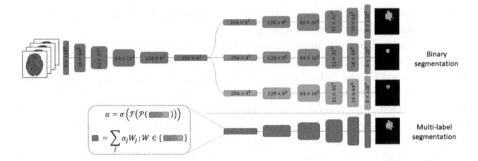

Fig. 1. Overview figure of Cerberus. Our approach uses a shared encoder that learns a common representations which can be then exploited by specialized decoders. Then, a combination module is applied to combine the kernels of each decoder to produce a multi-label segmentation. Skip connections between the encoder and the decoders are omitted for simplicity.

2 Method

We introduce Cerberus, a unified encoder-decoder architecture designed to simultaneously solve multiple binary semantic segmentation tasks and a multi-label task. Our method uses a single encoder to extract rich features at different resolutions of the input image, and then inputs the resulting information to three different decoders that specialize in specific structures. This strategy allows the decomposition of a multi-label segmentation problem into various simpler binary tasks. Finally, Cerberus learns to combine the parameters learned for the subtasks to solve a multi-label segmentation problem.

2.1 Cerberus Architecture

Our model is composed of two main stages: the first one encodes general features obtained from the multimodal volumes, and the second one specializes in solving three binary tasks and a multi-label task. We conduct exhaustive experiments to optimize every stage of our architecture and the training curriculum, and finally define the best configuration empirically.

Encoder. For the first stage we use an encoder with five blocks that contain Depthwise Separable Convolutions (DSC) and a residual connection. Every block duplicates the amount of feature maps and reduces to half the spatial resolution of the input. We define the output feature maps at block i as

$$e_i = \mathcal{E}_i(e_{i-1}) + e_{i-1}, i \in \{1, ..., 5\} \qquad (1)$$

where e_0 is the input image and $\mathcal{E}_i$ denotes a sequence of two DSC. Throughout the entire architecture, every convolutional operation is followed by a normalization layer and a ReLU activation. We use Group Normalization [17] instead of the standard Batch Normalization because the former is less affected by batches of small size. Additionally, similar to U-Net, we use skip connections to connect the features maps from each stage of the encoder with its corresponding block in the decoder.

Decoder. In the second stage, each block contains a single $3 \times 3 \times 3$ convolutional layer to combine the result of the next deepest decoder block with the input from the corresponding skip connection. We include an upsample module composed of a $1 \times 1 \times 1$ convolutional layer followed by a trilinear upsampling operation to double the spatial resolution and halve the number of feature maps of the input from the decoder. The output of block i within the jth branch is calculated as

$$d_{j,i} = \mathcal{D}_{j,i}(U_{j,i}(d_{j,i+1}) \parallel e_i), j = \{1, 2, 3\} \tag{2}$$

where $\mathcal{D}$ denotes the $3 \times 3 \times 3$ convolution, U represents the upsample module and $\parallel$ is a concatenation operation. These decoders are specialized in segmenting the different regions of the tumor.

To obtain a final multi-label segmentation we include a fourth path that combines the kernels learned in the other decoders as follows:

$$W_{ml} = \sum_j \alpha_j W_j$$
$$\alpha = \sigma(\mathcal{F}(\mathcal{P}(W_1 \parallel W_2 \parallel W_3))) \tag{3}$$

where α denotes the weight given to each kernel W. To calculate α we introduce an attention module that concatenates the kernels, performs an average pooling $\mathcal{P}$ along the spatial dimensions, and assign the weights using a fully connected layer $\mathcal{F}$ and a softmax activation σ. Since we use the kernels learned from the binary branches, the only additional parameters incurred in the multi-label path come from the attention modules.

2.2   Loss Function

Cerberus produces three binary segmentation outputs (WT, TC, ET) and a final multi-label (ML) output. We define a combination of the Dice loss and the Cross-entropy loss calculated for each generated mask, and perform a weighted sum of all four to optimize the network, as shown in Eq. 4.

$$Loss = \sum_\ell \beta_\ell \left(L_\ell^{Dice} + L_\ell^{CE} \right); \ell \in \{WT, TC, ET, ML\} \tag{4}$$

We empirically found that the weights that maximize our results are $\beta = [0.1, 0.3, 0.2, 0.4]$.

2.3 Pre-processing

Normalization of MRI intensity values is crucial for processing different modalities in Neural Networks. Hence, we follow a standard normalization per patient: for each modality, we subtract the mean and divide by the standard deviation of the brain region, while setting the intensities outside the brain to zero.

2.4 Implementation Details

We train our models with Adam optimizer with an initial learning rate of $1e-3$ and include an L_2 regularization coefficient of $1e-5$. We reduce the initial learning rate by a factor of 0.1 whenever the validation loss has not decreased for 30 epochs. We adopt the on-the-fly data augmentation strategy proposed by [12]. Our transformations include rotations, scaling, mirroring and gamma correction. To address data imbalance, we define a patch-based sampling strategy such that the center voxel in a patch has equal probability of belonging to any category. Given the memory limitations of processing 3D images, we use patches of size $128 \times 128 \times 128$ and set the batch size to 6 to maintain the memory consumption under 12 GB

2.5 Inference

During inference, the multi-label output corresponds to the segmentation of the tumor and the three binary outputs are used to obtain the uncertainty of the predictions for each region evaluated. For the uncertainty, we apply a sigmoid function to the predictions, calculate the complement of the probabilities and re-scale the values between 0 and 100. The result of this process is a pixel-wise map with values close to zero where the network predicted a category with high confidence. Also, we use patches extracted from the images at uniform intervals, insuring that all pixels are covered. To reconstruct the final image, we assign higher weights to the central voxel of each patch and combine all the predictions. In the ablation studies no further processing is made. For the final results in the official validation server, we train the models in a 5-fold cross-validation fashion and perform an additional test time augmentation (TTA) step that consists of flipping the patches along all axes and averaging the predictions. Finally, we define a simple final post-processing step consisting on the elimination of any component smaller than a threshold by assigning them to the nearest label.

3   Experiments and Results

3.1 Database

We develop our model on the BraTS 2020 dataset, which comprises MRI scans of high grade glioblastomas and low grade gliomas. The annotation labels manually provided by one to four raters are edema, necrosis and non-enhancing tumor (NCR/NET), and enhancing tumor (ET). The challenge evaluates the following overlapping regions: whole tumor (WT), which includes all the three labels; tumor core (TC) that comprises the ET and NCR/NET; and enhancing tumor (ET). The training and validation sets contain 369 and 125 patients respectively, each with four MRI modalities available: T1 weighted, post-contrast T1 weighted, T2 weighted and FLAIR. The challenge provides an official server for evaluation. Besides, we compare our performance to recent published papers with results on the official validation set of BraTS 2018.

To conduct the ablation experiments, we split the training dataset into training and validation subsets with the 80% and 20% of the patients, respectively. We choose the model's weights that achieved the best Dice score on our validation subset to obtain results on the official validation sets.

3.2   Evaluation

The BraTS challenge evaluates the performance of segmentation algorithms using the Dice score (DS), Hausdorff distance (H95%), sensitivity (recall) and specificity. We report the average Dice score and Hausdorff distance over the patients in the corresponding evaluation set. For the uncertainty evaluation, the Dice score is calculated at different confidence measurements and the area under the Dice vs. uncertainty threshold curve is reported as the Dice AUC. An additional integrated score considers the Dice AUC and the area under the curve of the filtered true negatives and true positives for different thresholds.

3.3   Ablation Experiments

We describe in detail the empirical choices within our model. We analyze the advantages of our approach by testing three types of architectures: multi-category networks (shared encoder and decoder), independent binary networks (separated encoder and decoder), and our proposed method with a shared encoder and separate decoders. We report the results in our validation subset from the BraTS 2020 patients in Table 1.

In the first setting we optimize the annotations provided by the challenge (edema, NCR/NET and ET) instead of the regions given that our loss function is designed for non-overlapping labels. Table 1 shows that this model obtains lower performance for ET, the fine grained region. This phenomenon is probably because the tasks are highly unbalanced, and a single decoder is not capable of learning a proper representation for the smaller categories.

In the second setting we train independent networks optimizing the annotations and the regions. The results demonstrate that directly optimizing the regions results in better predictions, specially for the WT and TC regions. In addition, we train cascaded networks by using the output from the coarsest regions as input to the following networks. In this case the results for ET, the smallest region. This happens because the coarsest regions guide the segmentation of smaller structures by limiting the search space within the image, but comes at the cost of training as many models as categories.

Table 1. Ablation experiments on our validation set. Parenthesis indicate the labels used for training. The models with * was trained optimizing the annotated labels (edema, NCR/NET and ET).

Model	Dice			Hausdorff95		
	ET	WT	TC	ET	WT	TC
Single network*	0.770	0.889	0.823	9.24	13.82	13.21
Separate networks*	0.773	0.884	0.796	12.85	8.20	15.89
Separate networks	0.771	0.896	0.811	6.64	12.47	11.02
Cascaded networks	0.782	0.896	0.808	13.35	12.47	10.73
Separate decoders	0.783	0.895	0.836	6.42	15.07	14.08
Communication modules	0.773	0.888	0.844	7.44	16.34	10.07
Cerberus	**0.794**	**0.897**	**0.845**	**3.59**	**6.19**	**6.47**

In the last setting, we train three models with shared encoders and different specialized decoders. The first approach has completely independent decoders that specialize to their corresponding task using only information from the encoder. In this case there is an improvement with respect to using separate networks, which proves that learning a unique representation space results in richer features. In the second approach we share information during the decoding stage as well by adding communication modules between the separate decoders. These modules take information from the three paths and combine it as presented in [8]. Finally, our Cerberus outperforms all methods in TC and ET and has results comparable to the cascaded networks in WT.

3.4 Comparison with the State-of-the-Art

In Table 2, we compare our best model with similar methods that have results in the official BraTS validation set and a published paper. Since all the methods are evaluated in the 2018 challenge, we retrain our method using this dataset. The metrics of the other methods are retrieved from the original papers.

Table 2. Comparison of Cerberus performance against competitive methods on the 2018 validation set.

Model	Patch	Params (M)	FLOPS (G)	Dice score			Hausdorff95		
				ET	WT	TC	ET	WT	TC
No new-Net [12]	128	10.36	202.25	0.796	0.908	0.843	3.12	4.79	8.16
Rev. U-Net [6]	Full	12.37	–	0.803	0.910	0.862	2.58	4.58	6.84
DMFNet [7]	128	3.88	27.04	0.801	0.906	0.845	3.06	4.66	6.44
DCAN [18]	128	–	–	0.817	0.912	0.862	3.57	4.26	6.13
Cerberus (ours)	128	4.02	49.82	0.797	0.895	0.835	4.22	7.77	10.3

Table 2 shows the competitive performance of our method in comparison to the state-of-the-art. Cerberus obtains similar results to four top performing methods, even using a single network for the evaluation, against the five fold cross-validation used by most of the other methods. If we compare Cerberus' performance to the No New-Net results, the Dice scores for the three categories are remarkably similar.

Cerberus presents minor differences in performance with respect to Reversible U-Net, a single model trained on the 80% of the training data without model ensembles. The major difference in performance is on the TC category, 3.1%, and minor for ET (0.7%) and WT (1.6%). However, note that Reversible U-Net has three times more parameters than our method.

We achieve comparable Dice scores in the three categories with respect to DMFNet, presenting differences of 0.4%, 1.2% and 1.2% for ET, WT and TC, respectively. Finally, compared to DCAN, the major performance gap is for TC category (3.1%), and our model does not require training with cross-validations and ensembles across different data partitions to achieve competitive scores.

Results on BraTS 2020 Validation and Test Sets. We further address the competitiveness of Cerberus by evaluating its performance on the BraTS 2020 validation and test sets. Tables 3 and 4 show the evaluation metrics in both sets for the Segmentation and Uncertainty Tasks, respectively.

Table 3. Cerberus performance in the Segmentation task on the official Validation and Test sets of the BraTS 2020 Challenge.

	Dice			Hausdorff95		
BraTS 2020 set	ET	WT	TC	ET	WT	TC
Validation	0.748	0.898	0.828	31.82	5.50	9.68
Test	0.807	0.867	0.826	12.34	6.14	21.20

Table 4. Cerberus performance in the Uncertainty task on the official Validation and Test sets of the BraTS 2020 Challenge.

	Dice AUC			Score		
BraTS 2020 set	ET	WT	TC	ET	WT	TC
Validation	0.912	0.826	0.762	–	–	–
Test	0.883	0.841	0.819	0.947	0.935	0.929

3.5   Qualitative Results

In Fig. 2, we show qualitative examples of the predictions using Cerberus for patients in our validation subset to allow a visual comparison with the annotations. These examples show the accurate localization of brain tumors, specially the largest region (edema in green) and its internal structures with our single model.

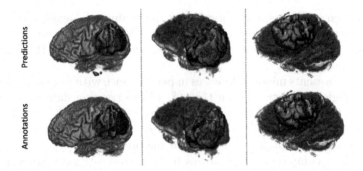

Fig. 2. Qualitative results of Cerberus' predictions over T1 in our validation subset. Top: predictions made by Cerberus; Bottom: annotations made by experts. Each column corresponds to a different patient. Edema is shown in green, ET in red and NCR/NET in blue. (Color figure online)

4   Conclusions

We present Cerberus, a method for brain tumor segmentation that uses a single encoder to learn a shared representation space, independent decoders to solve multiple binary tasks, and learns to combine the decoders parameters to solve a multi-label segmentation task. Cerberus achieves competitive performance against the state-of-the-art in the BraTS Challenge 2018, proving the advantages of sharing a model for the feature extraction stage, and training separate reconstruction networks to obtain specialized segmentations. We also demonstrate the superiority and advantages of our approach by comparing it with similar architectures that do not share the encoder or use a single decoder.

References

1. Ronneberger, O., Fischer, P., Brox, T.: U-Net: convolutional networks for biomedical image segmentation. In: Navab, N., Hornegger, J., Wells, W.M., Frangi, A.F. (eds.) MICCAI 2015. LNCS, vol. 9351, pp. 234–241. Springer, Cham (2015). https://doi.org/10.1007/978-3-319-24574-4_28
2. Bakas, S., et al.: Segmentation labels and radiomic features for the pre-operative scans of the TCGA-LGG collection. Cancer Imaging Arch. **286** (2017)
3. Bakas, S., et al.: Segmentation labels and radiomic features for the pre-operative scans of the TCGA-GBM collection. Cancer Imaging Arch. Nat. Sci. Data **4**, 170117 (2017)
4. Bakas, S., et al.: Advancing the cancer genome atlas glioma MRI collections with expert segmentation labels and radiomic features. Sci. Data **4**, 170117 (2017)
5. Bakas, S., et al.: Identifying the best machine learning algorithms for brain tumor segmentation, progression assessment, and overall survival prediction in the BRATS challenge. CoRR abs/1811.02629 (2018). http://arxiv.org/abs/1811.02629
6. Brügger, R., Baumgartner, C.F., Konukoglu, E.: A partially reversible u-net for memory-efficient volumetric image segmentation. CoRR abs/1906.06148 (2019). http://arxiv.org/abs/1906.06148

7. Chen, C., Liu, X., Ding, M., Zheng, J., Li, J.: 3D dilated multi-fiber network for real-time brain tumor segmentation in MRI. In: Shen, D., Liu, T., Peters, T.M., Staib, L.H., Essert, C., Zhou, S., Yap, P., Khan, A. (eds.) Medical Image Computing and Computer Assisted Intervention - MICCAI 2019–22nd International Conference, Shenzhen, China, 13–17 October 2019, Proceedings, Part III. Lecture Notes in Computer Science, vol. 11766, pp. 184–192. Springer (2019). https://doi.org/10.1007/978-3-030-32248-9_21

8. Chen, Y., Kalantidis, Y., Li, J., Yan, S., Feng, J.: Multi-fiber networks for video recognition. CoRR abs/1807.11195 (2018). http://arxiv.org/abs/1807.11195

9. Cheng, G., Cheng, J., Luo, M., He, L., Tian, Y., Wang, R.: Effective and efficient multitask learning for brain tumor segmentation. J. Real-Time Image Proc. **17**(6), 1951–1960 (2020). https://doi.org/10.1007/s11554-020-00961-4

10. Gomez, A.N., Ren, M., Urtasun, R., Grosse, R.B.: The reversible residual network: Backpropagation without storing activations. CoRR abs/1707.04585 (2017). http://arxiv.org/abs/1707.04585

11. Imai, H., Matzek, S., Le, T.D., Negishi, Y., Kawachiya, K.: High resolution medical image segmentation using data-swapping method. In: Shen, D., et al. (eds.) MICCAI 2019. LNCS, vol. 11766, pp. 238–246. Springer, Cham (2019). https://doi.org/10.1007/978-3-030-32248-9_27

12. Isensee, F., Kickingereder, P., Wick, W., Bendszus, M., Maier-Hein, K.H.: No new-net. In: Brainlesion: Glioma, Multiple Sclerosis, Stroke and Traumatic Brain Injuries - 4th International Workshop, BrainLes 2018, Held in Conjunction with MICCAI 2018, Granada, Spain, September 16, 2018, Revised Selected Papers, Part II, pp. 234–244 (2018). https://doi.org/10.1007/978-3-030-11726-9_21

13. Li, X., Luo, G., Wang, K.: Multi-step cascaded networks for brain tumor segmentation. CoRR abs/1908.05887 (2019). http://arxiv.org/abs/1908.05887

14. Menze, B.H., et al.: The multimodal brain tumor image segmentation benchmark (BRATS). IEEE Trans. Med. Imaging **34**(10), 1993–2024 (2014)

15. Simpson, A.L., et al.: A large annotated medical image dataset for the development and evaluation of segmentation algorithms. CoRR abs/1902.09063 (2019). http://arxiv.org/abs/1902.09063

16. Wang, G., Li, W., Ourselin, S., Vercauteren, T.: Automatic brain tumor segmentation using cascaded anisotropic convolutional neural networks. In: Crimi, A., Bakas, S., Kuijf, H.J., Menze, B.H., Reyes, M. (eds.) Brainlesion: Glioma, Multiple Sclerosis, Stroke and Traumatic Brain Injuries - Third International Workshop, BrainLes 2017, Held in Conjunction with MICCAI 2017, Quebec City, QC, Canada, 14 September 2017, Revised Selected Papers. Lecture Notes in Computer Science, vol. 10670, pp. 178–190. Springer (2017). https://doi.org/10.1007/978-3-319-75238-9_16

17. Wu, Y., He, K.: Group normalization. CoRR abs/1803.08494 (2018). http://arxiv.org/abs/1803.08494

18. Xu, H., Xie, H., Liu, Y., Cheng, C., Niu, C., Zhang, Y.: Deep cascaded attention network for multi-task brain tumor segmentation. In: Shen, D., et al. (eds.) MICCAI 2019. LNCS, vol. 11766, pp. 420–428. Springer, Cham (2019). https://doi.org/10.1007/978-3-030-32248-9_47

An Automatic Overall Survival Time Prediction System for Glioma Brain Tumor Patients Based on Volumetric and Shape Features

Lina Chato[(✉)], Pushkin Kachroo, and Shahram Latifi

Department of Electrical and Computer Engineering, University of Nevada, Las Vegas (UNLV),
Las Vegas, USA
{lina.chato,pushkin.kachroo,shahram.latifi}@unlv.edu

Abstract. An automatic overall survival time prediction system for Glioma brain tumor patients is proposed and developed based on volumetric, location, and shape features. The proposed automatic prediction system consists of three stages: segmentation of brain tumor sub-regions; features extraction; and overall survival time predictions. A deep learning structure based on a modified 3 Dimension (3D) U-Net is proposed to develop an accurate segmentation model to identify and localize the three Glioma brain tumor sub-regions: gadolinium (GD)-enhancing tumor, peritumoral edema, and necrotic and non-enhancing tumor core (NCR/NET). The best performance of a segmentation model is achieved by the modified 3D U-Net based on an Accumulated Encoder (U-Net AE) with a Generalized Dice-Loss (GDL) function trained by the ADAM optimization algorithm. This model achieves Average Dice-Similarity (ADS) scores of 0.8898, 0.8819, and 0.8524 for Whole Tumor (WT), Tumor Core (TC), and Enhancing Tumor (ET), respectively, in the train dataset of the Multimodal Brain Tumor Segmentation challenge (BraTS) 2020. Various combinations of volumetric (based on brain functionality regions), shape, and location features are extracted to train an overall survival time classification model using a Neural Network (NN). The model classifies the data into three classes: short-survivors, mid-survivors, and long-survivors. An information fusion strategy based on features-level fusion and decision-level fusion is used to produce the best prediction model. The best performance is achieved by the ensemble model and shape features model with accuracies of (55.2%) on the BraTS 2020 validation dataset. The ensemble model achieves a competitive accuracy (55.1%) on the BraTS 2020 test dataset.

Keywords: Artificial intelligence · Brain functionality regions · Deep learning · MRI images · Neural Network · U-Net

1 Introduction

Glioma is a common type of primary brain tumor that can be identified into two categories: High-Grade Glioma (HGG), and Low-Grade Glioma. Unfortunately, the HGG is recognized as overly aggressive type of brain tumor, and overall survival time does not

© Springer Nature Switzerland AG 2021
A. Crimi and S. Bakas (Eds.): BrainLes 2020, LNCS 12659, pp. 352–365, 2021.
https://doi.org/10.1007/978-3-030-72087-2_31

exceed two years [1]. Several factors are used by doctors to determine a suitable brain tumor treatment plan and a patient's prognosis, such as, tumor histology, age, symptoms, tumor location, and molecular features [2]. Magnetic Resonance Imaging (MRI) is the best method to identify and localize a Glioma brain tumor. Different MRI modalities are used to define Glioma brain tumor sub-regions: the Whole Tumor (WT) can be recognized in the T2 Fluid Attenuated Inversion Recovery (T2-FLAIR); the Tumor Core (TC) can be recognized in the T2-weighted (T2); and the Enhancing Tumor (ET) and the cystic/necrotic components of the core can be recognized in the post-contrast T1-weighted (T1Gd) [3].

Recently, Machine Learning (ML) including Deep Learning (DL) techniques have been used in medical field to enhance diagnosis and treatment processes by developing automatic prediction systems for classification, regression, and segmentation applications [4–11]. However, automatic prediction in the medical field is a complicated task due to limits in data availability. In addition, many factors control diseases' behaviors, and some of these factors are still unclear or unknown. Moreover, an unbalanced dataset is a common problem in training a prediction model for segmentation in medical images; for example, the tumor region is small compared to a brain's healthy region and an MRI image's background. In general, developing a proper DL structure is not an easy process, but it is still required for a large size dataset. Especially, a deep structure can develop numerous types of features without requiring a researcher's specifications. However, ML methods are suitable for small size datasets, but they mostly require robust features to define a pattern in the data, especially with images data. Many strategies and methods have been proposed and used to overcome the ML/DL weaknesses. Weighted Cross Entropy Loss (WCEL) function, Generalized Dice Loss (GDL) function, and balanced patch selection are common methods used to overcome the unbalanced data problem for segmentation tasks [12]. In addition, useful preprocessing methods have been used to improve performance, such as patch (input) normalization, data denoising, and data augmentation.

The BraTS challenge focuses on evaluation the state-of-the-art methods for Glioma brain tumor segmentation in MRI data. The BraTS challenge 2020 releases three tasks: the segmentation of the Glioma brain tumor's sub-regions; overall survival prediction; and the algorithmic uncertainty in the tumor's segmentation. An accurate overall survival time prediction system can help doctors to understand a tumor's behavior and activity. The volume, location, and shape of a brain tumor are factors related to overall survival time after diagnosis [2]. Experts are needed to annotate brain tumor sub-regions in MRI images, not only to segment brain tumors automatically, but also to study and test the above factors. However, the annotation process is time consuming and expensive. Therefore, DL methods based on segmentation tasks can be alternative methods to extract these features for overall survival time prediction models.

Various types of radiomics features have been proposed and examined by researchers to develop overall survival time prediction models, such as volumetric, location, shape, texture, and Tractographic Feature [12–16]. Feng et al. used shape, volume, and non-clinical imaging features to train a logistic regression model for overall survival prediction [13]. Weninger et al. combined the volumetric and location features with patient's age to train a model using linear regression [14]. Statistical and texture features with a decision tree classifier were proposed by Sun et al. [15]. A multi-layer perceptron was trained by a combination of texture and volumetric features for overall survival predictions [16]. The performance of a prediction model depends not only on the type of features, but also the accuracy of the DL segmentation model. Therefore, an accurate brain tumor segmentation model is required as a pre-processing stage to extract valuable volumetric, location, and shape features.

This paper proposes new methods to extract the volumetric, location, and shape features to develop an automatic overall survival time prediction system for Glioma brain tumor patients using the BraTS 2020 dataset. This paper is organized as follows: Sect. 2 presents the materials to develop the proposed prediction system. Section 3 explains suitable methods to develop the prediction system. Section 4 presents and discusses the experimental results. Finally, Sect. 5 concludes this work and highlights future work to improve the performance of the proposed system.

2   Materials

2.1   Dataset

Segmentation model: a multimodal MRI BraTS 2020 training dataset was used to train a DL model based on a modified 3D U-Net in order to segment Glioma brain tumor sub-regions into four categories: edema, enhanced tumor, non-enhanced tumor, and brain healthy region, as well as the image background. This data consists of 369 samples of HGG and LGG Glioma. Each sample consists of four MRI modalities: T1, T1Gd, T2-Flair, T2, and a segmentation Ground Truth (GT) file [3, 17–20], as shown in Fig. 1. To develop an accurate segmentation model, the data is divided into three groups: training data (305 samples) to train a DL segmentation model; validation data (33 samples) to tune the parameters of the proposed DL segmentation model; and test data (31 samples) to evaluate the DL model on unseen data.

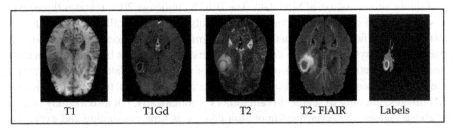

| T1 | T1Gd | T2 | T2- FlAIR | Labels |

Fig. 1. A specific slice of four MRI modalities and labels file from the BraTS 2020 training dataset; the labels are: red for the necrotic and non-enhancing tumor core, yellow for enhancing the tumor core, and green for peritumoral edema (Color figure online)

Overall survival prediction model: BraTS 2020 provides survival information for 236 samples of HGG to train an overall survival time prediction system. This information includes non-imaging information (patients' age, survival time in days, and resection status), as well as imaging data (four MRI modalities with GT). There are 119 samples with Gross Total Resection (GTR) status, and the remains are with Subtotal Resection (STR) and unknown resection status. To avoid over fitting, the training data is divided into three groups: training (166 samples) to train an accurate prediction model; validation (35 samples) to tune the parameters of the prediction model; and test (35 samples) to test a trained model on unseen data.

2.2 Software and Hardware

Matlab 2019b, which contains deep learning, image processing, machine learning, and computer vision toolboxes, is used to train the proposed DL segmentation model, as well as the overall survival time prediction model using NN. The TITAN RTX GPU 24 GB with 3.8 GHZ CPU is used to train the proposed DL structure, which is presented in Sect. 3.1.

3   Methods to Develop Overall Survival Time Prediction System

Three steps are required to design an automatic overall survival time prediction system based on volumetric, shape, and location features: a preprocessing, which is represented by a DL segmentation method; features extraction methods; and an NN prediction model. The structure of the proposed automatic prediction system is shown in Fig. 2.

3.1   DL Segmentation Model

A modified 3D U-net is designed to identify and localize the three Glioma brain tumor sub-regions (WT, TC, and ET) in an MRI volume of voxels. The proposed modified 3D-U-Net structure consists of 3 levels of Encoder/Decoder units with concatenation paths, as shown in Fig. 3. In general, the U-Net structure consists of a contracting path and expanding path [7]. The contracting path consists of the encoder units, which are used to increase the features map but reduce the size of the input patch. The expanding path consists of the decoder units, which are used to back up the input patch size in the output with reducing the size of features map. The robustness of the U-net is represented by the concatenation path, which is proposed to combine low- and high-level features. Low-level features are responsible for defining the contents of the input patch, while high level features are developed to identify the spatial information of the contents. The training patch (size 64 × 64 × 64 × 4) is cropped based on the availability of all classes, and then normalized (channel normalization) to avoid overfitting and overcome an unbalanced data problem. A mini-batch technique is used to speed up the training process in terms of a model convergence and improve performance. Each encoder unit consists of two convolutional layers (filter size [3 × 3×3], stride 1, same padding) and a maxpooling layer. Both convolutional layers are followed by Batch Normalization (BN) and a Rectifier Linear Unit (ReLU), but each has a different number of filters, as shown in Fig. 3-a.

A new version of modified 3D U-Net based on an Accumulated Encoder (U-Net AE) is proposed to improve segmentation results. The AE consists of Accumulated Block (AB), followed by ReLU, and maxpooling. The AB adds two features maps

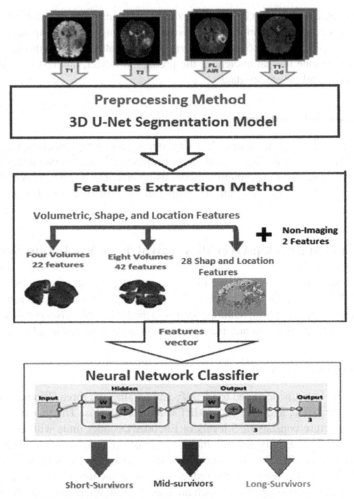

Fig. 2. Automatic overall survival time prediction system.

based on elements-wise addition to improve the quality of low-level features. These features are developed by two branches, after a previous maxpooling layer: the first branch increases the features map by using a convolutional layer (filter size [1 × 1 × 1], stride 1, same padding); and the second branch increases the features map by using another convolutional layer (filter size [3 × 3 × 3], stride 1, same padding), as shown in Fig. 3-b.

Because this data is highly unbalanced, the Generalized Dice Loss (GDL) [7] function is used with an ADAM optimizer to train the segmentation models. The GDL formula is defined as follows:

$$GDL = 1 - \frac{2\sum_{k=1}^{K} w_k \sum_{m=1}^{M} GT_{km} Pr_{km}}{\sum_{k=1}^{K} w_k \sum_{m=1}^{M} GT_{km}^2 + Pr_{km}^2} \tag{1}$$

where *GT* refers to the ground truth image, *Pr* refers to the prediction image, *K* is the number of the classes, *M* is the number of elements in the volume image, and w_k refers to a specific class weighting factor used to reverse the influence of larger classes, which helps to improve learning of smaller classes.

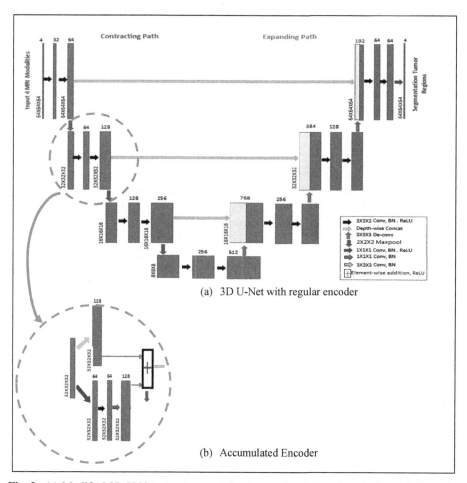

(a) 3D U-Net with regular encoder

(b) Accumulated Encoder

Fig. 3. (a) Modified 3D U-Net structure, and (b) Accumulated Encode unit (instead of regular encoder).

3.2 Features Extraction Methods

Volumetric Features

Each brain functionality region controls different functions and activity, as shown in Fig. 4. The volume and location information of Glioma brain tumor sub-regions, based on a specific brain functionality region, are proposed to develop an overall survival time prediction system. We did not find a program (software) to automatically bound each of these brain functionality regions in MRI brain images. Therefore, the cross of brain mid planes (mid-Sagittal, mid-Coronal, and mid-Horizontal) are used to divide a brain volume into, either four small sub-volumes using the crossing of the mid-Coronal plane and mid-Sagittal plane, or eight small sub-volumes using the crossing of the three mid planes, as shown in Fig. 4. The number of extracted volumetric features are 22 and 42 for the four small brain sub-volumes and eight small sub-volumes, respectively. Five volumetric features are collected from each small volume, which represent volume of a brain region, volume of the whole tumor, volume of the gadolinium (GD)- enhancing tumor, volume of the NCR/NET tumor, and volume of the Edema, with two additional volumetric features (volume of the whole brain and volume of the whole tumor). In addition, the availability of each brain tumor sub-region at a specific small volume can describe location information.

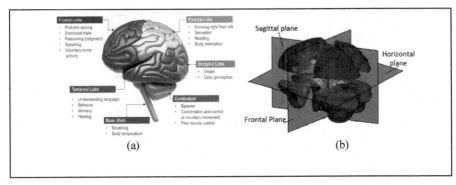

Fig. 4. (a) Brain Functionality regions [21], and (b) Brain structural planes.

Shape and Location Features

Twenty-eight shape and location features are proposed to describe the shape and location information of the whole brain tumor region and the tumor core region in MRI images. There are fourteen features to describe each of a brain tumor region (WT and TC), which include: *centroid,* which represents the center of the mass of a specific tumor region (three features: x, y and z); *volume* of a specific tumor region; *equivalent dimeter,* which represents the diameter of a sphere with the same volume as a specific tumor region; *orientation* (three orientation angle features: x, y, and z), which represents the orientation of the ellipsoid that has the same normalized second central moments as a specific tumor region, calculated by using eigen vectors; *principle axis length,* which represents a length of the major axes of the ellipsoid with the same normalized second central moments to a specific tumor region (three axes length features: x, y, and z); *surface area extent,* which represents the ratio of voxels in a specific tumor region to the whole region of a brain volume; and *solidity,* which represents the portion of the smallest convex polygon that can contain a tumor region that is also in a brain volume.

Non-imaging Features

The BraTS 2020 evaluates only the GTR resection status. The number of samples with GTR resection status in the BraTS 2020 training survival dataset is 119 of 236 samples. A dataset with a size of 119 samples is not enough to train a robust overall survival time prediction model. Therefore, the resection status is used as non-imaging feature with the value "1" for the GTR resection, and "0" for STR and unknown resection statuses. As the patient's age is one of the factors used by doctors to determine a suitable treatment plan for Glioma brain tumor patients [2], it is used as another non-imaging feature.

3.3 Overall Survival Time Prediction Model

This is the last stage of developing the automatic overall survival time prediction system for Glioma brain tumor patients. A simple NN is used as a ML method to train an overall survival time classification model. This model is used to classify the data into three survivor categories: short-survivors (<10 months), mid-survivors (10–15 months), and long-survivors (>15 months). The simple NN structure consists of three layers: the input layer, hidden layer, and output layer. The size of the input layer is equal to the number of training features. The size of the output layer is three, which represents the three classification categories. Different values are used to tune the size of the hidden layer to produce an accurate prediction model. A Sigmoid function is used in the hidden layer, and SoftMax is used in the output layer.

3.4 Evaluation Methods

Average Dice Similarity (ADS) is used to evaluate the 3D modified U-Net segmentation models. The Dice Similarity (DS) score calculates the similarity between the Ground Truth image (GT) and Prediction image (Pr). Confusion matrix and accuracy (ACC) are used to evaluate the performance of the overall survival time classification model. The confusion matrix is preferred to evaluate a classifier's performance with unbalanced data. In addition, it provides positive and negative classification rates for each class. ACC and ADS formulas are represented in Eqs. 2 and 3, respectively.

$$ACC = \frac{TP + TN}{P + N} \tag{2}$$

$$ADS = \sum_{i=1}^{n} \frac{2|GT_i.Pr_i|}{n|GT_i| + |Pr_i|} \tag{3}$$

where *TP* refers to the true positive, *TN* refers to true negative, *P* is the number of positive elements, *N* is the number of negative elements, and *n* is total number of samples in a dataset.

4   Experiments and Results

In the first experiment, three DL models based on a 3D modified U-Net were trained and tested to develop an accurate segmentation model for Glioma brain tumor sub-regions. The segmentation models were trained by using ADAM optimizer with a GDL function. The following parameters were specified to train the model: the initial Learning Rate (LR) parameter was 10^{-4}; the maximum number of epochs was 100; and the learning rate drop factor was 0.95 with a period drop of 5 epochs. Two different mini-batch sizes were used (4 and 8) to enhance the performance. The training time for the 3D U-Net AE was 52.6 h. The evaluation performance based on ADS for the three trained models in the training dataset is reported in Table 1. The segmentation results were improved

Table 1. Performance of the modified 3D U-Net segmentation models in terms of average dice similarity score.

Model	# Mini batch	ADS WT	ADS TC	ADS ET
3D U-Net	4	0.85809	0.87336	0.85511
3D U-Net	8	0.86306	0.86752	0.85054
3D U-Net RE	8	0.8898	0.88197	0.85248

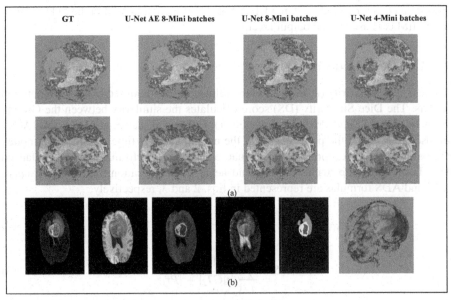

Fig. 5. (a) 3D segmentation results of two samples from the BraTS 2020 training dataset (unseen in the training process); The colored regions represent three brain tumor sub-regions as follows: red for the necrotic and non-enhancing tumor core; green for enhancing the tumor core; and yellow for the peritumoral edema. (b) Four MRI modalities and prediction tumor for slice 89 of sample BraTS20_Validation_008, with 3D prediction tumor.

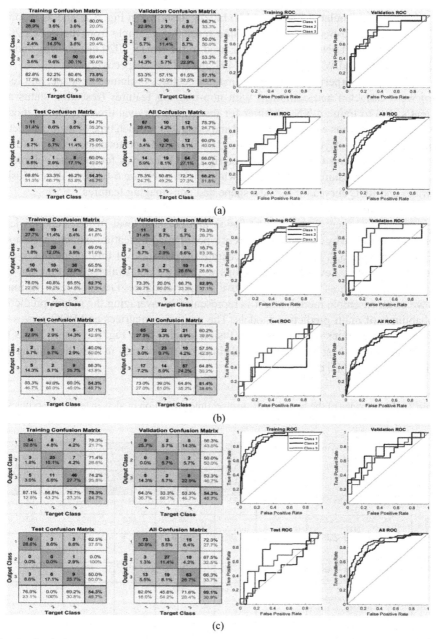

Fig. 6. Confusion matrices and ROC curves for: a) features fusion model based on volumetric features (4 sub-volumes) and non-imaging features; b) features fusion model based on shape, location, and non-imaging features; and c) features fusion based on volumetric (4 sub-volumes), shape, location and non-imaging features

by using the Accumulated Encoder (AE) block, as shown in Fig. 5-a, and as reported in Table 1. Figure 5-b shows the prediction results of a sample from the BraTS 2020 validation set. The U-Net AE model was used in the BraTS 2020 test dataset, and the ADS for ET, TC and WT are 0.6851, 0.7431 and 0.7708, respectively.

In the second experiment, different imaging features (volumetric, shape, and location) were extracted and combined with non-imaging features (resection status and patient's age) to train an NN classifier for the overall survival time prediction model. Several configurations were implemented and evaluated for each type of feature to determine the best hidden layer size that produced the best model performance. The confusion matrices and Receiver Operating Characteristic (ROC) curves for the best volumetric and non-imaging features model are shown in Fig. 6-a. Figure 6-b shows the confusion matrices and ROC curves for the best shape and non-imaging features.

For the third experiment, two information fusion strategies were used to enhance classifier performance: features-level fusion and decision-level fusion. Features-level fusion is represented by combining different types of imaging feature extraction methods, presented in Sect. 3.2, to develop an accurate overall survival time prediction model using a simple NN. Several configurations were tested and evaluated to tune the size of the hidden layer, as well as to improve the performance. Decision-level fusion is known as an ensemble model, and it is represented by averaging the output prediction probabilities of multi-prediction models. The confusion matrices and ROC curves for the best configurations of features-level fusion are shown in Fig. 6-c. Table 2 lists the performance of the best prediction models in terms of accuracy in both the training and validation BraTS 2020 datasets.

Table 2. Performance of the best overall survival time prediction models in terms of accuracy in BraTS Train 2020/Validation2020 datasets.

	Model name: Features type (# Features)	# Hidden Nodes	BraTS 2020 Train Dataset				BraTS 2020 Validation Dataset
			Train ACC%	Validation ACC%	Test ACC %	Total ACC %	ACC %
Features fusion	M1: 4V+ Non-Imaging (24)	150	73.5	57.1	54.3	68.2	51.7
	M2: 8V+ Non-Imaging (44)	200	68.1	68.6	62.1	67.4	48.3
	M3: Shape+ Non-Imaging (30)	60	62.7	62.9	54.3	61.4	**55.2**
	M4: 4V+8V+ Non-Imaging (66)	400	74.7	48.6	65.7	69.5	48.3
	M5: 4V+ shape+ Non-Imaging 52	100	75.3	54.3	54.3	69.1	51.7
Ensemble	M3&M5: (V4+shape+Non-Imaging) & (shape)		-				**55.2**

The best accuracy achieved in the validation BraTS 2020 dataset was 55.2%, by using 30 shape and location features. In addition, the ensemble model achieved an accuracy of 55.2% in the validation BraTS 2020 dataset. The ensemble model is based on averaging the prediction probabilities of two models: the first model is the best features fusion model based on volumetric (four sub-volumes), shape, and location features, as well as non-imaging features; the second model is the best features fusion model based on shape, location, and non-imaging features. The ensemble model archives accuracy 55.1% in the BraTS 2020 test dataset.

5 Conclusions and Future Work

An automatic overall survival time prediction system for Glioma brain tumor patients was proposed and evaluated. The proposed modified 3D U-Net AE outperformed the 3D U-Net with a regular encoder in segmenting the three-brain tumor sub-regions. The segmentation results were used to extract the proposed volumetric, shape, and location features. The volumetric features based on a four small sub-volumes model performed better than an eight small sub-volume model. The shape and location features achieved the best validation accuracy (55.2%) compared to the volumetric features based on four volumes (51.8%) in the BraTS 2020 validation dataset. However, the training accuracy for the volumetric features model was better than the shape features. The ensemble model achieved the same accuracy of the shape and location features model (55.2%) in the BraTS 2020 validation data. The ensemble model is based on two models: model 1 was represented by a combination of 22 volumetric features (four brain sub-volumes) and 28 shape features with 2 non-imaging features; model 2 was based on 28 shape features with 2 non-imaging features. As an ensemble model is used to reduce generalization errors and prediction variance, especially with NN models, we decided to use the ensemble model in the BraTS 2020 test dataset. The Ensemble model achieved a competitive accuracy 55.1% in the BraTS 2020 test dataset. where the first top rank methods achieved accuracy 62% in the BraTS 2020 test dataset but accuracy 41.3% in the validation dataset. The total number of corrected predictions was 16 out of 29, and 59 out of 107 in the BraTS 2020 validation and test datasets, respectively. This means 75 out of 136 samples of the survival data were classified correctly to one of the three classes (short-survivors, mid-survivors, and long-survivors) in both the test and the validation dataset for GTR resection status. These results indicate that our method maybe it is achieved one of the three top rank places if the validation dataset was used with the test dataset in the final evaluation. base on the reported results in the BraTS challenge 2020, the three top rank teams achieves accuracy less than our method in the validation dataset. Therefore, our other trained model probably would achieve better performance in the BraTS test dataset if they were used and evaluated.

For future work, two strategies are proposed to improve the performance of the overall survival time prediction system. The first is to improve the preprocessing stage by using data denoising techniques, increasing the complexity of the DL model, and using multi loss functions. The second is to improve the classification results by using other ML methods, and adding histogram texture features, based on the co-occurrence matrix, to the proposed features.

Acknowledgments. This research was supported by HOWE Foundation Fellowship/College of Engineering-UNLV and Summer Doctoral Research Fellowship/Graduate College-UNLV. We would like to thank **Dr. Edwin Oh**, Associate Professor, Neurogenetics and Precision Medicine Lab/UNLV, for his valuable help by providing a computer with Titan RTX GPU to train and test the Deep Learning segmentation models.

References

1. Suchorska, B., et al.: Complete resection of contrast-enhancing tumor volume is associated with improved survival in recurrent glioblastoma - results from the DIRECTOR trial. Neuro-Oncology **18** (4), 549–556 (01 2016). https://doi.org/10.1093/neuonc/nov326
2. Brain Tumor: Grades and Prognostic Factors, Approved by the Cancer.Net Editorial Board. https://www.cancer.net/cancer-types/brain-tumor/grades-and-prognostic-factors. Accessed 13 Aug 2020
3. Menze, B., et al.: The multimodal brain tumor image segmentation benchmark (BRATS), IEEE Trans. Med. Imaging **34**(10), 1993–2024 (2015). https://doi.org/10.1109/TMI.2014.2377694
4. Litjens, G., et al.: A survey on deep learning in medical image analysis. Med. Image Anal. **42**, 60–88 (2017)
5. Esteva, A., et al.: Dermatologist-level classification of skin cancer with deep neural networks. Nature **542**, 115–118 (2017)
6. Esteva, A., et al.: A guide to deep learning in healthcare. Nat. Med. **25**, 24–29 (2019)
7. Ronneberger, O., Fischer, P., Brox, T.: U-Net: convolutional networks for biomedical image segmentation. In: Navab, N., Hornegger, J., Wells, W.M., Frangi, A.F. (eds.) MICCAI 2015. LNCS, vol. 9351, pp. 234–241. Springer, Cham (2015). https://doi.org/10.1007/978-3-319-24574-4_28
8. Milletari, F.; Navab, N.; Ahmadi, S.A.:V-Net: fully convolutional neural networks for volumetric medical image segmentation, In: 3D Vision Fourth International Conference (3DV), Stanford, California, USA, October 25–28 (2016)
9. Chen, H., Dou, Q., Yu, L., Qin, J., Heng, P.A.: VoxResNet: deep voxelwise residual networks for brain segmentation from 3D MR Images, J. NeuroImage, **170**, 446–455 (2018)
10. Liu, S., et al.: Prostate cancer diagnosis using deep learning with 3D multiparametric MRI, In: Medical Imaging Computer-Aided Diagnosis Conference, vol. 10134, Orlando, Florida, United States March 2017
11. Myronenko, A.: 3D MRI brain tumor segmentation using autoencoder regularization. In: Crimi, A., Bakas, S., Kuijf, H., K, Farahani, Reyes, M., van Walsum, Theo (eds.) BrainLes 2018. LNCS, vol. 11384, pp. 311–320. Springer, Cham (2019). https://doi.org/10.1007/978-3-030-11726-9_28
12. Sudre, C.H., Li, W., Vercauteren, T., Ourselin, S., Jorge Cardoso, M.: Generalised dice overlap as a deep learning loss function for highly unbalanced segmentations. In: Cardoso, M.J., et al. (eds.) DLMIA/ML-CDS -2017. LNCS, vol. 10553, pp. 240–248. Springer, Cham (2017). https://doi.org/10.1007/978-3-319-67558-9_28
13. Feng, X., Tustison, N., Meyer, C.: Brain tumor segmentation using an ensemble of 3d u-nets and overall survival prediction using radiomic features. BrainLes 2018, pp. 279–28876, Springer LNCS11384 (2019)
14. Weninger, L., Rippel, O., Koppers, S., Merhof, D.: Segmentation of brain tumors and patient survival prediction: methods for the brats 2018 challenge. In: Crimi, A., Bakas, S., Kuijf, H., Keyvan, F., Reyes, M., van Walsum, T. (eds.) BrainLes 2018. LNCS, vol. 11384, pp. 3–12. Springer, Cham (2019). https://doi.org/10.1007/978-3-030-11726-9_1

15. Sun, L., Zhang, S., Luo, L.: Tumor segmentation and survival prediction in glioma with deep learning. In: Crimi, A., Bakas, S., Kuijf, H., Keyvan, F., Reyes, M., van Walsum, T. (eds.) BrainLes 2018. LNCS, vol. 11384, pp. 83–93. Springer, Cham (2019). https://doi.org/10.1007/978-3-030-11726-9_8

16. Baid, U., et al.: Deep Learning Radiomics Algorithm for Gliomas (DRAG) Model: A Novel Approach Using 3D UNET Based Deep Convolutional Neural Network for Predicting Survival in Gliomas. In: Crimi, A., Bakas, S., Kuijf, H., Keyvan, F., Reyes, M., van Walsum, T. (eds.) BrainLes 2018. LNCS, vol. 11384, pp. 369–379. Springer, Cham (2019). https://doi.org/10.1007/978-3-030-11726-9_33

17. Bakas, S., Akbari, H., Sotiras, A., Bilello, M., Rozycki, M., Kirby, J.S., et al.: Advancing the cancer genome atlas glioma MRI collections with expert segmentation labels and radiomic features. Nat. Sci. Data **4**, 170117 (2017). https://doi.org/10.1038/sdata.2017.117

18. Bakas, S., Reyes, M., Jakab, A., Bauer, S., Rempfler, M., Crimi, A., et al.: Identifying the Best Machine Learning Algorithms for Brain Tumor Segmentation, Progression Assessment, and Overall Survival Prediction in the BRATS Challenge, arXiv preprint arXiv:1811.02629 (2018)

19. Bakas, S., Akbari, H., Sotiras, A., Bilello, M., Rozycki, M., Kirby, J., et al.: Segmentation Labels and Radiomic Features for the Pre-operative Scans of the TCGA-GBM collection. Cancer Imaging Arch. (2017). https://doi.org/https://doi.org/10.7937/k9/tcia.2017.klxwjj1q

20. Bakas, S., Akbari, H., Sotiras, A., Bilello, M., Rozycki, M., Kirby, J., et al.: Segmentation labels and radiomic features for the pre-operative scans of the TCGA-LGG collection. Cancer Imaging Arch. (2017). https://doi.org/10.7937/K9/TCIA.2017.GJQ7R0EF

21. "Parts of the Brain & Function", Website: Anatomy info. https://anatomyinfo.com/parts-of-the-brain/. Accessed 20 July 2020

Squeeze-and-Excitation Normalization for Brain Tumor Segmentation

Andrei Iantsen[✉], Vincent Jaouen, Dimitris Visvikis, and Mathieu Hatt

LaTIM, INSERM, UMR 1101, University Brest, Brest, France
andrei.iantsen@inserm.fr

Abstract. In this paper we described our approach for glioma segmentation in multi-sequence magnetic resonance imaging (MRI) in the context of the MICCAI 2020 Brain Tumor Segmentation Challenge (BraTS). We proposed an architecture based on U-Net with a new computational unit termed "SE Norm" that brought significant improvements in segmentation quality. Our approach obtained competitive results on the validation (Dice scores of 0.780, 0.911, 0.863) and test (Dice scores of 0.805, 0.887, 0.843) sets for the enhanced tumor, whole tumor and tumor core sub-regions. The full implementation and trained models are available at https://github.com/iantsen/brats.

Keywords: Medical imaging · Brain tumor segmentation · SE norm · U-Net

1 Introduction

Glioma is a group of malignancies that arises from the glial cells in the brain. Nowadays, gliomas are the most common primary tumors of the central nervous system [1,2]. The symptoms of patients presenting with a glioma depend on the anatomical site of the glioma in the brain and can be too common (e.g. headaches, nausea or vomiting, mood and personality alterations) to give an accurate diagnosis in early stages of the disease. The primary diagnosis is usually confirmed by magnetic resonance imaging (MRI) or computed tomography (CT) that provide additional structural information about the tumor.

Gliomas usually consist of heterogeneous sub-regions (edema, enhancing and non-enhancing tumor core, etc.) with variable histologic and genomic phenotypes [1]. Presently, multimodal MRI scans are used for non-invasive tumor evaluation and treatment planning, due to its ability to depict the tumor sub-regions with different intensities. However, segmentation of brain tumors in multimodal MRI scans is one of the most challenging tasks in medical imaging because of the high heterogenity in tumor appearances and shapes.

The brain tumor segmentation challenge (BraTS) [3–6] is aimed at development of automatic methods for the brain tumor segmentation. All participants of the BraTS are provided with a clinically-acquired training dataset of pre-operative MRI scans (4 sequences per patient) and segmentation masks for

A. Crimi and S. Bakas (Eds.): BrainLes 2020, LNCS 12659, pp. 366–373, 2021.
https://doi.org/10.1007/978-3-030-72087-2_32

three different tumor sub-regions, namely the GD-enhancing tumor, the peritu-moral edema, and and the necrotic and non-enhancing tumor core. The MRI scans were acquired with different clinical protocols and various scanners from multiple 19 institutions. Each scan was annotated manually by one to four raters and subsequently approved by expert raters.

The performance of proposed algorithms was evaluated by the Dice score, sensitivity, specificity and the 95th percentile of the Hausdorff distance.

2   Materials and Methods

2.1   SE Normalization

Normalization layers have become an integral part of modern deep neural net-works. Existing methods, such as Batch Normalization [7], Instance Normal-ization [8], Layer Normalization [9], etc., have been shown to be effective for training different types of deep learning models. In essence, any normaliza-tion layer performs the following computations. First, for a n-dimensional input $X = (x^{(1)}, x^{(2)}, \ldots, x^{(n)})$, we normalize each dimension

$$x'^{(i)} = \frac{1}{\sigma^{(i)}}(x^{(i)} - \mu^{(i)}) \tag{1}$$

where $\mu^{(i)} = \mathrm{E}[x^{(i)}]$ and $\sigma^{(i)} = \sqrt{\mathrm{Var}[x^{(i)}] + \epsilon}$ with ϵ as a small constant. Normalization layers mainly differ in terms of the dimensions chosen to com-pute the mean and standard deviation [10]. Batch Normalization, for example, uses the values calculated for each channel within a batch of examples, whereas Instance Normalization - within a single example. Second, a pair of parameters γ_k, β_k are applied to each channel k to scale and shift the normalized values:

$$y_k = \gamma_k x'_k + \beta_k \tag{2}$$

The parameters γ_k, β_k are fitted in the course of training and enable the layer to represent the identity transform, if necessary. During inference, both parameters are *fixed* and *independent* of the input X. In this paper, we propose to apply *instance-wise normalization* and design each parameter γ_k, β_k as *functions* of the input X, i.e.

$$\gamma = f_\gamma(X) \tag{3}$$
$$\beta = f_\beta(X) \tag{4}$$

where $\gamma = (\gamma_1, \gamma_2, \ldots, \gamma_K)$ and $\beta = (\beta_1, \beta_2, \ldots, \beta_K)$ - the scale and shift parameters for all channels, K is a number of channels. We represent the function f_γ using the original Squeeze-and-Excitation (SE) block with the sigmoid [11], whereas f_β is modeled with the SE block with the tanh activation function to enable the negative shift (see Fig. 1a). This new architectural unit, that we refer to as *SE Normalization (SE Norm)*, is the major component of our model.

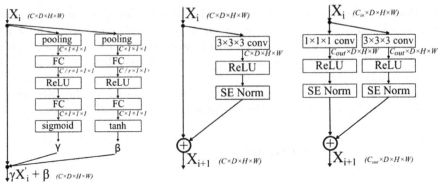

(a) SE Normalization layer

(b) Residual layer with the shortcut connection

(c) Residual layer with the non-linear projection

Fig. 1. Proposed layers. Output dimensions are depicted in brackets.

2.2   Network Architecture

The widely used 3D U-Net [12,13] serves as the basis to design our model. The basic element of the model, a convolutional block comprised of a $3 \times 3 \times 3$ convolution followed by the ReLU activation function and the SE Norm layer, is used to construct the decoder (Fig. 2, blue blocks). In the encoder, we utilize residual layers [14] consist of convolutional blocks with shortcut connections (see Fig. 1b). If numbers of input / output channels in a residual layer are different, we perform a non-linear projection by adding the $1 \times 1 \times 1$ convolutional block to the shortcut in order to match the dimensions (see Fig. 1c).

In the encoder, we perform downsampling applying max pooling with the kernel size of $2 \times 2 \times 2$. To linearly upsample feature maps in the decoder, we use $3 \times 3 \times 3$ transposed convolutions. In addition, we supplement the decoder with three upsampling paths to transfer low-resolution features further in the model by applying the $1 \times 1 \times 1$ convolutional block to reduce the number of channels, and utilizing trilinear interpolation to increase the spatial size of the feature maps (Fig. 2, yellow blocks).

The first residual layer placed after the input is implemented with the kernel size of $7 \times 7 \times 7$ to increase the receptive field of the model without significant computational overhead. The softmax layer is applied to output probabilities for four target classes.

To regularize the model, we add Spatial Dropout layers [15] right after the last residual block at each stage in the encoder and before $1 \times 1 \times 1$ convolution in the decoder tail (Fig. 2, red blocks).

2.3   Data Preprocessing

Intensities of MRI scans are not standardized and typically exhibit a high variability in both intra- and inter-image domains. In order to decrease the intensity

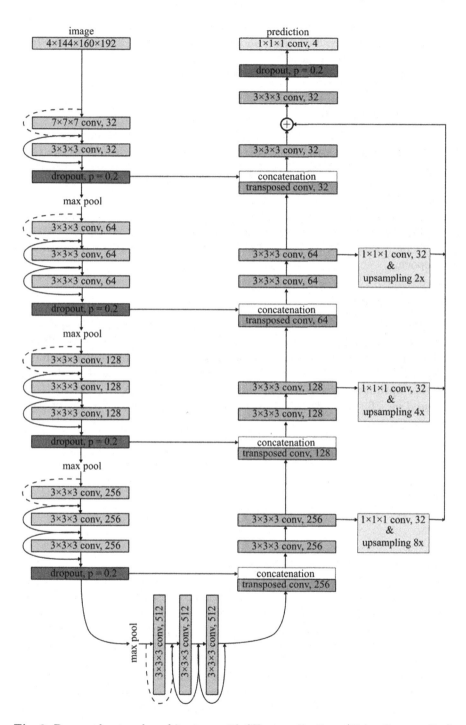

Fig. 2. Proposed network architecture with SE normalization. (Color figure online)

inhomogeneity, we perform Z-score normalization for each MRI sequence and each patient separately. The mean and standard deviation are calculated based on non-zero voxels corresponding to the brain region. All background voxels remain unchanged after the normalization.

2.4 Training Procedure

Due to the large size of provided MRI scans, we perform training on random patches of the size $144 \times 160 \times 192$ voxels (*depth* $\times$ height $\times$ *width*) on two GPUs NVIDIA GeForce GTX 1080 Ti (11 GB) with a batch size of 2 (one sample per worker).

We train the model for 300 epochs using Adam optimizer with $\beta_1 = 0.9$ and $\beta_2 = 0.99$ for exponential decay rates for moment estimates, and apply a cosine annealing schedule gradually reducing the learning rate from $lr_{max} = 10^{-4}$ to $lr_{min} = 10^{-6}$ within 25 epochs and performing the learning rate adjustment at each epoch.

2.5 Loss Function

We utilize the unweighted sum of the Soft Dice Loss [16] and the Focal Loss [17] as the loss function in the course of training. The Soft Dice Loss is the differentiable surrogate to optimize the Dice score that is one of the evaluation metrics used in the challenge. The Focal Loss, compared to the Soft Dice Loss, has much smoother optimization surface that ease the model training.

Based on [16], the Soft Dice Loss for one training example can be written as

$$L_{Dice}(y, \hat{y}) = 1 - \frac{1}{C} \sum_{c=1}^{C} \frac{2 \sum_i^N y_i^c \hat{y}_i^c + 1}{\sum_i^N y_i^c + \sum_i^N \hat{y}_i^c + 1} \tag{5}$$

The Focal Loss is defined as

$$L_{Focal}(y, \hat{y}) = -\frac{1}{N} \sum_i^N \sum_{c=1}^{C} y_i^c (1 - \hat{y}_i^c)^\gamma \ln(\hat{y}_i^c) \tag{6}$$

In both definitions, $y_i = \left[y_i^1, y_i^2, \ldots, y_i^C\right]^\top$ - the one-hot encoded label for the i-th voxel, $\hat{y}_i = \left[\hat{y}_i^1, \hat{y}_i^2, \ldots, \hat{y}_i^C\right]^\top$ - predicted probabilities for the i-th voxel. N and C are the total numbers of voxels and classes for the given example, respectively. Additionally we apply Laplacian smoothing by adding +1 to the numerator and denominator in the Soft Dice Loss to avoid the zero division in cases when one or several labels are not represented in the training example. The parameter γ in the Focal Loss is set at 2.

The training data in the challenge has labels for three tumor sub-regions, namely the necrotic and non-enhancing tumor core (NCR & NET), the peritumoral edema (ED) and the GD-enhancing tumor (ET). However, the evaluation is done for the GD-enhancing tumor (ET), the tumor core (TC), which is comprised of NCR & NET along with ET, and the whole tumor (WT) that combines all provided sub-regions. Hence, during training we optimize the loss directly on these nested tumor sub-regions.

2.6   Ensembling

To reduce the variance of the model predictions, we build an ensemble of models that are trained on different splits of the train set and use the average as the ensemble prediction. At each iteration, the model is built on 90%/10% splits of the train set and subsequently evaluated on the online validation set. Having repeated this procedure multiple times, we choose 20 models with the highest performance on the online validation set and combine them into the ensemble. Predictions on the test set are produced by averaging predictions of the individual models and applying a threshold operation with a value equal to 0.5.

2.7   Post-processing

The Dice score used for the performance evaluation in the challenge is highly sensitive to cases wherein the model predicts classes that are not presented in the ground truth. Therefore, a false positive prediction for a single voxel leads to the lowest value of the Dice score and might significantly affect the average model performance on the whole evaluation dataset. This primarily refers to patients without ET sub-regions. To address this issue, we add a post-processing step to remove small ET regions from the model outcome if their area is less than a certain threshold. We set its value at 32 voxels since it is the smallest ET area among all patients in the train set.

3   Results and Discussion

The results of the BraTS 2020 segmentation challenge are presented in Table 1 and Table 2. The Dice score, Sensitivity and Hausdorff distance (HD) were utilized for the evaluation. Results in Table 1 were obtained on the online validation set with 125 patients without publicly available segmentation masks. The U-Net model was used as a baseline for comparison purposes. Final results on the test set consisted of 166 patients are shown in Table 2.

For all cases, the lowest average Dice score was obtained for the ET sub-region. This can be partially explained by the relatively small size of the ET class compared to the other tumor sub-regions that made segmentation of this class more challenging. The proposed model outperformed U-Net in all evaluation metrics except for the Dice score for the ET class. It is mainly caused by cases wherein the ET sub-regions were not presented. Combining multiple models into the ensemble allowed to address this issue since it reduced the chance to receive false positive predictions for the ET class as well as led to the better performance in terms of HD.

Table 1. Performance on the online validation set ($n = 125$). Average results are provided for each evaluation metrics.

Metrics	Dice score			Sensitivity			HD		
Class	ET	WT	TC	ET	WT	TC	ET	WT	TC
U-Net	0.772	0.899	0.825	0.794	0.896	0.813	5.813	5.973	6.576
Best Model	0.740	0.908	0.862	0.816	0.909	0.854	3.841	4.602	5.339
Ensemble	0.761	0.911	0.863	0.814	0.908	0.850	3.695	4.475	4.816
Ensemble + pp	0.780	0.911	0.863	0.815	0.908	0.850	3.717	4.475	4.816

Table 2. Performance on the test set ($n = 166$).

Metrics	Dice score			Sensitivity			HD		
Class	ET	WT	TC	ET	WT	TC	ET	WT	TC
Ensemble + pp	0.805	0.887	0.843	0.854	0.909	0.866	15.429	4.535	19.589

References

1. Bakas, S.: Advancing the cancer genome atlas glioma MRI collections with expert segmentation labels and radiomic features. Nature Sci. Data **4**(1), 1–13 (2017)
2. Upadhyay, N., Waldman, A.: Conventional MRI evaluation of gliomas. British J. Radiol. **84**(2), Special Issue 2 S107–S111 (2011)
3. Menze, B.H., et al.: The multimodal brain tumor image segmentation benchmark (BRATS). IEEE Trans. Med. Imaging **34**(10), 1993–2024 (2015)
4. Bakas, S. et al: Segmentation labels and radiomic features for the pre-operative scans of the TCGA-GBM collection. TCIA (2017)
5. Bakas, S., et al.: Segmentation labels and radiomic features for the pre-operative scans of the TCGA-LGG collection. TCIA (2017)
6. Bakas, S., Reyes, M., Jakab, A., Bauer, S., Rempfler, M., Crimi, A. et al.: Identifying the Best Machine Learning Algorithms for Brain Tumor Segmentation, Progression Assessment, and Overall Survival Prediction in the BRATS Challenge. arXiv preprint arXiv:1811.02629 (2018)
7. Ioffe, S., Szegedy, C.: Batch Normalization: Accelerating Deep Network Training by Reducing Internal Covariate Shift. arXiv preprint arXiv:1502.03167 (2015)
8. Ulyanov, D., Vedaldi, A., Lempitsky, V.: Instance normalization: The missing ingredient for fast stylization. arXiv preprint arXiv:1607.08022 (2016)
9. Ba, J.L., Kiros, J.R., Hinton, G.E.: Layer Normalization. arXiv preprint arXiv:1607.06450 (2016)
10. Wu, Y., He, K.: Group normalization. In: European Conference on Computer Vision (ECCV) (2018)
11. Hu, J., Shen, L., Sun, G.: Squeeze-and-excitation networks, CoRR, vol. abs/1709.01507 (2017). http://arxiv.org/abs/1709.01507
12. Ronneberger, O., Fischer, P., Brox, T.: U-net: convolutional networks for biomedical image segmentation. In: Navab, N., Hornegger, J., Wells, W.M., Frangi, A.F. (eds.) MICCAI 2015. LNCS, vol. 9351, pp. 234–241. Springer, Cham (2015). https://doi.org/10.1007/978-3-319-24574-4_28

13. Çiçek, Ö., Abdulkadir, A., Lienkamp, S.S., Brox, T., Ronneberger, O.: 3D U-net: learning dense volumetric segmentation from sparse annotation. In: Ourselin, S., Joskowicz, L., Sabuncu, M.R., Unal, G., Wells, W. (eds.) MICCAI 2016. LNCS, vol. 9901, pp. 424–432. Springer, Cham (2016). https://doi.org/10.1007/978-3-319-46723-8_49

14. He, K., Zhang, X., Ren, S., Sun, J.: Identity mappings in deep residual networks. In: Leibe, B., Matas, J., Sebe, N., Welling, M. (eds.) ECCV 2016. LNCS, vol. 9908, pp. 630–645. Springer, Cham (2016). https://doi.org/10.1007/978-3-319-46493-0_38

15. Tompson, J., Goroshin, R., Jain, A., LeCun, Y., Bregler, C.: Efficient Object Localization Using Convolutional Networks. arXiv preprint arXiv:1411.4280 (2014)

16. Milletari, F., Navab, N., Ahmadi, S.-A.: V-net: fully convolutional neural networks for volumetric medical image segmentation. In: International Conference on 3D Vision, pp. 565–571. IEEE (2016)

17. Lin, T.-Y., Goyal, P., Girshick, R., He, K., Dollár, P.: Focal Loss for Dense Object Detection. arXiv preprint arXiv:1708.02002 (2017)

Modified MobileNet for Patient Survival Prediction

Agus Subhan Akbar[1,2]([envelope]) [ID], Chastine Fatichah[1] [ID], and Nanik Suciati[1] [ID]

[1] Institut Teknologi Sepuluh Nopember, Surabaya, Indonesia
{chastine,nanik}@if.its.ac.id
[2] Universitas Islam Nahdlatul Ulama Jepara, Jepara, Indonesia
agussa@unisnu.ac.id

Abstract. Glioblastoma is a type of malignant tumor that varies significantly in size, shape, and location. The study of this type of tumor, one of which is about predicting the patient's survival ability, is beneficial for the treatment of patients. However, the supporting data for the survival prediction model are minimal, so the best methods are needed for handling it. In this study, we propose an architecture for predicting patient survival using MobileNet combined with a linear survival prediction model (SPM). Several variations of MobileNet are tested to obtain the best results. Variations tested include modification of MobileNet V1 with freeze or unfreeze layers, and modification of MobileNet V2 with freeze or unfreeze layers connected to SPM. The dataset used for the trial came from BraTS 2020. A modification based on the MobileNet V2 architecture with the freezing layer was selected from the test results. The results of testing this proposed architecture with 95 training data and 23 validation data resulted in an MSE Loss of 78374.17. The online test results with the validation dataset 29 resulted in an MSE loss value of 149764.866 with an accuracy of 0.345. Testing with the testing dataset resulted in increased accuracy of 0.402. These results are promising for better architectural development.

Keywords: MobileNet · Survival prediction model · BraTS 2020 · MobileNet feature extractor · Glioblastoma

1 Introduction

Glioblastoma (GBM) is a malignant type of brain tumor that varies in size, shape, and location. Sufferers have a minimal ability to survive, generally around 14 months [12]. So that efforts to identify markers from MRI images of these tumors can be helpful for subsequent medical action.

Brain Tumor Segmentation (BraTS) is a challenge that provides MRI data from several patients with this type of tumor. BraTS starts from 2012 until this year 2020, by providing a dataset that is continuously updated [2–5]. In BraTS, available challenges include segmenting the tumor area, predicting patient survival, and evaluating the uncertainty of the tumor area segmentation algorithm used (this task began in 2019).

© Springer Nature Switzerland AG 2021
A. Crimi and S. Bakas (Eds.): BrainLes 2020, LNCS 12659, pp. 374–387, 2021.
https://doi.org/10.1007/978-3-030-72087-2_33

The provided dataset includes training data, validation, and testing. For training data, the results of the label segmentation and patient survival data have been provided. Meanwhile, validation data and testing data are without segmentation labels. Facilities are provided on the web to upload results and determine the accuracy of processing validation and testing data.

Participants of this BraTS challenge have produced many of the best algorithms, as listed on [13]. Some of the participants followed all the existing tasks, but some only completed two [6,9,11,17] or one tasks [1,7,8,15]. One challenging task is the prediction of a patient's survival. With the data provided in the form of patient age, surgery status, and a number of days of survival, the task that must be done is to predict the number of days the patient will survive with these data and/or supplement with features that can be taken from the 4 provided MRI images. For validation, the data provided consists of 4 MRI images without segmentation labels and data on the patient's age and surgical status. The available online validation facility will validate patients with Gross Total Resection (GTR) surgical status only.

For this BraTS 2020 dataset, we conducted research to utilize the MobileNet V1 [10] and V2 [14] architectures to predict patient survival. This choice is based on its best performance in object classification and detection and its flexibility to be modified while maintaining its pre-trained weight. The architecture is modified so that it can process MRI image data and produce feature maps. The features of the extracted architecture plus the patient's age are used as input for the predictive architecture. So we propose a modified MobileNet architecture as a model for feature extraction that will be input into a predictive model for predicting patient survival.

2   Methodology

The methodology used in this research includes the use of a modified MobileNet V1 architecture and a modified MobileNet V2 for feature extraction. The extracted features are used as input for prediction models. The prediction model not only gets input from the extracted features but also the patient's age data. The methodology used in more detail is shown in Fig. 1. The final output of this system is the survival number of days predicted.

2.1   Dataset

The dataset used in this study is BraTS 2020. In this dataset, there are 369 MRI image data for patients. Each patient has 4 MRI images consisting of T1, T1CE, T2, and Flair. Furthermore, each patient is provided with annotation images for recognized tumor areas. The tumor area is marked with three labels. Label 1 for necrosis/non-enhancing tumor (NEC) areas, Label 2 for areas of edema (ED), and label 4 for tumor enhancing (ET) areas.

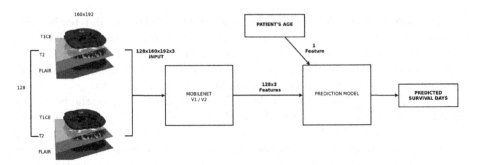

Fig. 1. Proposed methodology

For patient survival data, the data included were 236 patient data with columns, including patient ID, age, survival rate (in days), and surgery status. Of the 236 data, 119 had GTR surgery status, while the rest had STR or NA surgery status. Of 119 data with GTR status, one patient data is known to be alive, so the data used for survival training is 118.

For validation data not accompanied by a tumor area segmentation label, there are 125 MRI image data of the patient. Of the 125 data, validation data for survival were available, and there were 29 data. The online validation facility is provided to check the accuracy of the results of this validation data.

2.2   Proposed Architecture

The architecture proposed in this research includes modification of MobileNet V1 and MobileNet V2 as feature extraction architectures interchangeable. Extracted features combined with patient age data are input to the predictive architecture, as seen in Fig. 1.

MobileNet V1 Feature Extractor. We modified the MobileNet V1 architecture by removing part of the classification and connecting it with several additional layers to produce three features (the ratio of NEC, ED, and ET area to the overall area of the tumor), as shown in Table 1. The input received by MobileNet V1 is a 2D image with three channels, and the size of the 2D image is a multiple of 32. In contrast, the available dataset is four 3D images with cropped size $160 \times 192 \times 128$ from the center of the image. To adjust, the images used are T1CE, T2, and Flair. 3D MRI images were taken axially sliced to obtain 128 2D images of 160×192. The three types of images are stacked into three channels, while the 128 slices are made into batch numbers for one epoch.

The use of a dropout with a factor of 0.1 is used to prevent overfitting [16]. This model's output is a feature of size 3. Since the batch used is 128, the final output of this architecture is a feature of size 128×3.

Table 1. Modified MobileNet V1 as feature extractor

Layer	Kernel size	Feature shape	
		Input	Output
MobileNetV1	–	(None, 160, 192, 3)	(None, 5, 6, 1024)
GlobalAveragePooling2D	–	(None, 5, 6, 1024)	(None, 1024)
Reshape	–	(None, 1024)	(None, 1, 1, 1024)
Dropout(0.1)	–	(None, 1, 1, 1024)	(None, 1, 1, 1024)
Conv2D	3×3	(None, 1, 1, 1024)	(None, 1, 1, 128)
BatchNormalization	–	(None, 1, 1, 128)	(None, 1, 1, 128)
Activation(relu)	–	(None, 1, 1, 128)	(None, 1, 1, 128)
Conv2D	3×3	(None, 1, 1, 128)	(None, 1, 1, 16)
BatchNormalization	–	(None, 1, 1, 16)	(None, 1, 1, 16)
Activation(relu)	–	(None, 1, 1, 16)	(None, 1, 1, 16)
Conv2D	3×3	(None, 1, 1, 16)	(None, 1, 1, 3)
Reshape	–	(None, 1, 1, 3)	(None, 3)
Activation(Sigmoid)	–	(None, 3)	(None, 3)

MobileNet V2 Feature Extractor. Modifications to the MobileNet V2 architecture are carried out as in the MobileNet V1 architecture. The difference between the two is the output feature after deleting the classification layer. The size of the output feature of MobileNet V2 is 1280. Additional layers are made as shown in Table 2. For handling 3D MRI image input to this architecture, the same strategy is done to input the MobileNet V1 architecture.

Table 2. Modified MobileNet V2 as feature extractor

Layer	Kernel size	Feature shape	
		Input	Output
MobileNetV1	–	(None, 160, 192, 3)	(None, 5, 6, 1280)
GlobalAveragePooling2D	–	(None, 5, 6, 1280)	(None, 1280)
Reshape	–	(None, 1280)	(None, 1, 1, 1280)
Dropout(0.1)	–	(None, 1, 1, 1280)	(None, 1, 1, 1280)
Conv2D	3×3	(None, 1, 1, 1280)	(None, 1, 1, 128)
BatchNormalization	–	(None, 1, 1, 128)	(None, 1, 1, 128)
Activation(relu)	–	(None, 1, 1, 128)	(None, 1, 1, 128)
Conv2D	3×3	(None, 1, 1, 128)	(None, 1, 1, 16)
BatchNormalization	–	(None, 1, 1, 16)	(None, 1, 1, 16)
Activation(relu)	–	(None, 1, 1, 16)	(None, 1, 1, 16)
Conv2D	3×3	(None, 1, 1, 16)	(None, 1, 1, 3)
Reshape	–	(None, 1, 1, 3)	(None, 3)
Activation(Sigmoid)	–	(None, 3)	(None, 3)

Survival Prediction Model (SPM). The proposed SPM is a linear architecture using input from the output of the Feature Extractor (MobileNetV1/MobileNetV2) measuring 128 × 3, followed by the patient's age feature. The layer arrangement in this prediction model is shown in Table 3. Connecting patient age feature input is done using the Concatenate layer. A dropout by a factor of 0.1 is also used in this model to prevent overfitting [16]. The output of this model at the activation layer uses a linear function because it is a regression of values.

Table 3. Details of the survival prediction model layer

Layer	Kernel size	Feature shape	
		Input	Output
InputLayer1(FeatureExtractor)	–	(None, 128, 3)	(None, 128, 3)
Conv1D	3	(None, 128, 3)	(None, 128, 16)
MaxPooling1D	–	(None, 128, 16)	(None, 64, 16)
Activation(relu)	–	(None, 64, 16)	(None, 64, 16)
BatchNormalization	–	(None, 64, 16)	(None, 64, 16)
Conv1D	3	(None, 64, 16)	(None, 64, 32)
MaxPooling1D	–	(None, 64, 32)	(None, 32, 32)
Activation(relu)	–	(None, 32, 32)	(None, 32, 32)
BatchNormalization	–	(None, 32, 32)	(None, 32, 32)
Conv1D	3	(None, 32, 32)	(None, 32, 64)
MaxPooling1D	–	(None, 32, 64)	(None, 16, 64)
Activation(relu)	–	(None, 16, 64)	(None, 16, 64)
BatchNormalization	–	(None, 16, 64)	(None, 16, 64)
Conv1D	3	(None, 16, 64)	(None, 16, 128)
MaxPooling1D	–	(None, 16, 128)	(None, 8, 128)
Activation(relu)	–	(None, 8, 128)	(None, 8, 128)
BatchNormalization	–	(None, 8, 128)	(None, 8, 128)
Flatten	–	(None, 8, 128)	(None, 1024)
Dropout(0.1)	–	(None, 1024)	(None, 1024)
Dense	–	(None, 1024)	(None, 128)
Dense	–	(None, 128)	(None, 64)
Dense	–	(None, 64)	(None, 32)
Dense	–	(None, 32)	(None, 16)
Dense	–	(None, 16)	(None, 8)
InputLayer2(Age)	–	(None, 1)	(None, 1)
Concatenate	–	(None, 8)(None, 1)	(None, 9)
Dense	–	(None, 9)	(None, 4)
Dense	–	(None, 4)	(None, 1)
Activation(Linear)	–	(None, 1)	(None, 1)

2.3   Preprocessing

The MRI image data used for both training and inference is pre-truncated to the size $160 \times 192 \times 128$ from the center point of the original image, which is $240 \times 240 \times 128$. Cropping is done from the center point of the image. The next step is to normalize the cropped image using the Eq. 1 where X_{norm} and X_{orig} are the normalized images and the original image, $\bar{X}$ is the average voxels's intensity value of the cropped image, and σ is its standard deviation.

$$X_{norm} = \frac{X_{orig} - \bar{X}}{\sigma} \qquad (1)$$

2.4   Experiment Settings

Experiments carried out include training and inference using a modified variation of MobileNet V1 and V2 and the SPM. The variations include:

1. Modified MobileNet V1 with pre-trained weight and freezing layers and the SPM
2. Modified MobileNet V1 with pre-trained weight and unfreezing layers and the SPM
3. Modified MobileNet V2 with pre-trained weight and freezing layers and the SPM
4. Modified MobileNet V2 with pre-trained weight and unfreezing layers and the SPM
5. Online validation of one of the best architectural variations.

The training dataset used for experiment number 1–4 is 369 patient MRI image data for training on MobileNet modified architecture and 118 survival data for SPM training. Each data is divided into two parts, one part for training, and the other as local validation data. For this reason, the MRI image data of patients used for training in experiments 1–4 uses 118 data related to survival data. So that from 118 data, 95 data are used as training data, and the remaining 23 data are used as local validation data.

For experiment number 5, 369 MRI image data were used for training in selected architectures from experimental architectures 1–4, which resulted in a prediction of survival with the lowest error rate. For the SPM training, 118 survival data were used. The predicted data for the online validation process were 29 validation data with MRI images without a segmentation label.

2.5   Performance Evaluation

The model is trained using the loss means squared error (MSE) function and measured using the mean absolute error (MAE) metric. This function is available in the Keras library used in this study. For an online validation of the prediction results of survival, the metrics obtained include accuracy, MSE, Median Square Error (Median SE), Standard Deviation Error (StdSE), and SpearmanR.

3   Result

Experiment number 1–4 was carried out by producing the performance as shown in the Figs. 2,3,4,5 for MSE Losses and Figs. 6,7,8,9 for MAE Metrices .

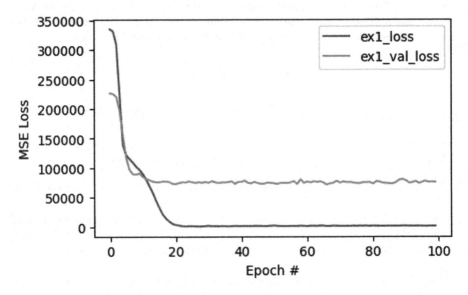

Fig. 2. MSE loss from Experiment 1

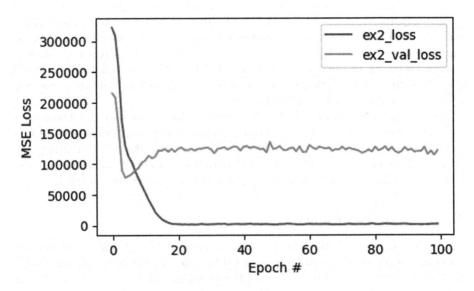

Fig. 3. MSE loss from Experiment 2

From these figures, it can be concluded that the first and third experiments gave better results than the second and fourth experiments. Furthermore, from the performance of the training results in Table 4, the two architectures are candidates for the 5th experiment. For this reason, both were trained with 369 MRI image data and 118 survival data to get training performance. At the end of the training, the training performance results are shown in Fig. 10 and Table 5

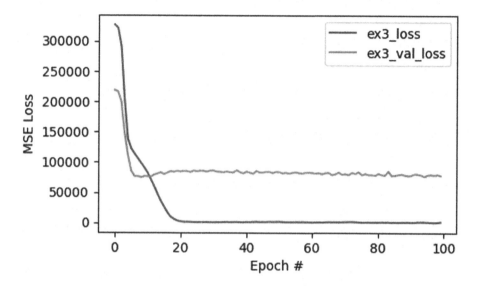

Fig. 4. MSE loss from Experiment 3

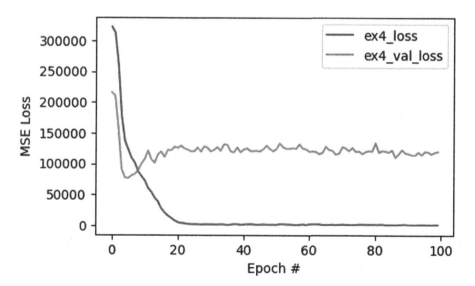

Fig. 5. MSE loss from Experiment 4

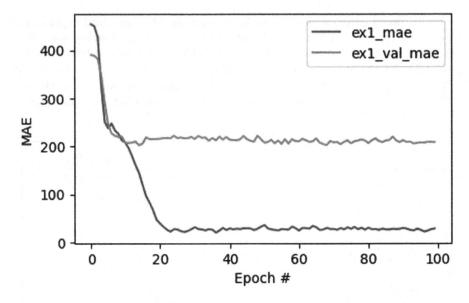

Fig. 6. MAE metric from Experiment 1

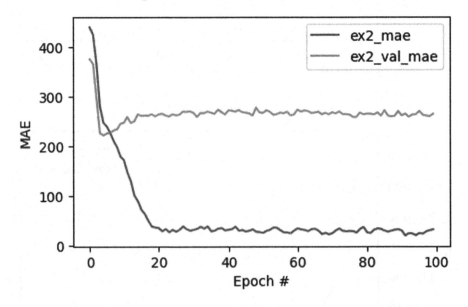

Fig. 7. MAE metric from Experiment 2

for corresponding last values. With this result, the architectural variation chosen for the 5th experiment is the 3rd variation.

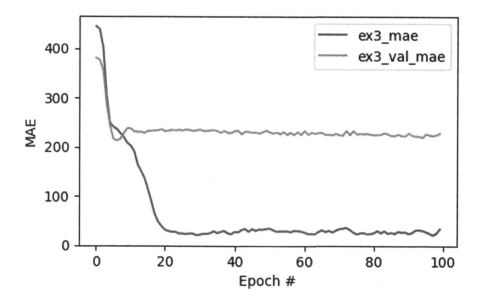

Fig. 8. MAE metric from Experiment 3

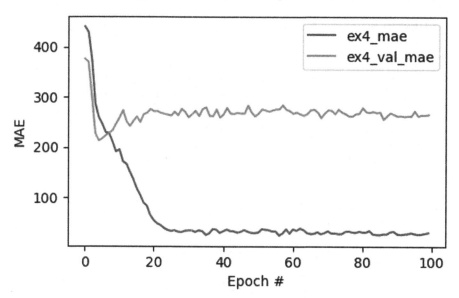

Fig. 9. MAE metric from Experiment 4

The 29 data for online validation were processed using the third experimental architecture. The results of online validation are shown in Table 6. This architecture is good enough from these results, although it still has to be developed to produce even better performance values.

Table 4. Training and local validation result

Experiment	Training		Validation	
	MSE	MAE	MSE	MAE
1	1236.90	28.20	**75644.20**	**208.12**
2	1753.93	32.47	121137.47	264.85
3	1655.59	33.83	**78374.17**	**228.16**
4	1385.52	30.46	120483.23	266.01

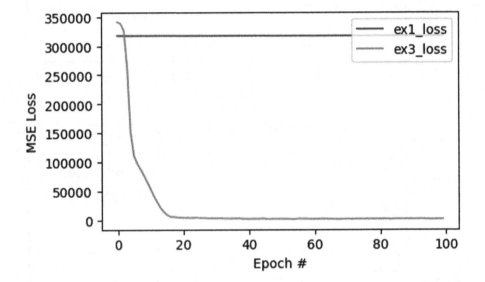

Fig. 10. MSE Loss in training two candidates architecture

Table 5. Training performance comparison

Experiment	Training	
	MSE	MAE
1	316680.45	445.50
3	**1073.13**	**25.41**

Table 6. Online validation result of validation dataset

Features	CasesEvaluated	Accuracy	MSE	MedianSE	StdSE	SpearmanR
With age	29	0.31	155272.714	66862.505	240390.717	0.036
Without age	29	0.345	149764.866	60314.09	190758.941	0.123

Based on the online validation results' performance, we used experimental model 3 with no age feature to process the testing dataset. The results of processing the testing dataset, amounting to 107 data, are sent to the online validation tool. The validation results obtained are shown in Table 7.

Table 7. Online validation result of testing dataset

NumberOfCases	Accuracy	MSE	MedianSE	StdSE	SpearmanR
107	0.402	470903.861	79348.037	1393608.343	0.086

4 Conclusion

This study proposes a variation of the MobileNet V1 and V2 architectures connected to an SPM so that it can combine the features of the MRI image and the features of the patient's age to obtain a survival prediction. Four architectural variations have been tested. Of the four architectures, the best candidates are first and third architecture. However, when trained with all the data, the first architecture becomes unable to converge, even though local validation is better than the third. This needs to be studied more deeply. On the other hand, the third one, which had an MSE loss of 78374.17 in local validation, was tested with online validation and varied with the use of the age feature, resulting in a more significant MSE loss (155272.714 for using age feature and 149764.866 for not using age feature). However, when tested with the test dataset, the MSE value is getting bigger. Nevertheless, the accuracy is better than the validation result of the validation dataset, which is 0.402, so that we need some improvements to the architecture to improve its performance.

Acknowledgements. This research was supported by the Ministry of Education and Culture and the Ministry of Research and Technology/BRIN, Indonesia. We are deeply grateful for both the Beasiswa Pendidikan Pascasarjana Dalam Negeri(BPPDN) and Penelitian Disertasi Doktor(PDD) 2020–2021 grant, which enabled this research to be carried out.

References

1. Agravat, R.R., Raval, M.S.: Prediction of overall survival of brain tumor patients. In: IEEE Region 10 Annual International Conference, Proceedings/TENCON, vol. 2019, pp. 31–35. Institute of Electrical and Electronics Engineers Inc., Oct 2019. https://doi.org/10.1109/TENCON.2019.8929497
2. Bakas, S., Akbari, H., Sotiras, A., et al.: Segmentation labels for the pre-operative scans of the tcga-gbm collection (2017). https://doi.org/10.7937/K9/TCIA.2017.KLXWJJ1Q
3. Bakas, S., Akbari, H., Sotiras, A., et al.: Segmentation labels for the pre-operative scans of the tcga-lgg collection (2017). https://doi.org/10.7937/K9/TCIA.2017.GJQ7R0EF

4. Bakas, S., Akbari, A., Sotiras, H., et al.: Advancing the cancer genome atlas glioma MRI collections with expert segmentation labels and radiomic features. Sci. Data **4**, p. 170117 (2017). https://doi.org/10.1038/sdata.2017.117, http://www.nature.com/articles/sdata2017117

5. Bakas, S., Reyes, M., Jakab, A., et al.: Identifying the Best Machine Learning Algorithms for Brain Tumor Segmentation, Progression Assessment, and Overall Survival Prediction in the BRATS Challenge. Technical report, November 2018. https://doi.org/10.17863/CAM.38755

6. Carver, E., et al.: Automatic brain tumor segmentation and overall survival prediction using machine learning algorithms. In: Crimi, A., Bakas, S., Kuijf, H., Keyvan, F., Reyes, M., van Walsum, T. (eds.) BrainLes 2018. LNCS, vol. 11384, pp. 406–418. Springer, Cham (2019). https://doi.org/10.1007/978-3-030-11726-9_36

7. Chato, L., Latifi, S.: Machine learning and deep learning techniques to predict overall survival of brain tumor patients using MRI images. In: Proceedings - 2017 IEEE 17th International Conference on Bioinformatics and Bioengineering, BIBE 2017, vol. 2018, no. 1, pp. 9–14. Institute of Electrical and Electronics Engineers Inc., July 2017. https://doi.org/10.1109/BIBE.2017.00-86

8. Dai, L., Li, T., Shu, H., Zhong, L., Shen, H., Zhu, H.: Automatic brain tumor segmentation with domain adaptation. In: Crimi, A., Bakas, S., Kuijf, H., Keyvan, F., Reyes, M., van Walsum, T. (eds.) BrainLes 2018. LNCS, vol. 11384, pp. 380–392. Springer, Cham (2019). https://doi.org/10.1007/978-3-030-11726-9_34

9. Feng, X., Tustison, N., Meyer, C.: Brain tumor segmentation using an ensemble of 3D U-Nets and overall survival prediction using radiomic features. In: Crimi, A., Bakas, S., Kuijf, H., Keyvan, F., Reyes, M., van Walsum, T. (eds.) BrainLes 2018. LNCS, vol. 11384, pp. 279–288. Springer, Cham (2019). https://doi.org/10.1007/978-3-030-11726-9_25

10. Howard, A.G., Zhu, M., Chen, B., et al.: MobileNets: efficient convolutional neural networks for mobile vision applications, April 2017. http://arxiv.org/abs/1704.04861

11. Lefkovits, S., Szilágyi, L., Lefkovits, L.: Brain tumor segmentation and survival prediction using a cascade of random forests. In: Crimi, A., Bakas, S., Kuijf, H., Keyvan, F., Reyes, M., van Walsum, T. (eds.) BrainLes 2018. LNCS, vol. 11384, pp. 334–345. Springer, Cham (2019). https://doi.org/10.1007/978-3-030-11726-9_30

12. Louis, D.N., Perry, A., Reifenberger, G., et al.: The 2016 world health organization classification of tumors of the central nervous system: a summary. Acta Neuropathol. **131**(6), 803–820 (2016). https://doi.org/10.1007/s00401-016-1545-1

13. Menze, B.H., Jakab, A., Bauer, S., et al.: The multimodal brain tumor image segmentation benchmark (BRATS). IEEE Trans. Med. Imaging **34**(10), 1993–2024 (2015). https://doi.org/10.1109/TMI.2014.2377694

14. Sandler, M., Howard, A., Zhu, M., et al.: MobileNetV2: inverted residuals and linear bottlenecks. In: Proceedings of the IEEE Computer Society Conference on Computer Vision and Pattern Recognition, pp. 4510–4520. IEEE Computer Society (2018). http://arxiv.org/abs/1801.04381

15. Shboul, Z.A., Alam, M., Vidyaratne, L., Pei, L., Iftekharuddin, K.M.: Glioblastoma survival prediction. In: Crimi, A., Bakas, S., Kuijf, H., Keyvan, F., Reyes, M., van Walsum, T. (eds.) BrainLes 2018. LNCS, vol. 11384, pp. 508–515. Springer, Cham (2019). https://doi.org/10.1007/978-3-030-11726-9_45

16. Srivastava, N., Hinton, G., Krizhevsky, A., et al.: Dropout: a simple way to prevent neural networks from overfitting. J. Mach. Learn. Res. **15**(1), 1929–1958 (2014)
17. Yang, H.-Y., Yang, J.: Automatic brain tumor segmentation with contour aware residual network and adversarial training. In: Crimi, A., Bakas, S., Kuijf, H., Keyvan, F., Reyes, M., van Walsum, T. (eds.) BrainLes 2018. LNCS, vol. 11384, pp. 267–278. Springer, Cham (2019). https://doi.org/10.1007/978-3-030-11726-9_24

Memory Efficient 3D U-Net with Reversible Mobile Inverted Bottlenecks for Brain Tumor Segmentation

Mihir Pendse[(⊠)], Vithursan Thangarasa[(⊠)], Vitaliy Chiley, Ryan Holmdahl,
Joel Hestness[(⊠)], and Dennis DeCoste

Cerebras Systems, Los Altos, CA 94022, USA
`mihir@cerebras.net`

Abstract. We propose combining memory saving techniques with traditional U-Net architectures to increase the complexity of the models on the Brain Tumor Segmentation (BraTS) challenge. The BraTS challenge consists of a 3D segmentation of a $240 \times 240 \times 155 \times 4$ input image into a set of tumor classes. Because of the large volume and need for 3D convolutional layers, this task is very memory intensive. To address this, prior approaches use smaller cropped images while constraining the model's depth and width. Our 3D U-Net uses a reversible version of the mobile inverted bottleneck block defined in MobileNetV2, MnasNet and the more recent EfficientNet architectures to save activation memory during training. Using reversible layers enables the model to recompute input activations given the outputs of that layer, saving memory by eliminating the need to store activations during the forward pass. The inverted residual bottleneck block uses lightweight depthwise separable convolutions to reduce computation by decomposing convolutions into a pointwise convolution and a depthwise convolution. Further, this block inverts traditional bottleneck blocks by placing an intermediate expansion layer between the input and output linear 1×1 convolution, reducing the total number of channels. Given a fixed memory budget, with these memory saving techniques, we are able to train image volumes up to 3x larger, models with 25% more depth, or models with up to 2x the number of channels than a corresponding non-reversible network.

Keywords: Depthwise separable convolution · Inverted residual · Reversible network

1 Introduction

Gliomas are a type of tumor affecting the glial cells that support the neurons in the central nervous system including the brain [8]. Gliomas are associated with

---

M. Pendse and V. Thangarasa—Equal contribution.

© Springer Nature Switzerland AG 2021
A. Crimi and S. Bakas (Eds.): BrainLes 2020, LNCS 12659, pp. 388–397, 2021.
https://doi.org/10.1007/978-3-030-72087-2_34

hypoxia which causes them to invade and deprive healthy tissue of oxygen leading to necrosis. This can result in a range of symptoms including headaches, nausea, and vision loss. Brain gliomas are typically categorized into low grade glioma and high grade glioma based on their size and rate of growth with high grade gliomas having a much poorer prognosis and higher likelihood of recurrence after treatment. Diagnosing and treating gliomas early before they become serious is essential for improving the prognosis of the disease.

Magnetic resonance imaging (MRI) is one of the most commonly used imaging techniques used to identify neurological abnormalities including brain gliomas [5]. One of the strengths of MRI is the ability to measure several different properties of tissue by adjusting the settings of the scan, namely the echo time and repetition time. For example, a scan with a short echo time and a short repetition time will result in a T1-weighted image that is sensitive to a property of tissues called spin-lattice relaxation which can help to differentiate between white and grey matter. A scan with longer echo and repetition times will result in a T2-weighted image that is sensitive to the spin-spin relaxation property of tissues and can be used to highlight the presence of fat and water. Another type of image called fluid attentuated inversion recovery (FLAIR) can be obtained by applying an inversion radiofrequency pulse that has the effect of nulling the signal from water making it easier to visualize lesions near the periphery of ventricles. Addition- ally, it is possible to inject a paramagnetic constrast agent such as gadolinium into the blood stream prior to the scan which will amplify signal from blood and make the vessels easier to visualize. Typically, for diagnosis of gliomas an MRI exam consists of T1-weighted, T1-weighted with gadolinium, T2-weighted, and FLAIR scans.

Based on an MRI exam, it is possible to identify four regions associated with the glioma [1]. At the center of the tumor is the region that is most affected and consists of a fluid filled necrotic core that is associated with high grade gliomas. The necrotic core is surrounded by a region called the enhancing region that hasn't undergone necrosis but still exhibits enhanced signal in T1-weighted images. Surrounding the enhancing region, is a region of the tumor that has reduced signal on T2-weighted images and is thus called the non-enhancing region. The core tumor consisting of the enhancing and non-enhancing regions is surrounded by peritumoral edematous tissue that is characterized by hyperintense signal on T2-weighted images and hypointense signal on T1-weighted images. Because the necrotic core is difficult to distinguish from the surrounding enhancing region the two can be grouped into the same class.

Because of the heterogeneous nature of the composition and morphology of gliomas segmentation of these tumors on the MRI is time consuming even for experienced radiologists [17]. For a human, the task of tracing an outline of various tumor classes on an imaging volume is limited by the two dimensional nature of human vision which requires iterating through several 2D slices in order view the entire volume. Furthermore, manual segmentation is subject variability between different radiologists and even between the same radiologist across multiple attempts. Computer vision algorithms could potentially help reduce the time needed for segmentation while also improving accuracy and reducing variance.

Convolutional neural networks have been shown to be a powerful class of machine learning models for extracting features from images to perform tasks such as classification, detection, and segmentation. For segmentation, the feature extraction stage of the network (encoder) is followed by a decoder that outputs a score for each output class. One of the first models proposed for segmentation called the Fully Convolutional Network (FCN) [13] consists of 7 convolutional layers in the encoder each of which reduces the image size while increasing the number of features followed by a single deconvolutional layer consisting of either transposed convolution or bilinear interpolation for upsampling. Because the FCN does not consist of any fully connected layers, it can be used with images of any size. While the FCN had strong performance on segmentation of natural images, the drawback of the FCN architecture is that the encoder layers lose local information through several layers of filtering that cannot be recovered through the single decoder layer. The U-Net [15] was shown to perform better on medical image segmentation. Instead of a single layer in the decoder, the U-Net uses the same number of layers as in the encoder resulting in a symmetric U shaped architecture. The U-Net introduced skip connections between the output of each encoder layer and the input of the corresponding decoder layer. The advantage of the skip connections is that precise local information is retained and can be used by the decoder in achieving sharp segmentation outlines. The U-Net model is very popular in biomedical image segmentation due to its ability to segment images efficiently with a very limited amount of labeled training data. In addition, several variants of U-Net models have also been successfully implemented in various kinds of computer vision applications [10,18,23].

Although, U-Net models have been used successfully for many vision tasks, they are difficult to scale to high resolution images or 3D volumetric datasets. Activation memory requirements, which scale with network depth and mini-batch size, quickly become prohibitive. Thus, one of the main challenges with 3D segmentation of high-resolution MRIs is that the large volumetric images result in a high memory footprint to store the activations at the intermediate layers of the U-Net, which in effect limits the size of the network that can be used within the memory budget of modern deep learning accelerators. One approach for addressing this limitation is to crop the image volume into patches sampled at different scales to reduce activation memory [12]. However, this strategy has limitations since it requires stitching together several cropped regions during inference which can be problematic at the border of these regions. Furthermore, cropping discards contextual information due to the lack of global context that can be used to increase the accuracy of the segmentation [11]. Another memory saving technique is to use multiple 2D slices and a less memory intensive 2D network but this also prevents full utilization of the entire context and can limit the power of the model.

2   Methods

2.1   Reversible Layers

An alternative memory saving approach that does not compromise the expressive power of the model is to use reversible layers [7,21] that reduce the memory requirements in exchange for additional computation. If certain restrictions are imposed on a residual layer, namely that the input dimensions are identical to the output dimensions, it is possible to recover the input of that layer from the output. Therefore, the input activations do not need to be stored during the forward pass and can be reconstructed on-the-fly during backward pass to compute the gradients of the weights.

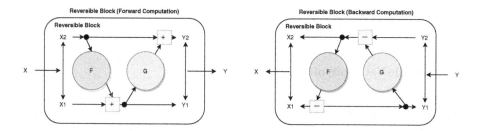

Fig. 1. The forward and backward computations of a reversible block.

The specific mechanism of a reversible layer is illustrated in Fig. 1. During the forward computation, the input to the reversible layer is split across the channel dimension into two equally sized tensors x_1 and x_2. The F and G blocks represent two identical blocks (e.g., Convolution $\rightarrow$ Normalization $\rightarrow$ Non-linear Activation). The two output tensors y_1 and y_2 can be concatenated to get a tensor with same dimensions as the input. This can be expressed with the following equations:

$$y_1 = x_1 + F(x_2) \qquad y_2 = x_2 + G(y_1). \tag{1}$$

On the backward pass, the input to the layer can be computed from the output as illustrated in Fig. 1. The gradients of the weights of F and G, as well as the reversible block's original inputs are calculated. The design of the reversible block allows to reconstruct x_1 and x_2 given only y_1 and y_2 using Eq. 2, thus making the block reversible.

$$x_2 = y_2 - G(y_1) \qquad x_1 = y_1 - F(x_2). \tag{2}$$

It has been shown that for many tasks reversible layers maintain the same expressive power and achieve the same model accuracy as traditional layers with approximately same number of parameters. Reversible layers have been combined with the U-Net architecture to achieve memory savings by replacing a portion of the blocks in both the encoder and decoder with a reversible variant [6].

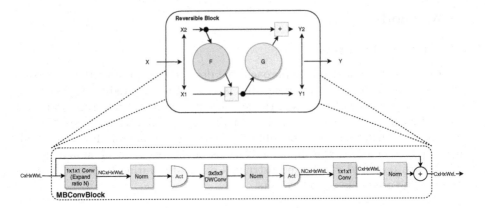

Fig. 2. The reversible MBConv block with inverted residual bottleneck and depthwise separable convolutions.

2.2   MobileNet Convolutional Block

We introduce another memory saving technique that can be combined with reversibility to achieve additional performance by replacing traditional convolutional layers found in the standard U-Net with mobile inverted bottleneck convolutional block (MBConvBlock) introduced in MobileNetV2 [16] and later used in neural architecture search (NAS) based models such as MnasNet [19] and EfficientNet [20]. The MBConvBlock consists of two important features. The components of this block are shown in Fig. 2. It replaces standard convolutions with depthwise separable convolutions consisting of a depthwise convolution (in which each input channel is convolved with a single convolutional kernel producing an output with same number of channels as the input) followed by a pointwise $1 \times 1 \times 1$ convolution (where for each voxel a weighted sum of the input channels is computed to get the value of the corresponding voxel in the output channel). In the case of separable convolutions such as the Sobel filter for edge detection, it is possible to find values for the kernels of the depthwise and pointwise convolutions that make it mathematically identical to a standard convolution. More generally, even when the kernel of standard convolution is not separable, the loss in accuracy with a depthwise separable convolution is minimal and compensated for with reduction in total amount of computation [20].

The second important feature of the MBConvBlock is the inverted residual with linear bottleneck block. In a conventional bottleneck block found in residual architectures such as ResNet-50 [9], the input to the block has a large number of channels and undergoes dimensionality reduction from convolutional layers with reduced number of channels before the final convolutional layer restores the original dimensionality. In an inverted residual block, the input has low dimensionality but the first convolutional layer consists of a pointwise convolution that results in expansion to a higher number of channels where the increase in dimensionality is given by a parameter called the expand ratio. This is followed by a depthwise separable convolution with the depthwise convolution occurring in the high dimensional space and the subsequent pointwise convolution projecting back into the lower dimensional space. This inverted bottleneck results in fewer number of parameters than a standard bottleneck block but also reduces the representational capacity of the network. To compensate for this, the nonlinear ReLU activation after the final convolutional layer is eliminated which was shown to improve accuracy in [16].

2.3   Architecture

Our architecture (Fig. 3) consists of a U-Net with multiple levels of contraction in the encoder (through $2 \times 2 \times 2$ max pooling) and the same number of levels of expansion in the decoder (through trilinear interpolation for upsampling instead of transposed convolutions as was shown to be preferable in [6]). Each level consists of two convolutional blocks. In the encoder, the first block is a pointwise convolution that increases the number of channels and the second block is a reversible block where each of the components (F and G in Fig. 1) is a MBConvBlock with half the number of channels. We use additive instead of concatenated skip connections as in [6]. Because this memory intensive task requires using a batch size of 1, we use group normalization [22] after the convolution instead of batch normalization.

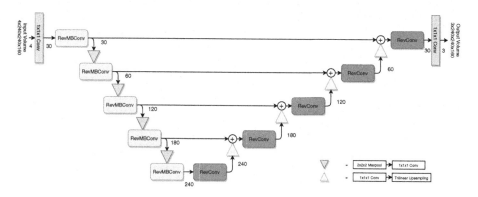

Fig. 3. Our reversible U-Net architecture with MBConv blocks in the encoder and regular convolutional blocks in the decoder. The downsampling and upsampling stages are depicted by red and yellow arrows, respectively. (Color figure online)

2.4   Training Procedure

Training was done using Nvidia V100 GPUs for 500 epochs with initial learning rate of 0.0001 and learning rate drop by 5x at epoch 250 and 400. To speed up training, mixed precision and data parallel training with 4 GPUs (effective batch size of 4) was used resulting in a net speedup of about 5x compared to single GPU full precision training.

Dataset: The provided BraTS [1–4,14] training dataset consists of 370 total examples each consisting of an MRI exam with 4 240 × 240 × 155 images (T1-weighted, Gadolinium enhanced T1-weighted, T2-weighted, and FLAIR) and a ground truth segmentation map grouping each voxel into one of four categories. We split this dataset into 330 examples for training and keep the remaining 40 examples as the hold out set for validation.

Augmentation: Because of the limited amount of data, we make extensive use of data augmentation to prevent overfitting. The augmentation applied includes the following: random rotation of the volume along the longitudinal axis by a random value between −20 and +20°, random scaling up or down (resizing) of the image by at most 10%, random flipping about each axis, randomly increasing or decreasing the intensity of the image by at most 10%, and random elastic deformation.

3   Experiments and Results

We compare four types of reversible U-Net architectures each with a constant 14 GB of memory usage. The baseline consists of standard convolutional blocks for F and G in the reversible layers of the encoder. In the MBConv variants, F and G in the reversible layers of the encoder are replaced with the MBConv block. To make use of the additional memory, we explore using the full image volume (MBConv-Base), using a deeper model with cropped images (MBConv-Deeper), and a wider model with cropped images (MBConv-Wider) (Table 1).

Table 1. Summary of experiments.

Experiment name	Conv block	Image size	Channels	Expand ratio
Baseline	Standard	256 × 256 × 160	60, 120, 180, 240, 480	NA
MBConv-Base	MB	256 × 256 × 160	30, 60, 120, 180, 240	2
MBConv-Deeper	MB	128 × 128 × 128	30, 60, 120, 180, 240, 480	2
MBConv-Wider	MB	128 × 128 × 128	30, 60, 120, 180, 240	8

As seen in Table 2, our best MBConv reversible architecture was found to be the MBConv-Base variant which achieves a mean Dice score (averaged over all classes) above 0.7317 on hold out set after 50 epochs of training and Dice score of 0.7513 after convergence. The rate of convergence is faster than the baseline which only reaches a Dice score of 0.7184 after 50 epochs of training although the final score after convergence is slightly higher (0.7513). In Fig. 4, a sample segmentation for an example from the training set and an example from the holdout set indicate a close match between the prediction and the ground truth.

After identifying that the MBConv-Base variant performed the best, we trained three different models of this architecture to convergence using different initializations. We used following procedure to ensemble the three models to make the final prediction on the validation and test sets. For each image in the test set, a histogram of the pixel values was computed and the chisquared distance was computed with the histogram of each image in the training set. A weighted sum was computed across the training set for each model where the Dice score on each image was weighted by the chisquared distance of that image to the test image. The model with the lowest weighted sum was used to make the prediction for that particular test image.

Table 2. Experimental results.

Experiment name	Dice score after 50 epochs	Dice score after convergence
Baseline	0.7184	0.7513
MBConv-Base	0.7317	0.7501
MBConv-Deeper	0.7129	0.7483
MBConv-Wider	0.7092	0.7499

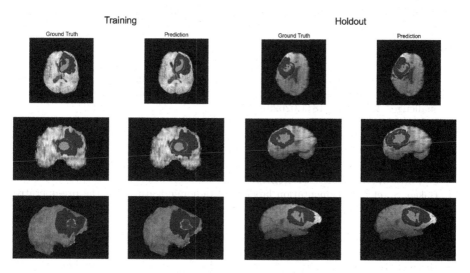

Fig. 4. Segmentation result for subject ID BraTS20_Training_210 (left) from training set (Dice_ET = 0.88, Dice_WT = 0.93, DICE_TC = 0.92) and subject ID BraTS20_Training_360 (right) from holdout set (Dice_ET = 0.92, Dice_WT = 0.91, Dice_TC = 0.95). Blue = whole tumor (WT), red = enhancing tumor (ET), green = tumor core (TC). (Color figure online)

4   Discussion

We demonstrated the benefits of replacing a standard convolutional block with a MobileNet inverted residual with linear bottlneck block inside the reversible block of the encoder. This more parameter efficient MBConvBlock results in faster convergence while still fitting in a 16 GB GPU. For the same computational budget, the MBConvBlock gives more expressive power by replacing a single convolution with multiple convolutions in the form of a bottleneck block which has shown to improve accuracy on image classification tasks with architectures such as ResNet-50. When comparing the Dice score for an equal number of training steps, the MBConv-Basic variant is higher than the baseline. This is despite the fact that hyperparameters were tuned on the baseline model and the same values were used on the MB-Conv variant without further tuning. A significant drawback however is that the depthwise separable convolutions that are the dominant computation in the MB-Conv Block are slow on GPU. This is because standard convolutions are optimized to make use of the reuse of a convolutional kernel's weights on different inputs whereas in the depthwise separable convolutions does not have this optimization since each convolutional kernel is only applied to a single input. Therefore even though the MB-Conv block has fewer FLOPs than the standard one it is slower and results in longer wall clock time for each epoch. The fact that fewer epochs were needed for convergence suggests that the MB-Conv architecture is powerful and motivates optimizations to hardware that make depthwise separable convolutions efficient.

References

1. Bakas, S., et al.: Identifying the best machine learning algorithms for brain tumor segmentation, progression assessment, and overall survival prediction in the brats challenge. ArXiv abs/1811.02629 (2018)
2. Bakas, S., et al.: Advancing the cancer genome atlas glioma MRI collections with expert segmentation labels and radiomic features. Sci. Data 4, 1–13 (2017)
3. Bakas, S., et al.: Segmentation labels and radiomic features for the pre-operative scans of the TCGA-GBM collection. Cancer Imaging Arch. (2017). https://doi.org/10.7937/K9/TCIA.2017.KLXWJJ1Q
4. Bakas, S., et al.: Segmentation labels and radiomic features for the pre-operative scans of the TCGA-LGG collection. Cancer Imaging Arch. (2017). https://doi.org/10.7937/K9/TCIA.2017.GJQ7R0EF
5. Bauer, S., Wiest, R., Nolte, L.P., Reyes, M.: A survey of MRI-based medical image analysis for brain tumor studies. Phys. Med. Biol. 58, R97–R129 (2013)
6. Brügger, R., Baumgartner, C.F., Konukoglu, E.: A partially reversible U-net for memory-efficient volumetric image segmentation. In: Shen, D. (ed.) MICCAI 2019. LNCS, vol. 11766, pp. 429–437. Springer, Cham (2019). https://doi.org/10.1007/978-3-030-32248-9_48
7. Gomez, A.N., Ren, M., Urtasun, R., Grosse, R.B.: The reversible residual network: Backpropagation without storing activations. In: Advances in Neural Information Processing Systems 30, pp. 2214–2224. Curran Associates, Inc. (2017)
8. Hanif, F., Muzaffar, K., Perveen, k., Malhi, S., Simjee, S.: Glioblastoma multiforme: a review of its epidemiology and pathogenesis through clinical presentation and treatment. Asian Pac. J. Cancer Prev. 18(1), 3–9 (2017)
9. He, K., Zhang, X., Ren, S., Sun, J.: Deep residual learning for image recognition. In: 2016 IEEE Conference on Computer Vision and Pattern Recognition, CVPR, pp. 770–778 (2016)

10. Iglovikov, V., Shvets, A.: Ternausnet: U-net with vgg11 encoder pre-trained on imagenet for image segmentation (2018)
11. Isensee, F., et al.: No new-net. In: Crimi, A., van Walsum, T., Bakas, S., Keyvan, F., Reyes, M., Kuijf, H. (eds.) Brainlesion. pp. 234–244. Lecture Notes in Computer Science (including subseries Lecture Notes in Artificial Intelligence and Lecture Notes in Bioinformatics), Springer (January 2019), 4th International MICCAI Brainlesion Workshop, BrainLes 2018 held in conjunction with the Medical Image Computing for Computer Assisted Intervention Conference, MICCAI 2018; Conference date: 16–09-2018 Through 20–09-2018
12. Kamnitsas, K., et al.: Deepmedic for brain tumor segmentation. In: MICCAI Brain Lesion Workshop (October 2016)
13. Long, J., Shelhamer, E., Darrell, T.: Fully convolutional networks for semantic segmentation. In: 2015 IEEE Conference on Computer Vision and Pattern Recognition (CVPR), pp. 3431–3440 (2015)
14. Menze, B.H., Jakab, A., Bauer, S., Kalpathy-Cramer, J., Farahani, K., et al.: The multimodal brain tumor image segmentation benchmark (brats). IEEE Trans. Med. Imaging **34**(10), 1993–2024 (2015). https://doi.org/10.1109/TMI.2014.2377694
15. Ronneberger, O., Fischer, P., Brox, T.: U-net: convolutional networks for biomedical image segmentation. In: Navab, N., Hornegger, J., Wells, W.M., Frangi, A.F. (eds.) MICCAI 2015. LNCS, vol. 9351, pp. 234–241. Springer, Cham (2015). https://doi.org/10.1007/978-3-319-24574-4_28
16. Sandler, M., Howard, A.G., Zhu, M., Zhmoginov, A., Chen, L.: Mobilenetv 2: inverted residuals and linear bottlenecks. In: 2018 IEEE Conference on Computer Vision and Pattern Recognition, CVPR, pp. 4510–4520 (2018)
17. Simi, V., Joseph, J.: Segmentation of glioblastoma multiforme from MR images - a comprehensive review. Egyptian J. Radiol. Nucl. Med. **46**(4), 1105–1110 (2015)
18. Sun, T., Chen, Z., Yang, W., Wang, Y.: Stacked u-nets with multi-output for road extraction. In: 2018 IEEE/CVF Conference on Computer Vision and Pattern Recognition Workshops (CVPRW), pp. 187–1874 (2018)
19. Tan, M., et al.: MnasNet: platform-aware neural architecture search for mobile. In: 2019 IEEE Conference on Computer Vision and Pattern Recognition, CVPR, pp. 2820–2828. Computer Vision Foundation/IEEE (2019)
20. Tan, M., Le, Q.V.: EfficientNet: rethinking model scaling for convolutional neural networks. In: Proceedings of the 36th International Conference on Machine Learning, ICML, vol. 97, pp. 6105–6114 (2019)
21. Thangarasa, V., Tsai, C.Y., Taylor, G.W., Köster, U.: Reversible fixup networks for memory-efficient training. In: NeurIPS Systems for ML (SysML) Workshop (2019)
22. Wu, Y., He, K.: Group normalization. In: Proceedings of the European Conference on Computer Vision (ECCV) (2018)
23. Yao, W., Zeng, Z., Lian, C., Tang, H.: Pixel-wise regression using u-net and its application on pansharpening. Neurocomputing **312**, 364–371 (2018)

Brain Tumor Segmentation and Survival Prediction Using Patch Based Modified 3D U-Net

Bhavesh Parmar[1,2]([⊠]) and Mehul Parikh[1,2]

[1] Gujarat Technological University, Ahmedabad, India
[2] L. D. College of Engineering, Ahmedabad, India
bhparmar@ldce.ac.in

Abstract. Brain tumor segmentation is a vital clinical requirement. In recent years, the developments of the prevalence of deep learning in medical image processing have been experienced. Automated brain tumor segmentation can reduce the diagnosis time and increase the potential of clinical intervention. In this work, we have used a patch selection methodology based on modified U-Net deep learning architecture with appropriate normalization and patch selection methods for the brain tumor segmentation task in BraTS 2020 challenge. Two-phase network training was implemented with patch selection methods. The performance of our deep learning-based brain tumor segmentation approach was done on CBICA's Image Processing Portal. We achieved a Dice score of 0.795, 0.886, 0.827 in the testing phase, for the enhancing tumor, whole tumor, and tumor core respectively. The segmentation outcome with various radiomic features was used for Overall survival (OS) prediction. For OS prediction we achieved an accuracy of 0.570 for the testing phase. The algorithm can further be improved for tumor inter-class segmentation and OS prediction with various network implementation strategies. As the OS prediction results are based on segmentation, there is a scope of improvement in the segmentation and OS prediction thereby.

Keywords: Brain tumor segmentation · Deep learning · Survival prediction · Uncertainty · Medical imaging

1 Introduction

An abnormal mass of tissue with uncontrolled growth and seemingly unchecked by the control mechanism, that to normal cells in the brain is known as a brain tumor. Based on aggresiveness and malignancy, tumors are classified as benign or malignant. The later is having a high malignancy and infiltrative nature. Tumor originates from the same tissue or immediate surrounding are primary tumors,

Computational support provided by Keepsake Welding Research and Skill Development Center, Center of Excellence - Welding, L. D. College of Engineering, Ahmedabad.

A. Crimi and S. Bakas (Eds.): BrainLes 2020, LNCS 12659, pp. 398–409, 2021.
https://doi.org/10.1007/978-3-030-72087-2_35

whereas tumors arise elsewhere and migrate to other tissues are secondary or metastatic tumors. Primary tumors can be malignant or benign. Metastatic tumors are malignant and fast metastasize.

Over 120 types of brain tumors, gliomas are the most prevalent type of brain tumor. Gliomas are intra-axial brain tumor, which includes astrocytomas, ependymomas, glioblastoma multiform, medulloblastomas, and oligodendrogliomas. Based on malignancy, aggressiveness, infiltration, reoccurrence, and Necrosis prone, World Health Organization (WHO) has developed a grading system on I to IV scale. Grade I-II are Low-Grade Gliomas and grade III-IV are High-Grade Gliomas. High-Grade Gliomas can spread rapidly and have wide infiltration.

Magnetic Resonance Imaging (MRI) gives detailed pictures and preferred modality for brain tumor imaging. MRI is highly sensitive to pathologic alterations of normal parenchyma and has been an important diagnostic tool in the evaluation of intracranial tumors [1,2]. The multimodal nature of MRI can be useful in clinical diagnosis and treatment.

The clinical expert uses MRI images for the diagnosis and locating the boundaries of the tumor. Due to the large size of data volume, the task is challenging and time consuming [3]. Automated or semi-automated segmentation methods of tumor segmentation are motivated to overcome such challenges. Broadly the automated brain tumor segmentation methods are classified into; generative, and discriminative [4]. In the last few years, the development of deep learning with a convolutional neural network (CNN) has performed par excellaence for various medical image segmentation tasks [5].

Various researchers have addressed the brain tumor segmentation task at subregion like; enhancing tumor (ET), tumor core (TC) and whole tumor (WT) [6]. The size of the tumor substructures help find the tumor extent and aggression. The volume of brain tumor and its location play a vital role in overall survival (OS) prediction.

This paper is arranged as follows. Section 2 discusses various state-of-the-art methods in brain tumor segmentation and overall survival prediction. Section 3 comprises the information of the brain tumor segmentation (BraTS) dataset used in the proposed work along with methodology used for brain tumor segmentation and overall survival prediction. In Sect. 4, experimental results are presented and discussed. Finally, Sect. 5 concludes the paper, with suggestions to further improve in adopted the methodology and future projections.

2   Related Works

Kamnitsas et al. [7] obtained a brain tumor mask using averaged probability maps from integrated models derived from seven different 3D neural networks with different parameters. Anisotropic convolutional neural network based hierarchical pipeline used by Wang et al. [8] to segment the different tumor subregions. Isensee et al. [9] suggested additional residual connections on context

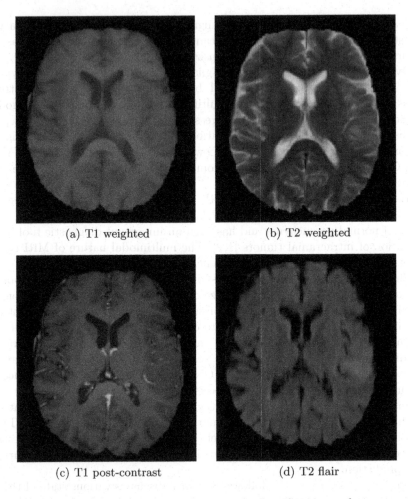

(a) T1 weighted

(b) T2 weighted

(c) T1 post-contrast

(d) T2 flair

Fig. 1. Example of image modalities in the BraTS2020 test dataset.

pathway and multi-scale aggregation on localization pathways in 3D U-Net architecture. The tumor characterized by image based features from segmentation mask and radiomic features are used for overall survival prediction. Random Forest Regressor (RFR) with 1000 trees and an ensemble of small multilayer perceptrons (MLP) are trained on these features. The produced accuracy 52.6 % on the test dataset for overall survival. Gates et al. implemented tumor segmentation using DeepMedic CNN architecture and overall survival prediction using the cox model [10]. They achieved 80%, 68%, and 67% Dice Similarity Coefficient (DSC) for WT, TC, and ET respectively. The reported OS prediction accuracy 38.2%. Liang et al. use Dense-Res-Inception Net (DRINet) for with reported accuracy of 83.47%, 73.41%, and 64.98% DSC for WT, TC and ET respectively [11]. Fully convolution neural network (FCNN) architecture is used for tumor segmentation and the derived features are fed to SVM linear kernel for OS prediction [12]. They reported OS prediction accuracy 60%. In an attempt by Kao PY et al., the author implemented an ensemble using 19 variations of DeepMedic and 7 variations of 3D U-net. The amalgamation of features like morphological, tractographical, volumetric, spatial, and age are used to train the SVM classifier. They achieved 63% accuracy in OS prediction [13]. IN another attempt at OS prediction, Agravat R. et al. used 2D U-Net architecture for brain tumor segmentation. The derived feature from segmentation outcome along with age feature is considered for OS prediction with 51.7% accuracy [14].

All methods in available literature use large network parameters or more features in training the segmentation model. U-Net based segmentation architecture gives good segmentation results in medical images, as suggested by the literature. The paper used U-Net like architecture as proposed by [15] with modifications.

The proposed work shrinks network depth to lessen network parameters. For the brain tumor segmentation task, we propose a methodology to modified U-Net as suggested by Isensee et al. [9] and patch-based 3D U-Nets adjusted from [15] with different parameters and different training strategies to get a robust brain tumor segmentation from multi-modal structural MR images. The volumetric and shape features are explored from the segmentation results. Along with these radiomic features, clinical (age) feature is used to train random forest classifier for OS prediction.

3   Methods

3.1   Datasets Description

Multimodal Brain Tumor Segmentation (BraTS) Challenge 2020 [16] provides good multimodal MRI scans of subjects with gliomas. The routine clinically acquired multimodal MRI, with tumor segmentation ground truth labels annotations by expert neuroradiologists, with pathological determined diagnosis and available OS, are provided for training and validation. BraTS multimodal scans provide a) native (T1) and b) post-contrast T1-weighted (T1ce), c) T2-weighted (T2), and d) T2 Fluid Attenuated Inversion Recovery (T2-FLAIR) volumes, and were acquired from multiple institutions with non-identical clinical protocols and distinct scanners from [5,6,17]. The annotations defined by the non-enhancing tumor core and necrotic (NET/NCR - label 1), the peritumoral edema (ED - label 2), and the GD-enhancing tumor (ET - label 4) [5,18,19]. The OS, defined in days, with corresponding subject images, includes the subject age and the resection status [6].

3.2   Preprocessing

Magnetic field intensity variation in multiple scanner, institute, and protocols causes bias in MRI scans. The structural MRI bias field was corrected with

N4ITK Insight Toolkit [20] to reduce bias. Z-score normalization with min-max scaling is performed on each modality. Mean value μ and the standard deviation σ are estimated on all training images by accumulating the voxel values inside the brain. For the given original image $\boldsymbol{P} \in \mathbb{R}^{240 \times 240 \times 155}$;

$$\hat{P}_{ijk} = \begin{cases} 100 \times \left(\dfrac{\hat{P}_{ijk} - \hat{P}_{\min}}{\hat{P}_{\max} - \hat{P}_{\min}} + 0.1 \right) & \text{if } P_{ijk} \neq 0 \\ 0 & \text{else,} \end{cases} \quad (1)$$

$$\tilde{P}_{ijk} = \begin{cases} (P_{ijk} - \mu)/\sigma & \text{if } P_{ijk} \neq 0 \\ 0 & \text{else,} \end{cases} \quad (2)$$

in which the minimum and maximum values for all the $\hat{P}_{ijk}$ with corresponding $P_{ijk} \neq 0$, $i, j \in [0, 239]$, $k \in [0, 154]$ indicated by $\hat{P}_{\min}$ and $\hat{P}_{\max}$. Given a voxel value P_{ijk} from the raw MRI, $\hat{P}_{ijk}$ is calculated for each voxel. Mean value μ, the standard deviation σ, $\hat{P}_{\min}$ and $\hat{P}_{\max}$ values are estimated from the claculated $\hat{P}_{ijk}$ values of all the accumulated voxels values of image under consideration as indicated in Eq. 1. Normalized values of each voxel $\tilde{P}_{ijk}$ is calculated by earlier estimated values of $\hat{P}_{ijk}$, $\hat{P}_{\min}$ and $\hat{P}_{\max}$ using Eq. 2. The final value of each voxel pertaining to brain would range from 10 to 110, differentiating the background voxels with value 0. Same set of values, derived in training phase for μ and σ are considered for validation and testing phase also in the z-score normalization process.

3.3 Patch Selection

Larger batch size could increase the optimization but leads to smaller patches that give less contextual information. On the contrary, a larger patch could give more contextual information but leads to the smaller batch size, which increases the variance of stochastic gradient and decreases optimization. We take advantage of both patch sizes by constructing a training batch pool with batches of different patch sizes as preset for this work. The model can learn global information from the largest a patch and informative texture from the small patch with the same parameter, by using the different numbers of cropping and padding layers between the convolution layers. Further, patching strategy would make it possible for a less powerful GPU to deal with a large image. In this work we applied two types of patch selection strategies, both patch selection methods are based on the cuboid boundary of the brain with a patch size is $128 \times 128 \times 128$. In the first patch selection method, a cubic patch initiated with a random starting point of origin, distance between 0 to 6 voxels away from each of the border. For the next patch, patch windows will move 96 voxels and keeping 32 voxels overlap with the neighboring patches. This patching method will also generate the background values. In the second patch selection method, each patch starts from one of the corner of quadrilateral boundary of image. Next patch will be selected in the similar manner to earlier method. In the second patch selection method all the patches arranged to the wide extent inside the brain.

3.4   Segmentation Pipeline and Network Structure

We have modified the U-Net framework suggested by Isensee et al. [9]. The input to the network is $4 \times 128 \times 128 \times 128$ matrices is a stack of 4 modalities input patch of size $128 \times 128 \times 128$. The downblock reduce the patch size and increase the channel length. The convolution block follows instant normalization and leaky ReLU [21]. To eliminate overfitting, downblock pass information from front to end. The upblock reconstructs the location information by joining corresponding downblock outputs. The patch size and number of channels are recovered with a probability matrix with the confidence of each voxel belongs to a particular sub-region of the tumor.

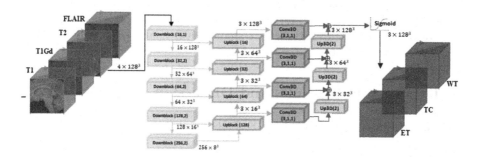

Fig. 2. Modified U-net structure for tumor segmentation

Loss Function. Weighted multi-class Dice loss function has been used to address the class imbalance problem. The weighted multi-class Dice loss function have been uesd effectively in previous methodologies for deep learning based brain tumor segmentation in earlier BraTS [6].

$$\mathcal{L} = -\sum_{c=0}^{2} \frac{\sum_{i,j,k=0}^{127} A_{cijk} B_{cijk}}{\sum_{i,j,k=0}^{127} A_{cijk} + \sum_{i,j,k=0}^{127} B_{cijk}}, \tag{3}$$

In the Eq. 3, c represent the channel number, A is the matrix of the ground truth image of size $3 \times 128 \times 128 \times 128$. Output of the proposed network is B. The total output channels are 3 $c \in [0,2]$ and voxel location values $i, j, k \in [0, 127]$.

All output channels of patch size $128 \times 128 \times 128$ are combined to make one patch. The values of each voxels in that patch will be one of the tumor sub-region or background. This can be achieved by setting up a suitable threshold values for each class. To simplify the thresholding approch in line with the voxel values and loss function, decreasing order priorities are assiged for each tumor sub-region as ET, TC, and, WT. At last, all patches are concatenated to form a brain size patch with sub-region labels to earlier patch boundary, where maximum values are taken in case of overlapping. This can be further aim to recover the original image size $240 \times 240 \times 155$ with priory knowledge.

3.5 Overall Survival Prediction

Features are extracted from the whole brain and tumor substructures for training using a random forest classifier.

Tumor segmentation is used to extract various volumetric and shape features. The volume of the tumor with respect to the whole brain, and the volume of tumor sub-structures, WT, ET, and TC are considered for volumetric features. Major axis dimension, minor axis dimension, elongation, maximum 2d diameter, maximum 3D diameter mess volume, spherical index, flatness for the whole brain, and tumor sub-structure are extracted to find interrelationship as shape features. Clinical feature (age) is given for corresponding segmentation labels.

Five volumetric features, thirty-two shape features, and age features are used to train the multi-fold random forest classifier.

4 Experimental Results and Discussion

4.1 Data and Implementation Details

BraTS2020 [16] contains 369 cases with four MRI sequences named, T1, T1ce, T2, and Flair, with manual segmentation. Each MRI sequences are normalized in 0 to 1 range of voxel value before feeding for training. We also applied a random axis mirror along the horizontal axis for augmentation during training. We have implemented this work using python and Keras with tensorflow on NVIDIA Quadro P5000 GPU for training and validation of our deep learning CNN model and random forest training. The modified U-Net is trained on 295 (80%) data (divided in training (280 images) and validation (15 images) set) and tested on 74 (20%) data.

We have trained the network with the proposed patching methods phase wise. In first phase, we have trained the network for 200 epochs, followed by second phase with 500 epochs. The saved trained model of first phase is provided as pre-trained input to the second phase in network. Initial learning rate of Adam optimizer was set to 5×10^{-4}, weight decay 5×10^{-7}, and batch size 4.

4.2 Task 1: Brain Tumor Segmentation in MRI Scans

The evaluation of brain tumor segmentation is done by CBICA's Image Processing Portal, for the multiple submission for both tasks. Though we have done multiple submissions, the result from best training model are not included for training and validation phases. Here, we are presenting the quantitative results as evaluated by organizer from our submission for training phase in Table 1, validation phase in Table 2, and testing phase in Table 3. The results obtained from CBICA's Image Processing Portal in the training, validation, and testing phase shows consistency. Looking at the table, the results seem to be consistent between all datasets, even though the training set provides slightly better results. This might be related to the fact that the presented results are based on limited epoch in training and smaller batch size. Further, testing datasets are likely to have higher variability.

Table 1. Result obtained from the evaluation on CBICA's Image Processing Portal for training phase indicating Mean (Standard Deviation) values for all 369 subjects.

Subregion	Dice	Sensitivity	Specificity	Hausdorff95
ET	0.81500 (0.2346)	0.82725 (0.2380)	0.99976 (0.0003)	24.74498 (88.0060)
WT	0.92869 (0.0387)	0.93327 (0.0601)	0.99922 (0.0007)	03.63437 (04.3263)
TC	0.90759 (0.0954)	0.91444 (0.0930)	0.99958 (0.0008)	03.16978 (04.2244)

Table 2. Result obtained from the evaluation on CBICA's Image Processing Portal for validation phase indicating Mean (Standard Deviation) values for all 125 subjects.

Subregion	Dice	Sensitivity	Specificity	Hausdorff95
ET	0.74494 (0.2917)	0.74673 (0.3087)	0.99972 (0.0004)	36.04162 (105.2888)
WT	0.90570 (0.0635)	0.92315 (0.0784)	0.99894 (0.0011)	05.08991 (07.1852)
TC	0.84668 (0.1568)	0.84191 (0.1771)	0.99957 (0.0007)	08.69096 (34.3819)

Table 3. Result obtained from the evaluation on CBICA's Image Processing Portal for testing phase indicating Mean (Standard Deviation) values for all 166 subjects.

Subregion	Dice	Sensitivity	Specificity	Hausdorff95
ET	0.79495 (0.2158)	0.82885 (0.2324)	0.99964 (0.0004)	18.04116 (74.8080)
WT	0.88645 (0.1119)	0.91949 (0.11023)	0.99895 (0.0011)	04.91966 (05.9886)
TC	0.82719 (0.2571)	0.84984 (0.2383)	0.99960 (0.0008)	021.98884 (79.5659)

Looking at the differences between the mean results of subregion segmentation scores for datasets, we observe that the enhanced tumor segmentation performed inferior. That also suggests, that while the overall performance is comparatively good, increase patch size and improved validation can betterment the performance. The median values across the various phases for all the class indicates better performance compare to mean values. That indicates that data cleaning can enhance the results here. Further, there are some outlier cases where the performance has deteriorated. In order to show the improvement of the approach, Fig. 3 present some qualitative examples, it is observed in the figure that the overall ROI of the tumor is similar between both ground truth and our prediction. However, the modified U-net under-performed to properly differentiate between tumor subregions, specifically enhanced tumor, the training parameter improvement is capable of better delineate these subregions.

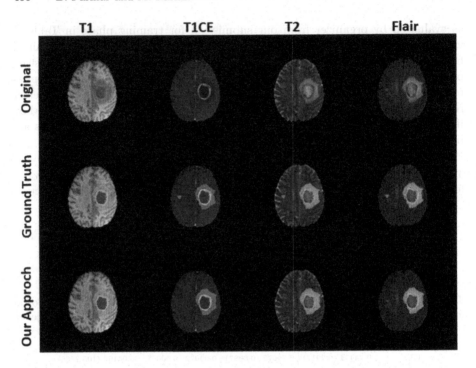

Fig. 3. Comparison of qualitative results of our approach to that of original modalities (i. e. t1, t1ct, t2, flair) and with ground truth. (Whole Tumor (WT) - Green, Tumor Core (TC) - Yellow, Enhancing Tumor (ET) - Red). (Color figure online)

4.3 Task 2: Overall Survival (OS) Prediction from Pre-operative Scans

We use subregion segmentation volumes as features along with the clinical data (age) as features for the survival task. We also include image features based on the surrounding area of the tumor. Five volumetric features, fourteen shape features, and clinical (age) features are used to train random forest classifier with multi-fold cross-validation (Table 4).

Table 4. Summary of the results obtained for evaluation from the on CBICA's Image Processing Portal in the training, validation and testing phase for the survival task.

Dataset	Cases	Accuracy	Mean SE	Median SE	Std. SE	Spearman R
Training	118	0.492	98649.25	24968	185575.446	0.392
Validation	29	0.414	101278.00	49248	132237.663	0.228
Testing	107	0.570	389943.33	50157	1182490.234	0.394

We realize from the evaluated survival prediction that the estimated values are in a narrow range of 150–400, whereas, training dataset has a wide range of 5–1767. The numbers of evaluated subjects for OS are fairly lower than that of the segmentation task. The strategies are not capable of extracting proper information in absence of a large training dataset. The network was not able to generalize, using cross-validation, even though less number of parameters has been considered. Further, normalization has less effect due to a large number of network features. By changing the strategies in training and increasing the feature set, may leads to more accuracy.

5 Conclusion

In this work, we have used a modified U-Net architecture to first segment the tumor and its subregions and then the estimation of overall survival prediction using this segmentation, clinical data, and multimodal image information from T1, T1ce, T2, and Flair MRI scans. Our results on BraTS2020 evaluated through CBICA's IPP. Brain tumor segmentation achieves a Dice score of 0.815, 0.928, 0.907 in the training phase and 0.744, 0.905, 0.846 in the validation phase for the enhancing tumor, whole tumor, and tumor core respectively. OS prediction achieved 0.492 and 0.414 accuracy during the training, and validation phase respectively. In the testing phase we achieved a Dice score of 0.795, 0.886, 0.827 for the enhancing tumor, whole tumor, and tumor core respectively, whereas OS prediction achieved 0.570 accuracy rate.

The OS prediction results are based on segmentation, there is a scope of improvement in the segmentation and OS prediction thereby. We plan to further improve our network and OS prediction strategy for the testing phase.

References

1. Leung, D., Han, X., Mikkelsen, T., Nabors, L.B.: Role of MRI in primary brain tumor evaluation. J. Natl. Compr. Canc. Netw. **12**(11), 1561–1568 (2014) https://doi.org/10.6004/jnccn.2014.0156
2. Beets-Tan, R.G.H., Beets, G.L., Vliegen, R.F.A., Kessels, A.G.H., Van Boven, H., De Bruine, A., et al.: Accuracy of magnetic resonance imaging in prediction of tumour-free resection margin in rectal cancer surgery. Lancet **357**(9255), 497–504 (2001). https://doi.org/10.1016/S0140-6736(00)04040-X
3. Bankman, I., (ed.) Handbook of Medical Image Processing and Analysis. Elsevier, Amsterdam (2008)
4. Işın, A., Direkoğlu, C., Şah, M.: Review of MRI-based brain tumor image segmentation using deep learning methods. Procedia Comput. Sci. **102**, 317–324 (2016). https://doi.org/10.1016/j.procs.2016.09.407
5. Menze, B.H., Jakab, A., Bauer, S., Kalpathy-Cramer, J., Farahani, K., Kirby, J., et al.: The multimodal brain tumor image segmentation benchmark (BRATS). IEEE Trans. Med. Imaging **34**(10), 1993–2024 (2015). https://doi.org/10.1109/TMI.2014.2377694

6. Bakas, S., Reyes, M., Jakab, A., Bauer, S., Rempfler, M., Crimi, A., et al.: Identifying the Best Machine Learning Algorithms for Brain Tumor Segmentation, Progression Assessment, and Overall Survival Prediction in the BRATS Challenge", arXiv preprint arXiv:1811.02629 (2018)

7. Kamnitsas, K., et al.: Efficient multi-scale 3D CNN with fully connected CRF for accurate brain lesion segmentation. Med. Image Anal. **36**, 61–78 (2017). [PubMed: 27865153]

8. Wang, G., Li, W., Ourselin, S., Vercauteren, T.: Automatic brain tumor segmentation using cascaded anisotropic convolutional neural networks. In: Crimi, A., Bakas, S., Kuijf, H., Menze, B., Reyes, M. (eds.) BrainLes 2017. LNCS, vol. 10670, pp. 178–190. Springer, Cham (2018). https://doi.org/10.1007/978-3-319-75238-9_16

9. Isensee, F., Kickingereder, P., Wick, W., Bendszus, M., Maier-Hein, K.H.: Brain tumor segmentation and radiomics survival prediction: Contribution to the brats 2017 challenge. In: International MICCAI Brainlesion Workshop, pp. 287–297. Quebec City, Quebec, Canada (2017). https://doi.org/10.1007/978-3-319-75238-9_25

10. Gates, E., Pauloski, J.G., Schellingerhout, D., Fuentes, D.: Glioma segmentation and a simple accurate model for overall survival prediction. In: Crimi, A., Bakas, S., Kuijf, H., Keyvan, F., Reyes, M., van Walsum, T. (eds.) BrainLes 2018. LNCS, vol. 11384, pp. 476–484. Springer, Cham (2019). https://doi.org/10.1007/978-3-030-11726-9_42

11. Liang, C., Bentley, P., Mori, K., Misawa, K., Fujiwara, M., Rueckert, D.: DRINet for medical image segmentation. IEEE Trans. Med. Imaging **37**(11), 2453–2462 (2018)

12. Varghese, A., Mohammed, S., Ganapathy, K.: Brain Tumor Segmentation from Multi Modal MR images using Fully Convolutional Neural Network. BRATS proceedings, MICCAI (2017)

13. Kao, P.-Y., Ngo, T., Zhang, A., Chen, J.W., Manjunath, B.S.: Brain tumor segmentation and Tractographic feature extraction from structural MR images for overall survival prediction. In: Crimi, A., Bakas, S., Kuijf, H., Keyvan, F., Reyes, M., van Walsum, T. (eds.) BrainLes 2018. LNCS, vol. 11384, pp. 128–141. Springer, Cham (2019). https://doi.org/10.1007/978-3-030-11726-9_12

14. Agravat, R., Raval, M.,: Prediction of overall survival of brain tumor patients. In: TENCON 2019–2019 IEEE Region 10 Conference (TENCON), Kochi, India, pp. 31–35 (2019) https://doi.org/10.1109/TENCON.2019.8929497

15. Wang, F., Jiang, R., Zheng, L., Meng, C., Biswal, B.: 3D U-net based brain tumor segmentation and survival days prediction. In: Crimi, A., Bakas, S. (eds.) BrainLes 2019. LNCS, vol. 11992, pp. 131–141. Springer, Cham (2020). https://doi.org/10.1007/978-3-030-46640-4_13

16. https://www.med.upenn.edu/cbica/brats2020/ MICCAI BRATS - The Multimodal Brain Tumor Segmentation Challenge. https://www.med.upenn.edu/cbica/brats2020/ (2020)

17. Bakas, S. et al.: advancing the cancer genome atlas glioma MRI collections with expert segmentation labels and radiomic features. Nature Sci. Data **4** 170117 (2017) https://doi.org/10.1038/sdata.2017.117

18. Bakas, S., Akbari, H., Sotiras, A., Bilello, M., Rozycki, M., Kirby, J., et al.: Segmentation labels and radiomic features for the pre-operative scans of the TCGA-GBM collection. Cancer Imaging Arch. (2017). https://doi.org/10.7937/K9/TCIA.2017.KLXWJJ1Q

19. Bakas, S., Akbari, H., Sotiras, A., Bilello, M., Rozycki, M., Kirby, J., et al.: Segmentation labels and radiomic features for the pre-operative scans of the TCGA-LGG collection. Cancer Imaging Archive (2017). https://doi.org/10.7937/K9/TCIA.2017.GJQ7R0EF
20. Tustison, N.J., et al.: N4ITK: improved N3 bias correction. IEEE Trans. Med. Imaging **29**(6), 1310–1320 (2010). https://doi.org/10.1109/TMI.2010.2046908
21. Ulyanov, D., Vedaldi, A., Lempitsky, V.: Instance normalization: The missing ingredient for fast stylization. arXiv preprint arXiv:1607.08022. (2016)

DR-Unet104 for Multimodal MRI Brain Tumor Segmentation

Jordan Colman[1,2(✉)], Lei Zhang[2(✉)], Wenting Duan[2(✉)], and Xujiong Ye[2(✉)]

[1] Ashford and St Peter's Hospitals NHS Foundation Trust, Surrey, UK
jordan.colman@nhs.net
[2] School of Computer Science, University of Lincoln, Lincoln, UK
{LZhang,wduan,XYe}@lincoln.ac.uk

Abstract. In this paper we propose a 2D deep residual Unet with 104 convolutional layers (DR-Unet104) for lesion segmentation in brain MRIs. We make multiple additions to the Unet architecture, including adding the 'bottleneck' residual block to the Unet encoder and adding dropout after each convolution block stack. We verified the effect of including the regularization of dropout with small rate (e.g. 0.2) on the architecture, and found a dropout of 0.2 improved the overall performance compared to no dropout, or a dropout of 0.5. We evaluated the proposed architecture as part of the Multimodal Brain Tumor Segmentation (BraTS) 2020 Challenge and compared our method to DeepLabV3+ with a ResNet-V2–152 backbone. We found the DR-Unet104 achieved a mean dice score coefficient of 0.8862, 0.6756 and 0.6721 for validation data, whole tumor, enhancing tumor and tumor core respectively, an overall improvement on 0.8770, 0.65242 and 0.68134 achieved by DeepLabV3+. Our method produced a final mean DSC of 0.8673, 0.7514 and 0.7983 on whole tumor, enhancing tumor and tumor core on the challenge's testing data. We produce a competitive lesion segmentation architecture, despite only using 2D convolutions, having the added benefit that it can be used on lower power computers than a 3D architecture. The source code and trained model for this work is openly available at https://github.com/jordan-colman/DR-Unet104.

Keywords: Deep learning · Brain tumor segmentation · BraTS · ResNet · Unet · Dropout

1 Introduction

Lesion segmentation is an important area of research necessary to progress the field of radiomics, using imaging to infer biomarkers, that can be used to aid prognosis prediction and treatment of patients [1]. Segmentation of gliomas, the most common form of primary brain malignancy [2], is a highly useful application of lesion segmentation. Accurate brain tumor segmentation in MRI produces useful volumetric information, and in the future may be used to derive biomarkers to grade gliomas and predict prognosis. Manual brain tumor segmentation, however, is a skilled and time-consuming task. Therefore, automated brain tumor segmentation would be of great benefit to progress

© Springer Nature Switzerland AG 2021
A. Crimi and S. Bakas (Eds.): BrainLes 2020, LNCS 12659, pp. 410–419, 2021.
https://doi.org/10.1007/978-3-030-72087-2_36

this area. However, accurate segmentation remains a challenging task due to the high variation of brain tumor size, shape, position and inconsistent intensity and contrast in various image modalities. This has contributed to the development of automatic segmentation methods, for which several methods have been proposed. The Multimodal Brain Tumor Segmentation (BraTS) challenge is an annual challenge set to act as a benchmark of the current state-of-the-art brain tumor segmentation algorithms.

In the recent years deep neural networks have achieved the top performance in the BraTS challenge. Many existing methods consider the Unet [3] as a base architecture, a basic but effective form of the encoder-decoder network design, with at least two winners of the BraTS challenge utilising a variation of the Unet in 2017 and 2019 [4, 5]. Other current state-of-the-art segmentation algorithms use the ResNet [6, 7] as an encoder in such architecture, such as the DeepLabV3+, which uses ResNet-101 as the initial encoder and spatial pyramid pooling module in the final layer [8]. The ResNet uses identity mapping, a residual connection which skips multiple network layers to aid back propagation and allows deeper networks to be made. Additionally, the ResNet uses a 'bottleneck' residual block which uses a 1×1 convolution to reduce the number of image features prior to a spatial, or 3×3 convolution, and then uses another 1×1 convolution to increase the number of features. This is done to increase computational efficiency and at the same time increase the number of image features represented in the network. The ResNet and DeeplabV3+ do not use random dropout, a commonly used regularizer, which randomly removes the signal of a given proportion of neurons in a layer in order to reduce overfitting of training data [9].

A common approach of improving the performance of current architectures in medical image segmentation is by extending 2D image segmentation to 3D, using 3D convolutional networks on whole images as opposed to a single 2D MRI slice. An example is the BraTS 2019 1st place solution for the brain tumor segmentation task, which used a two-staged cascaded 3D Unet [4]. The paper used a 3D Unet architecture with the addition of residual connections in convolutional blocks, in stacks of 1, 2 or 4. The first cascade has fewer total feature channels and detects 'coarse' features, the second cascade detects finer details as it has more total feature channels. This architecture achieved a mean dice score coefficient (DSC) of 0.8880 for whole tumor segmentation of the testing data, 0.8370 for tumor core and 0.8327 for enhancing tumor. However, this architecture required cropping 3D images to run in a batch size of 1 on a graphics card with 12Gb of memory due to the high memory usages of 3D processing and the large image size.

In this paper, given the success of the Unet in BraST challenge [4, 5] and inspired by the 'bottleneck' residual block from the ResNet [6, 7] we proposed a new architecture to couple the strengths of the 'bottleneck' residual block and the Unet for brain tumor segmentation in the BraTS 2020 challenge. The proposed network has a total of 104 convolutional layers, so is named, deep residual Unet 104 (DR-Unet104). We additionally include dropout and investigate if this improves architecture performance, as we mimic the ResNet in our encoder and it is suggested by its creators additional regularization may improve performance [7].

2 Methods

2.1 Architecture

The proposed architecture DR-Unet104 overview is shown below in Fig. 1 and 2. The detailed architecture description is shown in Fig. 3, in order to display the number of image feature channels in each layer. It comprises of three main components: encoder, decoder and bridge that forms a typical U-shape network. In this design, five stacked residual block levels with convolution layers and identify mapping are deployed in both encoder and decoder components, which are connected by the bridge component. The feature representations are encoded by the encoder path, which are recovered in the decoder path to a pixel-wise classification. Figure 2 shows the outline of the residual blocks in the encoder and decoder path. In the encoder path and bridge connection, the bottleneck design is applied, which consists of a 1 × 1 convolution layer for reducing the depth of feature channel, a 3 × 3 convolution layer, and then a 1 × 1 convolution layer for restoring dimension [6]. In the decoder path, the typical residual block consists of two stacked 3 × 3 2D convolutions. The batch normalisation (BN) and rectified linear unit (ReLU) activation are used in all residual blocks (Fig. 2), we use 'pre-activation' residual connections as used in ResNet-V2 and described in He *et al.* 2016 [7].

Given a deep architecture has many layers, the issues regarding overfitting and dead neurons in activation need to be considered for training the network. In our method, following the work [9], we employ a regularization using dropout with a small rate (e.g. dropout rate of 0.2) after each level. In our method, the input of each level (after upsampling) of the decoder is added with a concatenation connection from the output of the encoder to aid feature mapping. Downsampling is performed with a stride of 2 in the first convolution of each level (except for level 1). Upsampling is performed with 2D transposed convolution with a kernel of 2 × 2 and stride of 2. The final layer is convoluted with a 1 × 1 kernel and generating pixel-wise classification scores to represent the 3 tumor classes, and background class, the class of the pixel is decided by the channel with the largest output (argmax), softmax is used during training. The input is a 2D image slice with 4 channels representing each MRI modality. The proposed network code is publicly available at https://github.com/jordan-colman/DR-Unet104.

2.2 Loss Function

The loss function used was sparse categorical cross entropy (CE), which is calculated with Eq. (1). This was chosen for simplicity, however, better performance may have been produced using other loss functions such as 'soft Dice loss' [4].

$$CE = -\frac{1}{N}\sum_{n}^{N}\sum_{c}^{C} Y_{true}{}_{c}^{n} \times \log\left(Y_{pred}{}_{c}^{n}\right) \tag{1}$$

Where N is number of examples and C represents the classes, Y_{true} is the truth label and Y_{pred} the softmax of the prediction [10].

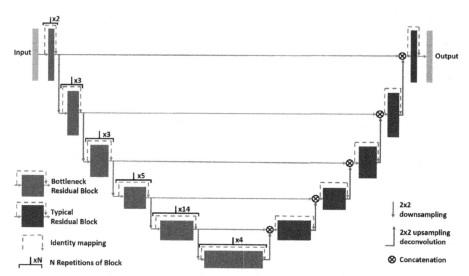

Fig. 1. Overview of the DR-Unet104 architecture showing the bottleneck residual block in the encoder and typical residual block in the decoder. The number of stacks of the bottleneck block is also shown. The downsampling is performed by a step of 2 in the initial 1×1 convolution in the first bottleneck block of the level.

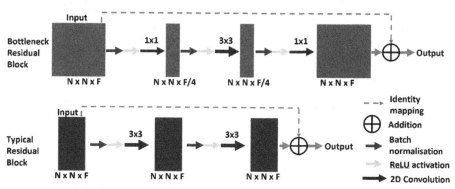

Fig. 2. The outline of the typical residual blocks in the decoder and the bottleneck residual block of the encoder path. The typical residual block can be seen to be formed of two 3×3 2D convolutions with batch normalization and rectified linear unit (Relu) activation before each convolution. The Bottleneck residual block has a 1×1 2D convolution, which reduces the number of image feature channels (F) to ¼ of the number. This is followed by a 3×3 2D convolution then a 1×1 convolution which increases the feature channel number by 4 times to the input number. Both blocks input is added to the output for the identity mapping to aid backpropagation [6, 7].

DR-Unet104

	Image dimensions	Encoder	Decoder
Input / output & Level 1	240 x 240	Input $\begin{bmatrix} 1x1, 16 \\ 3x3, 16 \\ 1x1, 64 \end{bmatrix} x2$ (Dropout 0.2)	Output Argmax $1 \times 1, n\_class$ $\begin{bmatrix} 3 \times 3, 32 \\ 3 \times 3, 32 \end{bmatrix} x1$ $\rightarrow$ $\uparrow$(Dropout 0.2)
Level 2	120 x 120	$\downarrow$ $\begin{bmatrix} 1x1, 32 \\ 3x3, 32 \\ 1x1, 128 \end{bmatrix} x3$ (Dropout 0.2)	$\begin{bmatrix} 3 \times 3, 64 \\ 3 \times 3, 64 \end{bmatrix} x1$ $\rightarrow$ $\uparrow$(Dropout 0.2)
Level 3	60 x 60	$\downarrow$ $\begin{bmatrix} 1x1, 64 \\ 3x3, 64 \\ 1x1, 256 \end{bmatrix} x3$ (Dropout 0.2)	$\begin{bmatrix} 3 \times 3, 128 \\ 3 \times 3, 128 \end{bmatrix} x1$ $\rightarrow$ $\uparrow$(Dropout 0.2)
Level 4	30 x 30	$\downarrow$ $\begin{bmatrix} 1x1, 128 \\ 3x3, 128 \\ 1x1, 512 \end{bmatrix} x5$ (Dropout 0.2)	$\begin{bmatrix} 3 \times 3, 256 \\ 3 \times 3, 256 \end{bmatrix} x1$ $\rightarrow$ $\uparrow$(Dropout 0.2)
Level 5	17 x 17	$\downarrow$ $\begin{bmatrix} 1x1, 256 \\ 3x3, 256 \\ 1x1, 1024 \end{bmatrix} x14$ (Dropout 0.2)	$\begin{bmatrix} 3 \times 3, 512 \\ 3 \times 3, 512 \end{bmatrix} x1$ $\rightarrow$ $\uparrow$(Dropout 0.2)
Bridge	8 x 8	$\downarrow$ $\begin{bmatrix} 1x1, 512 \\ 3x3, 512 \\ 1x1, 2048 \end{bmatrix} x4$	

Fig. 3. Our Proposed architecture for the DR-Unet104. The [] brackets denote the typical or bottleneck residual block with 3×3, 64 representing a 2D convolution with a 3×3 kernel and 64 layers. *xN* denotes the number, N, of stacked residual blocks in that layer. $\downarrow$ denotes reduction in spatial resolution performed by a stride of 2 in the first 1×1 2D convolution of the initial bottleneck residual block of that level. $\uparrow$ denotes upsampling via the 2D transposed convolution with a kernel of 3×3 and stride of 2. $\rightarrow$ denotes a skip connection between the output of the encoder with the input of the decoder at the same level, joined via concatenation to the unsampled output of the previous level.

3 Experiment

3.1 Dataset, Pre-processing and Data Augmentation

The data used for training and evaluation of our model consisted of the BraTS 2020 dataset including training data with 369 subjects, validation data with 125 subjects and testing data with 166 subjects, all contained low and high grade gliomas [11–15]. Each subject has a T2 weighted FLAIR, T1 weighted, T1 weighted post contrast, and a T2 weighted MRI. Each image is interpolated to a $1 \times 1 \times 1$ mm voxel sized giving images sized $240 \times 240 \times 155$ voxels. All subject images are aligned into a common space and skull stripped prior to data sharing. The training data additionally contains manually drawn tumor segmentation mask, annotated by experts and checked by a neuroradiologist, the segmentation is labelled with numbers denoting edema, tumor core and enhancing tumor. Figure 4 shows an example of a subjects imaging modalities and tumor mask.

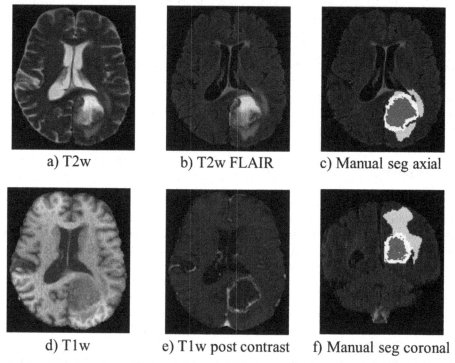

a) T2w	b) T2w FLAIR	c) Manual seg axial
d) T1w	e) T1w post contrast	f) Manual seg coronal

Fig. 4. Shows the different MRI modalities provided on one subject with a glioma, a) shows a T2 weighted MRI, b) a T2 weighted FLAIR MRI, d) a T1 weighted MRI, e) a T1 weighted MRI post contrast. c) and f) show the manual tumor segmentation in an axial and coronal slice respectively on a T2 weighted FLAIR MRI. Green is edema, red tumor core and yellow enhancing tumor.

We pre-processed the images using data standardisation on a whole 3D MRI modality wise basis, normalised using Eq. (2) where v is a given voxel and the *mean* and standard deviation (*SD*) are of the image of voxels > 0. This is so the image was rescaled prior to conversion to 2D image.

$$v = \begin{cases} 254 & v > mean + 3*SD \\ \frac{(v-mean)+127}{3*SD/128} & otherwise \\ 0 & v < mean - 3*SD \end{cases} \qquad (2)$$

Each slice of the image is then saved as individual png image, with each channel representing an MRI modality. The only image augmentation applied was randomly flipping the images in the left-right and anterior-posterior orientation with a 50% probability.

3.2 Setup and Training

For training, we used ADAM optimiser [16] with a learning rate of 1e-4 and He initialization [17]. We used a batch sizes of 10 and ran the training for 50 epochs. All subjects' images were converted to single RGBA images for each slice as described in the methods and inputted to the model in a random order. We evaluated our method with varying dropout rates and compared our proposed architecture to DeeplabV3+ as the architecture is openly available online and a current state-of-the-art segmentation architecture. We use a ResNet-V2–152 backbone for the BraTS evaluation and not the originally suggested ResNet-V1–101 in order to improve the performance and make the architecture more analogous to our proposed method [6]. The architectures are implemented in Keras with a tensorflow backend [18, 19]. This was performed using a NVIDIA GTX 2080Ti graphics card with 12 GB of memory.

4 Results

We initially evaluated our proposed DR-Unet104 architecture on the 125 validation subjects, pre-processed in the same way as the training data, however, without data augmentation. The resulting 2D masks are reconstructed into 3D nii images and evaluated on the IPP website [https://ipp.cbica.upenn.edu/], which computed and evaluated the lesion masks outputting the metrics, DSC, sensitivity, specificity and Hausdorff distance 95% (HD95). We evaluated our proposed architecture with varying random dropout rate after each level, the rate we trailed being 0.2, 0.5 and no dropout. We can observe from the Table 1, the architecture applying the dropout of rate 0.2 has superior performance than the architecture with a rate of 0.5 or without dropout, with all tumor components' DSC and HD95 being greater. These observations verify that dropout in our architecture allows the network to learn informative features from the data, but only with the small rate (e.g. 0.2), as the network with a large (0.5) rate performs poor and even worse than the setting without dropout. Comparing our proposed architecture with a droupout of 0.2 to DeepLabV3+ shows improvement of segmentation results in all areas except the mean tumor core DSC.

Table 1. Table showing the mean results of the Dice Score coefficient (DSC) and Hausdorff distance 95% (HD95) with the Standard deviation (SD) in brackets results on the BraTS20 validation data. We show results for the whole tumor (WT), enhancing tumor (ET) and tumor core (TC). We compare the results of our own architecture, the deep residual Unet 104 (DR-Unet104) with no dropout, a dropout of 0.2 and of 0.5 after each level. We additionally compare our model to DeepLabV3+ with a ResNet101 backbone. We show the best results in bold.

| | Validation results | | | | | |
| | DSC | | | HD95 | | |
Architecture	WT (SD)	ET (SD)	TC (SD)	WT (SD)	ET (SD)	TC (SD)
DR-UNET104 dropout (0)	0.8763 (0.0859)	0.6549 (0.3256)	0.6693 (0.3357)	18.39 (24.87)	53.61 (115.9)	16.19 (20.88)
DR-UNET104 dropout (0.2)	**0.8862 (0.0886)**	**0.6756 (0.3171)**	0.6721 (0.3462)	**12.11 (20.82)**	**47.62 (112.7)**	**15.74 (36.04)**
DR-UNET104 dropout (0.5)	0.8723 (0.0977)	0.6701 (0.3245)	0.6489 (0.3651)	23.80 (27.58)	51.53 (108.5)	28.56 (51.69)
DeepLabV3+	0.8771 (0.0853)	0.6524 (0.3101)	**0.6813 (0.3213)**	14.87 (23.20)	49.10 (112.6)	17.96 (37.54)

We finally evaluated our model on the 166 BraTS testing subjects as part of the 2020 challenge with the final results shown in Table 2. Our mean whole tumor DSC is lower than the validation results at 0.8673, however the enhancing tumor and tumor core DSC are much higher than validation results achieving 0.7514 and 0.7983 respectively. The proposed methods overall better performance on the testing data shows it generalizability aided by random dropout. Our model's performance is additionally competitive with 3D models.

Table 2. Table showing the mean results of the Dice Score coefficient (DSC) and Hausdorff distance 95% (HD95) with the Standard deviation (SD) in brackets results on the BraTS20 Testing data. We show results for the whole tumor (WT), enhancing tumor (ET) and tumor core (TC) evaluated using the deep residual Unet 104 (DR-Unet104) with dropout of 0.2.

| | Testing data | | | | | |
| | DSC | | | HD95 | | |
Architecture	WT (SD)	ET (SD)	TC (SD)	WT (SD)	ET (SD)	TC (SD)
DR-UNET104 dropout (0.2)	0.8673 (0.1279)	0.7514 (0.2478)	0.7983 (0.2757)	10.41 (16.59)	24.68 (83.99)	21.84 (74.35)

5 Discussion

In this paper, we present a 2D deep residual Unet with 104 convolution layers for automated brain tumor segmentation in multimodal MRI. The proposed network couples the strengths of deep residual blocks and the Unet with encoder-decoder structure. The regularization of dropout is included into the network, allowing it to learn more representative features than the plain architectures without regularization, producing improved validation results as shown in Table 1.

The results show that our method achieves promising performance when comparing to a state-of-the-art method (i.e. DeeplabV3+), and performs reasonably well despite being a 2D architecture, having minimal training data augmentation and being trained for only 50 epochs without any complex learning rate scheduling. The 2D architecture has the added benefit of meaning the model can be evaluated on lower powered computers/GPUs, only needing a GPU with 1–2 GBs of memory (when evaluated with a batch size of 1), unlike many other 3D architectures [4]. A limitation is that for simplicity we used the commonly used cross-entropy loss function, but would likely have received a better performance using a 'soft Dice loss' function, as used by the BraTS 2019 1st place method [4], due to DSC being used for the evaluation, and this could be included in future work.

Unusually, on the testing data set our architecture performed worse on whole tumor segmentation, but much better on enhancing tumor and tumor core labelling, compared to the validation data set. This is likely due a greater number of difficult to segment enhancing areas in the validation data set, which would also affect the tumor core DSC if mislabelled as one another. A greater number of difficult to label enhancing tumor areas in the validation data set is supported by a larger standard deviation of the mean DSC, 0.25 in testing vs. 0.32 on validation data.

6 Conclusion

We propose a variant of the Unet taking advantage of bottleneck residual block to produce a deeply stacked encoder. We additionally show the benefit of using dropout in our architecture. Our method has a competitive performance despite being a 2D architecture and having simple and limited training.

References

1. Lambin, P., Leijenaar, R., Deist, T., et al.: Radiomics: the bridge between medical imaging and personalized medicine. Nat. Rev. Clin. Oncol. **14**, 749–762 (2017). https://doi.org/10.1038/nrclinonc.2017.141
2. Ricard, D., Idbaih, A., Ducray, F., et al.: Primary brain tumours in adults. Lancet **379**(9830), 1984–1996 (2012). https://doi.org/10.1016/S0140-6736(11)61346-9
3. Ronneberger, O., Fischer, P., Brox, T.: U-Net: convolutional networks for biomedical image segmentation. In: Navab, N., Hornegger, J., Wells, W.M., Frangi, A.F. (eds.) MICCAI 2015. LNCS, vol. 9351, pp. 234–241. Springer, Cham (2015). https://doi.org/10.1007/978-3-319-24574-4_28
4. Jiang, Z., Ding, C., Liu, M., Tao, D.: Two-stage cascaded U-Net: 1st place solution to BraTS challenge 2019 segmentation task. In: Crimi, A., Bakas, S. (eds.) BrainLes 2019. LNCS, vol. 11992, pp. 231–241. Springer, Cham (2020). https://doi.org/10.1007/978-3-030-46640-4_22
5. Kamnitsas, K., et al.: Ensembles of multiple models and architectures for robust brain tumour segmentation. In: Crimi, A., Bakas, S., Kuijf, H., Menze, B., Reyes, M. (eds.) BrainLes 2017. LNCS, vol. 10670, pp. 450–462. Springer, Cham (2018). https://doi.org/10.1007/978-3-319-75238-9_38
6. He, K., Zhang, X., Ren, S., Sun, J.: Identity mappings in deep residual networks. In: Leibe, B., Matas, J., Sebe, N., Welling, M. (eds.) ECCV 2016. LNCS, vol. 9908, pp. 630–645. Springer, Cham (2016). https://doi.org/10.1007/978-3-319-46493-0_38

7. He, K., Zhang, X., Ren, S., Sun, J.: Deep residual learning for image recognition. In: Proceedings of the IEEE Conference on Computer Vision and Pattern Recognition (CVPR), pp. 770–778 (2016). https://doi.org/10.1109/CVPR.2016.90

8. Chen, L.-C., Zhu, Y., Papandreou, G., Schroff, F., Adam, H.: Encoder-decoder with atrous separable convolution for semantic image segmentation. In: Ferrari, V., Hebert, M., Sminchisescu, C., Weiss, Y. (eds.) ECCV 2018. LNCS, vol. 11211, pp. 833–851. Springer, Cham (2018). https://doi.org/10.1007/978-3-030-01234-2_49

9. Park, S., Kwak, N.: Analysis on the dropout effect in convolutional neural networks. In: Lai, S.-H., Lepetit, V., Nishino, Ko., Sato, Y. (eds.) ACCV 2016. LNCS, vol. 10112, pp. 189–204. Springer, Cham (2017). https://doi.org/10.1007/978-3-319-54184-6_12

10. Ho, Y., Wookey, S.: The real-world-weight cross-entropy loss function: modeling the costs of mislabeling. IEEE Access **8**, 4806–4813 (2020). https://doi.org/10.1109/ACCESS.2019.296 2617

11. Menze, B.H., Jakab, A., Bauer, S., Kalpathy-Cramer, J., Farahani, K., Kirby, J., et al.: The multimodal brain tumor image segmentation benchmark (BRATS). IEEE Trans. Med. Imaging **34**(10), 1993–2024 (2015). https://doi.org/10.1109/TMI.2014.2377694

12. Bakas, S., Akbari, H., Sotiras, A., Bilello, M., Rozycki, M., Kirby, J.S., et al.: Advancing the cancer genome atlas glioma MRI collections with expert segmentation labels and radiomic features. Nat. Sci. Data **4**, 170117 (2017). https://doi.org/10.1038/sdata.2017.117

13. Bakas, S., Reyes, M., Jakab, A., Bauer, S., Rempfler, M., Crimi, A., et al.: identifying the best machine learning algorithms for brain tumor segmentation, progression assessment, and overall survival prediction in the BRATS challenge. arXiv preprint arXiv:1811.02629 (2018)

14. Bakas, S., Akbari, H., Sotiras, A., Bilello, M., Rozycki, M., Kirby, J., et al.: Segmentation labels and radiomic features for the pre-operative scans of the TCGA-GBM collection. Cancer Imaging Arch. (2017). https://doi.org/10.7937/K9/TCIA.2017.KLXWJJ1Q

15. Bakas, S., Akbari, H., Sotiras, A., Bilello, M., Rozycki, M., Kirby, J., et al.: Segmentation labels and radiomic features for the pre-operative scans of the TCGA-LGG collection. Cancer Imaging Arch. (2017). https://doi.org/10.7937/K9/TCIA.2017.GJQ7R0EF

16. Kingma, D., Ba, J.: Adam: a method for stochastic optimization. arXiv 1412.6980 (2014)

17. He, K., Zhang, X., Ren, S., Sun, J.: Delving deep into rectifiers: surpassing human-level performance on ImageNet classification. Int. Conf. Comput. Vis. (2015). https://doi.org/10.1109/ICCV.2015.123

18. Chollet, F., et al.: Keras. GitHub (2015). https://github.com/fchollet/keras

19. Abadi, et al.: TensorFlow: large-scale machine learning on heterogeneous systems (2015). https://tensorflow.org

Glioma Sub-region Segmentation on Multi-parameter MRI with Label Dropout

Kun Cheng[✉], Caihao Hu, Pengyu Yin, Qianlan Su, Guancheng Zhou, Xian Wu, Xiaohui Wang, and Wei Yang

Beijing University of Posts and Telecommunications, Beijing, China
kcheng@bupt.edu.cn

Abstract. Gliomas are the most common primary brain tumor, the accurate segmentation of clinical sub-regions including enhancing tumor (ET), tumor core (TC) and whole tumor (WT) has great clinical importance throughout the diagnosis, treatment planning, delivery and prognosis. Machine learning algorithms particularly neural network based methods have been successful in many medical image segmentation applications. In this paper, we trained a patch based 3D UNet model with a hybrid loss between soft dice loss, generalized dice loss and multi-class cross-entropy loss. We also proposed a label dropout process that randomly discards inner segment labels and their corresponding network output during training to overcome the heavy class imbalance issue. On the BraTs 2020 final test data, we achieved 0.823, 0.886 and 0.843 for ET, WT and TC respectively.

Keywords: Brain tumor segmentation · 3D UNet · Deep learning · Multi-parameter MRI

1 Introduction

Medical image interpretation has drawn much attention past few years, leveraging the fast advances in machine learning, especially with regard to deep learning methodologies [8]. The delineation of glioma and its sub-regions is still a very challenging task due to large variance in lesion shape and image appearance, heterogeneous histological features of tumor sub-regions and the complex anatomical structure within the human brain. Since 2012, the brain tumor segmentation challenge (BraTS) has been held in conjunction with the Medical Image Computing and Computer-Assisted Intervention (MICCAI), which provides two great public data sets as benchmark for evaluating the state-of-art solution of glioma segmentation [1,2].

Multi-parameter Magnetic Resonance Imaging (mpMRI) sequences are provided in order to obtain complementary clinical information to distinguish the glioma-subregions, including tumor core (TC), enhancing tumor (ET) and whole

© Springer Nature Switzerland AG 2021
A. Crimi and S. Bakas (Eds.): BrainLes 2020, LNCS 12659, pp. 420–430, 2021.
https://doi.org/10.1007/978-3-030-72087-2_37

tumor (WT). In common clinical practices, T1-weighted (T1) and T2-weighted (T2) MRIs are used for structural and lesion screening respectively. In addition to T1 and T2, post-contrast T1-weighted (T1Gd) is commonly used for the investigation at the TC boundary, on which the contrast agent breaks through the blood-brain barrier and accumulates. Fluid Attenuation Inversion Recovery (FLAIR) sequence shows little clinical relevant information about TC inner structure, while it is very helpful in defining the peritumoral edema volume [15].

The BraTS 2020 challenge has three tracks, brain tumor segmentation, prediction of overall survival, and evaluation of uncertainty measure evaluation. The three tracks almost covered the image-guided clinical procedures throughout the diagnosis, planning, and treatment delivery [3,4,12]. This paper focuses on the segmentation task.

2   Related Work

The encoder and decoder based UNet architecture and its variants have been widely implemented in many segmentation related medical applications, for example in brain glioma, prostate cancer and kidney cancer [9,14]. The segmentation task has been dominated by UNet and its variants since Brats2018 [5,13]. In BraTs 2019 challenge, McKinley et al. [10] proposed a triplanar ensembles of DeepSCAN structure, which embedded with a shallow encoder and decoder network. It is worth mentioning that, McKinley et al. ensemble 30 different model outputs with respect to their uncertainty. Zhao et al. [17] presented comprehensive ablation experiments with a bag of learning tricks, including hard example mining, random patch and batch size, warm-up and learning rate decay and model ensemble. Zhao et al. also used a multitask learning optimization scheme, which basically added on the clinical correlation between ET, TC and WT segments as a constraint to the model. They achieved the significantly highest score on 0.86 TC segment among the top 3 teams. The 1st place winner of BraTs 2019, Jiang et al. [6] proposed a two-stage cascaded U-Net, which they claim as a variant of the work by Myronenko [13]. The second UNet produces two outputs using deconvolution and trilinear decoder path, in addition to the first UNet output, soft Dice loss (SDL) is computed against the training labels. They also applied post-processing to replace small ET prediction with necrosis, their model outperforms the rest teams on the testing data, particularly for the ET segments.

The top 3 solutions in BraTs 2019 more or less used model ensemble and other tricks to improve their accuracy and robustness. However, from their scores on testing data, similar performance on WT segment can be observed, while much lower scores and larger variance on ET and TC segments, which we believe is caused by the class imbalance between segments. Recall the clinical definition of sub-regions, TC actually entails the ET segments, the ET segment suffers from false positives as well as false negative on small lesions. False-positive removal or replacement into other segments indeed improves the ET scores, while compromising the TC scores. Focal loss, over-sample, generalized Dice loss (GDL), and other widely used learning tricks have not completely resolved the class imbalance issue during training. Without a fundamental breakthrough in the supervised learning methodology, the UNet variants, the class imbalance issue should be given priority in training.

3 Method

3.1 Preprocessing

N4ITK bias correction firstly applied to training sequences to remove possible artifacts caused on data acquisition [16], followed by case-wise intensity normalization and re-sampling into [1, 1, 1] isotropic pixel spacing.

3.2 Cropping and Data Augmentation

Whole image volume were randomly cropped into widely used patch size [128, 128, 128] to fit in our 3D UNet model. Data augmentation method including random rotation ($15°$ maximum), scaling and elastic deformations are implemented both in training and predicting.

3.3 Model Architecture

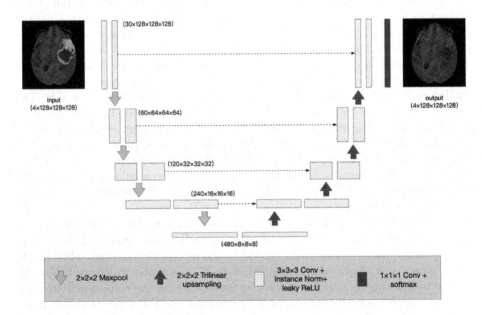

Fig. 1. The modified UNet structure.

We firstly tried Unet structure with residual block and dense connection with no significant difference with vanilla UNet structure. Given the fact, we chose Vanilla UNet structure with minor modifications tailored for the BraTS data set as our baseline, as shown in Fig. 1. We replaced batch normalization and Rectified Linear Unit (ReLU) with instance normalization and Leaky ReLU.

3.4 Loss Function

SDL and GDL inherently take advantage of the tasks evaluated using Dice, compared with multi-class cross entropy (CE) loss [11]. Consequently, SD loss and GDL are used along with CE loss as two hybrid losses for training the model, as in Eq. 1 and Eq. 2.

$$L_{total} = w_{SDL} * L_{SDL} + w_{CE} * L_{CE} \qquad (1)$$

$$L_{total} = w_{GDL} * L_{GDL} + w_{CE} L_{CE} \qquad (2)$$

3.5 Label Dropout

It is obvious that the pixels at segments' boundaries are the golden keys to solve the puzzle. The inner homogeneous region of a sub-region, for example the edema, is likely to have lower local gradient or less discriminate features to be learned, in some extend we can treat them as duplicates of their surrounding pixels. We propose the label dropout process which discards a few training label pixels and the corresponding feature map activation at the last layer, which means these pixels do not contribute to the hybrid loss in Eq. 1. The label dropout is designed to be like an regularization method to reduce the contribution of these duplicates to the network weights updating, prevent potential over-fitting on non-boundary pixels and naturally drive the network's attention to hard examples on the ET and TC segment boundaries.

As shown in Fig. 3, training label (top left) are filtered with Gaussian blur with a kernel size of [5, 5], followed by Sobel edge detector (top right), the random drop matrix (bottom right) is binomially distributed with probability p = 0.5, the new label is obtained by pixel-wise multiplication between sub-region interior (including background) and the drop matrix. The zeros on the label drop matrix indicate those pixels will be discarded in calculating the loss. Namely, the label dropout process discards 50% of the contribution of a patch to the network updating, except on the segment boundary.

As we expected from the fluctuated training loss of the full image label dropout scheme, the performance is getting poorer than the vanilla Unet baseline. The focal loss [7] intends to put a larger weight on the class with fewer samples, while we redesign the label drop out process here to solve the class imbalance in the opposite way. The label dropout now only discards sub-region interior pixels, as shown in Fig. 2 , in the first row we preserve pixels within both

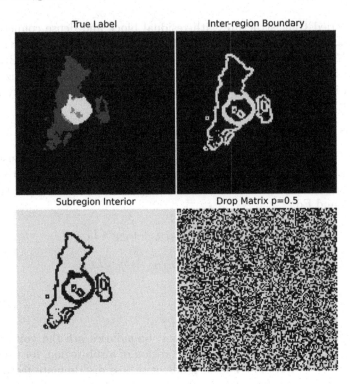

Fig. 2. Full image label dropout. All pixels and their corresponding model output are discarded at a probability p–0.5, except for those at sub-region boundaries.

ET and TC segment, in the second and third row we preserve pixels in ET and TC respectively, in all three columns WT segment interior pixels are randomly dropped as shown in Fig. 3 right column. After a few experiments, we found that WT only interior dropout significantly outperforms the rest 2 schemes, it is easy to understand since WT sub-region has the largest amount of pixels over the training data set. Consequently, in the rest of this paper, WT only pixel dropout is used as default.

3.6 Implementation Details

We implemented the proposed method using Pytorch on an 11 GB NVIDIA RTX 2080TI GPU with batch size 2. The model is trained using the ADAM optimizer with learning rate decay, initial learning rate 3e–4. We used 5-fold cross-validation to split our training data.

3.7   Postprocessing

We observed that, the trained models constantly produce false positives for around 10 cases or more, which significantly reduce the ET segment dice score. Therefore, we used morphological opening with kernel size [3, 3, 3] to remove small prediction and replace with its neighboring network prediction class.

4   Results

Dice, sensitivity, specificity and Hausdorff distance (95%) are used to evaluate the segmentation methods for each tumor sub-region. Our preliminary results on the validation data are summarized in Table 1 and Table 2. SDL and CE loss without label dropout achieved the best score on both WT and TC, label dropout significantly improved the ET sub-region score to 0.763, sadly WT and TC score dropped as a side effect. GDL and CE loss with label dropout further increase the ET score to 0.780 while compromising TC score. The training using label dropout experienced loss oscillation on the validation fold, the learning rate of 3e–4 is reduced to 1e–4 for a more stable training process. In the test phase, we used an ensemble of all three models submitted for the validation phase, the final test phase results are shown in Table 3, with promising dice score on ET and TC sub-region of test phase, and the lowest Hausdorff95 score on ET sub-region.

Table 1. Dice and Hausdorff95

Model	Dice			Hausdorff95		
	ET	WT	TC	ET	WT	TC
SDL+CE Loss	0.719	**0.903**	**0.842**	37.892	7.316	8.693
SDL+CE with Label Dropout	0.763	0.890	0.820	35.474	**5.803**	**7.215**
GDL+CE with label Dropout	**0.780**	0.894	0.814	**24.423**	7.071	12.655

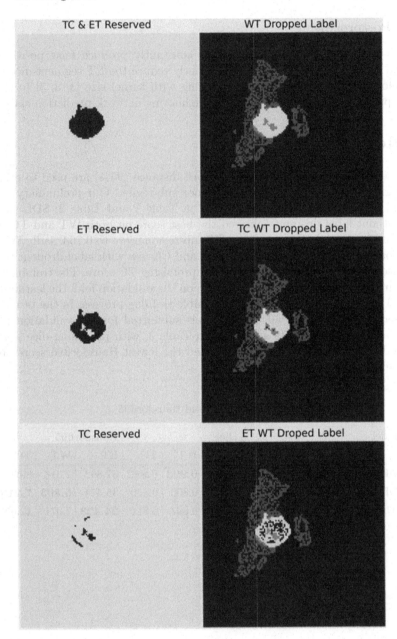

Fig. 3. Sub-region interior label dropout. Only pixel within specific sub-region are dropped to solve the class imbalance issue.

Table 2. Sensitivity and specificity

Model	Sensitivity			Specificity		
	ET	WT	TC	ET	WT	TC
SDL+CE Loss	0.749	**0.909**	**0.82**	0.999	0.999	0.999
SDL+CE+LD with Label Dropout	0.762	0.901	0.799	0.999	0.999	0.999
GDL+CE+LD with Label Dropout	**0.785**	0.908	0.791	0.999	0.999	0.999

Table 3. Test phase

Model	Dice			Sensitivity			Specificity			Hausdorff95		
	ET	WT	TC	ET	WT	TC	ET	WT	TC	ET	WT	TC
Ensemble	**0.823**	0.886	**0.843**	0.827	0.909	0.850	0.999	0.999	0.999	**13.100**	6.71	15.64

Results on the validation data set, trained using hybrid GDL and CE loss with label dropout enabled, is shown in Fig. 4. We visualize our results using validation data 001. Considering the nature of the four modalities, we superimpose edma on T2 as edma region generally is brighter on T2 weighted. Similarly, ET is plotted on T1ce for better contrast at the blood-brain barrier, TC is shown on FLAIR and Non-tumor healthy tissues are shown on T1. It can be observed that, the performance on ET and TC segments are promising, the Dice results for this case are 0.938, 0.865 and 0.946 on ET, WT and TC respectively.

428   K. Cheng et al.

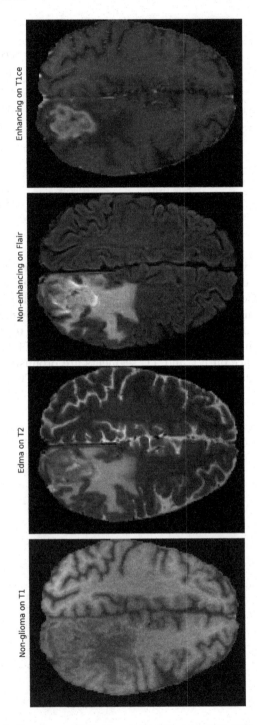

Fig. 4. 120th axial slice from validation data 001 with ET, TC and WT prediction superimposed on T1ce, FLAIR and T2 respectively.

5 Conclusion

In this paper, we reported our solution to BraTs 2020 segmentation track, a 3D patch based UNet was trained using hybrid loss functions of SDL and GDL in addition to commonly used CE loss. Label dropout method is proposed to conquer the heavy class imbalance between sub-regions, especial the ET segment. We successfully improved the segmentation on ET and TC sub-region while slightly compromising the performance for WT segment. We acknowledge that model ensemble might be helpful in dealing with the loss of information caused by the discard back-propagation. We empirically picked the hyper-parameters such as edge filter kernel size and the drop probability (0.5) without deep quantitative investigation.

Acknowledgment. We want to thank Beijing University of Posts and Telecommunication (BUPT) Undergraduate Innovation and Entrepreneurship Competition Program (UIEC), and the great effort of the UIEC team members during the Covid-19 pandemic period.

References

1. Bakas, S., Akbari, H., Sotiras, A., Bilello, M., Rozycki, M., Kirby, J., et al.: Segmentation labels and radiomic features for the pre-operative scans of the TCGA-GBM collection. Cancer Imaging Arch. (2017). https://doi.org/10.7937/K9/TCIA.2017. KLXWJJ1Q
2. Bakas, S., Akbari, H., Sotiras, A., Bilello, M., Rozycki, M., Kirby, J.: et al.: Segmentation labels and radiomic features for the pre-operative scans of the TCGA-LGG collection. Cancer Imaging Arch. (2017). https://doi.org/10.7937/K9/TCIA.2017. GJQ7R0EF
3. Bakas, S., et al.: Advancing the cancer genome atlas glioma MRI collections with expert segmentation labels and radiomic features. Sci. Data **4**, 1–13 (2017). https://doi.org/10.1038/sdata.2017.117
4. Bakas, S., et al.: Identifying the Best Machine Learning Algorithms for Brain Tumor Segmentation, Progression Assessment, and Overall Survival Prediction in the BRATS Challenge (2018). http://arxiv.org/abs/1811.02629
5. Isensee, F., Kickingereder, P., Wick, W., Bendszus, M., Maier-Hein, K.H.: No new-net. In: Crimi, A., Bakas, S., Kuijf, H., Keyvan, F., Reyes, M., van Walsum, T. (eds.) BrainLes 2018. LNCS, vol. 11384, pp. 234–244. Springer, Cham (2019). https://doi.org/10.1007/978-3-030-11726-9_21
6. Jiang, Z., Ding, C., Liu, M., Tao, D.: Two-stage cascaded u-net: 1st place solution to brats challenge 2019 segmentation task. In: Crimi, A., Bakas, S. (eds.) BrainLes 2019. LNCS, vol. 11992, pp. 231–241. Springer, Cham (2020). https://doi.org/10. 1007/978-3-030-46640-4_22
7. Lin, T.Y., Goyal, P., Girshick, R., He, K., Dollár, P.: Focal loss for dense object detection. In: Proceedings of the IEEE International Conference on Computer Vision, pp. 2980–2988 (2017)
8. Litjens, G., et al.: A survey on deep learning in medical image analysis. Med. Image Analy. **42**, 60–88 (2017)

9. Liu, J., Yin, P., Wang, X., Yang, W., Cheng, K.: Glioma subregions segmentation with a discriminative adversarial regularized 3D Unet. In: Proceedings of the Third International Symposium on Image Computing and Digital Medicine, pp. 269–273 (2019)
10. McKinley, R., Rebsamen, M., Meier, R., Wiest, R.: Triplanar ensemble of 3D-to-2D CNNs with label-uncertainty for brain tumor segmentation. In: Crimi, A., Bakas, S. (eds.) BrainLes 2019. LNCS, vol. 11992, pp. 379–387. Springer, Cham (2020). https://doi.org/10.1007/978-3-030-46640-4_36
11. Mehrtash, A., Wells, W.M., Tempany, C.M., Abolmaesumi, P., Kapur, T.: Confidence calibration and predictive uncertainty estimation for deep medical image segmentation. IEEE Trans. Med. Imaging 39(12), 3868–3878 (2020)
12. Menze, B.H., et al.: The multimodal brain tumor image segmentation benchmark (BRATS). IEEE Trans. Med. Imaging 34(10), 1993–2024 (2015). https://doi.org/10.1109/TMI.2014.2377694
13. Myronenko, A.: 3D MRI brain tumor segmentation using Autoencoder regularization. In: Crimi, A., Bakas, S., Kuijf, H., Keyvan, F., Reyes, M., van Walsum, T. (eds.) BrainLes 2018. LNCS, vol. 11384, pp. 311–320. Springer, Cham (2019). https://doi.org/10.1007/978-3-030-11726-9_28
14. Ronneberger, O., Fischer, P., Brox, T.: U-Net: convolutional networks for biomedical image segmentation. In: Navab, N., Hornegger, J., Wells, W.M., Frangi, A.F. (eds.) MICCAI 2015. LNCS, vol. 9351, pp. 234–241. Springer, Cham (2015). https://doi.org/10.1007/978-3-319-24574-4_28
15. Shukla, G., et al.: Advanced magnetic resonance imaging in glioblastoma: a review. JHN J. 13(1), 5 (2018)
16. Tustison, N.J., et al.: N4ITK: improved n3 bias correction. IEEE Trans. Med. Imaging 29(6), 1310–1320 (2010)
17. Zhao, Y.-X., Zhang, Y.-M., Liu, C.-L.: Bag of tricks for 3D MRI brain tumor segmentation. In: Crimi, A., Bakas, S. (eds.) BrainLes 2019. LNCS, vol. 11992, pp. 210–220. Springer, Cham (2020). https://doi.org/10.1007/978-3-030-46640-4_20

Variational-Autoencoder Regularized 3D MultiResUNet for the BraTS 2020 Brain Tumor Segmentation

Jiarui Tang[1], Tengfei Li[1], Hai Shu[2], and Hongtu Zhu[1]([✉])

[1] University of North Carolina, Chapel Hill, NC, USA
`htzhu@email.unc.edu`
[2] Department of Biostatistics, School of Global Public Health, New York University, New York, NY 10003, USA

Abstract. Tumor segmentation is an important research topic in medical image segmentation. With the fast development of deep learning in computer vision, automated segmentation of brain tumors using deep neural networks becomes increasingly popular. U-Net is the most widely-used network in the applications of automated image segmentation. Many well-performed models are built based on U-Net. In this paper, we devise a model that combines the variational-autoencoder regularuzed 3D U-Net model [10] and the MultiResUNet model [7]. The model is trained on the 2020 Multimodal Brain Tumor Segmentation Challenge (BraTS) dataset and predicts on the validation set. Our result shows that the modified 3D MultiResUNet performs better than the previous 3D U-Net.

Keywords: Brain tumor segmentation · 3D U-Net · MultiResUNet · Variational-autoencoder regularization

1 Introduction

Multimodal MRI scans are useful to identify brain tumors, and the brain tumor segmentation is an important task before the diagnosis. The major goal of the medical image segmentation is to extract the areas of interest in an image, such as tumor regions. It allows doctors to focus on the most important areas for diagnosis or monitoring [7,11]. For a long time, manual tumor segmentation from MRI scans conducted by physicians is a time-consuming task. In order to speed up the process of image segmentation, many automated methods were developed [2–5,7,9,10]. With the rapid development of deep learning technologies in the field of computer vision, deep-learning based automated brain tumor segmentation becomes increasingly popular [14]. In particular, convolutional neural networks (CNNs) show great successes in image segmentation tasks [12], and especially the U-Net [13] earns the most credits.

Multimodal Brain Tumor Segmentation Challenge (BraTS) is an annual challenge aims at gathering state-of-the-art methods for the segmentation of

© Springer Nature Switzerland AG 2021
A. Crimi and S. Bakas (Eds.): BrainLes 2020, LNCS 12659, pp. 431–440, 2021.
https://doi.org/10.1007/978-3-030-72087-2_38

432   J. Tang et al.

brain tumors. Participants are provided with clinically acquired training data
to develop their own models and produce segmentation labels of three glioma
sub-regions: enhancing tumor (ET), tumor core (TC), and whole tumor (WT)
[2–5,9,10]. The BraTS 2020 training dataset contains 369 cases, including 293
high-grade gliomas (HGG) and 76 low-grade gliomas (LGG) cases, each with
four 3D MRI modalities: the native (T1) and the post-contrast T1-weighted
(T1CE) images, and the T2-weighted (T2) and the T2 Fluid Attenuated Inversion Recovery (FLAIR) images. The example images of the four modalities are
shown in Fig. 1. The validation dataset contains 125 cases and the test dataset
contains 166 cases. The dimension of each MRI image is 240 × 240 × 155.

In this paper, we propose a model that combines the variational-autoencoder
(VAE) regularized 3D U-Net model [10] and the MultiResUNet model [7], which
is used to train end-to-end on the BraTS 2020 training dataset. Our model follows the encoder-decoder structure of the 3D U-Net model of [10] used in BraTS
2018 Segmentation Challenge but exchanges the ResNet-like block in the structure with the "MultiRes block" and connects the feature maps from the encoder

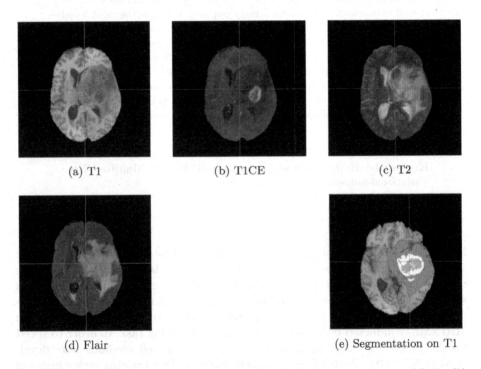

(a) T1   (b) T1CE   (c) T2

(d) Flair   (e) Segmentation on T1

Fig. 1. Visualization of the four modalities in BraTS 2020 training Dataset (a) T1, (b)
T1CE, (c) T2 and (d) Flair, and (e) the segmentation on T1. These images are from
the same patient. In (e), the blue area represents the Necrotic and Non-Enhancing
Tumor (NCR/NET), the orange area represents the peritumoral edema (ED) and the
white area represents the GD-enhancing tumor (ET). (Color figure online)

stages to the decoder stages with the "Res path", a chain of convolutional layers with residual connections [7].

2 Related Work

2.1 VAE-regularized 3D U-Net

In BraTS 2018, the top-1 winner Myronenko [10] proposed an encoder-decoder CNN model with an asymmetrically large encoder to extract deep image features and a VAE branch to regularize the encoder. In BraTS 2019, the champion team Jiang et al. [8] proposed a novel model, a two-stage cascaded U-Net. The first stage of the model is a variant of the asymmetrical U-Net in [10] and the second stage of the model doubles the number of filters in the initial 3D convolution to increase the network width and has two decoders, one using deconvolutions and another using the trilinear interpolation to facilitate regularizing encoders. The top ranks of the two models in the challenge have proved the power of VAE-regularized 3D U-Nets in the MRI brain tumor segmentation.

2.2 MultiResUNet

In early 2020, Ibtehaz and Rahman [7] proposed a modified U-Net, called MultiResUNet, which outperforms the classical U-Net. There are two important parts of the architecture: the MultiRes block and the Res path. In the MultiRes block, there are three 3×3 consecutive convolutional layers with gradually increasing numbers of filters. A residual connection and a 1×1 convolutional layer are added in each MultiRes block to gather more spatial information. The Res path passes the feature maps from the encoder stage through a chain of 3×3 filters with 1×1 filters residual connections and then concatenates them with the decoder features. This MultiResUNet can also be applied to 3D images. The structures of the MultiRes block and the Res path are shown in Figs. 2 and 3, respectively.

In this paper, we modify the VAE-regularized 3D U-Net model [10] by exchanging the initial $3 \times 3 \times 3$ 3D convolution block with a revised 3D MultiRes block based on the MultiRes block in [7] and adding the Res path between the encoder and decoder stages.

3 Method

The architecture of our proposed network is shown in Fig. 4. Due to the limited computational resource, each image is cropped from the size of $240 \times 240 \times 155$ voxels to $160 \times 192 \times 128$ and then resized to $80 \times 96 \times 64$. Each MRI modality is fed into the network as one channel and so we have totally four channels.

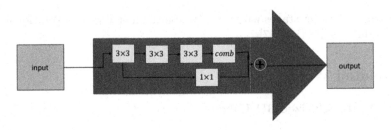

Fig. 2. The structure of the MultiRes block: the green boxes represent convolutional layers. The three 3 × 3 convolutional layers have increasing numbers of filters and then they are concatenated together, which can help the model capture different information from different scales. Then the result from the 1 × 1 convolutional layer following the input is added to the result from the three 3 × 3 convolutional layers to obtain the final output of the MultiRes block. (Color figure online)

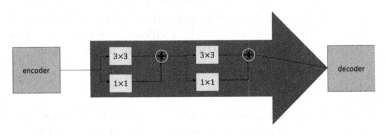

Fig. 3. The structure of the Res path: the green boxes represent convolutional layers. In many classic U-Nets, the encoder feature maps are simply sent to the decoder part. However, it is possible that there is semantic gap between the features of the decoder part and the encoder part, since the decoder part experiences more processing compared to the encoder part. Therefore, Ibtehaz and Rahman [7] proposed the Res path between the encoder and decoder before the features from the encoder part are concatenated to the decoder part. Within the Res path are series of 3 × 3 convolution layers with 1 × 1 residual connections. (Color figure online)

3.1 Encoder

The encoder part has multiple MultiRes blocks, each of which contains three 3 × 3 × 3 sequential convolution layers with a 1 × 1 × 1 residual connection to gather spatial features at different scales. After each MultiRes block, we use strided convolutions to downsize the image dimension by 2 and increase feature size by 2. The number of filters in the three 3 × 3 × 3 sequential convolution layers is 6, 12 and 18 respectively. Each MultiRes block contains the Group Normalization and the ReLU activation function. The output layer uses the sigmoid activation function.

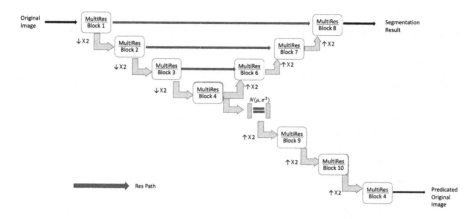

Fig. 4. The architecture of our MultiResUNet with a VAE branch. The size of the original input is $4 \times 80 \times 96 \times 64$. Each MultiRes block has three $3 \times 3 \times 3$ sequential convolution layers and a $1 \times 1 \times 1$ convolutional residual connection. Before each convolution layer, we use the group normalization and the ReLU activation function. In the Res path, there are two $3 \times 3 \times 3$ convolutional layers each with one $1 \times 1 \times 1$ residual connection. The output segmentation result has three channels and the shape of each channel is $4 \times 80 \times 96 \times 64$. Finally, we use the sigmoid activation function to obtain the segmentation for the three regions: WT (TC+ET), TC (ET+NCR/NET) and ET.

3.2   Decoder with a VAE Branch

The decoder part also has multiple MultiRes blocks. After the decoder, the output has three channels and the image of each channel has the same size as the input. We use strided convolutions to upsize the image dimension by 2 and decrease the feature size by 2.

For the VAE branch, we adopt the similar VAE structure used in [10]. The output of the fourth MultiRes block (MultiRes Block 4 in Fig. 4) is reduced to 256×1. We then draw a sample from a Gaussian distribution whose mean and standard deviation are both 128. Then, without connection to the encoder part through Res Path, we use the same structure as the decoder part to return to the original dimension $4 \times 80 \times 96 \times 64$. This step helps us to regularize encoder in training to prevent overfitting.

The VAE detail is shown in Table 1.

3.3   Res Path

The original skip connection between encoder and decoder parts is replaced by the Res path in our network model. Within each Res path, there is a chain of two $3 \times 3 \times 3$ convolutional layers our network architecture and each convolutional layer has one $1 \times 1 \times 1$ convoulational residual connection. The feature maps from the encoder parts is passed into the Res path and the output is concatenated

Table 1. The VAE part. Conv3 represents $3 \times 3 \times 3$ convolutional layer. GN is the group normalization and the group size is 8. Conv1 represents $1 \times 1 \times 1$ convolutional layer. Conv1 RC is the $1 \times 1 \times 1$ convolutional residual connection in the MultiRes block. AddId is the addition of the residual connection. Concate is the concatenation of the three consecutive Conv3. Step 5, 7 and 9 are the MultiRes Block 9, 10, 11 in Fig. 4.

Step	Details	Output size
1	GN, ReLU, Conv(16), Dense(256)	256×1
2	Sample $N(128, 128^2)$	128×1
3	Dense, ReLU, Conv1, UpLinear	$256 \times 10 \times 12 \times 8$
4	Conv1, UpLinear	$128 \times 20 \times 24 \times 16$
5	GN, ReLu, Conv3, Conv3, Conv3, Concate, Conv1 RC, AddId	$128 \times 20 \times 24 \times 16$
6	Conv1, UpLinear	$64 \times 40 \times 48 \times 32$
7	GN, ReLu, Conv3, Conv3, Conv3, Concate, Conv1 RC, AddId	$64 \times 40 \times 48 \times 32$
8	Conv1, UpLinear	$32 \times 80 \times 96 \times 64$
9	GN, ReLu, Conv3, Conv3, Conv3, Concate, Conv1 RC, AddId	$32 \times 80 \times 96 \times 64$
10	Conv1	$4 \times 80 \times 96 \times 64$

with the decoder features. The number of filters in the Res Path reduces as the image downsizes. All convolutions in the Res path have 32 filters.

3.4 Loss Function

We use a combined loss function with three components: the soft Dice loss (L_{dice}), the L2 loss on the VAE part (L_{L2}) and the VAE penalty term (L_{KL}), KL divergence between the estimated normal distribution $N(\mu, \sigma^2)$ and the prior distribution $N(0, 1)$, which was used in [10].

The equation of the soft dice loss is

$$L_{dice} = \frac{2 * \sum S * R}{\sum S^2 + \sum R^2 + \epsilon} \tag{1}$$

where S represents the true labels and R represents the predicted output by the model.

The equation of L_{L2} is

$$L_{L2} = ||I_{origin} - I_{VAE_{pred}}|| \tag{2}$$

where I_{origin} is the original input image and $I_{VAE_{pred}}$ is the predicted image from the VAE part.

The equation of L_{KL} is

$$L_{KL} = \frac{1}{N} \sum \mu^2 + \sigma^2 - log\sigma^2 - 1 \tag{3}$$

where N is the number of voxels in the image, μ and σ are the parameters of the estimated normal distribution.

3.5   Optimization

We use the Adam optimizer with an initial learning rate of $\alpha_0 = 10^{-4}$ and decrease the learning rate by

$$\alpha = \alpha_0 * (1 - \frac{e}{N_e})^{0.9}, \tag{4}$$

where e is an epoch counter, and N_e is a total number of epochs. The total epoch is set to 300 and the batch size is 1.

4   Experiment

We use Python [15] to implement the experiment. Particularly, the library Keras [6] with Tensorflow [1] as the backend is used to build the network model. The model is trained on the BraTS 2020 training dataset (total 369 cases) without additional in-house data.

For the training dataset, there are 369 cases, among which 293 cases are high-grade gliomas (HGG) and 76 are low-grade gliomas (LGG) cases. There are 125 cases in the validation dataset. The validation result is submitted to CBICA's Image Processing Portal (https://ipp.cbica.upenn.edu) for evaluation.

4.1   Data Preprocessing and Augmentation

All the original images are preprocessed to have mean 0 and standard deviation 1. The image is randomly cropped to the size of $160 \times 192 \times 128$. Due to the computational limitations, we resize the images from $160 \times 192 \times 128$ to $80 \times 96 \times 64$. For every input image, we randomly flip it across a random axis.

For consideration of robustness, we repeat the process of randomly flipping the input image 10 times for each case and use the flipped data as input to generate 10 training datasets. Using each of these training datasets, we train a model. Finally, we average the results from the 10 models to obtain a final segmentation.

4.2   Results

Our model is trained on NVIDIA Tesla V100 16 GB GPUs. If running on a single GPU, the approximate time for training 300 epochs is about 2.5 days. The best result of a single model on the validation dataset is provided in Table 2. The best final result on the validation dataset is reported in Table 3, which is obtained by averaging the results from 10 models trained on the 10 training datasets. The model ensemble improves our results by 0.4%. An example of our predicted segmentation is shown in Fig. 5.

Table 2. BraTS 2020 validation dataset result of the best single model.

	Dice			Hausdorff (mm)		
Validation	ET	WT	TC	ET	WT	TC
Best model	0.69800	0.88934	0.78357	34.29461	4.5323	10.06072

Table 3. BraTS 2020 validation dataset result of the best ensembled model.

	Dice			Hausdorff (mm)		
Validation	ET	WT	TC	ET	WT	TC
Best model	0.70301	0.89292	0.78977	34.30576	4.6287	10.07086

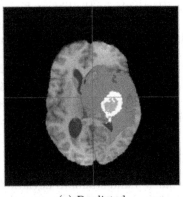 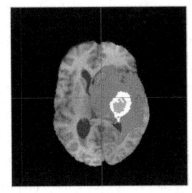

(a) Predicted   (b) Original

Fig. 5. An example of our predicted segmentation. The blue area represents the Necrotic and Non-Enhancing Tumor (NCR/NET), the orange area represents the peritumoral edema (ED) and the white represents the GD-enhancing tumor (ET) (Color figure online)

5   Discussion and Conclusion

In this paper, we modified a VAE-regularized 3D U-Net to the new proposed MultiResUNet architecture by replacing the classic ResNet with the MultiRes block and adding Res Path between the encoder and decoder parts to reduce the possible semantic gap. The MultiRes Block extracts more information on different scales of the image. We experimented our method on the BraTS 2020 training dataset and validation dataset. The result has shown that the architecture has satisfactory performance. We have tried to train the 3D U-Net model proposed by the top-1 winner Myronenko [10] in BraTS 2018 Challenge. However, we met a challenge of limited computational resource since Myronenko trained his model on the NVIDIA Volta V100 32 GB GPU. We tried to crop the original image from $240 \times 240 \times 150$ to $112 \times 126 \times 96$ which may largely affect the performance of that model. With the increasing amount of data, the model would expect to have higher requirements for GPUs. In comparison, the result of our model is very close to the result of that model but requires less computational resource, which makes our model more economically attractive.

References

1. Abadi, M., et al.: Tensorflow: a system for large-scale machine learning. In: 12th USENIX Symposium on Operating Systems Design and Implementation (OSDI 16), pp. 265–283. USENIX Association, Savannah, GA, November 2016. https://www.usenix.org/conference/osdi16/technical-sessions/presentation/abadi
2. Bakas, S., et al.: Advancing the cancer genome atlas glioma MRI collections with expert segmentation labels and radiomic features. Sci. Data **4** (2017). https://doi.org/10.1038/sdata.2017.117
3. Bakas, S., et al.: Segmentation labels and radiomic features for the pre-operative scans of the TCGA-LGG collection, July 2017. https://doi.org/10.7937/K9/TCIA.2017.GJQ7R0EF
4. Bakas, S., et al.: Segmentation labels and radiomic features for the pre-operative scans of the TCGA-GBM collection, July 2017. https://doi.org/10.7937/K9/TCIA.2017.KLXWJJ1Q
5. Bakas, S., et al.: Identifying the best machine learning algorithms for brain tumor segmentation, progression assessment, and overall survival prediction in the BRATS challenge. CoRR abs/1811.02629 (2018). http://arxiv.org/abs/1811.02629
6. Chollet, F., et al.: Keras (2015). https://github.com/fchollet/keras
7. Ibtehaz, N., Rahman, M.S.: Multiresunet: rethinking the u-net architecture for multimodal biomedical image segmentation. Neural Netw. Official J. Int. Neural Netw. Soc. **121**, 74–87 (2020)
8. Jiang, Z., Ding, C., Liu, M., Tao, D.: Two-stage cascaded U-net: 1st place solution to BraTS challenge 2019 segmentation task. In: Crimi, A., Bakas, S. (eds.) BrainLes 2019. LNCS, vol. 11992, pp. 231–241. Springer, Cham (2020). https://doi.org/10.1007/978-3-030-46640-4_22
9. Menze, B.H., Jakab, A., Bauer, S., Kalpathy-Cramer, J., Farahani, K., et al.: The multimodal brain tumor image segmentation benchmark (BRATS). IEEE Trans. Med. Imaging **34**(10), 1993–2024 (2015). https://doi.org/10.1109/TMI.2014.2377694
10. Myronenko, A.: 3D MRI brain tumor segmentation using autoencoder regularization. CoRR abs/1810.11654 (2018). http://arxiv.org/abs/1810.11654
11. Naik, S., Doyle, S., Agner, S., Madabhushi, A., Feldman, M., Tomaszewski, J.: Automated gland and nuclei segmentation for grading prostate and breast cancer histopathology, pp. 284–287, June 2008. https://doi.org/10.1109/ISBI.2008.4540988
12. Qayyum, A., Anwar, S.M., Majid, M., Awais, M., Alnowami, M.R.: Medical image analysis using convolutional neural networks: A review. CoRR abs/1709.02250 (2017). http://arxiv.org/abs/1709.02250
13. Ronneberger, O., Fischer, P., Brox, T.: U-net: convolutional networks for biomedical image segmentation. In: Navab, N., Hornegger, J., Wells, W.M., Frangi, A.F. (eds.) MICCAI 2015. LNCS, vol. 9351, pp. 234–241. Springer, Cham (2015). https://doi.org/10.1007/978-3-319-24574-4_28

14. Schindelin, J., Rueden, C., Hiner, M., Eliceiri, K.: The ImageJ ecosystem: an open platform for biomedical image analysis. Mol. Reprod. Dev. **82** (2015). https://doi.org/10.1002/mrd.22489
15. Van Rossum, G., Drake, F.L.: Python 3 Reference Manual. CreateSpace, Scotts Valley, CA (2009)

Learning Dynamic Convolutions for Multi-modal 3D MRI Brain Tumor Segmentation

Qiushi Yang[(✉)] and Yixuan Yuan

Department of Electrical Engineering, City University of Hong Kong, Hong Kong, China
qsyang2-c@my.cityu.edu.hk, yxyuan.ee@cityu.edu.hk

Abstract. Accurate automated brain tumor segmentation with 3D Magnetic Resonance Image (MRIs) liberates doctors from tedious annotation work and further monitors and provides prompt treatment of the disease. Many recent Deep Convolutional Neural Networks (DCNN) achieve tremendous success on medical image analysis, especially tumor segmentation, while they usually use static networks without considering the inherent diversity of multi-modal inputs. In this paper, we introduce a dynamic convolutional module into brain tumor segmentation and help to learn input-adaptive parameters for specific multi-modal images. To the best of our knowledge, this is the first work to adopt dynamic convolutional networks to segment brain tumor with 3D MRI data. In addition, we employ multiple branches to learn low-level features from multi-modal inputs in an end-to-end fashion. We further investigate boundary information and propose a boundary-aware module to enforce our model to pay more attention to important pixels. Experimental results on the testing dataset and cross-validation dataset split from the training dataset of BraTS 2020 Challenge demonstrate that our proposed framework obtains competitive Dice scores compared with state-of-the-art approaches.

Keywords: Brain tumor segmentation · Dynamic convolutional networks · Boundary-aware module

1 Introduction

As one of the most common and aggressive diseases, brain tumors draw a lot of researchers' attention to study. About 80% of brain tumors are a type of malignant tumors and they tend to be found among older people. Brain tumor segmentation is an important step to diagnosis, monitoring, treatment and prognosis. Manual analysis needs much anatomical knowledge with a great amount of human labor, which is difficult for segmenting brain tumors with MRI data accurately in limited time. In recent years, plenty of DCNNs based architectures [1] were proposed to segment brain tumor automatically and achieve remarkable performance.

The multimodal brain tumor segmentation challenge (BraTS) [11–15] provides a good 3D MRI dataset with pixel-wise ground truth labeled by experts. BraTS-2020 contains 369 patients' labeled examples as the training set. For each case, it comprises

© Springer Nature Switzerland AG 2021
A. Crimi and S. Bakas (Eds.): BrainLes 2020, LNCS 12659, pp. 441–451, 2021.
https://doi.org/10.1007/978-3-030-72087-2_39

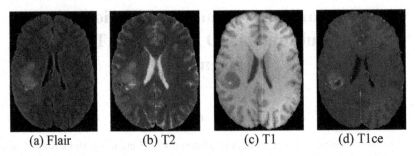

| (a) Flair | (b) T2 | (c) T1 | (d) T1ce |

Fig. 1. An example of four modalities image in the BraTS 2020 dataset.

four modalities data: T1, T1ce, T2 and FLAIR. Figure 1 shows an example of the four modalities. We aim at segmenting four kinds of target regions: the background of the brain image, the non-enhancing tumor (NET), the enhancing tumor (ET) and peritumoral edema (ED). Images of four modalities have different intensity distribution and their appearance shows different contrast on multiple kinds of target regions. Four common metrics: dice score, hausdorff distance (95%), specificity and sensitivity are adopted to evaluate the segmentation performances of methods proposed by participants.

In recent years, automated brain tumor segmentation has been widely investigated by using released MRI datasets. A lot of methods, especially deep learning-based frameworks are proposed and achieve state-of-the-art performance [2, 4, 10, 17–23]. Considering the spatial prior of various tumor regions, a series of approaches employed cascaded structures to segment different classes of tumors step-to-step in a multi-task fashion [20–22]. As the given brain tumor MR images usually have four modalities, many methods were presented to learn features from each modality separately and then fused them to get final segmentation results [18, 23]. In BraTS 2018, Myronenko [4] exhibited the best performance on the testing data. They leveraged a U-Net with an extra encoder to learn features and a decoder to reconstruct the ground truth with adding an additional variational autoencoder (VAE) branch as a regularization for encoder. Kamnitsas et al. [19] proposed an ensemble pipeline with integrating multiple deep learning models, in which the final brain tumor segmentation results were obtained by combining outputs of different models. In BraTS 2019, the first-place winner Jiang et al. [2] proposed a two-stage cascaded U-Net to learn features of tumor regions in a coarse-to-fine manner. They firstly employed a U-Net to get the coarse segmentation mask, and then utilized another U-Net to refine and decode the intermediate results.

Although those strategies improved the performance, their frameworks suffered from complexity problem due to the additional large networks. In addition, instead of working on multi-modal fashion, most of them tackled four modalities inputs in the same way, which ignored the differences among various modalities. With typical static neural networks, those methods cannot learn distinct representations adaptively for each example containing diverse localizations, scales and shapes of tumor regions.

In this work, inspired by the inherent properties of multi-modal inputs [6, 7], we consider enforcing the model to learn diverse features, especially low-level features from four modalities. Different from previous methods, we introduce four parallel branches consisting of small neural networks with dynamic convolutions [8], which could help

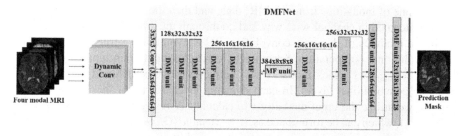

Fig. 2. The flowchart of our proposed framework. Each modality of inputs is fed into one dynamic convolutional module and then features maps are fused by concatenating to be put into the 3D Dilated Multi-Fiber Network (DMFNet) to obtain the final segmentation predictions. MF units and DMF units of DMFNet consist of multiple separated dilated residual units. Each Dynamic Conv module includes two dynamic convolutional layers following batch normalization and ReLU layers. Red dash lines indicate directions of gradient flow in the training phase.

to reduce the number of parameters while improve the capacity of the networks. Our pipeline is based on the 3D Dilated Multi-Fiber Network (DMFNet), which is also a highly efficient framework. Considering the highly diversity of tumor regions on localization, size and shape for each sequence, we propose to use dynamic convolutional networks in every branch to learn specific features adaptively. Consequently, the proposed dynamic convolutional networks not only can increase the capacity of the models in an input-adaptive way, but also maintaining efficient inference.

2 Methods

In order to tackle the multi-modal inputs adaptively and learn particular features from each modality, we propose a simple but efficient multi-branch dynamic network. We first design four branches to extract low-level features for four modal inputs separately. In each branch, we introduce two dynamic convolutional layers to learn discriminative information adaptively and then fuse the outputs of four branches to feed them to further layers. This is a modality-specific and sample-specific structure to capture distinct knowledge from different modalities respectively in different ways for various multi-modal inputs. To further learn the fused features and obtain the final segmentation maps, we utilize the segmentation networks 3D Dilated Multi-Fiber Network (DMFNet) as the feature extractor, which employs group convolution and 3D dilated convolution to learn multi-level features efficiently. We further propose a boundary-aware module to learn low-level cues using boundary supervision with a multi-level structure and deep supervision fashion. Figure 2 shows the framework of proposed model.

2.1 Multi-branch Dynamic Convolutional Networks

The multi-branch dynamic convolutional networks consist of four small parallel networks with two typically dynamic convolutional layers in every branch. Each branch is fed with one of modalities from the MRI data, and then the four series of feature maps are concatenated in channel-wise way and go through the downstream networks.

In the modules of dynamic convolution, 3D convolutional kernels are reweighted for convolutional computation. Specifically, given the input brain tumor images, $X = \{x_{t1}, x_{t1ce}, x_{flair}, x_{t2}\}$, each modality will be fed into one branch and go through two dynamic convolutional layers. In each dynamic layer, for all of the N convolutional kernels, we utilize the input modality x to calculate N reweighting coefficients $\alpha_i (i = 1, 2, ..., N)$, which can be formulated as: $\alpha = Sigmoid(GlobalAveragePool(x)R)$. Here, R is a learned matrix aiming at transforming pooled feature maps to coefficients and is implemented by one fully-connected layer. Relying on adaptively reweighted kernels to conduct dynamic convolution operations in a sample-specific way, the network can have the larger capacity than common static convolution networks. Detailed structures of the dynamic convolutional module are illustrated in Fig. 3.

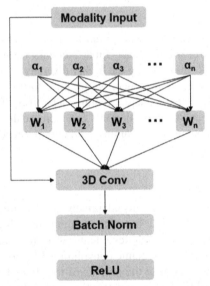

Dynamic Conv: $(\alpha_1 W_1 + \cdots + \alpha_n W_n) * x$

Fig. 3. The structure of one dynamic convolution layer. A series of weights learnt by features of inputs are added to reweight the 3D convolutional kernels adaptively. A batch normalization layer and a ReLU activation function layer are followed.

2.2 High-Level Features Learning Architecture

After learning low-level features via our proposed multi-branch dynamic networks from multi-modal inputs, we leverage 3D Dilated Multi-Fiber Network (DMFNet) to further extract high-level multi-scale features for segmentation.

The DMFNet introduces an efficient 3D fully convolutional neural network by leveraging multiple fiber units (MF units) that include ensemble simple 3D CNNs to learn high-level semantic features. It also combines to use 3D dilated convolutions (DMF units) to enlarge receptive field based on aforementioned MF units thus to learn multi-scale features in low-level layers.

To be specific, MF units consist of multiple paralleled residual layers with adding a $1 \times 1 \times 1$ convolutional bottleneck module in each layer to boost the flow of compact information. In order to learn abundant hierarchical features efficiently, DMF units add a group of dilated convolutional layers with multiple dilation rates at every residual layer. In both of MF units and DMF units, instead of common convolutional layers, it utilizes group convolutions to reduce the complexity. The general skeleton of DMFNet is showed in Fig. 2.

2.3 Loss Functions

We adopt softmax dice loss to train our neural networks. Compared with the common dice loss [3] that measures the overlap between predictions and labels explicitly. The softmax dice loss considers imbalance problem on multi-class segmentation and reweight the contributions of different categories in vanilla dice loss adaptively. It is formulated as following:

$$Loss_{GDL} = 1 - 2\frac{\sum_{i=1}^{c}\omega_i\sum_n y_{in}p_{in}}{\sum_{i=1}^{c}\omega_i\sum_n y_{in} + p_{in}}, \tag{1}$$

Where y, p, ω is prediction map, ground truth, and the weight of class i at pixel n respectively. c is the number of the target classes. Specifically, the weight ω is defined as following:

$$\omega_i = \frac{1}{(\sum_{n=1}^{N} y_{in})^2}. \tag{2}$$

2.4 Latest Further Works

Afterwards, we propose a boundary-aware module based on modified 3D-Unet with adding channel-wise attention units followed by convolutional layers as backbone to enforce the model pay more attention to the regions nearby the boundary of targeted foregrounds.

As a kind of low-level information, boundaries play an important role in the procedure of annotation. Doctors usually label pixels nearby boundaries of foreground regions and then label inner pixels [24]. Inspired by this real labeling process, we propose a novel boundary-aware module to help segmentation models learn semantic boundary positions in a unified framework. Learning semantic segmentation and semantic edge detection simultaneously at a multi-task regime can impose the model obtain high-level semantic features and low-level fine information in a complementary manner. Specifically, since multi-modal inputs perform specific boundary information towards to different classes, we employ the multi-path structures on the boundary detection modules to learn boundaries of various classes from different modalities. Moreover, we also employ multi-scale structure to learn low-level and high-level boundary features simultaneously and fuse them via a shared concatenated module [16]. Binary cross entropy (BCE) loss is added at multi-layers of the networks to optimize the boundary-aware module and it is trained with the segmentation branch together in an end-to-end manner.

3 Experiments

3.1 Datasets

BraTS-2020 offers training dataset with annotated ground truth of 369 patients, validation dataset with 125 unlabeled cases and testing dataset with 166 unlabeled cases [11–15]. Each data has four 3D MRI sequences consisting of T1, T1ce, T2, and FLAIR modalities, respectively, and the shape of each sequence is 240, 240, 155 on sagittal, coronal, axial view respectively.

In our experiments, in order to train the proposed models and evaluate the performance, we split a quarter dataset from the whole training dataset randomly as our own validation dataset and reported the results on it when the models were convergent. Then, we also showed the final results of our models trained on testing data, which were submitted via official evaluation system of BraTS Challenge.

3.2 Data Preprocessing

In our experiments, in the training stage, each MRI data of a patient was normalized before feeding the images to the models. To be specific, voxels of a MRI sequence were normalized as the average was 0 and the variance was 1. Considering the limited number of cases, we used random flipping along axis with a 0.5 ratio as data augmentation to avoid overfitting. In addition, due to the limited memory, we randomly cropped the MRI data from $240 \times 240 \times 155$ voxels to $128 \times 128 \times 128$ voxels. For testing, we adopted zero padding to original images to generate the resolution of input images as $240 \times 240 \times 160$ voxels.

3.3 Implementation Details

We conducted experiments to evaluate our proposed methods with PyTorch 1.6.0 and Python 3.6. The number of training iterations was set as 500 epochs. Adam optimizer was used to update the parameters, with the learning rate of 1e−3, weight decay of 1e−5 and batch size of 2. In each branch of dynamic convolutional networks, we adopted two dynamic 3D convolutional layers with $3 \times 3 \times 3$ kernel size followed by a batch normalization layer and a ReLU layer as the activation function. After the four branches, a simple concatenation operation was used to fuse the four feature maps to be fed into the following networks. In the MFNet, we used 9 MF units in the encoder and 3 MF units in the decoder. Each MF unit consists of two common 3D convolutional layers and two dilated 3D convolutional layers with $3 \times 3 \times 3$ kernel size. DMFNet was similar to MFNet. It added two $1 \times 1 \times 1$ convolutional layers in each MF unit as DMF units

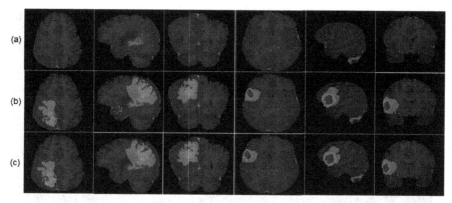

Fig. 4. The prediction results of proposed model on cross-validation of 2020 BraTS Challenge training dataset. (a) left three ones are Flair MRI, right three ones are T1 MRI; (b) our segmentation prediction; (c) ground truth.

and sum feature maps from multiple branches. In the decoder, trilinear interpolation was employed to upsample the feature maps. We applied softmax dice loss with square as the type of weighting. Experiments were conducted on two Nvidia GeForce RTX 2080Ti GPUs with 11 GB memory and the training time were about 30 h for 500 epochs.

3.4 Comparison with Previous Methods

We reported the results of our method on BraTS 2020 testing data and cross-validation data. Cross-validation data was split into three quarters dataset from original training datasets as our own training dataset and the other quarter as our validation dataset to be evaluated.

In order to verify the effectiveness of our approach, we compared it with state-of-the-art methods including vanilla 3D U-Net [1] and DMFNet [10]. Table 1 shows the detailed results of our model on the BraTS 2020 dataset. Our proposed framework with DMFNet as based structure performed average dice scores of 91.36%, 84.91% and 74.58% for regions of whole tumor (WT), tumor core (TC) and enhancing tumor core (ET), respectively, and our model achieved performance of 82.30%, 76.28% and 70.52% on those three regions respectively on testing dataset. We achieved the best dice score on the class of whole tumor and tumor core, with obtaining improvements of 1.84% and 1.37% respectively, and also got competitive performance on the other classes. These results solidly supported that the two difficult tumor regions can be better segmented by dynamic convolutions and multi-branch structures. We calculated the number of parameters for these models to verify the efficiency of our proposed networks. The original DMFNet had 3.88M parameters, after adding CondConv and multi-branch modules, total parameters increased slightly up to 5.92M and 4.11M respectively. To show the visualization results of our final methods on brain tumor segmentation, we presented our prediction maps with corresponding original inputs and ground truth in Fig. 4. Additionally, Fig. 5(b) and (d) showed some of segmentation results of our proposed model on the validation and testing datasets respectively.

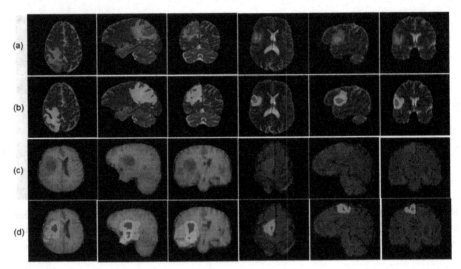

Fig. 5. Our segmentation results on validation and testing dataset. (a) validation MRI; (b) segmentation prediction on (a); (c) testing data; (d) segmentation prediction on (c).

Table 1. Dice score and Hausdorff distance of our segmentation approach on BraTS 2020 dataset. The bottom row shows our submitted results evaluated on testing dataset, other rows show results on cross-validation dataset split from training dataset. DSC: dice similarity coefficient (%), HD95: Hausdorff distance (95%).

Method	DSC			HD95		
Validation	WT	WT	ET	WT	TC	ET
Vanilla 3D U-Net	88.53	71.77	75.96	17.10	11.62	6.04
DMFNet	89.52	83.54	**76.2**	4.59	**9.09**	**13.52**
DMFNet + CondConv	88.92	84.71	72.1	4.26	15.72	14.2
DMFNet + Multi-branch	90.54	80.2	72.93	**4.07**	15.23	15.65
DMFNet + CondConv + Multi-branch (Ours)	**91.36**	**84.91**	74.58	4.81	10.26	14.43
Ours (on testing dataset)	82.30	76.28	70.52	10.33	34.13	28.35

Furthermore, we evaluated our latest work with boundary-aware module, named EdgeNet, to capture shape prior for each class and used the modified 3D U-net as backbone. The results of EdgeNet are presented in Table 2. For evaluating it, we also utilize aforementioned cross-validation dataset to conduct ablation experiments. Compared with our based model modified 3D Unet, our proposed boundary-aware branch delivers superior performance on the class of enhancing tumor of 80.06% dice score, 0.29% better than the baseline. It also exhibited competitive results on the region of whole tumor and tumor core, with dice score of 90.12% and 78.90% respectively. Performances of EdgeNet on three classes demonstrated that leveraging boundary information as prior

Table 2. Dice score of our latest boundary-aware segmentation approach on BraTS 2020 cross-validation dataset. DSC: dice similarity coefficient (%).

Method	DSC		
Validation	WT	TC	ET
DMFNet	89.52	**83.54**	76.2
3D modified U-net	**90.31**	81.10	79.77
EdgeNet (Ours)	90.12	78.90	**80.06**
EdgeNet + CondConv (Ours)	88.36	76.83	78.46
EdgeNet + Multi-branch (Ours)	88.18	77.35	76.63
EdgeNet + CondConv + Multi-branch (Ours)	86.1	76.44	77.09

can constrain the model learn better. We further showed the performance after adding dynamic layers and the multi-branch module on the EdgeNet, while found this decrease the final dice scores on three regions. We guessed it might be because our EdgeNet had similar ability as the dynamic networks and multi-branch scheme, which meant that EdgeNet framework and the two proposed modules were two duplicate approaches to improve the performance of brain tumor segmentation on the base models of DMFNet and modified 3D Unet.

3.5 Ablation Study

To further demonstrate the ability of each proposed module, we also conducted ablation experiments by adding every novel module one by one to get the results. We extensively studied the performance of the dynamic module and the multi-branch structure on DMFNet and our EdgeNet respectively. Built on DMFNet, dynamic networks boost the dice score on the region of tumor core mainly due to the adaptively learning ability and large capacity of dynamic layers. Multi-branch brought better dice score on the class of whole tumor, which could be attributed to the fact that multi-branch architecture can capture sufficient representations for the whole tumor from four modality inputs efficiently. Experiments on our developed EdgeNet showed that boundary-aware module seemed to fulfill the abilities of both dynamic networks and multi-branch module. The empirical insight was that our EdgeNet had already increased the model capacity and also learn difficult boundary regions well.

4 Conclusion

In this work, we propose a multi-branch network to extract low-level representations from four modalities MRI inputs respectively and employ dynamic convolutional neural networks to learn specialized convolutional kernels for each sample adaptively.

The four branches of networks help to fully exploit information via multi-modal images and the fusion of them is better for further segmentation. Dynamic networks are suitable for segmenting brain tumors with various localizations, scale and shape. This network can help to increase the models' capacity, meanwhile, reduce the complexity by conducting an efficient convolutional operation. The experimental performance on the released training data of 2020 BraTS Challenge demonstrates the effectiveness of our proposed framework with dice scores of 91.36%, 84.91% and 74.58% for regions of WT, TC and ET respectively.

References

1. Milletari, F., Navab, N., Ahmadi, S.: V-Net: fully convolutional neural networks for volumetric medical image segmentation. In: 2016 Fourth International Conference on 3D Vision, Stanford, CA, pp. 565–571 (2016)
2. Jiang, Z., Ding, C., Liu, M., Tao, D.: Two-stage cascaded U-Net: 1st place solution to BraTS challenge 2019 segmentation task. In: International MICCAI Brainlesion Workshop, pp. 231–241. Springer, Cham (2019). https://doi.org/10.1007/978-3-030-46640-4_22
3. Dice, L.R.: Measures of the amount of ecologic association between species. Ecology 26(3), 97–302 (1945)
4. Myronenko, A.: 3D MRI brain tumor segmentation using autoencoder regularization. In: International MICCAI Brainlesion Workshop, pp. 311–320. Springer, Cham (2018). https://doi.org/10.1007/978-3-030-11726-9_28
5. Kingma, D.P., Welling, M.: Auto-encoding variational Bayes. arXiv preprint arXiv:1312.6114 (2013)
6. Zhou, T., Canu, S., Ruan, S.: A review: deep learning for medical image segmentation using multi-modality fusion. arXiv preprint arXiv:2004.10664v2
7. Huo, Y., et al.: Splenomegaly segmentation on multi-modal MRI using deep convolutional networks. IEEE Trans. Med. Imaging 38(5), 1185–1196 (2019)
8. Yang, B., Bender, G., Le, Q.V., Ngiam, J.: CondConv: conditionally parameterized convolutions for efficient inference. In: Advances in Neural Information Processing Systems, pp. 1305–1316 (2019)
9. Yu, F., Koltun, V.: Multi-scale context aggregation by dilated convolutions. arXiv preprint arXiv:1511.07122 (2015)
10. Chen, C., Liu, X., Ding, M., Zheng, J., Li, J.: 3D Dilated Multi-Fiber Network for Real-time Brain Tumor Segmentation in MRI. arXiv Preprint arXiv:1904.03355v5 (2019)
11. Menze, B.H., Jakab, A., Bauer, S., Kalpathy-Cramer, J., Farahani, K., Kirby, J., et al.: The multimodal brain tumor image segmentation benchmark (BRATS). IEEE Trans. Med. Imaging 34(10), 1993–2024 (2015)
12. Bakas, S., Akbari, H., Sotiras, A., Bilello, M., Rozycki, M., Kirby, J.S., et al.: Advancing the cancer genome atlas glioma MRI collections with expert segmentation labels and radiomic features. Nat. Sci. Data 4, 170117 (2017)
13. Bakas, S., Reyes, M., Jakab, A., Bauer, S., Rempfler, M., Crimi, A., et al.: Identifying the Best Machine Learning Algorithms for Brain Tumor Segmentation, Progression Assessment, and Overall Survival Prediction in the BRATS Challenge. In: arXiv preprint arXiv:1811.02629 (2018)
14. Bakas, S., Akbari, H., Sotiras, A., Bilello, M., Rozycki, M., Kirby, J. et al.: Segmentation labels and radiomic features for the pre-operative scans of the TCGA-GBM collection. Cancer Imaging Arch. (2017)
15. Bakas, S., Akbari, H., Sotiras, A., Bilello, M., Rozycki, M., Kirby, J., et al.: Segmentation labels and radiomic features for the pre-operative scans of the TCGA-LGG collection. Cancer Imaging Arch. (2017)

16. Yu, Z., Feng, C., Liu, M., Ramalingam, S.: CASENet: deep category-aware semantic edge detection. In: IEEE Conference on Computer Vision and Pattern Recognition, Honolulu, HI, pp. 1761–1770 (2017)

17. Guo, X., Yang, C., Lam, P.L., Woo, P.Y.M., Yuan, Y.: Domain knowledge based brain tumor segmentation and overall survival prediction. In: International MICCAI Brainlesion Workshop, vol. 11993, pp. 285–295. Springer, Cham (2019). https://doi.org/10.1007/978-3-030-46643-5_28

18. Chen, L., Wu, Y., DSouza, A.M., Abidin, A.Z., Xu, C., Wismüller, A.: MRI tumor segmentation with densely connected 3D CNN. arXix Preprint arXiv:1802.02427 (2018)

19. Kamnitsas, K., et al.: Ensembles of multiple models and architectures for robust brain tumour segmentation. In: Crimi, A., Bakas, S., Kuijf, H., Menze, B., Reyes, M. (eds.) BrainLes 2017. LNCS, vol. 10670, pp. 450–462. Springer, Cham (2018). https://doi.org/10.1007/978-3-319-75238-9_38

20. Wang, G., Li, W., Ourselin, S., Vercauteren, T.: Automatic brain tumor segmentation using cascaded anisotropic convolutional neural networks. In: Crimi, A., Bakas, S., Kuijf, H., Menze, B., Reyes, M. (eds.) BrainLes 2017. LNCS, vol. 10670, pp. 178–190. Springer, Cham (2018). https://doi.org/10.1007/978-3-319-75238-9_16

21. Chen, X., Liew, J.H., Xiong, W., Chui, C.K., Ong, S.H.: Focus, segment and erase: an efficient network for multi-label brain tumor segmentation. In: Proceedings of the European Conference on Computer Vision, pp. 654–669 (2018)

22. Zhou, C., Ding, C., Wang, X., Lu, Z., Tao, D.: One-pass multi-task networks with cross-task guided attention for brain tumor segmentation. arXiv Preprint arXiv:1906.01796 (2019)

23. Dolz, J., Gopinath, K., Yuan, J., Lombaert, H., Desrosiers, C., Ayed, I.B.: HyperDense-Net: a hyper-densely connected CNN for multi-modal image segmentation. arXiv Preprint arXiv:1804.02967 (2018)

24. Hatamizadeh, A., Terzopoulos, D., Myronenko, A.: Edge-Gated CNNs for Volumetric Semantic Segmentation of Medical Images. arXiv Preprint arXiv:2002.0420 (2020)

Computational Precision Medicine:
Radiology-Pathology Challenge
on Brain Tumor Classification

Automatic Glioma Grading Based on Two-Stage Networks by Integrating Pathology and MRI Images

Xiyue Wang[1], Sen Yang[2(✉)], and Xiyi Wu[3]

[1] College of Computer Science, Sichuan University, Chengdu 610065, China
[2] College of Biomedical Engineering, Sichuan University, Chengdu 610065, China
[3] Shanghai Jiao Tong University, Shanghai 200240, China

Abstract. Glioma with a high incidence is one of the most common brain cancers. In the clinic, pathologist diagnoses the types of the glioma by observing the whole-slide images (WSIs) with different magnifications, which is time-consuming, laborious, and experience-dependent. The automatic grading of the glioma based on WSIs can provide aided diagnosis for clinicians. This paper proposes two fully convolutional networks, which are respectively used for WSIs and MRI images to achieve the automatic glioma grading (astrocytoma (lower-grade A), oligodendroglioma (middle-grade O), and glioblastoma (higher-grade G)). The final classification result is the probability average of the two networks. In the clinic and also in our multi-modalities image representation, grade A and O are difficult to distinguish. This work proposes a two-stage training strategy to exclude the distraction of the grade G and focuses on the classification of grade A and O. The experimental result shows that the proposed model achieves high glioma classification performance with the balanced accuracy of 0.889, Cohen's Kappa of 0.903, and F1-score of 0.943 tested on the validation set.

Keywords: Tumor classification · Digital pathology · Magnetic resonance image · Brain tumor

1 Introduction

Glioma is one common brain tumor, which occupies approximately 80% of malignant brain tumors [8]. According to its severity degree evaluated on the pathology images, world health organization (WHO) categorizes glioma as three grades: astrocytoma (lower grade), oligodendroglioma (middle grade), and glioblastoma (higher grade). The lower grade has more optimistic and more survival years. However, the higher grades with worse prognosis are usually life-threatening. In the clinic, the glioma is diagnosed depending on the histopathology technique on the microscopic examination environment. The accurate diagnosis of lesions for pathologists is very time-consuming, laborious, and expertise-dependent. The

© Springer Nature Switzerland AG 2021
A. Crimi and S. Bakas (Eds.): BrainLes 2020, LNCS 12659, pp. 455–464, 2021.
https://doi.org/10.1007/978-3-030-72087-2_40

computer-aided diagnosis is highly required to alleviate the difficulties of the pathologists.

The emergence of whole-slide images (WSIs) technology has realized the transformation from a microscopic perspective to a computer perspective and promoted the application of image processing technology to digital pathology. The digital pathology image analysis can help pathologists diagnose and provide quantitative information calculated from the WSIs, achieving the objectivity and reproducibility in the clinical diagnosis.

Magnetic resonance image (MRI) as a non-invasive imaging technique has been routinely used in the diagnosis of brain tumors. Multi-modalities with different brain tissue enhancement can be selected to make a clinical decision. Multi-modalities fusion technology can capture more abundant feature information to perform more precise tumor classification. MRI with the advantage of safety and non-invasion has been used to classify glioma [5,6,16,18]. However, pathological information acquired by invasive methods is adopted as the gold standard in the current clinical environment. The integration of the two types of images could achieve higher glioma grading performance.

The CPM-RadPath 2020 MICCAI challenge releases a multi-modalities dataset which contains paired MRI scans and histopathology images of brain gliomas collected from the same patients. Their glioma classification annotations (astrocytoma, oligodendroglioma, and glioblastoma) have been provided for the training set. Based on this dataset, this work trains one 3D Densenet for MRI images classification and 2D fully constitutional networks (EfficientNet-B2, EfficientNet-B3, and SE-ResNext101) for pathology classification. The final glioma grading is determined by the average of the two types of models.

2   Related Work

In recent years, automatic glioma classification and grading have attracted widespread attention using machine learning and deep learning techniques.

A majority of these methods adopt the MRI image as their experiment data since MRI modality is a non-invasive and fast imaging technique and is routinely used for glioma diagnosis in the clinic. These MRI-based glioma grading methods can be mainly divided into two categories: hand-crafted feature engineering and deep learning-based feature representation. These hand-crafted features are usually extracted based on the ROI (region of interest) region that is delineated by experienced radiologists or some automatic image segmentation techniques. These extracted features comprise histogram-based features [7], shape features [19], texture features [13], contour feature[13], and wavelet features [15]. Based on these features, some machine learning techniques including SVM (support vector machines), RF (random forest), NN (neural network), and DT (decision tree) are used to achieve the automatic glioma classification or grading. However, these traditional feature engineering techniques fail to generate a robust and general feature parameter and are easily affected by the data variations (MRI scanner, data collection protocol, and image noises), which limits their promise for clinical application. Deep learning has the ability to learn high-level feature representation from the raw data through the network training procedure. Due to the 3D nature of MRI images, 2D and 3D models are alternative. These 3D models could contact the context information during each MRI scan, such as the 3-D multiscale CNN model [7] and the 3D residual network [5]. It is known that the 3D model extremely increases the number of parameters and requires a large amount of computation resources to support network training. These 2D models, such as the VGG model [1,4] and Residual Network (ResNet) architecture [1,9], ignore the connection between MRI slices and treat each MRI slice as an independent image to complete the classification task.

The previously mentioned MRI-based studies have achieved limited performance since the gold standard for the glioma diagnosis is from the pathology images. Thus, a combination of the MRI and pathology images may provide complementary information and achieve finer classification results. The previous CPM challenge has reported four solutions for the glioma grading by combining MRI and pathology images [10,12]. Pei et al. [12] used an Unet-like model to segment ROI and a 3D CNN to classify the glioma types focusing on the extracted course tumor region. Their experimental results reported that using MRI sequences alone can realize better performance than using the pathology alone or the combination of MRI and pathology images. Ma et al. [10] applied a 2D ResNet-based model to classify glioma based on pathology and a 3D DenseNet-based model to classify glioma based on MRI images. Then, a simple regression model was used to achieve the ensemble of the two models. Chan et al. [2] extracted features by using two CNN networks (VGG16 and ResNet 50) and then classified three types of brain tumors based on several clustering methods (K-means and random forest). Xue et al. [17] trained a 2D ResNet18 and a 3D ResNet18 to classify three types of gliomas based on the pathology and MRI images, respectively. The features in fully connected layers of the two models were concatenated together as the input of the following softmax layer to achieve the classification of brain tumors.

Our work is different from the above-mentioned methods. We apply a two-stage classification algorithm to first detect the glioblastoma and then pay more attention to the distinction between the astrocytoma and oligodendroglioma. Sine the glioblastoma is defined as the Grade IV and astrocytoma and oligodendroglioma are both defined as the Grade II or III by WHO. Thus, astrocytoma and oligodendroglioma can be more difficult to separate. Our two-stage training has the ability to improve classification performance for astrocytoma and oligodendroglioma cases.

3   Method

This paper applies two fully convolutional networks to achieve a feature-independent end-to-end glioma grading. In the following, we introduce the image preprocessing and network framework in detail.

3.1 Data Preprocessing

The data used in this paper includes paired pathology and multi-sequence MRI (T2-FLAIR, T1, T1ce, and T2) images. Figure 1 and Fig. 2 illustrate the pathology and corresponding four modalities MRI images in terms of three glioma types (astrocytoma, glioblastoma, and oligodendroglioma).

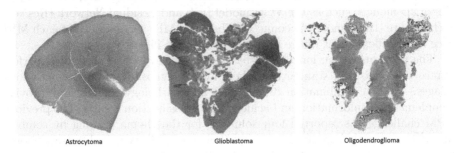

Astrocytoma   Glioblastoma   Oligodendroglioma

Fig. 1. Visualization of astrocytoma (A), glioblastoma (G), and oligodendroglioma (O) in pathology image

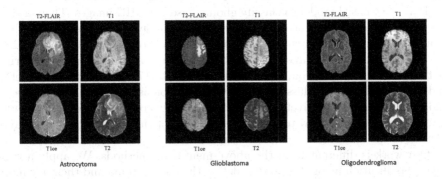

Astrocytoma   Glioblastoma   Oligodendroglioma

Fig. 2. Visualization of astrocytoma (A), glioblastoma (G), and oligodendroglioma (O) in MRI image. For each MRI case, four modalities are included (T2-FLAIR, T1, T1ce, and T2).

For the preprocessing procedure of the pathology images, the OTSU method is firstly adopted to remove the non-tissue region [10]. The original WSIs with large size (e.g., 50000 × 50000) can not be directly fed into the network as the input, we apply a sliding window with no overlap to segment the WSI into image patch with the size of 1024×1024 (Fig. 3). Not every path contains tumor tissue, so we set some rules for each path to exclude these useless patches. First, the mean value of the patch is limited from 100 to 200. Second, the standard deviation of each patch is limited to greater than 20. Third, in the color space HSV (hue, saturation, value), the mean of channel 0 is set as greater than 50.

In the preprocessing procedure of MRI images, the four MRI modalities images are all cropped to 128×192×192.

3.2   Model Details

Figure 4 illustrates the overall architecture of our proposed glioma grading system which is composed of two types of networks: 2D CNN for pathology image classification (Fig. 5) and 3D CNN for MRI image classification (Fig. 6). Each type of network performs the two-stage classification. Since the astrocytoma and oligodendroglioma appear as more similar, the first stage achieves the detection of glioblastoma, after that, the second stage distinguishes the astrocytoma and oligodendroglioma.

For the 2D pathology image classification network, the input image is a small WSI patch that is segmented from a large WSI in the preprocessing procedure. The backbone networks include EfficientNet-B2, EfficientNet-B3, and SE-ResNext101. Before the fully connected layer, generalized-mean (GEM) pooling is applied to the learned features, which is defined as

$$\mathbf{f}^{(g)} = \left[\mathbf{f}_1^{(g)} \dots \mathbf{f}_k^{(g)} \dots \mathbf{f}_K^{(g)}\right]^{\top}, \quad \mathbf{f}_k^{(g)} = \left(\frac{1}{|\mathcal{X}_k|} \sum_{x \in \mathcal{X}_k} x^{p_k}\right)^{\frac{1}{p_k}} \tag{1}$$

where $\mathcal{X}$ and $\mathbf{f}$ represent the input and output, respectively. When p_k equals to ∞ and 1, the equation denotes the max pooling and average pooling, respectively. Following [14], this work set p_k to 3.

After the GEM operation, the Meta info (age information) is added to the final classification feature vector. Then, a classification and a regression branches are appended as the cross-entropy (L_{BCE}) and smooth L1 (L1) losses, respectively, to achieve more robust brain tumor classification.

$$L_{BCE} = -\sum_l \left[(y_l \log \hat{y}_l) + (1 - y_l) \log (1 - \hat{y}_l)\right] \tag{2}$$

$$L_{\text{loc}} = \sum_l \text{smooth}_{L_1}(\hat{y}_l - y_l) \tag{3}$$

$$\text{smooth}_{L_1}(x) = \begin{cases} 0.5x^2 & \text{if } |x| < 1 \\ |x| - 0.5 & \text{otherwise} \end{cases} \tag{4}$$

where y_l and $\hat{y}_l$ denote the ground truth and predicted annotations.

For the 3D MRI image classification network, the input image is MRI images with four channels, which correspond to the four modalities. The backbone adopts 3D ResNet, following by global average pooling and fully connected layer to grade the brain tumor. In the MRI-based classification process, the loss function also adopts the cross-entropy. To minimize the loss function, Adam optimization algorithm is used.

The ensemble of the 2D pathology and 3D MRI classification models is the probability average of the two types of networks.

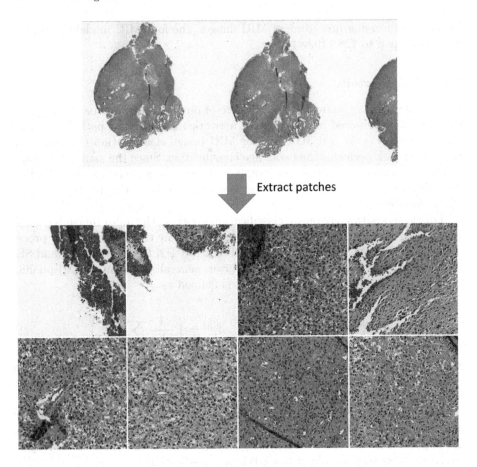

Fig. 3. The preprocessing process for the pathology image

4   Results and Discussion

4.1   Data and Evaluation Metrics

Dataset. The CPM-RadPath 2020 MICCAI challenge provides paired radiology scans and digitized histopathology images of brain gliomas, and image-level ground truth label as well. The goal of CPM2020 is classifying each case into three sub-types: Glioblastoma (grade IV), Oligodendroglioma (grade II or III), and Astrocytoma (grade II or III).

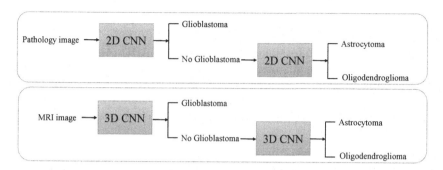

Fig. 4. The overall architecture of our glioma grading system. A two-stage classification strategy is applied to both the 2D pathology and 3D MRI images. The glioblastoma with more serious anatomy representation is detected in the first step. Then, in the second step, our algorithm focuses on the classification of astrocytoma and oligodendroglioma.

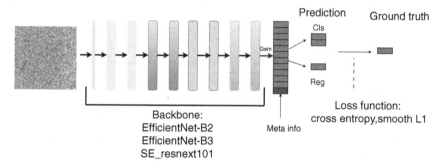

Fig. 5. The detailed 2D CNN network. The backbone includes EfficientNet-B2, EfficientNet-B3, and SE-ResNext101. In the final feature representation, the Meta info (age information) is included. A regression branch with a smooth L1 loss function is added to relieve the overfitting. The classification branch with the cross-entropy loss function is used to complete the classification procedure.

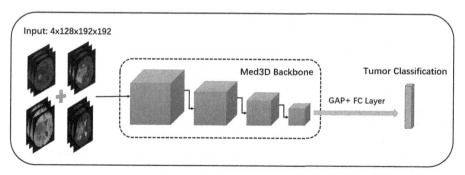

Fig. 6. The detailed 3D CNN network. The four MRI modalities are integrated as the network input. All the images are cropped to a fixed size of 128×192×192. The backbone adopts the 3D ResNet, following by global average pooling and fully connected layer to grade the brain tumor.

Specifically, for each patient, the provided 3D MRI images comprise four modalities: native (T1), post-contrast T1-weighted (T1Gd), T2-weighted (T2), and T2 Fluid Attenuated Inversion Recovery (T2-FLAIR). All MRI images were co-registered to the same anatomical template and interpolated to the same resolution (1 cubic mm) in three directions. Also, the contest provides one digitized whole slide tissue image for each case, which were scanned at 20x or 40x magnifications. The dataset is divided into 221, 35, 73 cases for training, validation, and testing respectively.

Evaluation Metrics. The algorithmic performance is evaluated from three aspects: F1-Score, Balanced Accuracy, and Cohen's Kappa. Suppose TP, FP, FN is the number of true positives, false positives, and false negatives respectively. Then the three metrics can be computed as below.

$$Sensitivity = \frac{TP}{TP + FN} \tag{5}$$

$$Precision = \frac{TP}{TP + FP} \tag{6}$$

$$F1\_Score = \frac{2 * (Precision * Sensitivity)}{(Precision + Sensitivity)} \tag{7}$$

$$Balanced\ Accuracy = \sum_{Class=1}^{K} Sensitivity/K \tag{8}$$

$$Kappa = \frac{p_o - p_e}{1 - p_e} \tag{9}$$

where K is the number of classes, p_o is the relative observed agreement among raters, and p_e is the hypothetical probability of chance agreement, using the observed data to calculate the probabilities of each observer randomly seeing each category [3]. The aforementioned evaluation metrics are suitable for imbalanced categories, avoiding depending on the proportion of each class.

4.2 Experiments and Discussion

In the training stage, we perform 5-fold cross validation for local validation. Adam optimizer is used with an initial learning rate of 0.001 and decreases by 10 times every 50 epochs. All models are implemented based on the Pytorch framework [11] and trained on a workstation with Intel(R) Xeon(R) CPU E5-2680 v4 2.40 GHz processors and four NVIDIA Tesla P40s (24 GB) installed.

In the inference stage, the multimodal MRI images and the whole-slide pathology images are pre-processed and sent to the classification network respectively. The average predictions of the 5-fold models are employed to get more accurate results. Then, we sum the probabilities of the two networks to obtain the final ensemble prediction.

As the online evaluation performances shown in Table 1, our proposed two-stage coarse-to-fine classification framework contributes to gaining higher accuracy on all three evaluation metrics. What's more, the classification model of MRI and pathological images can complement each other to obtain more robust and accurate results.

Table 1. Online results on CPM2020 validation data

Method	Balanced_acc	Kappa	F1_micro
ResNet50 only MRI (One stage)	0.700	0.665	0.800
ResNet50 only MRI (Two stage)	0.733	0.712	0.829
Efficientb2 only Pathology (One stage)	0.767	0.758	0.857
Efficientb2 only Pathology (Two stage)	0.822	0.808	0.886
Ensemble	0.889	0.903	0.943

5   Conclusion

This paper proposed a two-stage glioma grading algorithm to classify the brain tumor into three types: astrocytoma, glioblastoma, and oligodendroglioma. The classification algorithm is designed based on the feature representation difference between the severe and lower glioma grades. The more serious glioblastoma grade is separated out in the first stage, and the second stage eliminates the interference of glioblastoma and only focuses on learning the difference between astrocytoma and oligodendroglioma. Our two-stage strategy is applied on the classification networks for the pathology (2D CNN) and MRI images (3D ResNet), respectively. By testing on the validation data, we have achieved state-of-the-art performance by the ensemble of the 2D pathology and 3D MRI images classification networks. In the final submission of this challenge, we omit the 3D MRI image classification network, since the diagnosis based on pathology is adopted as the gold standard in the clinic. In this CPM-RadPath 2020 MICCAI challenge, the number of training samples is limited, which can greatly influence the robustness of designed network. In future work, small sample based deep learning could be developed to build a more general model.

Acknowledgement. This research was funded by the National Natural Science Foundation of China, grant number 61571314.

References

1. Banerjee, S., Mitra, S., Masulli, F., Rovetta, S.: Glioma classification using deep radiomics. SN Comput. Sci. **1**(4), 1–14 (2020)
2. Chan, H.-W., Weng, Y.-T., Huang, T.-Y.: Automatic classification of brain tumor types with the MRI scans and histopathology images. In: Crimi, A., Bakas, S. (eds.) BrainLes 2019. LNCS, vol. 11993, pp. 353–359. Springer, Cham (2020). https://doi.org/10.1007/978-3-030-46643-5_35

3. Cohen, J.: A coefficient of agreement for nominal scales. Educ. Psychol. Measur. **20**(1), 37–46 (1960)

4. Decuyper, M., Bonte, S., Van Holen, R.: Binary glioma grading: radiomics versus pre-trained CNN features. In: Frangi, A.F., Schnabel, J.A., Davatzikos, C., Alberola-López, C., Fichtinger, G. (eds.) MICCAI 2018. LNCS, vol. 11072, pp. 498–505. Springer, Cham (2018). https://doi.org/10.1007/978-3-030-00931-1_57

5. Decuyper, M., Holen, R.V.: Fully automatic binary glioma grading based on pre-therapy MRI using 3D convolutional neural networks (2019)

6. Fusun, C.E., Zeynep, F., Ilhami, K., Ugur, T., Esin, O.I.: Machine-learning in grading of gliomas based on multi-parametric magnetic resonance imaging at 3t. Comput. Biol. Med. **99**, 154–160 (2018)

7. Ge, C., Qu, Q., Gu, I.Y.H., Jakola, A.S.: 3D multi-scale convolutional networks for glioma grading using MR images. In: 2018 25th IEEE International Conference on Image Processing (ICIP), pp. 141–145. IEEE (2018)

8. Goodenberger, M.K.L., Jenkins, R.B.: Genetics of adult glioma. Cancer Genet. **205**(12), 613–621 (2012)

9. Ismael, S.A.A., Mohammed, A., Hefny, H.: An enhanced deep learning approach for brain cancer MRI images classification using residual networks. Artif. Intell. Med. **102**, 101779 (2020)

10. Ma, X., Jia, F.: Brain tumor classification with multimodal MR and pathology images. In: Crimi, A., Bakas, S. (eds.) BrainLes 2019. LNCS, vol. 11993, pp. 343–352. Springer, Cham (2020). https://doi.org/10.1007/978-3-030-46643-5_34

11. Paszke, A., et al.: Automatic differentiation in pytorch (2017)

12. Pei, L., Vidyaratne, L., Hsu, W.-W., Rahman, M.M., Iftekharuddin, K.M.: Brain tumor classification using 3D convolutional neural network. In: Crimi, A., Bakas, S. (eds.) BrainLes 2019. LNCS, vol. 11993, pp. 335–342. Springer, Cham (2020). https://doi.org/10.1007/978-3-030-46643-5_33

13. Pintelas, E., Liaskos, M., Livieris, I.E., Kotsiantis, S., Pintelas, P.: Explainable machine learning framework for image classification problems: case study on glioma cancer prediction. J. Imaging **6**(6), 37–50 (2020)

14. Radenović, F., Tolias, G., Chum, O.: Fine-tuning CNN image retrieval with no human annotation. IEEE Trans. Pattern Anal. Mach. Intell. **41**(7), 1655–1668 (2018)

15. Su, C., et al.: Radiomics based on multicontrast MRI can precisely differentiate among glioma subtypes and predict tumour-proliferative behaviour. Eur. Radiol. **29**(4), 1986–1996 (2019)

16. Wang, X., et al.: Machine learning models for multiparametric glioma grading with quantitative result interpretations. Frontiers Neurosci. **12**, 1–10 (2019)

17. Xue, Y., et al.: Brain tumor classification with tumor segmentations and a dual path residual convolutional neural network from MRI and pathology images. In: Crimi, A., Bakas, S. (eds.) BrainLes 2019. LNCS, vol. 11993, pp. 360–367. Springer, Cham (2020). https://doi.org/10.1007/978-3-030-46643-5_36

18. Yang, Y., et al.: Glioma grading on conventional MR images: a deep learning study with transfer learning. Frontiers Neurosci. **12**, 1–10 (2018)

19. Zhang, Z., et al.: Deep convolutional radiomic features on diffusion tensor images for classification of glioma grades. J. Digit. Imaging **33**(4), 826–837 (2020). https://doi.org/10.1007/s10278-020-00322-4

Brain Tumor Classification Based on MRI Images and Noise Reduced Pathology Images

Baocai Yin[1,2], Hu Cheng[2], Fengyan Wang[2], and Zengfu Wang[1(✉)]

[1] University of Science and Technology of China, HeFei, China
zfwang@ustc.edu.cn
[2] iFLYTEK Research, HeFei, China
{bcyin,hucheng,fywang6}@iflytek.com

Abstract. Gliomas are the most common and severe malignant tumors of the brain. The diagnosis and grading of gliomas are typically based on MRI images and pathology images. To improve the diagnosis accuracy and efficiency, we intend to design a framework for computer-aided diagnosis combining the two modalities. Without loss of generality, we first take an individual network for each modality to get the features and fuse them to predict the subtype of gliomas. For MRI images, we directly take a 3D-CNN to extract features, supervised by a cross-entropy loss function. There are too many normal regions in abnormal whole slide pathology images (WSI), which affect the training of pathology features. We call these normal regions as noise regions and propose two ideas to reduce them. Firstly, we introduce a nucleus segmentation model trained on some public datasets. The regions that has a small number of nuclei are excluded in the subsequent training of tumor classification. Secondly, we take a noise-rank module to further suppress the noise regions. After the noise reduction, we train a gliomas classification model based on the rest regions and obtain the features of pathology images. Finally, we fuse the features of the two modalities by a linear weighted module. We evaluate the proposed framework on CPM-RadPath2020 and achieve the first rank on the validation set.

Keywords: Gliomas classification · Noise-rank · Nucleus segmentation · MRI · Pathology

1 Introduction

Gliomas are the most common primary intracranial tumors, accounting for 40% to 50% of all cranial tumors. Although there are many systems for gliomas grading, the most common one is the World Health Organization (WHO) grading system, which classifies gliomas into grade 1 (least malignant and best prognosis) to grade 4 (most malignant and worst prognosis). According to the pathological malignancy of the tumor cells, brain gliomas are also classified into low-grade gliomas (WHO grade 1 to 2) and high-grade gliomas (WHO grade 3 to 4).

Magnetic resonance imaging (MRI) is the common examination method for glioma. In MRI images, different grades of gliomas show different manifestations: low-grade

© Springer Nature Switzerland AG 2021
A. Crimi and S. Bakas (Eds.): BrainLes 2020, LNCS 12659, pp. 465–474, 2021.
https://doi.org/10.1007/978-3-030-72087-2_41

gliomas tend to show T1 low signal, T2 high-signal. They are mainly located in the white matter of the brain and usually have a clear boundary with the surrounding tissue. However, high-grade gliomas usually show heterogeneous signal enhancement, and the boundary between tumor and surrounding brain tissue is fuzzy. As mentioned in previous works [4, 5], MRI is mainly used to identify low-grade gliomas and high-grade gliomas, while it alone cannot accurately identify astrocytoma (grade II or III), oligodendroglioma (grade II or III), and glioblastoma (grade IV).

The current standard diagnosis and grading of brain tumors are done by pathologists according to Hematoxylin and Eosin (H&E) staining tissue sections fixed on glass slides under an optical microscope after resection or biopsy. Different subtypes of stained cells often have unique characteristics that pathologists use to classify glioma subtypes. However, some gliomas with a mixture of features from different subtypes are difficult for pathologists to distinguish accurately. In addition, the entire process is time-consuming and is prone to sampling errors. The previous studies [6–8, 10] have shown that CNN can be helpful for improving the accuracy and efficiency of pathologists. The whole slide pathological image is too large for end-to-end training, hence, typical solutions are patch-based, in which they assign the whole image label directly to the randomly sampled patches. This process introduces quite a lot of noisy samples inevitably and affects the network training.

In this paper, we intend to reduce the noise samples in the training stage. Main contributions of this paper can be highlighted as follows:

(1) A nucleus segmentation model is adopted to filter patches with a small number of nuclei.
(2) We stack the nucleus segmentation mask with the original image as the input to increase the attention on the nucleus.
(3) The noise-rank module in study [9] is used to further suppress noise samples. In addition to CNN features, we also introduce the LBP feature as complementary.

2   Related Work

The goal of CPM-RadPath is to assess automated brain tumor classification algorithms developed on both radiology and histopathology images. Pei et al. [11] proposed to use the tumor segmentation results for glioma classification, and experimental results demonstrated the effectiveness. However, the pathological images were not employed in their work. Weng et al. [12] used the pre-trained VGG16 models to extract pathological features and a pyradiomics module [13] to extract radiological features, which were not updated by the CPM-RadPath data. Yang et al. [14] introduced a dual path residual convolutional neural network model and trained it by MRI and pathology images simultaneously. This work did not impose any constraints in the patch sampling process from the whole slide images, which could introduce noisy patches during training inevitably. [15] won the first-place in CPM-RadPath 2019, which set several constraints to prevent sampling in the background. Unfortunately, not all the patches sampled in the foreground contained diseased cells. These noisy patches have no discriminative features and will affect the training. In order to remove the noisy patches as much as possible, and dig more informative patches, we propose a delicate filtering strategy and a noise-rank module in this work.

3   Dataset and Method

In this section, we describe the dataset and our solution including preprocessing and networks.

3.1   Dataset

The training dataset of CPM-RadPath-2020 [1] consists of 221 paired radiology scans and histopathology images. The pathology data are in.tiff format, and there are four types of MRI data: Flair, T1, T1Ce, and T2. The corresponding pathology and radiology examples are shown in Fig. 1.

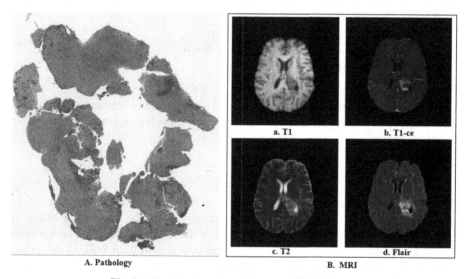

A. Pathology   B. MRI

Fig. 1. Visualization of pathology and MRI images.

The patients' age information was also provided in the dataset. The age distribution of different subtypes was shown in Fig. 2. Low-grade astrocytoma and glioblastoma have different age distributions. Low-grade astrocytoma is predominantly in the lower age groups, while glioblastoma is predominantly in the higher age groups.

CPM-RadPath2020 aims to distinguish between three subtypes of brain tumors, namely astrocytoma (Grade II or III), oligodendroglioma (Grade II or III) and glioblastoma (Grade IV). The number of each subtype in the training data is shown in Table 1.

Except for the CPM-RadPath2020 dataset, two public datasets are also utilized: the Multimodal Brain Tumor Segmentation Challenge 2019 (BraTS-2019) [2] and

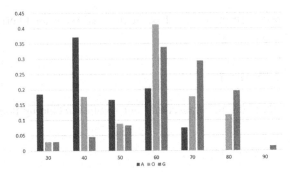

Fig. 2. Age distribution of different subtypes

Table 1. Data distribution of different subtypes in the training set

Subtype	A	O	G	Total
Nums	54	34	133	221

A = astrocytoma (Grade II or III)
O = oligodendroglioma (Grade II or III)
G = glioblastoma (Grade IV)

MoNuSAC [3]. The BraTS-2019 dataset is used to train a tumor segmentation model of MRI images and MoNuSAC dataset is used to train a nucleus segmentation model of pathology images.

3.2 Preprocessing

Pathology. Due to limited computational resources, it is not feasible to process the whole image directly, so we performed patch extraction on the whole slide image. In each WSI, only a few cells were stained, most of the regions were white background, we need to find an efficient way to extract valid information from WSI. In this paper, the WSI was first down-sampled to generate a thumbnail, and then these thumbnails were binarized to distinguish between foreground and background. We only sampled patches for training from the foreground. In addition, some of the data also had staining anomalies, we refer to the work [15] to constrain the extracted patch.

The above operations can filter out most of the background patches, but there are still some patches with less information or abnormal staining (e.g. Fig. 3c or d). Therefore, the nucleus segmentation model was added to suppress the noise patches by counting the proportion of nuclei in the patch. Figure 3 lists four different types of patch data.

The size of WSI in the training set has a wide range, which means the number of the extracted patches also differs from each WSI. To balance the training patches from different WSIs, we set a maximum number of extracted patches from one WSI as 2000. The extracted patch was assigned the same label as the corresponding WSI, as most studies [5] do. This process is unavoidable to bring in some noisy patches, which may affect the model training. We provide some solutions detailed in the next section.

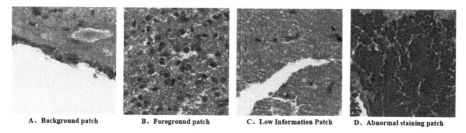

A、Background patch   B、Foreground patch   C、Low Information Patch   D、Abnormal staining patch

Fig. 3. Visualization of the extracted patches from pathological whole slide images

Due to differences in the staining process of the slices, WSIs have a big variance in color. The general practice is color normalization [17]. For the sake of simplicity, this paper adopted the same strategy as study [15] to directly convert RGB images into gray images.

The whole pipeline of the preprocessing step is shown in Fig. 4.

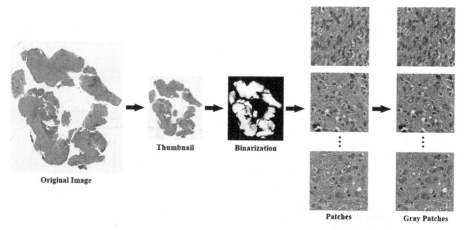

Original Image   Thumbnail   Binarization   Patches   Gray Patches

Fig. 4. Preprocessing of the pathological image

The pre-processing failed for one WSI (CPM19_TCIA06_184_1.tiff), depicted in Fig. 5. We could not sample any valid patches from it. Hence, this WSI was discarded during the training.

Radiology. For each patient, there are four types of MRI images, with a size of 240 * 240 * 155. We first used BraTS2019 dataset to train a segmentation model of the tumors and then extracted the tumor regions according to the segmentation mask. The extracted regions were resized to a fixed size of 160 * 160 * 160. The extracted regions from different MRIs of the same patient were concatenated forming a 4D data, with a size of 4 * 160 * 160 * 160. Z-score normalization was performed for a fast convergence.

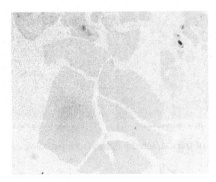

Fig. 5. Thumbnail of the discarded WSI (CPM19_TCIA06_184_1.tiff)

3.3 Classification Based on MRI Images and Noise Reduced Pathology Images

The pipeline of the proposed framework is shown in the Fig. 6, including a pathological classification network, a radiological classification network and a features fusion module. Next, we will detail each module.

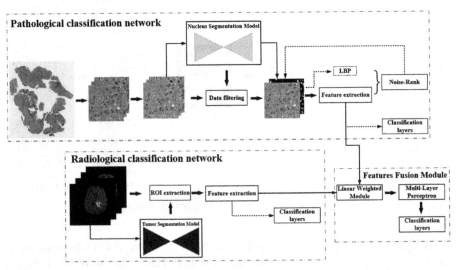

Fig. 6. Pipeline of the proposed framework. The black dotted lines indicate that these processes are only used in the training phase. (Color figure online)

Pathological Classification Network. The nucleus segmentation Model was trained on the MoNuSAC dataset. After the training, we used it to filter the patches whose total area of nuclei was less than 7% of the patch size. The rest patches were transformed into gray images and resized to 256 * 256. To force the classification model to focus on the nuclei, we stitched the original gray image with the nucleus segmentation mask. The final input size of the classification model was 2 * 256 * 256. We adopted a unified

Densenet network structure [16] for feature extraction, and set the numbers of dense blocks in different stages as 4, 8, 12, 24.

Directly assigning the whole image label to the patches inevitably introduced noisy samples, so we took a noise-rank [9] module to filter out those samples with high noise probability. In the noise-rank module, we first used K-means to cluster in the feature space for each subtype. Besides the CNN features, we also utilized LBP features [18] for the clustering, which is visually discriminative and a complementary to CNN features. The center of each cluster was regarded as a prototype, with an original label of its corresponding subtype. We can also obtain a predicted label by KNN for each prototype. Based on the original and predicted label of these prototypes, the posterior probability that indicates the label noise for all the samples was estimated. We then ranked all the samples according to the probability and dropped the top 20% samples in the training of classification model. The detail of noise-rank could be found in the study [9]. The noise-rank module and the classification model were trained alternatively.

The loss function was cross entropy. To avoid overfitting, the same augmentation methods as the study [19] were used, including Random-Brightness, Random-Contrast, Random-Saturation, Random-Hue, Flip and Rotation.

Radiological Classification Network. We extracted regions of interest (ROI) by a lesion segmentation model pre-trained on BraTS2019 [2] as the input to the classification network. The backbone of the network was a 3D-Densenet [16]. The numbers of dense blocks in different stages were 4, 8, 12. The loss function was cross entropy. Dropout and augmentation were also used to avoid over-fitting.

Features Fusion. After the training of the above two classification models, we extracted the features from the two models. To keep the same feature length of different pathological images, we only selected a fixed number of patches to represent the whole pathological image. Then we predicted a weight for each modality to fuse the two features. The fused features were sent to a multi-layer perception (MLP) for glioma classification.

4 Experiments

4.1 Implementation Details

The tumor segmentation model of radiology images was trained on TensorFlow [20] and all the other models were based on MXNet [21]. Parameters were optimized by SGD [22], and the weight decay and momentum were set to 1e−4 and 0.95 respectively. All of our models were trained for 200 epochs on a TeslaV100 GPU. The learning rate was initially set to 0.001 and was divided by 10 at 50% and 75% of the total training epochs. The Noise-Rank module was updated every 20 epochs during the training.

4.2 Results and Discussion

In this section, we will present our experimental results on CPM-RadPath-2020. Since the number of training data is small, we perform 3-forder cross validation on the training data for all the experiments.

Figure 2 shows that low-grade astrocytoma and glioblastoma have different age distributions. We tried to combine the age information with the CNN features after global average pooling. However, the results did not show any improvement brought by the age information. So we discarded the age information in the subsequent experiments.

We first show the classification results based on only pathology images in Table 2. The baseline model was trained on original gray image patches directly with a lot of noisy samples. When we filtered out some noisy patches with a small number of nuclei and concatenated the nuclei segmentation mask with the original gray image patches, all the indexes got a significant increase. And noise-rank module brought a further increase. These results demonstrated the effectiveness of the proposed noise reduction algorithm.

Table 2. Classification results based on pathology images.

Model	Data	Balanced-ACC	F1-micro	Kappa
Baseline	Cross validation	0.814	0.887	0.794
+ nuclei segmentation	Cross validation	**0.853**	**0.894**	**0.812**
+ noise-rank	Cross validation	**0.877**	**0.917**	**0.852**

The results of radiology images were quite lower than the pathology images, as shown in Table 3. We figure out that the reason is that MRI images have poor discriminative capacity between astrocytoma and oligodendroglioma. When fusing the features of the two modalities by the proposed linear weighted module, the F1-micro was increased from 91.7% to 93.2 on the cross-validation dataset. The results indicated that MRI images could provide some complementary features to pathology images, despite the big performance difference.

Table 3. Classification results on cross validation.

Model	Data	Balanced-ACC	F1-micro	Kappa
Pathology only	Cross validation	0.877	0.917	0.852
Radiology only	Cross validation	0.722	0.818	0.683
Fusion	Cross validation	**0.886**	**0.932**	**0.878**

Table 4 presents the results on the online validation set. Since the online validation sets are the same as the last year, we compared our results with the first-place methods [15] in CPM-RadPath-2019. We achieved higher performances on both pathology only and radiology only settings. The results further demonstrated the effectiveness of the proposed noise reduction in pathology images. The ROI extraction step in MRI images could make the model focus on tumor regions, so it brought improvements. We evaluated our solution on the validation set of CPM-RadPath-2020. The final F1-micro reached 0.971 and ranked 1st among 61 teams.

Table 4. Results on CPM-RadPath-2020

Model	Data	Balanced-ACC	F1-micro	Kappa
Pathology only [15]	Online validation set	0.833	0.914	0.866
MRI only [15]	Online validation set	0.711	0.829	0.748
Pathology only [ours]	Online validation set	**0.889**	**0.943**	**0.903**
MRI only [ours]	Online validation set	**0.820**	**0.857**	**0.767**
Ensemble [ours]	Online validation set	**0.944**	**0.971**	**0.952**

5   Conclusion

In this paper, we proposed a framework combining MRI Images and pathology Images to identify different subtypes of glioma. To reduce the noisy samples in pathology images, we leverage a nucleus segmentation model and a noise-rank module. With the help of noise reduction, we obtain a more precise classification model. Fusing the two modalities on feature space provides a more complete representation. Our results ranked in the first place on the validation set, which demonstrates the effectiveness of the proposed framework.

Acknowledgements. This work was supported by the National Natural Science Foundation of China under Grant 61472393.

References

1. Computational Precision Medicine Radiology-Pathology challenge on Brain Tumor Classification (2020). https://www.med.upenn.edu/cbica/cpm2020.html
2. Bakas, S., et al.: Identifying the Best Machine Learning Algorithms for Brain Tumor Segmentation, Progression Assessment, and Overall Survival Prediction in the BRATS Challenge (2018)
3. Kumar, N., et al.: A multi-organ nucleus segmentation challenge. IEEE Trans. Med. Imaging **39**, 5 (2017)
4. Choi, K.S., et al.: Prediction of IDH genotype in gliomas with dynamic susceptibility contrast perfusion MR imaging using an explainable recurrent neural network. NEURO-ONCOLOGY **21**, 9 (2019)
5. Decuyper, M., et al.: Automated MRI based pipeline for glioma segmentation and prediction of grade, IDH mutation and 1p19q co-deletion. MIDL 2020 (2020)
6. Li, Y., et al.: Classification of breast cancer histology images using multi-size and discriminative patches based on deep learning. IEEE Access **7**, 21400–21408 (2019)
7. Das, K., et al.: Multiple instance learning of deep convolutional neural networks for breast histopathology whole slide classification. In: 15th International Symposium on Biomedical Imaging (2018)
8. Hashimoto, N., et al: Multi-scale domain-adversarial Multiple-instance CNN for cancer subtype classification with unannotated histopathological images. In: CVPR 2020 (2020)

9. Sharma, K., Donmez, P., Luo, E., Liu, Y., Yalniz, I.Z.: NoiseRank: unsupervised label noise reduction with dependence models. In: Vedaldi, A., Bischof, H., Brox, T., Frahm, J.-M. (eds.) ECCV 2020. LNCS, vol. 12372, pp. 737–753. Springer, Cham (2020). https://doi.org/10.1007/978-3-030-58583-9_44

10. Kurc, T., et al.: Segmentation and classification in digital pathology for glioma research: challenges and deep learning approaches. Front. Neurosci. **14**, 1–15 (2020). https://doi.org/10.3389/fnins.2020.00027

11. Pei, L., et al.: Brain tumor classification using 3D convolutional neural network. In: MICCAI 2019 Workshop (2019)

12. Weng, Y.-T., et al.: Automatic classification of brain tumor types with the mri scans and histopathology images. In: MICCAI 2019 Workshop (2019)

13. Van Griethuysen, J.J.M., et al.: Computational radiomics system to decode the radiographic phenotype. Cancer Res. **77**, 21 (2017)

14. Yang, Y., et al.: Brain tumor classification with tumor segmentations and a dual path residual convolutional neural network from MRI and pathology images. In: MICCAI 2019 Workshop (2019)

15. Ma, X., et al.: Brain tumor classification with multimodal MR and pathology images. In: MICCAI 2019 Workshop (2019)

16. Huang, G., et al.: Densely connected convolutional networks. Computer Era. In: Proceedings of the IEEE Conference on Computer Vision and Pattern Recognition, pp. 4700–4708 (2017)

17. Tellez, D., et al.: Quantifying the effects of data augmentation and stain color normalization in convolutional neural networks for computational pathology. Med. Image Anal. **58**, 101544 (2019)

18. Chenni, W., et al.: Patch Clustering for Representation of Histopathology Images. Springer, Cham (2019). https://doi.org/10.1007/978-3-030-23937-4_4

19. Liu, Y., et al.: Detecting Cancer Metastases on Gigapixel Pathology Images. In: MICCAI Tutorial (2017)

20. Abadi, M., et al.: TensorFlow: a system for large-scale machine learning. Oper. Syst. Des. Implementation 2016 (2016)

21. Chen, T., et al.: MXNet: a flexible and efficient machine learning library for heterogeneous distributed systems. arXiv preprint arXiv:1512.01274 (2015)

22. Niu, F., et al.: HOGWILD!: a lock-free approach to parallelizing stochastic gradient descent. In: Advances in Neural Information Processing Systems, vol. 24 (2011)

Multimodal Brain Tumor Classification

Marvin Lerousseau[1,2](✉), Eric Deutsch[1], and Nikos Paragios[3]

[1] Paris-Saclay University, Gustave Roussy, Inserm, 94800 Villejuif, France
[2] Paris-Saclay University, CentraleSupélec, 91190 Gif-sur-Yvette, France
[3] TheraPanacea, 75014 Paris, France

Abstract. Cancer is a complex disease that provides various types of information depending on the scale of observation. While most tumor diagnostics are performed by observing histopathological slides, radiology images should yield additional knowledge towards the efficacy of cancer diagnostics. This work investigates a deep learning method combining whole slide images and magnetic resonance images to classify tumors. In particular, our solution comprises a powerful, generic and modular architecture for whole slide image classification. Experiments are prospectively conducted on the 2020 Computational Precision Medicine challenge, in a 3-classes unbalanced classification task. We report cross-validation (resp. validation) balanced-accuracy, kappa and f1 of 0.913, 0.897 and 0.951 (resp. 0.91, 0.90 and 0.94). For research purposes, including reproducibility and direct performance comparisons, our finale submitted models are usable off-the-shelf in a Docker image available at https://hub.docker.com/repository/docker/marvinler/cpm_2020_marvinler.

Keywords: Histopathological classification · Radiology classification · Multimodal classification · Tumor classification · CPM RadPath

1 Introduction

Gliomas are the most common malignant primary brain tumor. They start in the glia, which are non-neuronal cells from the central system that provide supportive functions to neurons. There are three types of glia, yielding different types of brain tumor: astrocytomas develop from astrocytes, tumors starting in oligodendrocytes lead to oligodendrogliomas, and ependymomas develop from ependymal cells. The most malignant form of brain tumors is glioblastoma multiforme, which are grade IV astrocytomas. Additionally, some brain tumors may arise from multiple types of glial cells, such mixed gliomas, also called oligo-astrocytomas.

In the clinical setting, the choice of therapies is highly influenced by the tumor grade [17]. In such a context, glioblastoma are grade IV, astrocytoma are grade II or III and oligodendroglioma are grave II. With modern pushes towards precision medicine, a finer characterization of the disease is considered for treatment strategies. Such therapeutic strategies can be surgery, radiation

© Springer Nature Switzerland AG 2021
A. Crimi and S. Bakas (Eds.): BrainLes 2020, LNCS 12659, pp. 475–486, 2021.
https://doi.org/10.1007/978-3-030-72087-2_42

therapy, chemotherapy, brachytherapy or their combinations such as surgery followed by neoadjuvant radiation therapy. Historically, the classification of brain tumors has relied on histopathological inspection, primarily characterized with light-microscopy observation of H&E-stained sections, as well as immunohistochemistry testing. This histological predominance has been contrasted with the recent progress in understanding the genetic changes of tumor development of the central nervous system. Additionally, while it is not used in neoplasm diagnostic, neuro-imaging can bring other information regarding the state of the disease.

This work deals with a novel strategy for classifying tumors with several image types. We present a strategy to classify whole slide images and/or radiology images extracted from a common neoplasm. Experiments are conducted on a dataset made of pairs of histopathological images and MRIs, for a 3-categories classification of brain tumor.

1.1   Related Work

Whole Slide Image Classification. Whole slide images (WSI) are often at gigapixel size, which drastically impede their processing with common computer vision strategies. By essence, a WSI is obtained at a certain zoom (called magnification), from which lower magnifications can be interpolated. Therefore, a first strategy for WSI classification would consist in downsampling a full-magnification image sufficiently to be processed by common strategies. However, since the constituent elements of WSIs are the biological cells, downsampling should be limited or the loss of information could completely prevent the feasibility of the task.

One first strategy therefore consists of first classification tiles extracted from a WSI, which is equivalent to WSI segmentation, and then combining those tile predictions into a slide prediction. This was investigated for classifying glioma in [8], breast cancer in [5], lung carcinoma in [3], and prostate cancer, basal cell carcinoma, and breast cancer metastases to axillary lymph nodes in [2]. More generally, the segmentation part of these works has been unified in [16]. For combining tile predictions into slide predictions, numerous strategies exist such as ensembling or many other techniques that fall under the umbrella of instance-based multiple instance learning [4].

One limitation with such a decoupled approach is that the tile classifier method does not learn useful features to embed tiles into a latent space which could be sampled by a slide classification method. A generalization consists in having a first model that converts tiles into an embedded (or latent) vector, and a second model which combines multiples such vectors from multiple tiles into a single WSI prediction. This is known as embedded-based multiple instance learning [4]. The function that maps the instance space to the slide space can be max-pooling, average-pooling, or more sophisticated functions such as noisy-or [24], noisy-and [13], log-sum-exponential [20], or attention-based [10].

Magnetic Resonance Imaging Classification. There are two major approaches for MRI classification. The first is called radiomics [15] and consists in first extracting a set of features for all images of a training set. Such features typically consist in clinical features such as age, first-order features such as descriptors of tumor shape or volume, and second order features describing textural properties of a neoplasm. These features are then used by common machine learning algorithms as a surrogate of crude MRIs. Examples of such approaches for brain tumor classification are illustrated in [12,25]. The major limitation of radiomics lies in the fact that tumor volumes must be annotated beforehand in order to extract meaningful features.

The other general approach of tumor classification from MRIs consists in using end-to-end deep learning systems. Deep learning bypasses the necessity of delineating the tumor volume and is more flexible than traditional machine learning since features are learned on-the-fly. Compared to machine learning methods, deep learning is known to require more training samples, but outcompetes the former when the number of data is sufficient. The majority of variability of brain tumor classification studies relying on deep learning lies in the architecture used. Examples of such studies are [21] or others identified in [23].

Combined Radiographic and Histologic Brain Tumor Classification. The 2018 instance of the Computational Precision Medicine Radiology-pathology challenge [14] is the first effort towards providing a publicly available dataset with pairs of WSIs and MRIs. It provided 32 training cases as well as 20 testing cases, balanced between the two brain tumor classes oligodendroglioma and astrocytoma. 3 methods were developed by the participating teams. The top-performing method [1] used a soft-voting ensemble based on a radiographic model and a histologic model. The MRI model relies on radiomics with prior automatic tumor delineation, while the histologic model classifies tiles extracted from WSIs which priorly are filtered using an outlier detection technique. The second-best performing team [18] also used two models for both modalities. The radiographic model is an end-to-end deep learning approach, while the histologic model uses a deep learning model to perform feature extraction by dropping its last layer, to further classify a set of extracted tiles into a WSI prediction. While these two models are trained separately, authors combine features extracted from both of them into an SVM model for case classification, that relies on dropout as regularization to counter the low number of training samples. Finally, the third best performing team used a weighted average of predictions obtained with two end-to-end deep-learning-based classification models for both histologic [19] and radiographic modalities, where weights are empirically estimated.

2   Methods

Our proposed method leverages both imaging and histological modalities through an ensemble. Specifically, a first deep learning model is intended to classify WSIs, while a second network classifies MRIs.

2.1   Whole Slide Image Classification with a Generic and Modular Approach

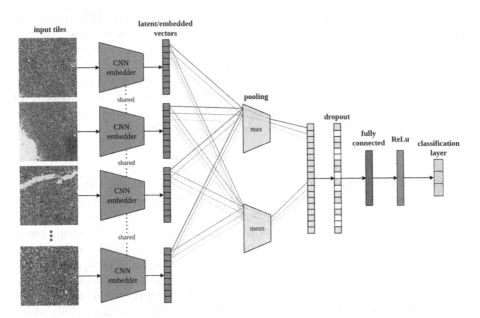

Fig. 1. Schematic representation of the end-to-end WSI classification strategy. A set of tiles is extracted from a WSI and are each inferred into a unique learnable CNN embedder (i.e. CNN classifier whose last layer has been discarded). Each of the tiles is thus converted into a latent vector (blue vectors) of size L, where $L = 9$ in this figure. Each latent dimension is then maxed, and simultaneously averaged, into two vectors of size L, which are then concatenated into a vector of size $2L$. This WSI latent vector is then forwarded into a dropout and a finale classification layer.

Our WSI classifier bears most concepts in multiple instance learning approaches as depicted in Fig. 1. At training and inference, a WSI is split into non-overlapping tiles. Each tile is forwarded into a standard 2D image classifier, such as a ResNet [7] or an EfficientNet [22], whose classifier layer has been removed. Each tile is thus embedded into a latent vector of size L. For a bag of n tiles, a latent matrix of size $n \times L$ is obtained. Then, a max-pooling operation is performed on each feature - that is, across the n dimension. Similarly, average pooling is performed for each feature. Both max-pooled and average-pooled are concatenated, making a resulting vector of size $L + L = 2L$. Finally, a classifier network ("head") processes this pooled latent vector into a 1 class prediction. This head is made of a dropout layer, followed by a final classification layer with softmax activation.

This system can be implemented end-to-end in a single network, for standard deep learning optimization. The head part can be more sophisticated, with the addition of an extra linear layer, followed by an activation function such as ReLu, and batch normalization. This implementation was kept simple due to the low number of training samples in our experiments. Of major interest, the slide embedding output of size $2L$ is independent of the size n of the input bag. Consequently, bags of any size can be fed into the network during training or inference.

During inference, it is common that the number of non-overlapping tiles is above the memory limit induced by the network, which is roughly the memory footprint of the embedding model. For instance, the number of tiles per slide is depicted in Fig. 4 for the training samples at magnification 20. Typically, for EfficientNet_b0 as an embedding model, only 200 tiles can be fitted in a 16Gb graphic card. Therefore, during inference, a random set of tiles is sampled from a WSI, each yielding one predicted class. The resulting finale predicted class is obtained by hard-voting the latter.

2.2 Magnetic Resonance Imaging Classification

Our MRI classification pipeline is straightforward and consists in a single network direct classification of the 4D volumes made of all 4 modalities. Specifically, we use a Densenet [9] made of 169 convolutional layers, whose architecture is displayed in Fig. 2. The Densenet family was used due to its low number of parameters, which seems appropriate to the low number of experiment training samples, as well as its high number of residual connections which alleviate much gradient issues. To accommodate with the 3D spatial dimensions of input volumes, the 2D convolutions, and the pooling operators have been modified to 3D. Furthermore, all convolutional kernels are cubic, i.e. with the same size in all 3 dimensions. Various kernel sizes are used throughout the architecture, as specified in Fig. 2. In particular, inputs MRI made of 4 3-dimensional modalities are stacked on their modality dimension such that the first convolution treats MRI modalities as channels. In practice, the first convolution is of kernel size 7 with stride 2, padding 3 and 64 channels, effectively converting an input MRI volume of size $(4, 128, 128, 128)$ into a output volume of size $(64, 64, 64, 64)$.

2.3 Multimodal Classification Through Ensembling

Ensembling was performed for multiple reasons. First of all, without any a priori, histopathological classification could detect different features at various scales, thus producing different or complementary diagnostics depending on the input magnification. In our implementation, multiple networks were trained at various magnifications. Secondly, ensembling allow to use both histopathological

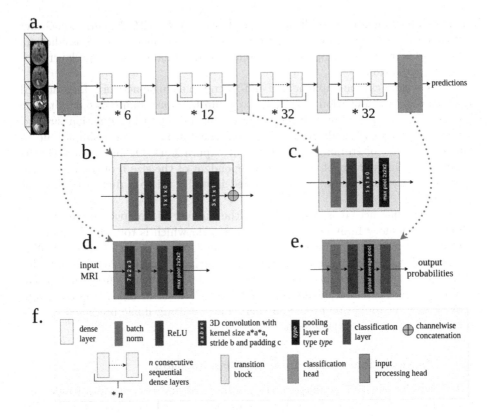

Fig. 2. Illustration of the Densenet169 architecture made of 169 convolutional layers. The macro representation (a.) shows a feature extractor module made of an input processing head (d.) followed by a combination of dense (b.) and transition (c.) layers. The extracted features are then processed into probabilities by a classification head (e.). All elements are detailed in the legend (f.).

and radiological modalities to determine a final diagnostic. Finally, ensembling several neural networks is known to significantly improve the performance and robustness of a neural network system [6]. For these reasons, our final classification uses an ensemble of multiple histopathological networks, radiological networks in a soft-voting way. Besides, as shown in [26], it may be advantageous to ensemble some of the at-hand neural networks rather than all of them. As discussed in 3.2, a portion of the trained networks is discarded in the finale decision system.

3   Experiments

Experiments were conducted during the 2020 Computational Precision Medicine Radiology-Pathology challenge (CPM-RadPath 2020). All the data is provided by the organizers, i.e. a training set and an online validation set. Additionally, final results are computed on a hidden testing set with only one try, such as to minimize testing fitting. Our team name was *marvinler*.

3.1 Data

Dataset. The training data consists in 221 cases extracted from two cohorts: The Cancer Image Archive (TCIA) and the Center for Biomedical Image Computing and Analytics (CBICA). For each case, one formalin-fixed, paraffin-embedded (FFPE) resection whole slide image was provided, along with one MRI, which consists in 4 modalities: T1 (spin-lattice relaxation), T1-contrasted, T2 (spin-spin relaxation), and fluid attenuation inversion recovery (FLAIR) pulse sequences.

Each case belongs to one of three diagnostic categories among astrocytomas, oligodendrogliomas, and glioblastoma multiforme. This information was provided for each training case. On top of that, a similar validation set made of 35 cases was available for generalization assessment. For these cases, ground-truth labels were hidden, although up to 50 submissions were possible, with feedback containing balanced accuracy, f1 score, and kappa score. Figure 3 shows an example from the validation set.

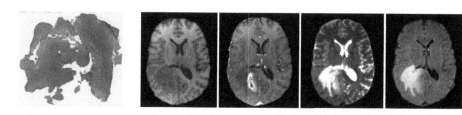

Fig. 3. Example of data point from the online validation set. The image on the left is a downscaled version of the whole slide image which is initially of width 96224 pixels and of height 82459 pixels. The four images on the right represent the MRI of the same case, and are, by order, the T1, the T1 contrast-enhanced, the T2, and the FLAIR modalities. For the MRI, the same slice extracted from each 3D volume is represented. Each 3D volume is initially of size $240 \times 240 \times 155$ pixels.

Pre-processing. No data pre-processing was performed on MRIs which have already been skull-stripped by data providers. Further MRI processing was subcontracted to the radiology data augmentation step, as detailed in Sect. 3.2.

Whole slide images were available in non-pyramidal .tiff format. They were first tiled in a pyramidal scheme using libvips, with a tile width of 512 pixel and no tile overlap. For each resulting magnification level, all tiles considered background were discarded. This was done by detecting tiles where at least 75% of pixels have both red, green, and blue channels above a value of 180 (where

255 is absolute white and 0 is black), which saved more than two third of disk space by discarding non informative tiles from further processing. After this filtering, the number of tiles per slide is depicted in the histogram of Fig. 4 for magnification 20. The complete WSI pre-processing was applied to both training, validation, and testing sets.

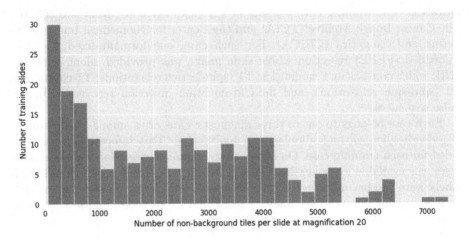

Fig. 4. Histogram of the number of tiles considered non-background for all training slides at magnification 20, or equivalently micrometer per pixel (mpp) of 0.5.

3.2   Implementation Details

A pre-trained Efficientnet_b0 was selected as the histopathological model embedder, converting 224 pixel-wide images into an embedded vector of size $L = 1280$. During training, 50 tiles were selected from 4 WSIs, resulting in a batch of size 200. Each tile was data augmented with random crop from 512 pixel width to 224 pixel width, color jitter with brightness, contrast, saturation of 0.1, and hue of 0.01, and was normalized by dividing the mean and standard deviation of each RGB channel as computed on the training dataset. To counter the low number of training samples, dropout of 0.5 was used in the head of the network. The final softmax activation was discarded during training, and cross-entropy loss was used to compute the error signal (which contains a log softmax operation for numerical stability). To handle class imbalance, weights were computed as the inverse of the frequency of each of the 3 classes and used in the error computation. Adam [11] optimizer was used to back-propagate the error signal, with default momentum parameters and learning rate of $5e - 5$ for 50 epochs. This process was applied for the 5 magnifications of 40, 20, 10, 5, and 2.5, or equivalently and respectively of micrometers per pixel of 0.25, 0.5, 1, 2, and 4. During inference, 200 tiles were randomly sampled from each WSI and for all magnifications, with data augmentation independently applied from tile to tile.

Due to poor results on the online validation set, the models from magnification 40 were discarded.

For the classification from MRIs, data augmentation first consisted of cropping the foreground, which roughly crops the MRI to their contained brain. These cropped volumes were resized to a unique size of $128 \times 128 \times 128$, and each modality was standard scaled such that 0 values (corresponding to background) were not computed in mean and standard deviation computations. Then, a random zoom between 0.8 and 1.2 was applied, followed by a random rotation of 10 degrees on both sides for all dimensions. Random elastic deformations with parameter values of sigma in range $(1, 10)$ and magnitude in range $(10, 200)$. All of the data augmentation was implemented using the Medical Open Network for AI (MONAI) toolkit[1], which is also the source of the Densenet169 architecture implementation. Error signal was computed with cross-entropy loss and back-propagated with Adam optimizer for 200 epochs at learning rate $5e - 4$ with a batch size of 3. The same class imbalance was used than the histopathological networks.

For both 5 models (4 histopathological and 1 radiological), rather than selecting the best performing models on cross-validation, two snapshots in a well performing region were selected. Specifically, the last epoch weights snapshot, as well as the snapshot from the 10 epochs to the end were collected for each model. Following [26], the 2 least performing (on the local validation) out of the 10 networks were discarding, which consisted in one radiology-based model and one model from magnification 5. Soft-voting was performed on the 8 resulting models for all slides, and the class with highest probability was assigned as the predicted class.

3.3   Results

After 3 online submissions, the hidden validation performance was a balanced accuracy of 0.911, a kappa score of 0.904, and an f1 score (with micro average) of 0.943 which ranked us second overall. Unseen testing results are still awaiting. Our cross-validation results are similar, with a balanced accuracy of 0.913, a kappa score of 0.897 and an f1 score of 0.951, denoting a certain high generalization capacity from our approach.

For each diagnostic category, a histogram of validation predicted probabilities was computed, resulting in 3 plots depicted in Fig. 5a. Notably, there seems to be less hesitation for the prediction of the glioblastoma multiforme class (G), where only 3 cases were predicted with a probability between 0.2 and 0.8. In comparison, 7 (resp. 9) cases are predicted with a probability between 0.2 and 0.8 for the astrocytoma (A) (resp. oligodendrogliomas (O)). Figure 5b further highlights that most hesitation comes from both classes A and O, with all probabilities of class A that are not below 0.2 or above 0.8 predicted mostly as class O compared to G. Besides, 6 out of 9 unsure probabilities for the O class were split with the A class compared to the G one. This could illustrate that some

---

[1]   https://github.com/Project-MONAI/MONAI.

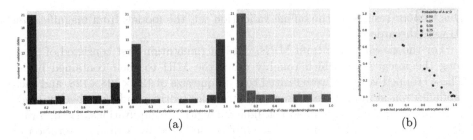

Fig. 5. Analysis of validation predicted probabilities. (a) Histogram of predicted probabilities for all 35 validation cases. Each plot (3 in total) refers to one of the 3 predicted classes, as indicated in the x-axis label. (b) Scatterplot of predictions of astrocytoma vs. oligodendrogliomas. Color and size of points are proportional to the sum of predictions of the latter class, or equivalently inversely proportional to the predicted probability of glioblastoma multiforme.

cases seem to exhibit both oligodendrogliomas and astrocytoma, known as mixed gliomas. Notably, while mixed gliomas are a valid diagnostic in clinical settings, this class was absent in the challenge data, with a decision between class A and O taken by the challenge annotators for such cases.

An ablation study was performed on the online validation set to understand the impact of both radiographic and histologic models. The MRI-only model obtained a balanced accuracy of 0.694, a kappa score of 0.662 and an f1 score of 0.800, which is significantly lower than the WSI-only model which had the same performance than our finale model combining both MRI and WSI modalities. However, upon inspection of class predictions, there were 2 differences between the predictions of the latter and the ensemble of WSI-only models. Specifically, one ensemble predicted an oligodendroglioma while the other predicted astrocytoma, and the opposite. Both WSI-only and WSI coupled with MRI models had a consensus on the glioblastoma class.

4   Conclusion and Discussion

Our work illustrates the feasibility of the classification of tumor types among 3 pre-defined categories. Although testing results are still pending, the proposed approach appears to generalize well to unseen data from the same distribution. The proposed pipeline heavily relies on histopathological slides, which have been the golden standard for tumor diagnosis for a hundred years. This system is an end-to-end trainable decision system that relies on a learnable embedding model (e.g. a convolutional neural network) and combines a set of any size of tiles embedded vectors into a slide latent vector for further classification. This generic deep learning pipeline can be taken off-the-shelf and applied to many other histopathological classification tasks, would it be grading, diagnostic, primary determination or prognosis purposes. For this reason, we open-source our complete whole slide image classification method in a Docker image at https://hub.docker.com/ repository/docker/marvinler/cpm_2020_marvinler, ensuring its use off-the-shelf on any platform. Such a resource can be used to perform direct comparisons of future research contributions on combined WSI and MRI brain tumor classification, while also ensuring reproducibility of results.

While our use of the radiology modality has not been heavily in this work, there should be improvement in classifying cases with more information than in a histopathological context only. We believe that, with more training data, diagnosis could be further improved with multimodal solutions that embed both radiographic images and histologic images into a common latent space.

References

1. Bagari, A., Kumar, A., Kori, A., Khened, M., Krishnamurthi, G.: A combined radio-histological approach for classification of low grade gliomas. In: Crimi, A., Bakas, S., Kuijf, H., Keyvan, F., Reyes, M., van Walsum, T. (eds.) BrainLes 2018. LNCS, vol. 11383, pp. 416–427. Springer, Cham (2019). https://doi.org/10.1007/978-3-030-11723-8_42
2. Campanella, G., et al.: Clinical-grade computational pathology using weakly supervised deep learning on whole slide images. Nat. Med. **25**(8), 1301–1309 (2019)
3. Coudray, N., et al.: Classification and mutation prediction from non-small cell lung cancer histopathology images using deep learning. Nat. Med. **24**(10), 1559–1567 (2018)
4. Dietterich, T.G., Lathrop, R.H., Lozano-Pérez, T.: Solving the multiple instance problem with axis-parallel rectangles. Artif. Intell. **89**(1–2), 31–71 (1997)
5. Gecer, B., Aksoy, S., Mercan, E., Shapiro, L.G., Weaver, D.L., Elmore, J.G.: Detection and classification of cancer in whole slide breast histopathology images using deep convolutional networks. Pattern Recogn. **84**, 345–356 (2018)
6. Hansen, L.K., Salamon, P.: Neural network ensembles. IEEE Trans. Pattern Anal. Mach. Intell. **12**(10), 993–1001 (1990)
7. He, K., Zhang, X., Ren, S., Sun, J.: Identity mappings in deep residual networks. In: Leibe, B., Matas, J., Sebe, N., Welling, M. (eds.) ECCV 2016. LNCS, vol. 9908, pp. 630–645. Springer, Cham (2016). https://doi.org/10.1007/978-3-319-46493-0_38
8. Hou, L., Samaras, D., Kurc, T.M., Gao, Y., Davis, J.E., Saltz, J.H.: Patch-based convolutional neural network for whole slide tissue image classification. In: Proceedings of the IEEE Conference on Computer Vision and Pattern Recognition, pp. 2424–2433 (2016)
9. Huang, G., Liu, Z., Van Der Maaten, L., Weinberger, K.Q.: Densely connected convolutional networks. In: Proceedings of the IEEE Conference on Computer Vision and Pattern Recognition, pp. 4700–4708 (2017)
10. Ilse, M., Tomczak, J.M., Welling, M.: Attention-based deep multiple instance learning. arXiv preprint arXiv:1802.04712 (2018)
11. Kingma, D.P., Ba, J.: Adam: a method for stochastic optimization. arXiv preprint arXiv:1412.6980 (2014)
12. Kotrotsou, A., Zinn, P.O., Colen, R.R.: Radiomics in brain tumors: an emerging technique for characterization of tumor environment. Magn. Reson. Imaging Clin. **24**(4), 719–729 (2016)
13. Kraus, O.Z., Ba, J.L., Frey, B.J.: Classifying and segmenting microscopy images with deep multiple instance learning. Bioinformatics **32**(12), i52–i59 (2016)

14. Kurc, T., et al.: Segmentation and classification indigital pathology for glioma research: challenges and deep learningapproaches. Front. Neurosci. **14** (2020)
15. Lambin, P., et al.: Radiomics: extracting more information from medical images using advanced feature analysis. Eur. J. Cancer **48**(4), 441–446 (2012)
16. Lerousseau, M., et al.: Weakly supervised multiple instance learning histopathological tumor segmentation. arXiv preprint arXiv:2004.05024 (2020)
17. Louis, D.N., et al.: The 2016 world health organization classification of tumors of the central nervous system: a summary. Acta Neuropathol. **131**(6), 803–820 (2016)
18. Momeni, A., Thibault, M., Gevaert, O.: Dropout-enabled ensemble learning for multi-scale biomedical data. In: Crimi, A., Bakas, S., Kuijf, H., Keyvan, F., Reyes, M., van Walsum, T. (eds.) BrainLes 2018. LNCS, vol. 11383, pp. 407–415. Springer, Cham (2019). https://doi.org/10.1007/978-3-030-11723-8_41
19. Qi, Q., et al.: Label-efficient breast cancer histopathological image classification. IEEE J. Biomed. Health Inform. **23**(5), 2108–2116 (2018)
20. Ramon, J., De Raedt, L.: Multi instance neural networks. In: Proceedings of the ICML-2000 Workshop on Attribute-value and Relational Learning, pp. 53–60 (2000)
21. Talo, M., Baloglu, U.B., Yıldırım, Ö., Acharya, U.R.: Application of deep transfer learning for automated brain abnormality classification using MR images. Cogn. Syst. Res. **54**, 176–188 (2019)
22. Tan, M., Le, Q.V.: Efficientnet: rethinking model scaling for convolutional neural networks. arXiv preprint arXiv:1905.11946 (2019)
23. Tandel, G.S., et al.: A review on a deep learning perspective in brain cancer classification. Cancers **11**(1), 111 (2019)
24. Zhang, C., Platt, J.C., Viola, P.A.: Multiple instance boosting for object detection. In: Advances in Neural Information Processing Systems, pp. 1417–1424 (2006)
25. Zhou, M., et al.: Radiomics in brain tumor: image assessment, quantitative feature descriptors, and machine-learning approaches. Am. J. Neuroradiol. **39**(2), 208–216 (2018)
26. Zhou, Z.H., Wu, J., Tang, W.: Ensembling neural networks: many could be better than all. Artif. Intell. **137**(1–2), 239–263 (2002)

A Hybrid Convolutional Neural Network Based-Method for Brain Tumor Classification Using mMRI and WSI

Linmin Pei[1]([✉]), Wei-Wen Hsu[2], Ling-An Chiang[2], Jing-Ming Guo[2], Khan M. Iftekharuddin[3], and Rivka Colen[1]

[1] Department of Diagnostic Radiology, University of Pittsburgh Medical Center, Pittsburgh, PA 15232, USA
peil@upmc.edu
[2] Department of Electrical Engineering, National Taiwan University of Science and Technology, Taipei, Taiwan
[3] Vision Lab, Electrical and Computer Engineering, Old Dominion University, Norfolk, VA 23529, USA

Abstract. In this paper, we propose a hybrid deep learning-based method for brain tumor classification using whole slide images (WSIs) and multimodal magnetic resonance image (mMRI). It comprises two methods: a WSI-based method and a mMRI-based method. For the WSI-based method, many patches are sampled from the WSI for each category as the training dataset. However, not all the sampling patches are representative of the category to which their corresponding WSI belongs without the annotations by pathologists. Therefore, some error tolerance schemes were applied when training the classification model to achieve better generalization. For the mMRI-based method, we firstly apply a 3D convolutional neural network (3DCNN) on the multimodal magnetic resonance image (mMRI) for brain tumor segmentation, which distinguishes brain tumors from healthy tissues, then the segmented tumors are used for tumor subtype classification using 3DCNN. Lastly, an ensemble scheme using the two methods was performed to reach a consensus as the final predictions. We evaluate the proposed method with the patient dataset from Computational Precision Medicine: Radiology-Pathology Challenge (CPM: Rad-Path) on Brain Tumor Classification 2020. The performance of the prediction results on the validation set reached 0.886 in f1_micro, 0.801 in kappa, 0.8 in balance_acc, and 0.829 in the overall average. The experimental results show that the performance with the consideration of both MRI and WSI outperforms the performance using single type of image dataset. Accordingly, the fusion from two image datasets can provide more sufficient information in diagnosis for the system.

Keywords: Brain tumor classification · Whole slide image · Magnetic resonance image · Convolutional neural network · Tumor subtype

© Springer Nature Switzerland AG 2021
A. Crimi and S. Bakas (Eds.): BrainLes 2020, LNCS 12659, pp. 487–496, 2021.
https://doi.org/10.1007/978-3-030-72087-2_43

1 Introduction

Brain tumor originates from glioma in central nervous system (CNS). Accord to a report, there are 23 out of 100,000 population diagnosed as brain tumor [1]. The brain tumor can be categorized into grade I, II, III, and IV (from least aggressive to most aggressive) based on its progressiveness [2]. The life expectancy of patient with brain tumors is highly related to the tumor grade [1]. In general, patients with lower-grade gliomas has a longer survival period. Even with treatment advancement, the median survival period of patients with glioblastoma (GBM), a progressive primary brain tumor, still remains 12–16 months [3]. Therefore, accurate tumor classification is imperative for proper prognosis and treatment planning. Prior to 2017, the standard criterion used by pathologists and neuropathologists for diagnosis and grading of gliomas were defined by WHO classification in 2007 [4]. Brain tumors are graded ranging from WHO grade I to IV according to the degrees of malignancy. Due to limitations of the histologic-based classification system, molecular information is introduced into the revised grading system [2]. There are several tumor subtypes, including diffuse astrocytoma, *IDH*-mutant/-wildtype, anaplastic astrocytoma *IDH*-mutant/-wildtype, Oligodendroglioma, *IDH*-mutant and 1p/19q-codeleted, Glioblastoma, *IDH*-mutant/-wildtype, etc. According to the new criterion, the tumor classification not only consider phenotype information, but also genotype information [2]. Conventionally, diagnosis and grading of brain tumor by pathologists, who examine tissue sections fixed on glass slides under a light microscope. However, the manual process of brain tumor classification is time-consuming, tedious, and susceptible to human errors. Therefore, a computer-aided automatic brain tumor subtype classification method is highly desirable.

The pathological images are still the main sources for tumor subtype classification. However, pathological examination with the format of WSI involves heavy computation due to its massive size. The pathological slides scanned at the magnification rate of 40 usually contain $100k \times 100k$ pixels. Several methods were proposed for tumor classification using WSI. Xiao *et al.* proposed CNN-based methods on WSI for tumor classification [5–7]. Kothari *et al.* utilized a multi-class model for the histological classification [8]. The authors use spatial pyramid matching framework (SPM) with a linear support vector machine (SVM) classifier for classifying glioblastoma multiforme (GBM) [9]. On the other hand, MRI is also widely used for glioma classification because of its non-invasive property. There are many methods are proposed in literature. Zacharaki *et al.* propose a SVM-based method for grading gliomas [10]. Other methods, such as hybrid of SVM and k-nearest neighbor (KNN) [11], and two layers feed forward neural network, are also used for glioma classification [12]. However, hand-crafted feature extraction is the prerequisite for all above methods, which is very challenging.

Deep learning becomes feasible because of availability of large dataset and hardware improvement. It has been successfully using in many domains, including computer vision [13], pattern recognition [14], natural language processing (NLP) [15], and medical image processing [16], etc. As such, deep learning-based methods for glioma classification are becoming prevalent in both WSI-based and MRI-based research [17–20].

In this paper, we investigate deep learning-based tumor classification method using digital histopathology data and mMRI. The CNN-based classification model is applied to recognize the tumor subtypes from the sampling patches. The biggest challenge is to reduce the impact of noisy labels when training the deep CNN models. On the other hand, for mMRI-based method, we first apply a brain tumor segmentation, then the segmented tumors are used for tumor subtype classification through a 3DCNN. In order to have brain tumor segmentation, we utilize the Multimodal Brain Tumor Segmentation Challenge 2019 (BraTS 2019) training data set [21–24]. Finally, we combine these results using majority voting to achieve a consistent outcome.

2 Method

In this section, we introduce both WSI-based and mMRI-based method for tumor subtype classification.

2.1 WSI-Based Method

The pathological assessment has been the gold standard for diagnosis in cancer. In digital pathology, glass slides with tissue specimens were digitized by the whole-slide scanner at high resolution as whole-slide images (WSIs). The analysis of WSIs is non-trivial because it usually involves massive data in the gigapixel level to process and visualize. One approach is to sample many patches from the lesion regions labeled by well-trained pathologists. However, for the classification task in the challenge, only the categorical labels of WSIs are given without having the precise annotations of lesion regions. Consequently, many patches sampled from the WSIs for each category became noisy labels in the dataset that should not be considered in the phase of training models.

For data acquisition, several patches were sampled from a WSI and labelled to the category to which their corresponding WSI belongs. Besides, sampling conditions are needed to screen out the meaningless sampling patches. In our experiment, the sampling patches should meet the requirements that the standard deviation of the pixels should be above 20 and the mean values should fall between 50 and 150 for RGB three channels. Figure 1 shows some qualified patches that were sampled from a WSI.

However, as it can be observed in Fig. 1, not all the qualified samples are representative of the tumor subtype that the corresponding pathological slide belongs to. Since we assign all the sampling patches from a WSI to the same category, the patches with normal cells and tissues sampled from different categorical slides will share the same characteristics but with conflicting labels. As a result, the noisy labels in the dataset are unfavorable for the deep learning approach since the models usually overfit the dataset, leading to low performance in testing. To overcome this problem, the open-source *clean-lab* that performs the confident learning approach [25] was applied to screen out the noisy samples from the training dataset. In addition, to further reduce the influence of the noisy labels and improve the training model's property of generalization, the modified teacher-student training scheme [26] was proposed to enhance the robustness of the predictions. The training process is as shown in Fig. 2. Firstly, a "teacher" model was trained on the original training data and the given noisy labels, and the inferencing results by the "teacher" model were used to update the labels for the original training data. Secondly, the original training dataset with the updated labels from the "teacher"

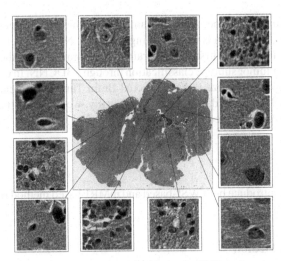

Fig. 1. Several patches were sampled from the WSIs in data acquisition.

model was modified by augmentation and dropout. Afterwards, the modified training dataset was used to train a "student" model. In the next iteration, the "student" model became a "teacher" model and the same steps were executed. After several iterations, the problem of the noisy data that caused conflictions in predictions can be alleviate while training; as a result, the trained model can achieve better robustness. In the testing phase, 300 patches are randomly sampled from each WSI under the same sampling conditions as in the training phase for inference, and the majority voting from all the predictions of sampling patches became the final predicted subtype for each case of WSI.

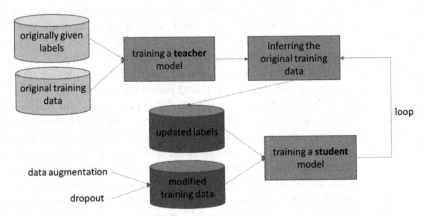

Fig. 2. The modified teacher-student training scheme was proposed to enhance the robustness of the predictions.

2.2 mMRI-Based Method

MRI is less invasive comparing to biopsy; therefore, MRI is also widely used as an alternative source for tumor classification. Since the intensity heterogeneity and other artificial facts, a single image modality is inadequate to distinguish all brain tumor tissues, such as peritumoral edema (ED), necrosis (NC), enhancing tumor (ET), and non-enhancing tumor (NET). However, multi-modal MRI (mMRI) offers a complementary information for all type of brain tumors, and has been widely used for brain tumor analysis. The mMRI contains T1-weighted image (T1), T1-contrast-enhanced weighted image (T1ce), T2-weighted image, and T2 Fluid Attenuated Inversion Recovery image (T2-FLAIR). Different image modality highlights different type of tumor: T1ce has good contrast in ET and NC, while T2 and T2-FLAIR emphasis in ED.

The MRI-based method consists of two part: tumor segmentation and tumor classification. In the first part, we utilize a 3DCNN model for the tumor segmentation, and then the segmented tumors are fed into another 3DCNN for tumor classification. The proposed framework of MRI-based method is shown in Fig. 3. Note that considering the non-radiomic age information is also a factor for certain type of tumor, mostly associated with GBM, we concatenate age information in a fully connected (FC) layer.

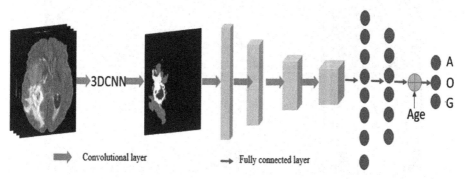

Fig. 3. The proposed framework of MRI-based method.

3 Materials and Pre-processing

3.1 Data

The training data is obtained from Computational Precision Medicine: Radiology-Pathology Challenge on Brain Tumor Classification 2020 (CPM-RadPath 2020) [27]. It has 221 cases with paired radiology images and digital pathology images. Radiology data consists of four image modalities: T1-weighted MRI (T1), T1-weighted MRI with contrast enhancement (T1ce), T2-weighted MRI (T2), and T2-weighted MRI with fluid-attenuated inversion recovery (T2-FLAIR). The mMRI is pre-processed with co-registration, skull stripping, and noise reduction. The size of each modality of each case is $240 \times 240 \times 155$. In the 221 cases, there are 54 cases, 34 cases, and 133 cases for lower grade astrocytoma, *IDH*-mutant (A), oligodendroblioma, 1p/191codeltion (O), and Glioblastoma and diffuse astrocytic glioma with molecular features of glioblastoma, *IDH*-wildtype (G).

In addition, to achieve tumor segmentation, we take Multimodal Brain Tumor Segmentation Challenge 2020 (BraTS 2020) [21–24] data for training. It has total 369 cases, including 293 high-grade glioma (HGG) cases and 76 low-grade glioma (LGG) cases. It also has four image modalities, as CPM2020, and corresponding ground truth.

3.2 Pre-processing for Radiology Images

Even though the mMRIs are pre-processed with co-registration, skull-stripping, and noise reduction, intensity inhomogeneity still exists across all cases. To reduce the impact of intensity heterogeneity, we first apply bias correction, and follow with intensity normalization. We simply apply z-score normalization, which ensures zero mean and unit standard deviation. Figure 4 shows an example with z-score normalization.

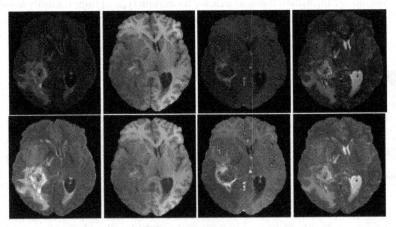

Fig. 4. An example of applying z-score normalization. Top row: raw images. Bottom row: z-score normalized images. From left to right: T2-FLAIR, T1, T1ce, and T2 image.

4 Experiments and Results

To reduce computational burden and contain all tumors information, we empirically crop size of all mMRIs to $160 \times 192 \times 128$. In tumor classification training phase, we use Adam [28] optimizer with initial learning rate of $R_0 = 0.001$, and the learning rate (R_i) is gradually decayed as follows:

$$R_i = R_0 \left(1 - \frac{i}{N}\right)^{0.9},\tag{1}$$

where i is epoch counter, and N is total number of epochs in training.

The corresponding cross-entropy loss function is computed as follows:

$$L_{ce} = -\sum_{c=1}^{M} y_{o,c} log(p_{o,c}),$$ (2)

where M is the total number of classes, y is a binary indictor (0 or 1) if class label c is the correct classification for observation o. p is predicted probability observation with o belonging to class c.

4.1 Brain Tumor Segmentation in Radiology Images

We use the BraTS 2020 data for tumor segmentation training. To evaluate the performance of segmentation model, we also submit the validation result through the BraTS 2020 portal. Note that the ground truths of the validation data are privately owned by the challenge organizer. The online evaluation shows our average dice score coefficient (DSC) is 0.752, 0.898, and 0.809 for ET, whole tumor (all tumor tissues), and tumor core (combination of ET and NC), respectively. There is a tumor segmentation example shown in Fig. 5.

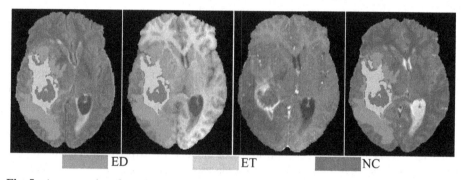

ED   ET   NC

Fig. 5. An example of tumor segmentation. From left to right: T2-FLAIR overlaid with segmentation, T1 overlaid with segmentation, T1ce, and T2 overlaid with segmentation.

4.2 Brain Tumor Classification Using Radiology and Pathology Images

For MRI-based method, we utilize another 3DCNN. The segmented tumors in the previous step are fed into the 3DCNN. In order to overcome the data imbalance issue, we manually duplicate the minority cases, to have tumor class having similar number of cases. We applied the proposed methods to CPM-RadPath 2020 validation data set, which includes 35 cases with radiology and pathology images. The online evaluation performance is reported in Table 1.

For WSI-based method, two models named "WSI-based 1" and "WSI-base 2" in Table 1 were obtained from the training phase with the iterations of 7 times and 5 time, respectively. Finally, we applied one MRI-based model and two WSI-based models on CPM2020 validation data for evaluation. At the end, we fused these predictions using majority voting method to form a consistent.

Table 1. Online evaluation results by using our proposed method.

Method	f1_micro	Kappa	Balance_acc	Average
MRI-based	0.771	0.627	0.698	0.699
WSI-based 1	0.8	0.657	0.7	0.719
WSI-based 2	0.886	0.798	0.777	0.821
Fusion	0.886	0.801	0.8	0.829

4.3 Discussion

There are two main issues for this work by using deep learning-based method: lack of data and data imbalance. Lack of data common exists in many deep learning applications. In general, deep learning-based method requires thousands/millions of data to training the network. In addition, data imbalance is another drawback in this work. In training data, GBM class takes more than 60% of the whole data. It easily dominates other class and result in overfitting issue. In doing so, we manual duplicate minority class to balance the data. However, it is an essential way to solve the problem.

For the deep learning approaches using the modality of WSIs, the task becomes very tricky without any precise annotations of lesions provided for each subtype of cancer. In the challenge, the label for each WSI was given but without indicating the representative lesions for each category. It is a classic scenario of inexact supervision in weakly supervised learning [29]. After sampling several patches from each WSI and labelling them to the same category as the WSI from which they were sampled, the task of inexact supervision became the case of inaccurate supervision [29]. Therefore, two strategies to tackle the problem of noisy dataset were performed to reduce the impact of noisy label and enhance the generalizability of models. The experimental results on the validation set show our proposed schemes can provide good predictions even with the noisy dataset. Besides, the fusion of the predictions from MRI-based model and WSI-based model achieve better accuracy rate on the metrics of Kappa and Balance_acc than the prediction results only with one type of medical images considered.

5 Conclusion

This work proposes a hybrid method of using deep learning approaches for brain tumor classification. It comprises two methods: WSI-based method and MRI-based method. For WSI-based method, we sampled several patches and trained a 2D CNN model to determine the subtype of brain tumor with some denoising schemes. For MRI-based method, we firstly applied tumor segmentation using a 3D CNN, and then performed the tumor classification on the segmented tumors using another 3D CNN. Finally, we combined the predictions using both methods to achieve an agreement. The online evaluating results show that the fusion offers a stable/better performance than each individual method.

References

1. Ostrom, Q.T., Gittleman, H., Truitt, G., Boscia, A., Kruchko, C., Barnholtz-Sloan, J.S.: CBTRUS statistical report: primary brain and other central nervous system tumors diagnosed in the United States in 2011–2015. Neuro Oncol. **20**(Suppl_4), iv1–iv86 (2018)
2. Louis, D.N., et al.: The 2016 World Health Organization classification of tumors of the central nervous system: a summary. Acta Neuropathol. **131**(6), 803–820 (2016)
3. Chen, J., McKay, R.M., Parada, L.F.: Malignant glioma: lessons from genomics, mouse models, and stem cells. Cell **149**(1), 36–47 (2012)
4. Louis, D.N., et al.: The 2007 WHO classification of tumours of the central nervous system. Acta Neuropathol. **114**(2), 97–109 (2007)
5. Ma, X., Jia, F.: Brain tumor classification with multimodal MR and pathology images. In: Crimi, A., Bakas, S. (eds.) BrainLes 2019. LNCS, vol. 11993, pp. 343–352. Springer, Cham (2020). https://doi.org/10.1007/978-3-030-46643-5_34
6. Chan, H.-W., Weng, Y.-T., Huang, T.-Y.: Automatic classification of brain tumor types with the MRI scans and histopathology images. In: Crimi, A., Bakas, S. (eds.) BrainLes 2019. LNCS, vol. 11993, pp. 353–359. Springer, Cham (2020). https://doi.org/10.1007/978-3-030-46643-5_35
7. Xue, Y., et al.: Brain tumor classification with tumor segmentations and a dual path residual convolutional neural network from MRI and pathology images. In: Crimi, A., Bakas, S. (eds.) BrainLes 2019. LNCS, vol. 11993, pp. 360–367. Springer, Cham (2020). https://doi.org/10.1007/978-3-030-46643-5_36
8. Kothari, S., Phan, J.H., Young, A.N., Wang, M.D.: Histological image classification using biologically interpretable shape-based features. BMC Med. Imaging **13**(1), 9 (2013)
9. Chang, H., Zhou, Y., Spellman, P., Parvin, B.: Stacked predictive sparse coding for classification of distinct regions in tumor histopathology. In: 2013 Proceedings of the IEEE International Conference on Computer Vision, pp. 169–176 (2013)
10. Zacharaki, E.I., et al.: Classification of brain tumor type and grade using MRI texture and shape in a machine learning scheme. Magn. Reson. Med. Off. J. Int. Soc. Magn. Reson. Med. **62**(6), 1609–1618 (2009)
11. Machhale, K., Nandpuru, H.B., Kapur, V., Kosta, L.: MRI brain cancer classification using hybrid classifier (SVM-KNN). In: 2015 International Conference on Industrial Instrumentation and Control (ICIC), pp. 60–65. IEEE (2015)
12. Zulpe, N., Pawar, V.: GLCM textural features for brain tumor classification. Int. J. Comput. Sci. Issues (IJCSI) **9**(3), 354 (2012)
13. LeCun, Y., Bengio, Y., Hinton, G.: Deep learning. Nature **521**(7553), 436–444 (2015)
14. Baccouche, M., Mamalet, F., Wolf, C., Garcia, C., Baskurt, A.: Sequential deep learning for human action recognition. In: Salah, A.A., Lepri, B. (eds.) HBU 2011. LNCS, vol. 7065, pp. 29–39. Springer, Heidelberg (2011). https://doi.org/10.1007/978-3-642-25446-8_4
15. Young, T., Hazarika, D., Poria, S., Cambria, E.: Recent trends in deep learning based natural language processing. IEEE Comput. Intell. Mag. **13**(3), 55–75 (2018)
16. Havaei, M., et al.: Brain tumor segmentation with deep neural networks. Med. Image Anal. **35**, 18–31 (2017)
17. Sajjad, M., Khan, S., Muhammad, K., Wu, W., Ullah, A., Baik, S.W.: Multi-grade brain tumor classification using deep CNN with extensive data augmentation. J. Comput. Sci. **30**, 174–182 (2019)
18. Pei, L., Vidyaratne, L., Hsu, W.-W., Rahman, M.M., Iftekharuddin, K.M.: Brain tumor classification using 3D convolutional neural network. In: Crimi, A., Bakas, S. (eds.) BrainLes 2019. LNCS, vol. 11993, pp. 335–342. Springer, Cham (2020). https://doi.org/10.1007/978-3-030-46643-5_33

19. Barker, J., Hoogi, A., Depeursinge, A., Rubin, D.L.: Automated classification of brain tumor type in whole-slide digital pathology images using local representative tiles. Med. Image Anal. **30**, 60–71 (2016)
20. Sultan, H.H., Salem, N.M., Al-Atabany, W.: Multi-classification of brain tumor images using deep neural network. IEEE Access **7**, 69215–69225 (2019)
21. Bakas, S., et al.: Identifying the best machine learning algorithms for brain tumor segmentation, progression assessment, and overall survival prediction in the BRATS challenge. arXiv preprint arXiv:1811.02629 (2018)
22. Bakas, S., et al.: Segmentation labels and radiomic features for the pre-operative scans of the TCGA-LGG collection. The Cancer Imaging Archive, vol. 286 (2017)
23. Menze, B.H., et al.: The multimodal brain tumor image segmentation benchmark (BRATS). IEEE Trans. Med. Imaging **34**(10), 1993–2024 (2014)
24. Bakas, S., et al.: Advancing the cancer genome atlas glioma MRI collections with expert segmentation labels and radiomic features. Sci. Data **4**, 170117 (2017)
25. Northcutt, C.G., Jiang, L., Chuang, I.L.: Confident learning: estimating uncertainty in dataset labels. arXiv preprint arXiv:1911.00068 (2019)
26. Xie, Q., Luong, M.-T., Hovy, E., Le, Q.V.: Self-training with noisy student improves imagenet classification. In: 2020 Proceedings of the IEEE/CVF Conference on Computer Vision and Pattern Recognition, pp. 10687–10698 (2020)
27. Kurc, T., et al.: Segmentation and classification in digital pathology for glioma research: challenges and deep learning approaches. Front. Neurosci. **14**(27) (2020). Original Research
28. Kingma, D.P., Ba, J.: Adam: a method for stochastic optimization. arXiv preprint arXiv:1412.6980 (2014)
29. Zhou, Z.-H.: A brief introduction to weakly supervised learning. Nat. Sci. Rev. **5**(1), 44–53 (2018)

CNN-Based Fully Automatic Glioma Classification with Multi-modal Medical Images

Bingchao Zhao[1,2], Jia Huang[1], Changhong Liang[1], Zaiyi Liu[1], and Chu Han[1(✉)]

[1] Department of Radiology, Guangdong Provincial People's Hospital, Guangdong Academy of Medical Sciences, Guangzhou, Guangdong 510080, China
[2] The School of Computer Science and Engineering, South China University of Technology, Guangzhou, Guangdong 510006, China

Abstract. The accurate classification of gliomas is essential in clinical practice. It is valuable for clinical practitioners and patients to choose the appropriate management accordingly, promoting the development of personalized medicine. In the MICCAI 2020 Combined Radiology and Pathology Classification Challenge, 4 MRI sequences and a WSI image are provided for each patient. Participants are required to use the multi-modal images to predict the subtypes of glioma. In this paper, we proposed a fully automated pipeline for glioma classification. Our proposed model consists of two parts: feature extraction and feature fusion, which are respectively responsible for extracting representative features of images and making prediction. In specific, we proposed a segmentation-free self-supervised feature extraction network for 3D MRI volume. And a feature extraction model is designed for the H&E stained WSI by associating traditional image processing methods with convolutional neural network. Finally, we fuse the extracted features from multi-modal images and use a densely connected neural network to predict the final classification results. We evaluate the proposed model with F1-Score, Cohen's Kappa, and Balanced Accuracy on the validation set, which achieves 0.943, 0.903, and 0.889 respectively.

Keywords: Glioma classification · Convolutional neural networks · Multiple modalities

1 Introduction

As the most common primary malignant tumor of the central nervous system, glioma comprise approximately 100, 000 newly diagnosed cases each year [2]. According to the 2016 World Health Organization (WHO) classification, diffuse glioma is categorized into five subtypes based on both its tumor histology and molecular alterations, among which glioblastoma, astrocytoma and oligodendroglioma are further designated with genetic subgroups, including promoter mutations in TERT, IDH mutations and chromosome arms 1p and 19q codeletion [14]. Glioblastoma accounts for 70–75% of all diagnoses in adults, and as

© Springer Nature Switzerland AG 2021
A. Crimi and S. Bakas (Eds.): BrainLes 2020, LNCS 12659, pp. 497–507, 2021.
https://doi.org/10.1007/978-3-030-72087-2_44

well has the poorest prognosis with a 5-year survival rate smaller than 5%. On the other hand, anaplastic astrocytoma and oligodendroglioma are rarer than glioblastoma, associated with relatively better overall survivals [1,9]. The treatment strategies, including adjuvant therapy selections after surgery, chemotherapy regimens and dosing schedules, vary by different subtypes [8]. Therefore, an accurate classification of glioma based on radiological and pathological images would be valuable for clinical practitioners and patients to choose the appropriate management accordingly, promoting the development of personalized medicine.

As a non-invasive clinical procedure, magnetic resonance image (MRI) reveals the characteristics of tumor phenotypes. Thus, it has been widely used for computer-aided diagnosis of glioma. Existing approaches usually require either manual or semi-automatic tumor segmentation before quantitatively analyzing 3D MRI volumes, which is labor intensive and time consuming. Besides considering a single MRI sequence, it is worth to associate multiple sequences due to the differences in signal intensities and patterns between different tissues on them. What's more, the enhancement patterns depending on various tumor subtypes would also greatly enrich the information of the tumors and their surrounding areas. Therefore, not only conventional non-enhanced T1-weighted and T2-weighted images, but also post-contrast images should be investigated.

Histopathology slide is the gold standard for the cancer diagnosis. It reflects the tumor microenvironment. The born of the digital whole slide scanner makes it possible for computers to quantitatively analyze diffuse gliomas at the microscopic level. However, even with the rapid development of computer hardware, the extreme resolution of whole slide image (WSI) is still the obstacle towards fully automatic clinical adoption.

In the real world scenario, it is meaningful and vital to read both radiological and pathological images for their diagnostic values in different subtypes of gliomas. However, manual quantification of multi-omics images is commonly subjective and experience dependent. Therefore, many researchers dedicate to find out an automatic and objective way to quantify multi-omics images. But how to fuse the features of the images acquired from different modals in a reasonable and interpretable manner is still unknown.

The CPM-RadPath 2020 MICCAI challenge is conducted for automatic brain tumor classification using two different modal images, including radiology and pathological images. Each case provides MRI of the four modalities of native (T1), post-contrast T1-weighted (T1Gd), T2-weighted (T2), and T2 Fluid Attenuated Inversion Recovery (T2-FLAIR), and an H&E stain digitized whole-slide images(WSI) which was scanned at 20x or 40x magnification. This challenge provides 388 glioma cases, including three types of gliomas, i.e., glioblastoma, oligodendroglioma, and astrocytoma, divided into a training set (70% of cases), validation set (20%), and test set (10%). In this paper, we proposed an effective pipeline for multi-modal images tumor classification. Firstly, a segmentation-free self-supervised feature extraction network is proposed for 3D MRI volume. Secondly, we proposed a feature extraction model for the H&E stained WSI by associating traditional image processing methods with convolutional neural

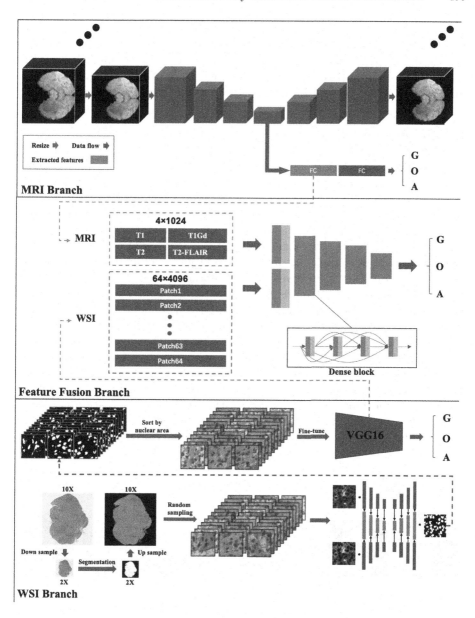

Fig. 1. Schematic diagram of the pipeline. MRI and WSI branches are proposed to extract radiology and pathology features. Feature fusion branch is proposed to aggregate multi-modal features and predict final classification results. G, O, and A denote Glioblastoma, Oligodendroglioma and Astrocytoma, respectively. The details of each branch are shown in corresponding subfigures.

network. These two feature extraction models are directly guided by the classification labels. Finally, we fused the extracted features from multi-modal images and used a densely connected neural network to predict the final classification results. Three evaluation measurements of F1-Score, Cohen's Kappa, and Balanced Accuracy achieve 0.943, 0.903, and 0.889 on the validation set.

2 Related Works

In the CPM-RadPath 2019 MICCAI challenge [6], several works have achieved great performance on automatic brain tumor classification with multi-modal images. Pei et al. [11] segmented the brain tumor from the MRI sequence, and then classified it by a regular 3D CNN model. But they did not discover the massive information in WSI. Ma et al. [7] used two convolutional neural networks for radiology and pathology images respectively. ResNet34 and ResNet50 were directly applied to extract features from WSI grayscale patches and classified them. 3D DenseNet was employed for MRI sequence. A further regression model was introduced for the inference. Chan et al. [3] grouped WSI tiles into several clusters in an unsupervised manner and applied a random forest for final prediction. Xue et al. [15] proposed a multi-modal tumor segmentation network by leveraging the information from four MRI sequences. Then a two-branch network for both MR images and pathology images was introduced for classification.

Different from the previous works, our proposed model is segmentation-free for MRI sequences. In the meanwhile, the area where tumor cells gather is regarded as the representative region of the tumor for the feature extraction of pathology images. A deep neural network is finally applied to aggregate the multi-modal image features and to make prediction.

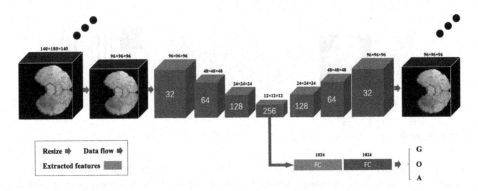

Fig. 2. Network architecture of MRI branch. It is designed as a multi-task learning model with a self-supervised feature reconstruction task and a classification task. The resolution of feature maps in each layer is shown on the top of each block. The number of the channels is shown inside the block.

3   Methods

Figure 1 demonstrates the complete classification model, which is constructed by three individual branches. The MRI branch and the WSI branch serve for radiology and pathological image feature extraction, respectively. A feature fusion branch is designed for aggregating multi-modal features and predicting the subtypes of the glioma.

3.1   Radiological Features Extraction

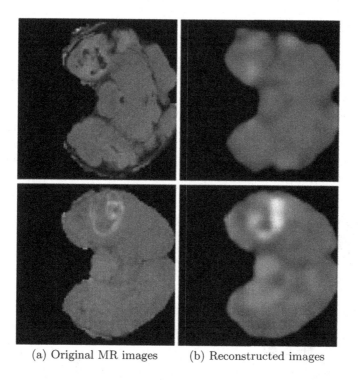

(a) Original MR images   (b) Reconstructed images

Fig. 3. The original MR images and the reconstructed images.

To extract the image features of the MRI volume, previous approaches commonly first segment the tumor lesion, and then extract the feature of the tumor. However, the accuracy of the segmentation result would directly affect the final prediction results. Moreover, since malignant tumor is invasive, the surrounding area of the tumor is also valuable for the assessment of the tumor. Therefore, we proposed a multi-task learning network to extract the features of MRI volume, which is segmentation-free. Figure 2 shows the network details. The first task is a self-supervised learning model with an autoencoder-decoder structure. It is

B. Zhao et al.

a common feature compression scheme by encoding the latent features of the input and reconstructing the original input. Figure 3 demonstrates the original images and the reconstructed images. However, this scheme tends to learn the pixel-wise features instead of high-level semantic features. So in the second task, we force the network to learn and predict the subtypes of the tumor from the latent features. L1 loss and cross entropy loss are used for the reconstruction and the classification tasks respectively.

$$\mathcal{L}_{L1} = \frac{1}{N} \sum_{i=1}^{N} |y_i - y_i'| \tag{1}$$

$$\mathcal{L}_{ce}(x) = -\sum_{i=1}^{K} p(x_i) \log p(x_i) \tag{2}$$

where y_i' and y_i denote the i-th pixel value of the reconstruction image and the input image respectively. N is the total number of pixels. K denote the number of tumor types.

The feature vector from the first fully connected layer of the second task are extracted as the feature representation of the MRI volume. Each MRI sequence (T1, T1Gd, T2, and T2-FLAIR) has a 1×1024 feature vector. The concatenated feature vectors of all the MRI sequence form the final feature representation of the radiological images.

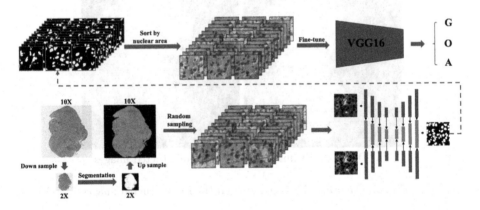

Fig. 4. Model architecture of WSI branch. The input WSI image was firstly downsampled to 2× magnification for a rough segmentation of the tissue. Then we randomly sampled a large among of patches from the segmented region. A nuclei segmentation approach [16] was applied for each patch sample. After sorting by the area of nuclei, 64 patched with the largest nuclei area were selected and passed into a classification network (VGG). The features of WSI image can be obtained from the fully connected layer of VGG network.

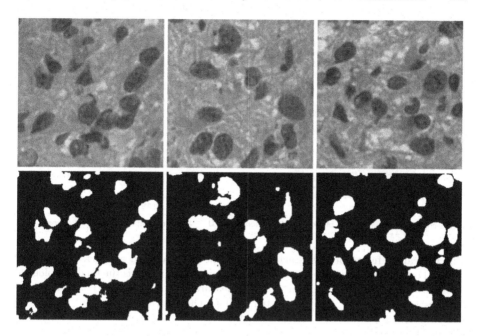

Fig. 5. Nuclei segmentation results by Triple U-net [16].

3.2 Pathological Features Extraction

Due to the gigapixel of WSI, we proposed a step-by-step procedure to quantify the pathological image, as shown in Fig. 4. We first performed a segmentation method under 2X magnification to roughly segment the tissue. Specifically, we introduced a color deconvolution algorithm [12] to map the WSI into H (Hematoxylin) channel and E (Eosin) channel, and then used the Otsu [10] to segment the foreground and background with the intersection area of the two channels as the rough segmentation mask of the tissue. Then we upsample the mask from 2X to 10X magnification to obtain the segmented region at 10X magnification. After that, we randomly sample 1000 patches (256 × 256) from the segmented region under 10X magnification and applied a nuclei segmentation network, Triple U-net [16]. Figure 5 shows the nuclei segmentation results. Note that, the nuclei segmentation network was trained on a public dataset MoNuSeg [5]. 64 patches with the largest nuclei area, which indicate the areas with dense tumor cells, they were selected to be the representative patches of the WSI. A VGG16 network [13] was introduced as the backbone of feature extraction and classification. The VGG16 was optimized by a cross entropy loss as Eq. 2.

3.3 Features Fusion Branch

Given the representative features extracted from the radiological and pathological images, we can get (1 × 4096) MRI features and (64 × 4096) WSI features.

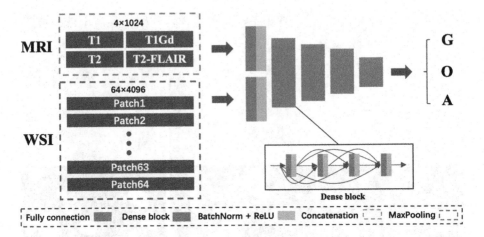

Fig. 6. Feature fusion branch. Each WSI sequence is quantified as a 1×1024 feature vector. Four MRI sequences form a 1×4096 feature vector which is the final radiological image features. We can obtain a 1×1024 feature vector of the pathological image features by maxpooling the 64 selected WSI patches features. Then the features of each modal are passed into a densely connected neural network [4] and predict the classification result.

Maxpooling operation was performed to the pathological features to downsample the features to the same dimension with the radiological features (1×4096). Then the features from two modalities were passed into the feature fusion branch, as shown in Fig. 6. Densely connected network [4] was applied to predict the subtypes of the tumor. Cross entropy loss was utilized to optimize the network.

3.4 Implementation and Training Details

All networks were implemented on Pytroch 1.5.0 and ran on a workstation equipped with an NVIDIA GeForce RTX 2080 Ti. The learning rates of the MRI branch, WSI branch, and densely connected network are 0.001, 0.0001, 0.001, respectively, and all with the learning rate decay of 0.96. All the training processes used adam optimizer without applying the dropout layer. Triple U-net [16] was trained on the public dataset from another MICCAI challenge MoNuSeg [5]. All the other models in the classification pipeline were not pre-trained.

4 Results

4.1 Quantitative Comparison

We evaluate the property of our pipeline with F1-Score, Cohen's Kappa, and Balanced Accuracy calculated on the validation set and compare with the top four models [3,7,11,15] in CPM-RadPath 2019 MICCAI challenge. Note that, all the quantitative results are from their respective papers. As is shown in Table 1.

Our pipeline achieved a classification result very close to the manual label with F1-Score of 0.943, Cohen's Kappa of 0.903, and Balanced Accuracy of 0.889. It outperforms three existing models and get the same performance with the champion of CPM-RadPath-2019 in the validation set.

Table 1. Quantitative evaluation of classification results.

Models	F1-Score	Cohen's Kappa	Balanced accuracy
Chan et al. [3]	–	–	0.780
Xue et al. [15]	–	–	0.849
Pei et al. [11]	0.829	0.715	0.794
Ma et al. [7]	**0.943**	**0.903**	**0.889**
Ours	**0.943**	**0.903**	**0.889**

4.2 Timing Statistics

We randomly select a patient from the test set with ID: CPM19_CBICA_ART_1 and provide comprehensive timing statistics of in Table 2. The resized resolution of MRI sequences from this patient is $96 \times 96 \times 96$. The resolution of the whole slide image is 108528×92767. In MRI branch, feature extraction for 4 MRI sequences takes 10 s. In WSI branch, because of the huge resolution of the whole slide image, it takes 701 s to read the WSI image and 227 s to sample patches, including nuclei segmentation. The feature extraction of 64 WSI patches takes 8 s. And the final prediction only takes 2 s. Note that, the variation of the resolutions of WSI images may lead to vibration of total running time. In the future, the timing performance can be further optimized if we can effectively decrease the time of reading WSI images.

Table 2. Timing statistics of each step (second).

MRI branch	WSI branch			Prediction
	Read image	Patch sampling	Feature extraction	
10	701	227	8	2

5 Conclusion

We propose an intuitive and fully automatic pipeline for glioma classification with the input 4 MRI sequences and a H&E stained whole slide image. The proposed pipeline can effectively extract and aggregate multi-modal image features and predict the subtypes of glioma without the necessity of any additional labels. Our model was examined on the validation set and gave a promising result.

Since our proposed pipeline has two feature extraction branches for both radiology and pathology images. And these two branches are supervised by the groundtruth labels of tumor classes. So even we omit one of the modality, the feature extraction models can still be utilized for classification. Each feature extraction branch itself can be utilized to predict the class of glioma. But lack of any modality will harm the classification performance.

Acknowledgements. This work was supported by the National Key R&D Program of China (No.2017YFC1309100), the National Science Fund for Distinguished Young Scholars (No.81925023), the National Natural Science Foundation of China (No. 81771912) and Science and Technology Planning Project of Guangdong Province (No.2017B020227012).

References

1. Annette, M.M., Jennie, W.T., John, K.W., Margaret, R.W.: Genetic and molecular epidemiology of adult diffuse glioma. Nat. Rev. Neurol. **15**, 405–417 (2019)
2. Bray, F., Ferlay, J., Soerjomataram, I., Siegel, R.L., Torre, L.A., Jemal, A.: Global cancer statistics 2018: globocan estimates of incidence and mortality worldwide for 36 cancers in 185 countries. CA Cancer J. Clin. **68**(6), 394–424 (2018)
3. Chan, H.-W., Weng, Y.-T., Huang, T.-Y.: Automatic classification of brain tumor types with the MRI scans and histopathology images. In: Crimi, A., Bakas, S. (eds.) BrainLes 2019. LNCS, vol. 11993, pp. 353–359. Springer, Cham (2020). https://doi.org/10.1007/978-3-030-46643-5_35
4. Huang, G., Liu, Z., Van Der Maaten, L., Weinberger, K.Q.: Densely connected convolutional networks. In: Proceedings of the IEEE Conference on Computer Vision and Pattern Recognition, pp. 4700–4708 (2017)
5. Kumar, N., Verma, R., Sharma, S., Bhargava, S., Vahadane, A., Sethi, A.: A dataset and a technique for generalized nuclear segmentation for computational pathology. IEEE Trans. Med. Imaging **36**(7), 1550–1560 (2017)
6. Kurc, T., et al.: Segmentation and classification in digital pathology for glioma research: challenges and deep learning approaches. Front. Neurosci. **14**, 27 (2020). https://doi.org/10.3389/fnins.2020.00027
7. Ma, X., Jia, F.: Brain tumor classification with multimodal MR and pathology images. In: Crimi, A., Bakas, S. (eds.) BrainLes 2019. LNCS, vol. 11993, pp. 343–352. Springer, Cham (2020). https://doi.org/10.1007/978-3-030-46643-5_34
8. Omuro, A., DeAngelis, L.M.: Glioblastoma and other malignant gliomas: a clinical review. JAMA **310**(17), 1842–1850 (2013)
9. Ostrom, Q.T., et al.: The epidemiology of glioma in adults: a "state of the science" review. Neuro-Oncol. **16**(7), 896–913 (2014)
10. Otsu, N.: A threshold selection method from gray-level histograms. IEEE Trans. Syst. Man Cybern. **9**(1), 62–66 (1979)
11. Pei, L., Vidyaratne, L., Hsu, W.W., Rahman, M.M., Iftekharuddin, K.M.: Brain tumor classification using 3D convolutional neural network. In: Crimi, A., Bakas, S. (eds.) BrainLes 2019. LNCS, vol. 11993, pp. 335–342. Springer, Cham (2020). https://doi.org/10.1007/978-3-030-46643-5_33

12. Ruifrok, A.C., Johnston, D.A., et al.: Quantification of histochemical staining by color deconvolution. Anal. Quant. Cytol. Histol. **23**(4), 291–299 (2001)
13. Simonyan, K., Zisserman, A.: Very deep convolutional networks for large-scale image recognition. Computer Science (2014)
14. Wesseling, P., Capper, D.: Who 2016 classification of gliomas. Neuropathol. Appl. Neurobiol. **44**(2), 139–150 (2018)
15. Xue, Y., et al.: Brain tumor classification with tumor segmentations and a dual path residual convolutional neural network from MRI and pathology images. In: Crimi, Alessandro, Bakas, Spyridon (eds.) BrainLes 2019. LNCS, vol. 11993, pp. 360–367. Springer, Cham (2020). https://doi.org/10.1007/978-3-030-46643-5_36
16. Zhao, B., et al.: Triple u-net: hematoxylin-aware nuclei segmentation with progressive dense feature aggregation. Med. Image Anal. **65**, 101786 (2020)

Glioma Classification Using Multimodal Radiology and Histology Data

Azam Hamidinekoo[1,2]([✉])[iD], Tomasz Pieciak[3,4][iD], Maryam Afzali[5][iD], Otar Akanyeti[6][iD], and Yinyin Yuan[1,2][iD]

[1] Division of Molecular Pathology, Institute of Cancer Research (ICR), London, UK
azam.nekoo@icr.ac.uk
[2] Centre for Evolution and Cancer, Institute of Cancer Research (ICR), London, UK
[3] LPI, ETSI Telecomunicación, Universidad de Valladolid, Valladolid, Spain
[4] AGH University of Science and Technology, Kraków, Poland
[5] CUBRIC, School of Psychology, Cardiff University, Cardiff, UK
[6] Department of Computer Science, Aberystwyth University, Ceredigion, UK

Abstract. Gliomas are brain tumours with a high mortality rate. There are various grades and sub-types of this tumour, and the treatment procedure varies accordingly. Clinicians and oncologists diagnose and categorise these tumours based on visual inspection of radiology and histology data. However, this process can be time-consuming and subjective. The computer-assisted methods can help clinicians to make better and faster decisions. In this paper, we propose a pipeline for automatic classification of gliomas into three sub-types: oligodendroglioma, astrocytoma, and glioblastoma, using both radiology and histopathology images. The proposed approach implements distinct classification models for radiographic and histologic modalities and combines them through an ensemble method. The classification algorithm initially carries out tile-level (for histology) and slice-level (for radiology) classification via a deep learning method, then tile/slice-level latent features are combined for a whole-slide and whole-volume sub-type prediction. The classification algorithm was evaluated using the data set provided in the CPM-RadPath 2020 challenge. The proposed pipeline achieved the F1-Score of 0.886, Cohen's Kappa score of 0.811 and Balance accuracy of 0.860. The ability of the proposed model for end-to-end learning of diverse features enables it to give a comparable prediction of glioma tumour sub-types.

Keywords: Glioma classification · Digital pathology · Multimodal MRI

1 Introduction

Gliomas are tumours of the brain parenchyma which are typically graded from I (low severity) to IV (high severity), and the five-year survival rates are 94% (for grade I) and 5% (for higher grades) [11]. Magnetic Resonance Imaging (MRI) has been widely used in examining gliomas during diagnosis, surgical

© Springer Nature Switzerland AG 2021
A. Crimi and S. Bakas (Eds.): BrainLes 2020, LNCS 12659, pp. 508–518, 2021.
https://doi.org/10.1007/978-3-030-72087-2_45

planning and follow-up. Clinical protocols include T1-weighted (T1w), T2-weighted (T2w), fluid-attenuated inversion recovery (FLAIR), and gadolinium-enhanced T1-Weighted (Gd-T1w) imaging. T1w, T2w and FLAIR show the tumour lesion and oedema whereas Gd-T1w shows regions of blood-brain-barrier disruption [1]. However, due to the complex tumour micro-environment and spatial heterogeneity, MRI itself is not sufficient for complete characterisation of gliomas (e.g. grading and sub-typing) and for high-grade cases histopathology examination is often required [6]. In histopathology, gliomas are classified based on the morphological features of the glial cells including increased cellularity, vascular proliferation, necrosis, and infiltration into normal brain parenchyma [11,17]. Oncologists examine patients' medical history, radiology scans, pathology slides and reports to provide suitable medical care for a person diagnosed with cancer. This decision making is often subjective and time-consuming. Machine learning offers powerful tools (e.g. deep learning methods) to support automated, faster and more objective clinical decision making. One active area of research is to design data processing pipelines that can effectively combine imaging data at different spatial domains (e.g. micro-scale histology and macro-scale MRI data). These pipelines would enable image-based precision medicine to improve treatment quality [7,9]. Recent efforts in automatic classification of gliomas using histology and radiology was reviewed by Kurc et al. [12].

In this study, we respond to the CPM-RadPath 2020 challenge and propose a data processing pipeline that classifies gliomas into three sub-types: oligodendroglioma, astrocytoma, and glioblastoma. In our approach, we analysed radiology and histopathology data independently using separate densely connected networks. We then combined the outcomes of each network using a probabilistic ensemble method to arrive at the final sub-type prediction. We used our pipeline to analyse data from a cohort of 35 patients and the results have been submitted to the challenge board.

2 Related Work

In this section, we summarise the work submitted by the participants for the previous similar challenge in CPM-RadPath 2019 (https://www.med.upenn.edu/cbica/cpm-rad-path-2019). Ma et al. [13], as the first ranked group, proposed two convolutional neural networks to predict the grade of gliomas from both radiology and pathology data: (i) a 2D ResNet-based model for pathology patch based image classification and (ii) a 3D DenseNet-based model for classifying the detected regions (using a detection model) on multi-parametric MRI (mp-MRI) images. To extract the pathology patches (512 × 512) they used mean and standard deviation with predefined thresholds to consider patches including cells and excluding patches with background contents. To avoid the effect of intensity variations, they converted the original RGB pathology images into gray-scale images. Then the labels of all extracted patches were set to the labels of the entire WSI image. On the radiology side, they used a detection model that was trained on the BraTS2018 (https://www.med.upenn.edu/sbia/brats2018.html) dataset and used the output of this model (detected abnormality) as the input for the 3D DenseNet model to perform the classification on the MRI volumes. Finally, the prediction results from these two modalities were concatenated using the multinomial logistic regression, resulting in F1-score = 0.943 for the validation set. Pei et al. [14] only used the mp-MRI data. They implemented two regular 3D convolutional neural networks (CNN) to analyse MRI data: (i) the first CNN was trained on the BraTS 2019 (https://www.med.upenn.edu/cbica/brats2019/data.html) dataset to differentiate tumourous and normal tissues; and (ii) the other CNN was trained on the CPM-RadPath 2019 dataset to do tumour classification. They performed z-score normalization to reduce the impact of intensity inhomogeneity and reported the F1-score = 0.764 for the validation set. Chan et al. [4] used two neural networks, including VGG16 and ResNet50, to process the patches extracted from the whole slide images. For the radiology analysis, they used the brain-tumour segmentation method (based on the SegNet) that was developed in the BraTS2018 challenge to obtain the core region of the tumour. Using the predicted regions, they used the pyradiomics module to extract 428 radiomic features. They also used the k-means clustering and random forest to only include relevant pathology patches. Finally, they calculated the sum of the probabilities to produce an ensemble prediction. They compared the prediction results with and without MRI features and found out that the classification models based on MRI radiomic features did not perform as accurate as the whole-slide features. Xue et al. [16] used a dual-path residual convolutional neural network to perform classification. For analysing radiology data, they used their custom designed U-Network to predict tumour region mask from MRI images. This network was trained on data from the BraTS2019 challenge. Then 3D and 2D ResNet18 CNN architectures were used jointly on detected MRI regions and selected pathology patches (extracted with the pixel resolution of 512 × 512), respectively. They reported the accuracy of 84.9% on the validation set, and concluded that combining the two image sources yielded a better overall accuracy.

Our proposed method does not use any external data and annotation, and does not require the pre-identification of the abnormality on the radiology data. We use the raw data (without changing them to gray scale or without doing the extensive sub-scaling that reduces the data quality) and feed this data to the models for sub-type prediction. In addition, this approach is not computationally expensive and is fast in the inference phase.

3   Methodology

3.1   Dataset

To train and validate our pipeline, we used the challenge dataset (https://zenodo.org/record/3718894). The dataset consists of multi-institutional paired mp-MRI scans and digital pathology whole slide images (WSIs) of brain gliomas, obtained from the same patients. For each subject, a diagnostic classification label, confirmed by pathology, is provided as 'A' (Lower grade astrocytoma, IDH-mutant (Grade II or III)), 'O' (Oligodendroglioma, IDH-mutant, 1p/19q codeleted (Grade II or III)) and 'G' (Glioblastoma and Diffuse astrocytic glioma, IDH-wildtype (Grade IV)). There are three data subsets: training, validation, and testing with 221, 35, and 73 co-registered radiology and histology subjects, respectively. Considering the training subset released by the challenge in the training phase, there were 133, 54, and 34 samples provided for the 'G', 'O' and 'A' classes, respectively. The radiology data consists of four modalities: T1w, T2w, Gd-T1w, and FLAIR volumes, acquired with different clinical protocols. The provided data are distributed after their pre-processing, co-registered to the same anatomical template, interpolated to the same resolution (1 mm^3), and skull-stripped. The histopathology data contains one WSI for each patient, captured from Hematoxylin and Eosin (H&E) stained tissue specimens. The tissue specimens were scanned with different magnifications and the pixel resolutions.

3.2   Deep Learning Model: Densely Connected Network (DCN)

Our deep learning classification model is primarily derived from DenseNet architecture [8]. DenseNet is a model notable for its key characteristic of bypassing signals from preceding layers to subsequent layers that enforce optimal information flow in feature maps. The DenseNet model comprises a total of L layers, while each layer is responsible for implementing a specific non-linear transformation. A dense-block consists of multiple densely connected layers and all layers are directly connected. At the end of the last dense-block, a global average pooling is performed to minimize over-fitting and reduce the model's number of parameters. We have used specific configurations of this model (called DCN). In the histology DCN (DCN1), the block configuration of $[2, 2, 2, 2]$ was used while in the radiology based DCN (DCN2), the block configuration was set to $[6, 12, 36, 24]$. The DCN architecture is shown in Fig. 1a. These configurations were selected by training more than 50 models with different block parameters and were empirically chosen. The network growth-rate, which defines the number of filters to add in each layer was set to 32 and 24 for DCN1 and DCN2, respectively. The initial convolution layer contained 64 filters in DCN1 and 48 filters in DCN2 to learn and the multiplicative factor for the number of bottle neck layers was set to 4 with zero dropout rate after each dense layer, as defined in [8].

3.3   Histological Phase Analysis

The process of selecting representative tiles to be used during the training phase is very important and can be labor-intensive. Weakly-supervised approaches [15] only use the slide labels as supervision during the training of the aggregation model. Some other approaches adopt the multiple instance learning (MIL) method, assuming that the slide label is represented by the existence of positive tiles [3]. To automatically classify WSIs, we have adopted a three-stage approach. In the first stage, multiple recognizable tiles/patches of size 2000×2000 pixels without any overlaps were selected form the WSI. Each tile was assigned with the slide's label if it included more than 80% cellularity information. These tiles were counted as positive tiles, which belonged to a specific class of 'A', 'O' or 'G'.

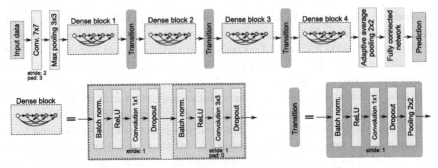

(a) The DCN architecture used for the glioma classification.

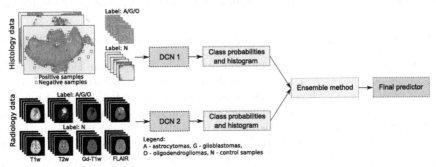

(b) Models trained on different modalities are combined with an ensemble method to predict the patient's glioma sub-types.

Fig. 1. The proposed pipeline using multi-modality data to classify glioma sub-types.

The histology images were representative of slides generated in a true pathology laboratory, including common slide preparation and digitisation artifacts, such as air bubbles, folds, cracks, striping, and blurred regions. To avoid the need for quality control and background removal, we added a new category in the classification problem called 'N' representing "None" and negative class. All the tiles (i) located on the tissue-background regions (with 80% background region), (ii) containing any artefacts addressed earlier, and (iii) representing hemorrhage were labeled as the class 'N'. So, the tiles with more than 95% of pixels exceeding 80% intensity on all three RGB channels were considered. The extracted tiles (positive and negative) were then investigated visually to reassure false tiles were not included in the training set. Due to the different number of samples provided in each class, we extracted various numbers of positive tiles from WSIs of a class considering the pixel resolution variation. This information is provided in Table 1. We selected a balanced number of samples for each class. For example, considering the samples with the 0.25 resolution, we selected 853, 921, 807, and 900 tiles for the 'A', 'G', 'O' and 'N' classes, respectively. In the second stage, we trained the explained DCN on small image tiles extracted from WSIs at high resolution (see Fig. 1b). Overall, two models were trained: a model for WSIs with pixel resolution of 0.50 and the other model for WSIs with pixel resolution of 0.25. Each tile was then encoded to a feature vector of a low dimension and a prediction score. The output of the DenseNet model was a tile/patch-wise probability prediction. In the third stage, an ensemble approach based on weighted probability prediction major voting was used to integrate the obtained tile level information for whole slide level prediction. For the second stage, we have used the provided slide level labels for the supervision without using any external, extensive, pixel-wise annotations from pathologists. It is important to mention that we assigned the slide label to the extracted patches. This idea of assigning the same label to all extracted patches from the WSI might be criticised since all parts of the WSI might not represent the same information which could lead to a correct classification of the tumour sub-type. To answer this concern, we emphasize that the final prediction was based on the abundance prediction of one of the three desired sub-types.

3.4 Radiological Phase Analysis

There are several pipelines proposed for automatic prediction of glioma sub-types using MRI data [12]. Most of these approaches are based on analysing radiomics in terms of high-dimensional quantitative features extracted from a large number of medical images. Several approaches initially perform segmentation, followed by the classification based on the detected bounding box of the region of interest [5]. The performance of these approaches is affected by the detection and segmentation outcome. We adopted the similar classification scheme used in Sect. 3.3 to analyse different MRI modalities. For each modality, the slices in each volume were considered either negative (without any lesions apparent) or positive (with a visible lesion). Each positive slice could belong to one of the target classes ('A', 'G', and 'O'). With this approach, we avoided the need for lesion segmentation in brain volumes. The initial preparation steps for the MRI slices included (i) visually categorising all the slices provided for each modality into negative and positive and (ii) preparing the training and validation subsets for training with the proportion of 90% and 10% of the prepared categorised data, respectively. Considering that there were fewer samples provided for the 'A' class, and the fact that we wanted to include a balanced number of samples in each class during the training, we selected 1500 samples for each class. Then we trained a DCN on image slices from each modality volume (see Fig. 1b). So, four

Table 1. Number of selected samples from each class using the histology data.

Class	Resolution = 0.25		Resolution = 0.50	
	Provided cases	**Extracted tiles**	Provided cases	**Extracted tiles**
O	91	**10**	42	**5**
A	46	**10**	8	**20**
G	31	**31**	3	**20**

models were trained. Each slice was encoded to the latent feature vector of a low dimension and a prediction score. The outputs of these models, trained for each modality, were slice-based probability predictions. Subsequently, the weighted average operation was applied to the probability predictions to integrate the obtained slice level information for the whole brain volume. In this classification approach, we have used the provided volume level labels for the supervision without using any external pixel-wise lesion masks from radiologists.

3.5   Training DCNs

We randomly split the provided training dataset (radiology and histology) into training and validation sets (90% and 10%, respectively), making sure that the validation samples did not overlap with the training samples. Data augmentation was performed to increase the number of training samples by applying random flipping, rotations, scaling and cropping. The final softmax classifier of the DCN contained 4 output nodes that would predict the probability for each class based on the extracted features in the network. The models were implemented in PyTorch and trained via the Adam optimisation algorithm [10] for the cross-entropy loss with mini-batches of size 128. We trained the histology model (DCN1) for 300 epochs while the radiology based models (DCN2) were trained for 500 epochs. All models were trained independently using an initial learning rate of 0.001, without any step-down policy and weight-decay. The loss function was calculated for each task on all samples in the batch with known ground truth labels and averaged to a global loss. Then the predicted loss for the slide label was back-propagated through the model. The training process in terms of loss decrease and accuracy improvement for each of these models is shown in Fig. 2.

3.6   Outcome Integration and Final Sub-type Prediction

It is common to use the combination of histology and MRI data in deep learning approaches to predict the class of gliomas. In this work, we used a deep learning-based method (DCN) for the classification of gliomas. The dataset in each modality was used for training an independent DCN model. Initially, the outputs of these models were either tile-wise or slice-wise probability predictions. Therefore, an ensemble approach based on probability prediction and major voting was used to integrate the obtained tile-level/slice-level information for the whole histology slide or MRI volume. Then the outcomes from five models (one for histology and four for radiology) were integrated with the same ensemble approach based on confidence prediction and major voting. All modality predictions were used for the final classification prediction of a patient.

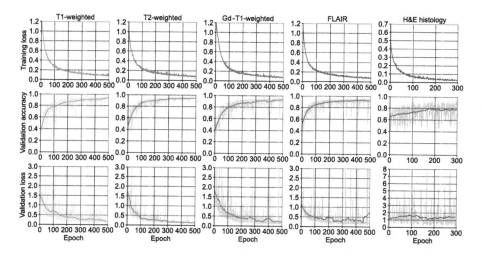

Fig. 2. The training process in terms of loss decrease and accuracy improvement for models trained on each modality. Pixel resolution for histology images = 0.25.

4   Results and Discussion

We trained five models (4 radiology based models and 1 histology based model when pixel resolution = 0.25). The F1-Score of the combined model on the training subset was 0.946. Further evaluation of these models were separately performed on the provided validation dataset and the evaluation metrics were calculated by the online web-portal of the challenge. Table 2 shows the classification performance of these models. These results show that among multiple MRI modalities, T2w was able to extract more salient latent features while hyperintense FLAIR did not reflect the morphological complexity underlying tumour accurately. The agreement between T1w and Gd-T1w was also observed with regards to the classification performance. However, the model trained on the histology tiles achieved the best performance of F1-Score = 0.771 compared to other MRI modalities. This illustrated the existence of an abundant amount of information in the histological images which led to the extraction of more distinct latent features to be used by the softmax classifier, which is crucial for high quality clinical decision support. Combining the outcomes from all the models reduced the F1-Score to 0.714. This was expected because including three modalities (T1w, Gd-T1w, and FLAIR) with more false positive and false negatives along with the major voting could deviate the prediction agreement and affect the final results. However, using the best predictions from histology, T2w and Gd-T1w achieved the best performance of F1-Score = 0.886. Apart from the quantified results, we visually investigated several histology samples that were correctly classified in the first validation set (derived from the provided training subset). Based on morphological features, we observed that *astrocytomas* contained hyperchromatic nuclei, with prominent nucleoli. *Oligodendrogliomas* were mostly represented as round nuclei surrounded by a clear cytoplasmic halo (like

Table 2. Classification performance of each model, evaluated on the validation set provided by the CPM-RadPath 2020 challenge. Pixel resolution for histology images = 0.25.

Modality	F1-Score	Kappa	Balance accuracy
MRI FLAIR	0.543	0.346	0.634
MRI T1w	0.400	0.143	0.429
MRI Gd-T1w	0.600	0.433	0.723
MRI T2w	0.714	0.519	0.632
Histology	0.771	0.606	0.682
Combined Histology + MRI T2w + MRI Gd-T1w	**0.886**	**0.811**	**0.860**
Combined Histology and MRIs	0.714	0.554	0.723

fried egg-shaped) and *glioblastomas* showed a densely cellular tumour with variation in the gross appearance of the viable tumour along with more necrosis. These observations meet the clinical observations addressed by Bertero et al. [2].

5   Conclusion

In this paper, we proposed using specific versions of DenseNet (called DCN) for sub-typing the glioma into astrocytoma, oligodendroglioma, and glioblastoma classes from the MRI and histology images provided by the CPM-RadPath-2020 organisers. The evaluations of the five trained models indicated that the combination of radiographic with histologic image information can improve glioma classification performance when careful training is performed in terms of curated data, model architecture, and ensemble approach.

Acknowledgements. Azam Hamidinekoo acknowledges the support by Children's Cancer and Leukaemia Group (CCLGA201906). Tomasz Pieciak acknowledges the Polish National Agency for Academic Exchange for project PN/BEK/2019/1/00421 under the Bekker programme. Maryam Afzali was supported by a Wellcome Trust Investigator Award (096646/Z/11/Z). Otar Akanyeti acknowledges Sér Cymru Cofund II Research Fellowship grant. Yinyin Yuan acknowledges funding from Cancer Research UK Career Establishment Award (CRUK C45982/A21808), CRUK Early Detection Program Award (C9203/A28770), CRUK Sarcoma Accelerator (C56167/A29363), CRUK Brain Tumour Award (C25858/A28592), Rosetrees Trust (A2714), Children's Cancer and Leukaemia Group (CCLGA201906), NIH U54 CA217376, NIH R01 CA185138, CDMRP BreastCancer Research Program Award BC132057, European Commission ITN (H2020-MSCA-ITN-2019), and The Royal Marsden/ICR National Institute of Health Research Biomedical Research Centre. This research was also supported in part by PLGrid Infrastructure. Tomasz Pieciak acknowledges Mr. Maciej Czuchry from CYFRONET Academic Computer Centre (Kraków, Poland) for outstanding support.

References

1. Abdalla, G., et al.: The diagnostic role of diffusional kurtosis imaging in glioma grading and differentiation of gliomas from other intra-axial brain tumours: a systematic review with critical appraisal and meta-analysis. Neuroradiology **62**(7), 791–802 (2020). https://doi.org/10.1007/s00234-020-02425-9
2. Bertero, L., Cassoni, P.: Classification of tumours of the central nervous system. In: Bartolo, M., Soffietti, R., Klein, M. (eds.) Neurorehabilitation in Neuro-Oncology, pp. 21–36. Springer, Cham (2019). https://doi.org/10.1007/978-3-319-95684-8_3
3. Campanella, G., et al.: Clinical-grade computational pathology using weakly supervised deep learning on whole slide images. Nat. Med. **25**(8), 1301–1309 (2019)
4. Chan, H.-W., Weng, Y.-T., Huang, T.-Y.: Automatic classification of brain tumor types with the MRI scans and histopathology images. In: Crimi, A., Bakas, S. (eds.) BrainLes 2019. LNCS, vol. 11993, pp. 353–359. Springer, Cham (2020). https://doi.org/10.1007/978-3-030-46643-5_35
5. Decuyper, M., Bonte, S., Deblaere, K., Van Holen, R.: Automated MRI based pipeline for glioma segmentation and prediction of grade, IDH mutation and 1p19q co-deletion. arXiv preprint arXiv:2005.11965 (2020)
6. Dhermain, F.G., Hau, P., Lanfermann, H., Jacobs, A.H., van den Bent, M.J.: Advanced MRI and PET imaging for assessment of treatment response in patients with gliomas. Lancet Neurol. **9**(9), 906–920 (2010)
7. Hamidinekoo, A., Denton, E., Rampun, A., Honnor, K., Zwiggelaar, R.: Deep learning in mammography and breast histology, an overview and future trends. Med. Image Anal. **47**, 45–67 (2018)
8. Huang, G., Liu, Z., Van Der Maaten, L., Weinberger, K.Q.: Densely connected convolutional networks. In: Proceedings of the IEEE Conference on Computer Vision and Pattern Recognition, pp. 4700–4708 (2017)
9. Jameson, J.L., Longo, D.L.: Precision medicine-personalized, problematic, and promising. Obstet. Gynecol. Surv. **70**(10), 612–614 (2015)
10. Kingma, D.P., Ba, J.: Adam: a method for stochastic optimization. arXiv preprint arXiv:1412.6980 (2014)
11. Kumar, V., Abbas, A.K., Aster, J.C.: Robbins Basic Pathology, 10th edn. Elsevier, Philadelphia (2018)
12. Kurc, T., et al.: Segmentation and classification in digital pathology for glioma research: challenges and deep learning approaches. Front. Neurosci. **14**, 27 (2020)
13. Ma, X., Jia, F.: Brain tumor classification with multimodal MR and pathology images. In: Crimi, A., Bakas, S. (eds.) BrainLes 2019. LNCS, vol. 11993, pp. 343–352. Springer, Cham (2020). https://doi.org/10.1007/978-3-030-46643-5_34
14. Pei, L., Vidyaratne, L., Hsu, W.-W., Rahman, M.M., Iftekharuddin, K.M.: Brain tumor classification using 3D convolutional neural network. In: Crimi, A., Bakas, S. (eds.) BrainLes 2019. LNCS, vol. 11993, pp. 335–342. Springer, Cham (2020). https://doi.org/10.1007/978-3-030-46643-5_33
15. Wang, X., et al.: Weakly supervised deep learning for whole slide lung cancer image analysis. IEEE Trans. Cybern. **50**(9), 3950–3962 (2019)

16. Xue, Y., et al.: Brain tumor classification with tumor segmentations and a dual path residual convolutional neural network from MRI and pathology images. In: Crimi, A., Bakas, S. (eds.) BrainLes 2019. LNCS, vol. 11993, pp. 360–367. Springer, Cham (2020). https://doi.org/10.1007/978-3-030-46643-5_36
17. Ye, Z., et al.: Diffusion histology imaging combining diffusion basis spectrum imaging (DBSI) and machine learning improves detection and classification of glioblastoma pathology. Clin. Cancer Res. (2020). https://doi.org/10.1158/1078-0432.CCR-20-0736

Author Index

Printed in the United States
by Baker & Taylor Publisher Services